ASTROPHYSICS II
INTERSTELLAR MATTER AND GALAXIES

OTHER JONES AND BARTLETT TITLES OF INTEREST

Introduction to Cosmology
Jayant V. Narlikar, Tata Institute of
Fundamental Research, Bombay, India

The Comet Book
Robert D. Chapman and **John C. Brandt**,
NASA, Goddard Space Flight Center, Maryland

Numerical Astrophysics
Edited by **Joan Centrella**, University of Texas,
James LeBlanc, Lawrence Livermore Laboratory,
Richard Bowers, Los Alamos National Laboratory

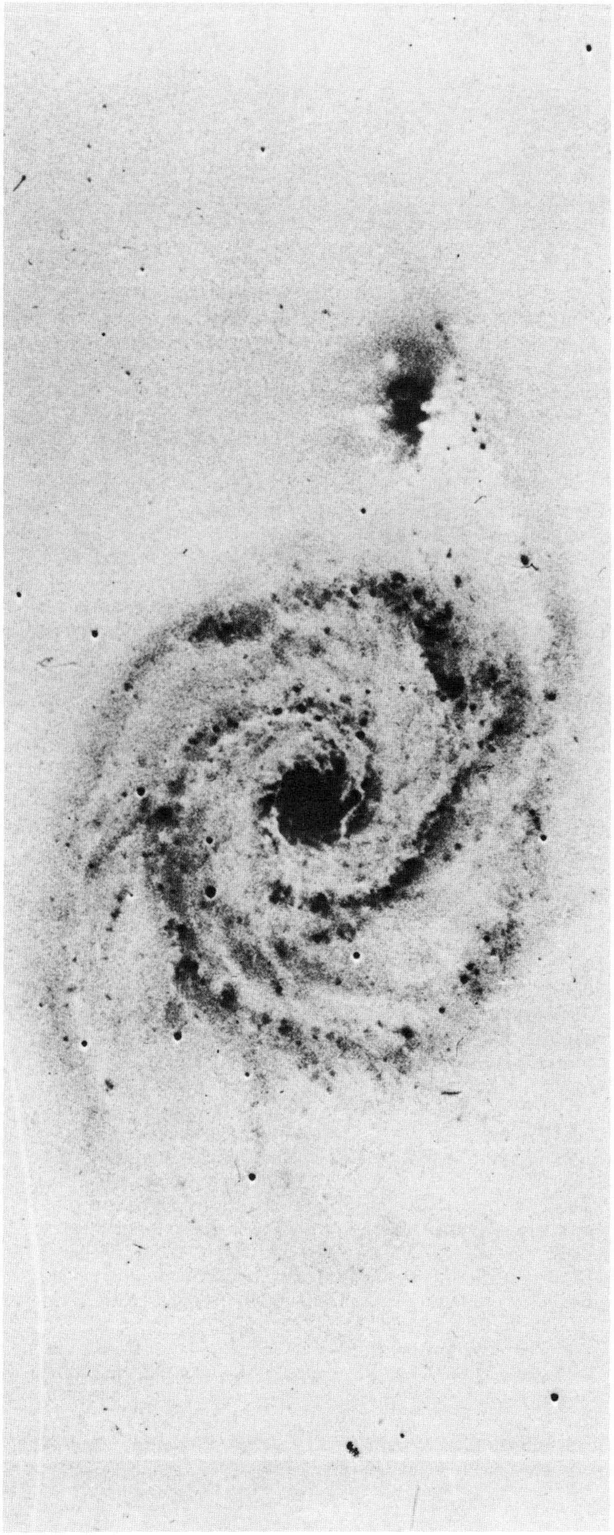

ASTROPHYSICS II

INTERSTELLAR MATTER AND GALAXIES

Richard L. Bowers
LOS ALAMOS NATIONAL LABORATORY

Terry Deeming
DIGICON GEOPHYSICAL CORPORATION

JONES AND BARTLETT PUBLISHERS
Boston   London

Editorial, Sales, and Customer Service Offices
Jones and Bartlett Publishers, One Exeter Plaza, Boston, MA 02116
Jones and Bartlett Publishers International, P. O. Box 1498, London, W6 7RS England

Copyright © 1984 by Jones and Bartlett Publishers, Inc. All rights reserved. No part of the material protected by this copyright notice may be reproduced or utilized in any form, electronic or mechanical, including photocopying, recording, or by any information storage and retrieval system, without written permission from the copyright owner.

Library of Congress Cataloging in Publication Data

Bowers, Richard, 1941–
  Astrophysics.
  Bibliography: p.
  Includes index.
  Contents: 1. Stars— 2. Interstellar matter and galaxies.
  1. Astrophysics. I. Deeming, Terry. II. Title.
QB461.B64   1984   523.01   83–17234

ISBN: 0-86720-047-2

Publisher: Arthur C. Bartlett
Production: Bookman Productions
Book and cover design: Hal Lockwood
Copyeditor: Aidan Kelly
Illustrator: Nancy Warner
Composition: Science Press

Printed in the United States of America

Printing number (last digit)
10 9 8 7 6 5 4

The authors gratefully acknowledge permission to use the following figures: **Figure 18.4:** L. Spitzer, *Physical Processes in the Interstellar Medium* (New York: Wiley, 1978), Fig. 6.2. **Figures 18.6(a) and 18.6(b):** R. N. Manchester and J. H. Taylor, *Pulsars* (San Francisco: W. H. Freeman, 1977), Figs. 7.2 and 7.5. **Figure 19.3:** D. S. Mathewson and V. L. Ford, *Mem. Royal Astron. Soc.* 74 (1970):139, Fig. 1(a). **Figure 20.4:** D. E. Osterbrock, *Astrophysics of Gaseous Nebulae* (San Francisco: W. H. Freeman, 1974), Fig. 3.2. **Figures 20.5–20.7:** D. R. Fowler, *Mon. Notices Royal Astron. Soc.* 146 (1969):243, Figs. 7, 8, 9, 10, 11, and 12. **Figure 24.6:** S. A. Kaplan, *Interstellar Gas Dynamics* (London: Pergamon Press, 1970), Fig. 9. **Figure 25.3:** I. S. Shklovskii, *Stars: Their Birth, Life and Death* (San Francisco: W. H. Freeman, 1978), Figs. 16.7 and 17.12. **Figure 25.6:** Mount Wilson and Las Campanas Observatories, Carnegie Institution of Washington, D.C. **Figure 25.7:** R. G. Conway, "The Crab Nebula," IAU Symposium No. 46, edited by R. D. Davies and F. G. Smith (Dordrecht, Holland: Reidel, 1971), Figs. 1 and 2. **Figure 26.1:** Adapted from A. Sandage, *Ap. J.* 178 (1972):1, Fig. 4. **Figure 26.6:** Adapted from A. Sandage, *Ap. J.* 173 (1972):494, Fig. 5. **Figure 26.7:** Schramm and Wagoner, *Phys. Today* 27 (1974):46, No. 12 **Figure 26.9:** Adapted from D. Woody and P. L. Richards, *Phys. Rev. Lett.* 92 (1979):925, Fig. 2. **Figure 27.1:** Kitt Peak National Observatory, AURA, Inc., Tucson, Ariz. **Figures 27.2–27.4:** Mount Wilson and Las Campanas Observatories, Carnegie Institution of Washington, D.C. **Figure 27.6:** Mount Wilson and Las Campanas Observatories, Carnegie Institution of Washington, D.C. **Figure 27.11:** G. de Vaucouleurs, *Ap. J. Suppl.* 5 (1961):233, Fig. 13 **Figure 27.12:** W. Baum, *Publ. Ast. Soc. Pacific* 71 (1959):106, Fig. 4. **Figure 27.15:** H. Spinrad and M. Peimbert, "Spiral and Gaseous Content of Galaxies," in *Galaxies and the Universe*, edited by A. Sandage, M. Sandage, and J. Kristian (Chicago: University of Chicago Press, 1975), Fig. 2. **Figure 27.19:** J. H. Oort, in *Galactic Astronomy*, edited by A. Blaauw and M. Schmidt (Chicago: University of Chicago Press, 1965), Fig. 3. **Figures 28.1 and 28.2:** I. R. King, *Astron. Journal* 71 (1966):67, Figs. 1 and 2. **Figures 28.3–28.5:** L. Spitzer, "Dynamical Theory of Spherical Stellar Systems with Large N," IAU Symposium, *Dynamics of Stellar Systems*, edited by A. Hayli (Dordrecht, Holland: Reidel, 1975), Figs. 3, 4, and 6. **Figure 28.6:** P. J. E. Peebles, *Physical Cosmology*, Princeton Series in Physics. Copyright © 1971 by Princeton University Press. Fig. IV-1, p. 69, reprinted by permission of Princeton University Press. **Figure 28.7:** Adapted from *Frontiers of Astrophysics*, edited by Eugene Avrett (Cambridge, Mass.: Harvard University Press, 1976), Fig. 12-1. **Figure 28.8:** S. S. Murray et al., *Ap. J.* 234 (1979), Plates L6 and L7. **Figure 29.1:** I. King, "The Structure of Round Stellar Systems: Observation and Theory," IAU Symposium 69, *Dynamics of Stellar Systems*, edited by A. Hayli (Dordrecht, Holland: Reidel, 1975), Fig. 10. **Figure 29.9:** J. H. Oort, "Stellar Dynamics," in *Galactic Structure*, edited by A. Blaauw and M. Schmidt (Chicago: University of Chicago Press, 1965), Fig. 1. **Figure 30.5:** C. C. Lin, C. Yuan, and F. Shu, *Ap. J.* 155 (1969):721, Fig. 1. **Figure 30.7:** Adapted from C. C. Lin, "Theory of Spiral Structure," in *Galactic Astronomy*, edited by H. Y. Chiu and A. Murriel (New York: Gordon and Breach, 1970), Fig. 6.3(a). **Figures 30.9 and 30.10:** W. W. Roberts, "Shock and Star Formation in Galactic Spirals," in *Galactic Astronomy*, edited by H. Y. Chiu and A. Murriel (New York: Gordon and Breach, 1970), Figs. 3.2 and 3.3. **Figure 31.5:** H. Spinrad, J. Stauffer, and H. Butcher, *Ap. J.* 244 (1981):382, Fig. 6. **Figures 31.8–31.10:** O. J. Eggen, D. Lynden-Bell, and A. R. Sandage, *Ap. J.* 136 (1962):748, Figs. 1, 2, and 3. **Table 28.6:** Martin Schwarzschild, *Structure and Evolution of the Stars*. Copyright © 1958 by Princeton University Press. Reprinted by permission of Princeton University Press.

Contents

VOLUME I

Part I
INTRODUCTION 1

Chapter 1
AN OVERVIEW OF STELLAR STRUCTURE AND EVOLUTION 2
- 1.1. Stars 2
- 1.2. Energy Transport and Generation in Stars 3
- 1.3. Stellar Time-scales 4
- 1.4. Static Configurations (Hydrostatic Equilibrium) 7
- 1.5. The Virial Theorem 9
- 1.6. Relativistic Effects 10
- 1.7. Star Formation 11
- 1.8. Stellar Evolution 13

Chapter 2
PROPERTIES OF MATTER 16
- 2.1. Equations of State 16
- 2.2. Ideal Gas 17
- 2.3. Mixtures of Ideal Gases: Mean Molecular Weight 19
- 2.4. Radiation and Matter 20
- 2.5. Degenerate Matter 22
- 2.6. Matter at High Temperatures 24
- 2.7. Real Fluids 25

Chapter 3
ASPECTS OF OBSERVATIONAL ASTRONOMY 28
- 3.1. Systems of Brightness Measurement 28
- 3.2. Interstellar Absorption and Reddening 31
- 3.3. Color Magnitude and Two-Color Diagrams 33
- 3.4. Stellar Populations and Stellar Evolution 35
- 3.5. Spectrum Analysis and Spectroscopy 38
- 3.6. Binary Systems 42
- 3.7. Pulsating Stars 45
- 3.8. Rotating Stars 47
- 3.9. Astronomical Statistics 48

Part 2
STELLAR STRUCTURE   53

Chapter 4
STATIC STELLAR STRUCTURE   54
4.1. Introduction to Stellar Structure   54
4.2. The Equation of Hydrostatic Equilibrium   55
4.3. Simplified Stellar Models   59

Chapter 5
RADIATION AND ENERGY TRANSPORT   65
5.1. Radiative Transport   65
5.2. Description of the Radiation Field   66
5.3. Opacity and Emissivity   69
5.4. Equation of Radiative Transfer   71
5.5. Black-Body Radiation   74
5.6. Radiative Equilibrium   75
5.7. Simple Stellar Atmospheres   76
5.8. True Absorption and Scattering   79
5.9. Radiation in the Solar Atmosphere   82
5.10. Summary of Results on Radiative Stellar Structure   86
5.11. Nonradiative Energy Transport   88

Chapter 6
ATOMIC PROPERTIES OF MATTER   97
6.1. The Hydrogen Atom   97
6.2. Thermal Excitation and Ionization   102
6.3. Detailed Balancing, Transition Probabilities, and Line Opacities   109
6.4. Continuous Opacity in Stars   115
6.5. Simplified Stellar Models   125
6.6. Line Broadening and Line Opacity   130
6.7. Line Intensities in Stellar Spectra   134
6.8. Line Broadening in Hydrogen and Helium   138

Part 3
STELLAR EVOLUTION   143

Chapter 7
NUCLEAR ENERGY SOURCES   144
7.1. Thermonuclear Energy Sources   144
7.2. Thermonuclear Energy Release   147
7.3. Nuclear Energy Generation Rates   149
7.4. Nuclear-Burning Stages   154
7.5. Homologous Stellar Models   158
7.6. Electron Screening in Nuclear Reactions   162

Chapter 8
INTRODUCTION TO STELLAR EVOLUTION   166
8.1. Phases of Stellar Evolution   166
8.2. Evolution of a Protostar   170

Chapter 9
THE MAIN SEQUENCE   175
9.1. The Zero-Age Main Sequence   175
9.2. Evolution on the Main Sequence   176
9.3. Lower Main Sequence   177
9.4. Upper Main Sequence   181
9.5. Isothermal Cores   182
9.6. Termination of the Main Sequence   184

Chapter 10
EVOLUTION AWAY FROM THE MAIN SEQUENCE   185
10.1. Post-Main-Sequence Evolution   185
10.2. Composition Inhomogeneities   185
10.3. Central Condensation   186
10.4. Characteristics of Shell-Burning Sources   187
10.5. Evolution of Shell Sources   189
10.6. Red Giants   195
10.7. Modifications: Composition and Mass Loss   203

Chapter 11
DEVIATIONS FROM QUASISTATIC EVOLUTION   208
11.1. Deviations from Hydrostatic Equilibrium   208
11.2. Adiabatic Stellar Pulsations   210
11.3. Stellar Stability   213
11.4. Pulsational Stability   215
11.5. Classical Cepheid and RR Lyrae Variables   218
11.6. Unstable Shell Sources   223
11.7. Mass Loss from Red Giants   225

Chapter 12
FINAL STAGES OF STELLAR EVOLUTION   229
12.1. Stellar Mass and the Final Stage   229
12.2. Advanced Stages of Nuclear Burning and Stellar Nucleosynthesis   230

Chapter 13
WEAK INTERACTIONS IN STELLAR EVOLUTION   239
13.1. Solar Neutrinos   241
13.2. Neutrino Energy-Loss Rates   243
13.3. Coherent Scattering Off Nuclei   249

Chapter 14
DEGENERATE STARS   252
- 14.1. Degenerate Matter in Stars   252
- 14.2. Degenerate Matter in Hydrostatic Equilibrium   254
- 14.3. White Dwarfs   256
- 14.4. Envelope Structure   261
- 14.5. Evolving White Dwarfs   265

Chapter 15
SUPERNOVAE   267
- 15.1. Observational Features   267
- 15.2. Stellar Core Collapse   274

Chapter 16
COMPACT STELLAR AND RELATIVISTIC OBJECTS   283
- 16.1. Compact Supernova Remnants   283
- 16.2. Neutron Stars   285
- 16.3. Gravitational Collapse and Black Holes   291
- 16.4. Pulsars   301

Chapter 17
CLOSE BINARY SYSTEMS   316
- 17.1. Mechanics of Binary Systems   316
- 17.2. Structure of Close Binary Systems   318
- 17.3. Evolving Binary Stars   322
- 17.4. X-Ray Sources   330
- 17.5. The Binary Pulsar   338
- 17.6. Novae   340

VOLUME II

Part 4
THE INTERSTELLAR MEDIUM   345

Chapter 18
INTERSTELLAR MATTER   346
- 18.1. Physical Processes in Interstellar Gas   346
- 18.2. Thermal States of Interstellar Gas   351
- 18.3. Interstellar Clouds   357
- 18.4. Interstellar Electron Density   360
- 18.5. Radio Emission and Absorption   363

Chapter 19
INTERSTELLAR DUST GRAINS   370
- 19.1. Interstellar Dust   370
- 19.2. Grain Properties   373
- 19.3. Infrared Excess   376
- 19.4. Grain Evolution   377
- 19.5. Dust Dynamics   378

Chapter 20
GASEOUS NEBULAE   382
- 20.1. Gaseous Nebulae   382
- 20.2. Ionization and Recombination   383
- 20.3. Energy Loss Mechanisms   386
- 20.4. Structure Equations   389
- 20.5. Model Nebulae   391
- 20.6. Relative Line Strengths   396
- 20.7. Thermal Radio Emission   398

Chapter 21
HYDRODYNAMICS   400
- 21.1. Reference Frames   400
- 21.2. Equations of Motion   402
- 21.3. Magnetohydrodynamic Effects   406
- 21.4. Cylindrical Coordinates   407

Chapter 22
THE VIRIAL THEOREM   409
- 22.1. General Form of the Virial Theorem   409
- 22.2. Stability (Macroscopic)   412

Chapter 23
STAR FORMATION   14
- 23.1. Matter Condensations and Star Formation   414
- 23.2. Linearized Hydrodynamic Equations   415
- 23.3. Effects of Rotation   419
- 23.4. Collapse of Isolated Clouds   420
- 23.5. Effects of Magnetic Fields   421
- 23.6. Fragmentation of Collapsing Clouds   425
- 23.7. Difficulties   429
- 23.8. Summary   430

Chapter 24
SUPERSONIC FLOW AND SHOCK WAVES   432
- 24.1. Shock Waves   434
- 24.2. Luminous Shock Waves   437
- 24.3. Ionization Fronts and Strömgren Spheres   440
- 24.4. Accretion onto Compact Objects   447

Chapter 25
DIFFUSE SUPERNOVA REMNANTS 451

- 25.1. Expanding Nebulae 451
- 25.2. Filamentary Structure 454
- 25.3. Nonthermal Radio Component 455
- 25.4. The Crab Nebula 463

Part 5
GALAXIES AND THE UNIVERSE 467

Chapter 26
THE EXPANDING UNIVERSE 468

- 26.1. Redshift and Expansion 468
- 26.2. Newtonian Cosmology 470
- 26.3. General Properties of Cosmological Models 473
- 26.4. Cosmological Redshifts 476
- 26.5. Cosmological Distances 478
- 26.6. The Primeval Fireball 481
- 26.7. The Mass Density of the Universe 487

Chapter 27
GALAXIES 489

- 27.1. Galactic Morphology 489
- 27.2. Surface Brightness of Galaxies 494
- 27.3. Galactic Masses 499
- 27.4. Stellar Content of Galaxies 504
- 27.5. General Characteristics of Galaxies 508

Chapter 28
DYNAMICS OF STELLAR SYSTEMS 512

- 28.1. Stellar Dynamics 512
- 28.2. Relaxation Times and Stellar Encounters 516
- 28.3. Globular Clusters 518
- 28.4. Clusters of Galaxies 524

Chapter 29
AXIALLY SYMMETRIC GALAXIES 532

- 29.1. Elliptical Galaxies 532
- 29.2. Spiral Galaxies 538
- 29.3. Rotation Curves 542
- 29.4. Force Laws 544
- 29.5. Galactic Mass Distribution 549
- 29.6. Noncircular Orbits 551

Chapter 30
SPIRAL STRUCTURE 555

- 30.1. Difficulties: Streaming Motions and the Winding Dilemma 556
- 30.2. Density-Wave Theory: Physical Picture 560
- 30.3. Spiral Density-Wave Theory: Formulation 561
- 30.4. Observational Consequences 571

Chapter 31
GALACTIC EVOLUTION 576

- 31.1. Formation of Galaxies 579
- 31.2. Stellar Populations 584

Appendix 1 Constants and Units A-1

Appendix 2 Atomic Mass Excesses A-3

Bibliography B-1

Index I-1

Preface to Volume II

Student interest in astronomy and astrophysics has grown dramatically during the past decade, and from it has sprung a need for modern texts reflecting the advances in theoretical and observational astronomy of the sixties and seventies. This need has previously been met largely at the introductory (descriptive) level and at the advanced level. *Astrophysics* is intended to fill the intermediate range. Much of the material developed from course material and lectures presented over a five-year period to upper-level undergraduate and graduate students in the departments of Astronomy and Physics at the University of Texas at Austin and from a one-semester course presented in the Department of Physics and Astronomy at Texas A & M University.

The second volume of *Astrophysics* emphasizes the nonstellar aspects of the field, including: the physics of interstellar matter; galactic dynamics; the structure and evolution of galaxies and systems of galaxies; and cosmology. This volume, like the first one, is intended for a senior-level or first-year graduate course in astrophysics. In this volume we present a relatively self-contained discussion of core topics that are established and basic to understanding advanced work in the field.

Some of the background material needed has been discussed in Volume I. In particular, Chapters 5 and 6 develop the physics of atomic and molecular spectroscopy and radiative transport, respectively. However, students who already have a background in these topics and who are primarily interested in nonstellar astrophysics can begin with Volume II.

Part I of Volume I contains an overview and a review of basic concepts in stellar astrophysics, and should be read by students who have no background in descriptive astronomy. In particular, Chapter 1 develops basic stellar parameters at the order-of-magnitude level that we will use throughout the present volume. Similar order-of-magnitude methods are used extensively in the discussions of nonstellar phenomena.

We acknowledge support by the Department of Astronomy and the Department of Physics at the University of Texas at Austin. We are indebted to the faculty and our colleagues for their criticisms and suggestions. In particular, we are grateful to Dimitri Mihalas and Austin Gleeson for valuable discussions and suggestions at the time the manuscript was being developed. We also thank Margaret Burbidge and William Kauffman, who read portions of the manuscript. We acknowledge Digicon Geophysical Corporation and Los Alamos National Laboratory for support during the final stages of manuscript preparation.

We would like to express special gratitude to Stewart Sharpless for the use of his composite (yellow positive on blue negative) photograph of NGC 5194-95, which appears on the title page of this volume; to T. D. Kinman for his assistance in selecting and obtaining photographs from Kitt Peak National Observatory; and to G. de Vaucouleurs for his help in locating photographs of elliptical galaxies. The final manuscript was completed largely through the encouragement and typing efforts of J. R. S. Bowers, to whom we owe a special debt of gratitude. Finally, we would like to thank the many students who read and patiently endured preliminary sets of notes.

January 1984

Richard L. Bowers
Terry Deeming

ASTROPHYSICS II
INTERSTELLAR MATTER AND GALAXIES

Part 4
THE INTERSTELLAR MEDIUM

Chapter 18

INTERSTELLAR MATTER

18.1. Physical Processes in Interstellar Gas

A major contribution to modern astrophysics has been the realization that the space between the stars is not devoid of interesting phenomena and challenging problems. It has long been guessed that some nebulae may be sites of active star formation, and that others are the vestiges of novae or supernovae eruptions. The matter in these objects is up to 10^6 times as dense as the average interstellar matter density, $\rho = n_0 m_H \approx 2 \times 10^{-24}$ g/cm^3. In the denser regions the discovery of dust grains, complex molecules, and evidence of large-scale (galactic) magnetic fields is yielding data about stars and how they form, and about the structure of the Galaxy itself.

Most of our knowledge about the interstellar medium is based on observations within the disk of our Galaxy, but observations of nearby spiral galaxies suggest that they contain similar interstellar matter. Excluding stars, the primary constituents of the medium are: gas (about 60 percent H and 30 percent He, with traces of heavier elements in approximately solar abundances), representing about 0.025 $M_\odot$/pc^3; dust or grains, about 0.002 M_C/pc^3; cosmic rays, 0.5 eV/cm^3; Galactic magnetic fields (H $\sim 10^{-6}$ gauss), 0.2 eV/cm^3; and starlight, 0.5 eV/cm^3. We have not included the rotational energy of the differential rotation of the Galactic disk, or turbulent energy resulting from disordered gas motions.

Many phenomena in the interstellar medium are hydrodynamic in nature, and some include magnetic fields, which, though weak, nevertheless have a major influence on the state of gas motion. Local temperatures vary from 10 K to over 10^4 K, and the radiation mean free path λ_ν may vary so much that a region may be transparent at some energies, but opaque at others. Furthermore, in some regions the gas contains electrons and ions, which, to be described thermodynamically, require different temperatures. Finally, the interstellar medium is continually evolving as new stars form out of it, and others eject matter (often enriched by stellar nucleosynthesis) back into it. The motion of spiral features in the Galactic disk induces shock waves in the interstellar gas; matter is continually being pumped into the disk by the Galactic nucleus, and small amounts may also be accreting onto the disk from the Galactic corona. Therefore, the large-scale dynamic behavior of the Galaxy itself contributes to the evolution of the medium.

Nuclear reactions, which are fundamental to stellar processes, play no role in the interstellar medium. Their role is replaced by atomic processes, including forbidden transitions, which do not occur in denser stellar matter. Molecular processes are also important, and are particularly useful for mapping out the nonstellar structure of the Galaxy, and as indicators of physical conditions in dense gas clouds.

The presence of diffuse matter between the stars was first deduced from stationary absorption features observed in spectra taken of bright stars in binary systems. The lines originating in the stellar atmospheres exhibit the expected periodic variation in wavelength about their natural value (Doppler shift). For some stars, however, the spectra also contain fixed absorption lines corresponding to optical transitions in CaI, CaII, and Na(D lines). Because these absorption lines show no evidence of periodic variation, they can not originate in the stellar atmosphere, but must originate between the binary system and the observer. The absorption lines show a complex structure, consisting of several lines corresponding to a single atomic transition. Each line appears displaced from its natural wavelength by an amount indicating that the absorbing gas is distributed in distinct clumps, or clouds, and that these clouds have random velocities of order 10 km/sec, though a few exceed 50 km/sec. Satellite-based observations that cover the ultraviolet spectrum show similar absorption lines for $H(L_\alpha)$, N, O, Mg, Si, S, Ar, Mn, and Fe. Measurements of relative line strengths for these elements can be used to estimate chemical abundances in interstellar gas. Abundances vary from region to region, but show a trend toward cosmic abundances.

Forbidden atomic transitions, which play no major role in stellar atmospheres or interiors (Section 6.1), are common in the interstellar medium. To appreciate their importance, consider a typical region where the number density of hydrogen atoms $n_H \approx 1$ cm^{-3}. Trace elements, such as Ca, O, or N, may be raised to excited levels as the result of collisions with other atoms. Collisional cross sections between neutral atoms are of order 10^{-15} cm^2; so the collisional mean free path is

$$l_c \approx (n_H \sigma_c)^{-1} = \frac{10^{15}}{n_H} \text{ cm}. \quad (18.1)$$

If the gas is at a temperature T, which is related to the mean atomic velocity by $3 m_H v^2 / 2 = kT$, then their collision rate is of order

$$\frac{1}{\tau_c} \approx \frac{v}{l_c} \approx \left(\frac{2kT}{3m_H}\right)^{1/2} n_H \sigma_0$$
$$= 7 \times 10^{-12} n_H T^{1/2} \text{ sec}^{-1}. \quad (18.2)$$

In nonionized H, $T \approx 80$ K, and, according to (18.2), the time between collisions is about 500 yrs. Many collisionally excited states in the interstellar medium are metastable; that is, decay to the ground state occurs only by forbidden transitions (Section 6.1), whose decay times vary from seconds to thousands of years. Because collisions are rare, atoms will persist in metastable states until spontaneous decay to the ground state occurs. At the relatively low temperatures common in much of the interstellar gas, nearly all atoms, especially H and He, will be in their ground states. An energy of about 10 eV is required to promote H to its first excited state (1s $\rightarrow$ 2s) and about 21 eV is required for He. These energies would require temperatures in excess of 10^5 K to be effected by collisions; so H and He scatter elastically on each other (the excitation of hyperfine transitions will be discussed later). The energies available from collisions in low-temperature regions lie in the range 10^{-2} to 10^{-3} eV, and can excite hyperfine transitions in ions such as CII. The excited state can decay to the ground state only by a forbidden transition. Since the collision rate is much smaller than the forbidden transition rate, the ion eventually decays, producing a photon in the radio or the infrared. Processes such as these convert thermal energy of the gas into radiation, and can act as cooling mechanisms.

Molecules also play an important role in the physics of interstellar matter, though they probably represent only a small fraction of its mass. Molecular transitions occur in emission and absorption from the far ultraviolet to radio frequencies. Electronic transitions involving electrons in molecular bound states extend from the far ultraviolet through the visible. Vibrational states result from harmonic oscillations of the nuclei about their equilibrium separation. For example, in a linear diatomic molecule whose two oscillating nuclear masses are m_1 and m_2, and whose instantaneous separation is r, the classical Coulomb force between the two nuclei can be written as

$$F_{\text{Coul}} = \frac{Z_1 Z_2 e^2}{r^2} \approx F_0 - \frac{2 Z_1 Z_2 e^2}{r_0^3} \delta r, \quad (18.3)$$

where $r = r_0 + \delta r$, and F_0 is $F_{\text{Coul}}(r_0)$, the force between $Z_1 e$ and $Z_2 e$ in equilibrium. What is neglected in this classical analysis (and can only be adequately treated by using quantum mechanics) is the counterbalancing effect of the molecular electron distribution. We will include it implicitly by assuming that an equilibrium state exists for finite r_0. The relative acceleration of the two oscillating nuclei is

$$\frac{d^2 r}{dt^2} = \frac{d^2 r_0}{dt^2} + \frac{d^2 (\delta r)}{dt^2}, \quad (18.4)$$

and the relative force must be $\mu d^2 r/dt^2$, where $\mu = m_1 m_2/(m_1 + m_2)$ is the reduced mass of the linear molecule. Equating the force obtained from (18.4) and F_{Coul},

$$\mu \frac{d^2 \delta r}{dt^2} = -\frac{2 Z_1 Z_2 e^2}{r_0^3} \delta r + \left(F_0 - \mu \frac{d^2 r_0}{dt^2} \right). \quad (18.5)$$

The quantity in parentheses vanishes by definition of the equilibrium state; so the displacement δr satisfies the differential equation for a simple harmonic oscillator with frequency

$$\omega = \left(\frac{2 Z_1 Z_2 e^2}{\mu r_0^3} \right)^{1/2}. \quad (18.6)$$

The energy levels for a quantum-mechanical oscillator with frequency $\nu = \omega/2\pi$ are described, for integer n, by

$$\begin{aligned} E_{\text{vib},n} &= (n + 1/2) \hbar \omega \\ &= (n + 1/2) 0.4 (Z_1 Z_2/r_0^3)^{1/2} \text{eV}, \end{aligned} \quad (18.7)$$

where r_0 is in Å. The energy of the ground state (zero-point energy) results for $n = 0$. The measured values of r_0 for H_2, CH^+, and CN are 0.74 Å, 1.12 Å, and 1.17 Å, respectively. Transitions between vibrational states produce radiation in the wavelength range $1 \lesssim \lambda \lesssim 20\mu$ ($1\mu = 10^{-4}$ cm). Vibrational dissociation of H_2 occurs for energies above $E_B = 4.48$ eV under laboratory conditions. Molecular photodissociation can generally be induced by photons of frequency E_B/h. The highly symmetric H_2 molecule is an exception, however, and generally requires photon energies of more than 14.7 eV.

Rotational degrees of freedom can also be excited in molecules. The corresponding radiation lies in the far infrared and radio spectrum. The rotational kinetic energy and angular momentum of a classical molecule are related by

$$\begin{aligned} E_{\text{rot}} &= \frac{1}{2} I \Omega^2 = \frac{(I\Omega)^2}{2I} = \frac{L^2}{2I} \\ &= j(j+1) \hbar^2/2I. \end{aligned} \quad (18.8)$$

In the last step we use the quantum-mechanical restriction on angular momentum $L = \sqrt{j(j+1)}\hbar$ (Section 6.1). For a linear diatomic molecule consisting of nuclear masses m_1 and m_2, lying at $\mathbf{r}_1$ and $\mathbf{r}_2$ relative to the center of mass, the moment of inertia is

$$I = m_1 r_1^2 + m_2 r_2^2 = \frac{m_1 m_2}{m_1 + m_2} r^2 = \mu r^2, \quad (18.9)$$

where $\mathbf{r} = \mathbf{r}_1 - \mathbf{r}_2$, and μ is the reduced mass. Denoting each mass by $A_i m_H$, we can combine (18.9) and (18.8) to yield

$$\begin{aligned} E_{\text{rot},j} &= \frac{\hbar^2 j(j+1)}{2 m_H r^2} \frac{A_1 + A_2}{A_1 A_2} \\ &= 2.08 \times 10^{-3} \left(\frac{A_1 + A_2}{A_1 A_2} \right) \\ &\quad \times \frac{j(j+1)}{r_0^2} \text{eV}, \end{aligned} \quad (18.10)$$

where r_0 is in Å. The energy change associated with a transition from state j to $j - 1$ follows from (18.10):

$$\begin{aligned} \Delta E_{\text{rot},j} &= \frac{\hbar^2 j}{I} \\ &= 4.16 \times 10^{-3} \left(\frac{A_1 + A_2}{A_1 A_2} \right) j \text{ eV}. \end{aligned} \quad (18.11)$$

Finally, we note that each rotational energy state is degenerate (Section 6.1) with multiplicity $g_j = (2j + 1)$.

Hyperfine transitions (Section 6.1) also occur in molecules, and can couple with electronic degrees of freedom in molecular ions such as OH^-. Figure 18.1 shows the ion in its lowest rotational state ($j = 1$); the shaded lobes represent the quantum-mechanical distribution of the extra electron that gives the ion its charge. In the absence of rotation, the two states (a) and (b) are indistinguishable. In the first rotational state, coupling between electron spin and molecular

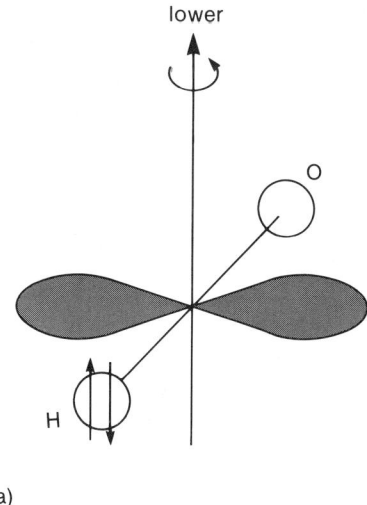
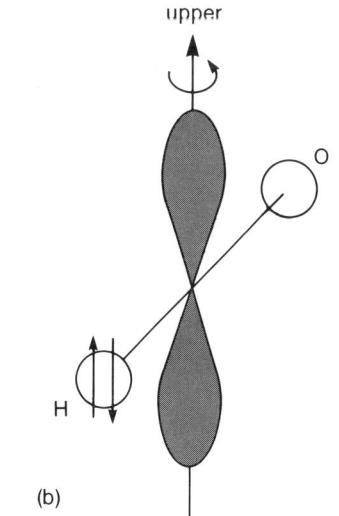
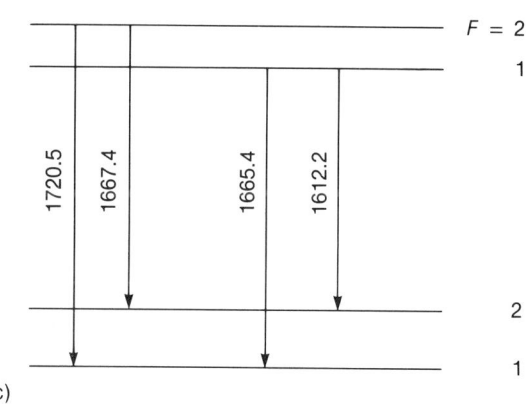

Figure 18.1. Electronic states of the OH^-. (a,b) The oxygen and hydrogen atoms making up the ion are shown on the x-axis. The shaded lobes represent the quantum-mechanical distribution of the electron that gives the ion its net charge (the electron is shared by the two atoms). The arrows show the two possible spin states of the proton. (c) Energy levels of the OH^- ion including rotational splitting.

rotation splits the energy level into upper and lower states. Finally, the spin of the proton in the hydrogen nucleus can have two possible values, and hyperfine interactions between the proton's spin and the total molecular angular momentum $J = 3/2$ split each level into a doublet. The doublet levels have total angular momentum $F = 2$ (when molecular and proton angular momenta are parallel) and $F = 1$ (when molecular and proton angular momenta are antiparallel). The frequencies in Hz of the resulting emission lines are shown in Figure 18.1(c).

Interstellar molecular absorption lines have been detected for many molecules such as CN, CH, and CO, as well as for radicals such as OH^-. Absorption of starlight by the first two rotational levels of CN has been used to measure the temperature of the cosmic background radiation. Transitions between vibrational states have also been observed at wavelengths of 0.04μ to 1μ, corresponding to energies of a few tenths eV. Vibrational excitation therefore requires higher temperatures, typically 10^3 K.

The gaseous component of the interstellar medium can be conveniently categorized by the ionization state of its dominant species, atomic hydrogen. Approximately 95 percent of the mass in interstellar gas is contained in the HI component made up of neutral hydrogen (HI), H_2, and neutral helium (HeI); less than 1 percent is heavy elements and molecules. Most of the data for the HI component comes from absorption of starlight in the ultraviolet and 21.1-cm emission by atomic hydrogen. Observations clearly indicate that most HI is concentrated in clouds confined to a narrow disk in the Galactic plane, whose scale height is about 250 pc. The clouds vary in size, mass, and temperature (see Table 18.1), though all are relatively cold. The most important difference between diffuse and dark clouds is their transparency to ultraviolet radiation. Ultraviolet photons readily penetrate diffuse clouds and play an important role in heating the gas. They also effectively dissociate molecules that may be present. Dark clouds, which contain relatively high concentrations of dust grains, are essentially opaque in the ultraviolet. Consequently they are cooler, and are found to be richer in molecules. The temperature of the denser clouds is estimated from molecular emission, particularly of CO molecules. The central regions of dark or molecular clouds probably have cooled to about 2.7 K, in equilibrium with the cosmic background radiation. Cloud sizes range from 10–100 pc, implying masses in the range 10^2–10^5 $M_\odot$. The composition of an HI cloud can be estimated by taking the ultraviolet spectrum of a bright O-type star located behind the cloud, and comparing it with the spectrum of a nearby star, of similar spectral type, whose line of sight does not traverse the cloud. The results (Figure 18.2) vary from cloud to cloud, but are in general agreement with the cosmic abundances for H and He. C, N, and O are underabundant by about 70 percent, and the refractory elements are almost entirely absent. For example, $N(Fe)/N(Fe)_{cosmic} \lesssim 10^{-2}$, but for Ca and Al the ratio is about 10^{-1}. HI clouds contain interstellar dust grains, which may consist primarily of refractory elements. The observed underabundances probably result because the grains formed from the available heavy elements in the cloud (see Chapter 19).

Table 18.1
Properties of interstellar clouds. Numbers in parentheses are most typical, or average, values. T_k is the kinetic temperature obtained from absorption line strengths.

Type of cloud	T_k(K)	n_H(cm^{-3})	Composition
Diffuse clouds	50–150 (80)	10–10^3 (20)	Mostly HI $n_e = 10^{-4} n_0$ $M/M_\odot \approx 400$ $R \approx 5$ pc
Clouds near HII regions	30		H, H$_2$, and some molecules (CO)
Dark clouds (molecular clouds)	3–10 (10)	10^3–10^6 (2×10^3)	Mostly H$_2$, molecules, and dust. $n_0 \gtrsim 10^4$ cm^{-3} $M/M_\odot \approx 300$ $R \approx 1$ pc

The abundance of free electrons n_e/n_H in diffuse clouds is estimated to be a few times 10^{-4}. They arise primarily from photoionization of atoms like C, and from low-energy cosmic rays. Thermal ionization of these elements would require temperatures far in excess of those observed.

The HI clouds are surrounded by a more nearly uniform, dilute hydrogen gas whose presence is revealed by ultraviolet absorption lines of H, deuterium, and O, indicating a temperature between 4 and 8×10^3 K, and a neutral-hydrogen number density $0.05 \lesssim n_H(HI) \lesssim 0.2$ cm^{-3}.

Problem 18.1. Find the number density of electrons and protons in a gas at 6,000 K if $n_H(HI) = 0.1$ cm^{-3}.

HII regions are generally located near O-type stars, whose spectra peak in the ultraviolet. Radiation at wavelengths below 912 Å will ionize essentially all the H within a cavity containing the bright star (Chapter 20). Optical and radio emissions from HII regions imply n_e up to 2×10^4 cm^{-3}, and temperatures between 7,000 and 10,000 K. The presence of dust grains is also indicated by scattered and polarized starlight, and infrared emission (Chapter 19); they probably account for a few percent of the mass of the HII region. Most of the mass is in ionized hydrogen and He, with trace amounts of N, O, Ne, and S in essentially cosmic abundances. As with HI clouds, there is a noticeable underabundance of refractory elements, which are presumably contained in dust grains.

The HI and HII components described in the preceding make up most of the interstellar gas. In addition to these components, a very hot coronal gas is believed to exist between HI regions. When matter escapes from stars into the interstellar medium at speeds in excess of 10^3 km/sec (from novae, supernovae, or extremely bright stars), a shock wave forms, heating the interstellar gas to 10^6 K or higher. At these temperatures H and He will be completely ionized, and the heavy elements will be highly ionized. Gas at 10^6 K will emit soft x rays in the .2 to 2 keV range, contributing to the Galactic x-ray background. Absorption in [OVI], which sets a lower limit on the coronal temperature, requires that the number density of atomic oxygen $n_0 \approx 10^{-8}$ cm^{-3} in the coronal gas. Assuming normal cosmic abundances, the implied hydrogen density is $n_H \approx 10^{-4}$ cm^{-3}.

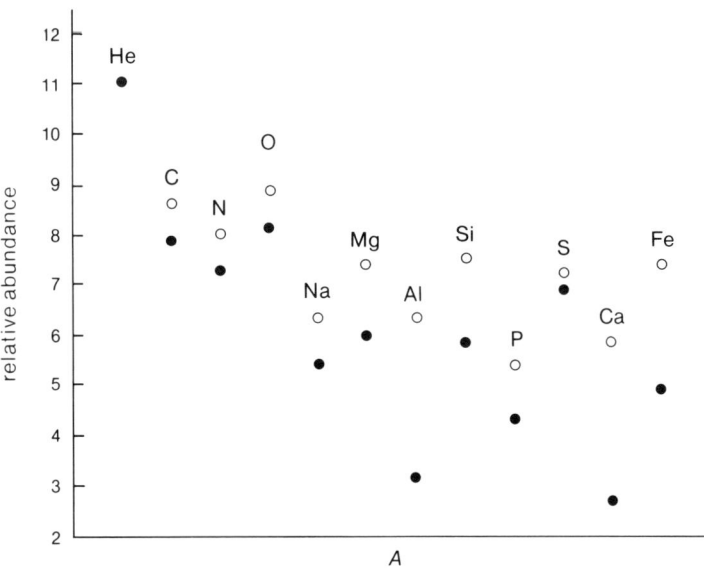

Figure 18.2. Relative abundances of chemical elements in HI clouds in the direction of ζ Oph (filled circles); open circles are cosmic abundances. Abundances are normalized to log $N(H) = 12$.

The picture that emerges from observations of the interstellar gas is far from simple. Since nearby regions often differ in temperature by several orders of magnitude, large-scale dynamic motions are to be expected. The extremely tenuous character of the gas and the presence of strong energy sources (typically stars) can result in thermal instabilities, or large-scale deviations from thermodynamic equilibrium. In addition to these problems, the effects of dust grains and of large-scale Galactic magnetic fields, cosmic rays, and variations in density caused by spiral structure must eventually be considered. Many of these topics will be considered in subsequent sections.

18.2. THERMAL STATES OF INTERSTELLAR GAS

For most purposes it can be assumed that the constituents in stars are in local thermodynamic equilibrium, because emission, absorption, scattering, and collision processes occur on time-scales that are many orders of magnitude smaller than the shortest dynamical or evolutionary time-scales, except (probably) for the neutrinos emitted during core collapse (Section 15.2). The dilute matter making up much of the interstellar medium is more complex to deal with, particularly when energy sources (hot stars, cosmic-ray or x-ray background, or shock waves) are present. This section will review the conditions that obtain for a gas in the thermodynamic equilibrium conditions encountered in interstellar space, and the approach to this state. Some of the arguments developed here can be modified slightly (replacing the Coulomb scattering by scattering in a gravitational field) to obtain results applicable to problems in stellar dynamics (Chapter 28).

Consider a gas containing a single atomic species (the arguments are readily extended to more species) and photons. We will have a reasonably complete specification of the state of the gas if we know: the radiation spectrum; the particle velocity distribution; and the distribution of atoms among excited states and among ionization states. All these characteristics can readily be defined without reference to thermodynamic equilibrium. The radiation spectrum can be integrated to supply the total-radiation energy density u, which can be used to define a temperature parameter T_R, called the *radiation temperature*, by the relation $u = aT_R^4$. If the radiation is in thermodynamic equilibrium, its spectrum will be Planckian, and a second parameter T_c, called the *color* or *spectral temperature*, may be defined. If the particle (atom, electron, or ion) velocity distribution is Maxwellian, then the *kinetic temperature* T_k may be defined (note that atoms, electrons, and ions may each have a different kinetic temperature). If the distribution of atomic excited states satisfies the Boltzmann relation, then an *excitation temperature* T_E is defined. Finally, if the ionization state is described by a Saha-type equation, we may define an *ionization temperature* T_I. In principle, some or all of these temperatures may be different; but if the gas is in local thermodynamic equilibrium, all the temperatures will be the same. In stellar interiors, local thermodynamic equilibrium is assumed to hold, and we have only one temperature, T. Deviations from thermodynamic equilibrium, though rare for stars, are not uncommon in nonstellar environments. For example, the inside of a terrestrial room on a sunny day is not in complete thermodynamic equilibrium. Typically, $T_k = T_I = T_E \approx 300$ K for all material particles. However, the sunlight filling the room, emitted by the solar surface, has a spectral temperature $T_c \approx 6{,}000$ K. Further, the energy emitted from the Sun at T_c expands and is diluted by a factor $(d/R_\odot)^2$ by the time it reaches Earth; so the radiation energy density near the Earth's orbit is

$$u_\oplus \approx aT_c^4 (R_\odot/d)^2 \approx 2 \times 10^{-4} \text{ erg/cm}^3$$

(d is the radius of the Earth's orbit). This implies a radiation temperature

$$T_R \approx (u_\oplus/a)^{1/4} \simeq 400 \text{ K}.$$

Evidently $T_c \gg T_R > T_k$, and the contents of the room are not in thermal equilibrium.

Often a steady state is reached in which the usual concept of LTE does not apply. For example, a number of dust grains of radius a suspended inside an evacuated glass bell jar through which sunlight passes will be heated to a temperature T_D as they absorb radiation. If the atoms in the grain are in thermal equilibrium, the grains will lose energy by emission of infrared radiation at a rate $4\pi a^2 \sigma T_D^4 \epsilon_E$, where ϵ_E is the grain's emissivity. A steady state results when the energy-loss rate equals the rate of energy gain. If the grain's absorptivity is ϵ_A, then

$$T_D = \left(\frac{\epsilon_A}{\epsilon_E} \frac{R_\odot^2}{d^2} \frac{T_\odot^4}{4} \right)^{1/4}, \qquad (18.12)$$

where $T_\odot = T_c$, the color temperature. For a gray body $\epsilon_A = \epsilon_E$, and (18.12) gives $T_D \simeq 300$ K. Most of the starlight in the interstellar medium comes from the relatively few bright stars whose effective temperatures are of order 10^4 K, and grain temperatures are about 10 K. We note that (18.12) does not include collisional effects.

Problem 18.2. Derive (18.12) for a spherical grain of radius a.

It is obvious from the preceding examples that one or more temperature parameters can often be assigned to matter distributions and radiation, even though a complete state of thermal equilibrium may not exist. When discussing "the temperature" of such a system, one must remember which parameter is meant, though this is often obvious from the context.

Problem 18.3. What temperature parameters were used in solving Problem 18.1, and what relation between these was implicitly assumed in the solution?

The approach to a steady state in a diffuse HI cloud is illustrative. The number of free electrons expected at $T \approx 80$ K and $n_H(\text{HI}) \approx 20$, assuming LTE, is entirely negligible. However, ionization will result from processes such as

$$\gamma + \text{H} \rightarrow \text{H}^+ + e^-, \qquad (18.13)$$

$$A_c + \text{H} \rightarrow \text{H}^+ + e^- + A_c, \qquad (18.14)$$

$$\gamma + A_i \rightarrow A_i^+ + e^-, \qquad (18.15)$$

where γ denotes a photon from the Galactic soft x-ray background ($0.2 \lesssim h\nu \lesssim 2$ keV), and A_c denotes a low-energy cosmic ray ($E \approx$ MeV). In (18.15) the most important nuclei A_i are those whose ionization potentials are low (see Table 6.1) and which are relatively common, for example, Si and Ca. If only these processes operated, then eventually the entire gas would become extremely hot. The photoelectrons produced by (18.13) and (18.15) have energies given by

$$\epsilon_e = \frac{1}{2} m_e v_e^2 = h\nu - \chi_i, \qquad (18.16)$$

where χ_i is the ionization energy of hydrogen or of the heavy nucleus A_i. Typically ϵ_e is comparable to χ_i. Elastic collisions between photoelectrons and ions

$$e^- + A(i) \to e^- + A(i) \tag{18.17}$$

gradually establish an isotropic electron-velocity distribution, while electron-electron scattering forces the electrons to a Maxwellian velocity distribution, which maximizes the electrons' entropy. The result is a thermal distribution of electrons with kinetic temperature $T_{k,e}$. The initial neutral gas is assumed to have a kinetic temperature $T_{k,H}$. Collisions between H and e^- will eventually heat the neutral gas. Given enough time, the entire system would reach a common temperature comparable to χ_i, of order 10^7 K or more. Clearly this does not happen in diffuse HI clouds (Table 18.1), which is hardly surprising, because cooling effects have not been taken into consideration.

In addition to the elastic process (18.17), the photoelectrons will inelastically scatter,

$$\begin{aligned} e^- + A(i) &\to A^*(i) + e^- \\ &\to A(i) + e^- + \gamma, \end{aligned} \tag{18.18}$$

and the excited nuclei will decay, emitting a photon. In HI clouds, a particularly important cooling process is the collisional excitation of CII:

$$e^- + C^+(^2P_{1/2}) \to C^+(^2P_{3/2}) + e^-. \tag{18.19}$$

The ground state of the C^+ atoms in HI clouds is the $^2P_{1/2}$ state. The excitation energy is 0.0079 eV, and the transition to the ground state is forbidden, since it involves a change in the electronic spin state, but no change in orbital angular momentum. Processes similar to (18.19) can also occur in P states of NII and OIII, and in D states of OII. The excitation temperature corresponding to (18.19), if all species are in equilibrium, is 92 K. However, the emitted photon easily escapes the cloud. The actual steady-state temperature results when the energy-gain rates Γ due to all heating processes [such as (18.13)] equal the energy-loss rates Λ due to all cooling processes [such as (18.19)]. It is notable that the final steady state attained in HI clouds is strongly influenced by the abundance of trace elements, particularly CII, which make up an extremely small fraction (less than 10^{-3}) of the cloud's mass. These issues will be treated in detail in Chapter 20.

Problem 18.4. Consider the elastic collision of two point masses m and M in one dimension, and show that the fractional change in kinetic energy and momentum of m is given by

$$\begin{aligned} \left(\frac{\Delta E}{E_i}\right)_m &= -\frac{4mM}{(m+M)^2}, \\ \left(\frac{\Delta p}{p_i}\right)_m &= -\frac{2M}{m+M}, \end{aligned} \tag{18.20}$$

where subscript i denotes initial values.

Collisional Relaxation Times

It should be clear from the preceding discussion that the rate of equilibration for gas and radiation in the interstellar medium will be strongly influenced by atomic, ionic, and electronic collisions. A crude estimate of the collision rate for electrically neutral atoms is given by (18.2). When the colliding particles are electrically neutral, τ_c is relatively easily obtained; the particles collide only if they touch in the classical sense. Electrons and ions are, however, charged and the forces between them are long range. Because interstellar gas is neutral on a macroscopic scale, even when partially or completely ionized, the interaction between two distant charges will be weakened (screened). It will be assumed that only Coulomb interactions occur between charged particles. In particular, quantum effects and excitation will be ignored. To the extent that this limitation is reasonable, the principal result of collisions (scattering) will be to deflect the incident particle, and to exchange momentum and energy with the scatterers. The primary result of deflection is to produce an isotropic particle-velocity distribution, which will generally be Maxwellian (2.16). Energy exchange simply redistributes the kinetic energy of the various particles. The rate for each process will be obtained in what follows. For a specific process denoted by subscript i, we can define the rate of change of the quantity Q by

$$\frac{dQ_i}{dt} = -\frac{f_i}{\tau_i}, \tag{18.21}$$

where f_i is a function of Q_i; the relaxation time τ_i for the process will depend on the state of the plasma, and may also depend on Q_i. The functional form of f_i and τ_i can be found by analyzing the change in Q_i caused by the scattering force.

First consider the change in momentum experienced by a group of test particles of mass m_A and charge $Z_A e$ incident on particles of mass m_B and charge $Z_B e$ (Figure 18.3). The change in momentum of a test particle scattering off a particle B is given by

$$\Delta \mathbf{p}_A = m_A \Delta \mathbf{v}_A = \mu \Delta \mathbf{u},$$
$$\mathbf{u} = \mathbf{v}_A - \mathbf{v}_B, \quad (18.22)$$

$$\mu = m_A m_B / (m_A + m_B). \quad (18.23)$$

The relative velocity of the two charges is $\mathbf{u}$, and μ is then reduced mass. For Coulomb forces, the scattering angle θ and the impact parameter s are related by

$$\sin^2 \theta/2 = (1 - \cos\theta) = \frac{s_\perp^2}{s^2 + s_\perp^2}, \quad (18.24)$$

$$s_\perp = \frac{Z_A Z_B e^2}{\mu u^2}, \quad (18.25)$$

where $s_\perp$ is the impact parameter for scattering at $\pi/2$ in the laboratory frame. Any particle of type A whose initial impact parameter and relative velocity at large distances from m_B are s and u, respectively, will pass through the annular area $2\pi s\, ds$ (Figure 18.3) and scatter into a cone of angular thickness $d\theta$ about the θ direction. The quantity $2\pi s\, ds = d\sigma$ is therefore a cross section. It follows from (18.24) that the differential cross section for Coulomb scattering is

$$\frac{d\sigma}{d\Omega} = 2\pi s \frac{ds}{d\Omega} = \left(\frac{Z_A Z_B e^2}{\mu u^2}\right)^2 \frac{1}{4 \sin^4 \theta/2}, \quad (18.26)$$

where $d\Omega = 4\pi \sin\theta\, d\theta$ is the solid angle. Now, consider a gas containing particles of type B, which will be called field particles, into which is introduced a particle of type A, which will be called the test particle (a photoelectron ejected into a partially ionized plasma is a typical example). The test particle A scatters off all the field particles, because the Coulomb force is long range. The rate of change in the momentum of a test mass (whose initial velocity is $\mathbf{v}_A$) caused by scattering off a field particle of velocity $\mathbf{v}_B$ is given by the product of the test-particle flux u (assuming one per unit volume), the momentum change $\mu \Delta \mathbf{u}$, integrated over the cross section $d\sigma$, and the distribution of field particles $f_B(v_B)$:

$$\frac{d\mathbf{p}_A}{dt} = -\int f_B(v_B)\, d^3 v_B \int_0^\infty u\mu \Delta \mathbf{u}\, d\sigma. \quad (18.27)$$

The first integral is over all values of $\mathbf{v}_B$. If we define spherical coordinates about the scattering center m_B with the Z axis as shown in Figure 18.3, the change in relative velocity has components

$$\Delta u_x = u \sin\theta \cos\phi,$$
$$\Delta u_y = u \sin\theta \sin\phi, \quad (18.28)$$
$$\Delta u_z = -u(1 - \cos\theta).$$

It should be obvious that physically there is no change in $\mathbf{p}_A$ in the x or y directions if the distribution of field particles is isotropic. Therefore $d\mathbf{p}_A/dt$ will have only a z component. Thus combining (18.28) and (18.27) yields

$$\left(\frac{dp_A}{dt}\right)_z = -\mu \int f_B(\mathbf{v}_B)\, dv_B$$
$$\times 2\pi \int_0^\infty u^2 (1 - \cos\theta) s\, ds$$
$$= -2\pi\mu \int f_B(v_B)\, d^3 v_B$$
$$\times u^2 \int_0^\infty \frac{s_\perp^2}{s^2 + s_\perp^2} s\, ds, \quad (18.29)$$

where (18.24) has been used to eliminate $\cos\theta$. The integral over the impact parameter s diverges logarithmically as $s \to \infty$. This is characteristic of Coulomb scattering where the range of the interaction is infinite, and simply means that no matter how large the impact parameter becomes, there will always be some deflection. However, in an electrically neutral plasma, the scattering charge $Z_B e$ will be screened by free electrons (see Section 7.6), and for impact parameters much greater than the Debye length λ_D given by (7.86), essentially no scattering occurs. Consequently, the upper limit in the integral over s in (18.29) may be replaced by λ_D, and

$$\int_0^{\lambda_D} \frac{s_\perp^2}{s^2 + s_\perp^2} s\, ds = s_\perp^2 \ln \frac{\lambda_D^2 + s_\perp^2}{s_\perp^2}$$
$$\approx 2 s_\perp^2 \ln \lambda_D / s_\perp. \quad (18.30)$$

In the last step we have assumed that $\lambda_D \gg s_\perp$, which is usually valid for interstellar gas. Substituting (18.30)

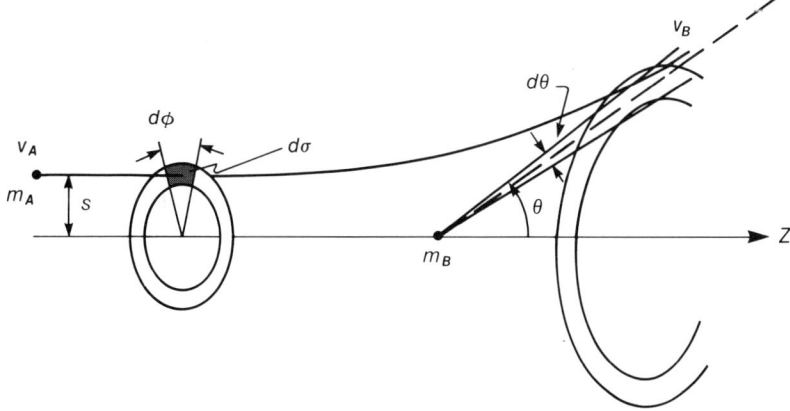

Figure 18.3. Coulomb scattering of test charge A incident on charge B. The impact parameter is s.

into (18.29) yields

$$\frac{d\mathbf{p}_A}{dt} = -4\pi\mu \left(\frac{Z_A Z_B e^2}{\mu}\right)^2$$
$$\times \int f_B(v_B) \frac{\mathbf{u}}{u^3} \ln \Lambda_c \, d^3 v_B$$
$$\simeq -4\pi \frac{Z_A^2 Z_B^2 e^4}{\mu} \ln \Lambda_c$$
$$\times \int f_B(v_B) \frac{\mathbf{u}}{u^3} d^3 v_B, \qquad (18.31)$$

where

$$\ln \Lambda_c = \ln \frac{\lambda_D}{s_\perp} = \ln \left(\frac{\mu u^2}{Z_A Z_B e^2} \lambda_D\right) \qquad (18.32)$$

is the Coulomb logarithm. Although Λ_c depends on $\mathbf{v}_B$ through the relative velocity u, $\ln \Lambda_c$ varies slowly, and has been taken outside the integral. For a plasma under interstellar conditions, we will set

$$\lambda_D = \left(\frac{kT}{4\pi n Z_A Z_B e^2}\right)^{1/2}. \qquad (18.33)$$

For most applications to interstellar plasma, T is the kinetic temperature of the electrons $Z_A = 1$, and $n = n_e$. We find $\lambda_D = 47.7 \, (T_e/n_e)^{1/2}$ cm.

Further evaluation of $d\mathbf{p}_A/dt$ requires knowledge of the field-particle distribution function. It is usually assumed to be Maxwellian, with kinetic temperature $T_{k,B}$:

$$f_B(\mathbf{v}_B) = \frac{4}{\sqrt{\pi^3}} \alpha_B^{3/2} v_B^2 e^{-\alpha_B v_B^2} n_B, \qquad (18.34)$$
$$\alpha_B = m_B/2kT_B,$$

where n_B is the number density of field particles. Substituting (18.34) into (18.32) produces an integral over the field-particle velocities that is straightforward though tedious to evaluate. Omitting intermediate details, we find

$$\int f_B(v_B) d^3 v_B \frac{\mathbf{u} \cdot \hat{\mathbf{v}}_A}{u^3}$$
$$= 2\pi n_B \left(\frac{m_B}{2\pi k T_B}\right)^{3/2} \int_0^\infty v_B^2 dv_B \exp\left(-\frac{m_B v_B^2}{kT_B}\right) I$$

$$I(v_B) = \int_{-1}^{1} \frac{(v_A - v_B x) \, dx}{(v_A^2 + v_B^2 - 2 v_A v_B x)^{3/2}}$$
$$= \frac{2\theta(v_A - v_B)}{v_A^2}.$$

The final integral over v_B cannot be evaluated in closed form; since $v_A \gg v_B$ for most field particles in the interstellar gas, the integral may be taken to infinity. We then find

$$\int f_B(v_B) \, d^3 v_B \frac{\mathbf{u} \cdot \hat{\mathbf{v}}_A}{u^3} \approx \frac{n_B}{4\pi},$$

and the momentum change is given by

$$\frac{d\mathbf{p}_A}{dt} = -4\pi \frac{Z_A Z_B e^4}{\mu v_A^2} n_B \ln \Lambda_c = -\frac{\mathbf{p}_A}{\tau^D_{AB}}. \quad (18.35)$$

The last expression defines the deflection time τ^D_{AB}. For photoelectrons incident on ions of charge Z_i,

$$\tau^D_{ei} \simeq \frac{m_e^2 v_e^3}{4\pi Z_i^2 e^4 n_i \ln \Lambda_c},$$

where

$$\ln \Lambda_c = \ln \frac{m_e v_e^2}{Z_i e^2} \left(\frac{kT}{4\pi n_e e^2 Z_i} \right)^{1/2}. \quad (18.36)$$

Near O-type stars, $n_e \approx 1 \text{ cm}^{-3}$, and $T_e \approx 10^4$ K, the electron velocity $v_e \approx (3kT_e/m_e)^{1/2}$ and $\ln \Lambda_c = 23$. The deflection time is $\tau_{ei} = 1.7 \times 10^4$ sec.

The deflection time τ^D_{AB} can also be applied to electron-electron scattering or ion-ion scattering. It is readily verified that for electrons and protons ($Z_i = 1$),

$$\tau_{ee} = \frac{n_p}{n_e} \tau_{ep} = \left(\frac{m_p T_{k,p}^3}{m_e T_{k,e}^3} \right)^{1/2} \tau_{pp}. \quad (18.37)$$

Since $n_e \simeq n_p$ in interstellar gas, $\tau_{ee} \approx \tau_{ep}$, and electrons are slowed down as rapidly by ions as by other electrons. If $T_{k,i} \approx T_{k,e}$, then $\tau_{ee} \approx (m_p/m_e)^{1/2}\tau_{pp} \approx 40\, \tau_{pp}$.

Problem 18.5. The Coulomb scattering cross section may be defined by

$$\sigma_{\text{Coul}} = \int_{\theta_m}^{\pi} (1 - \cos\theta) \frac{d\sigma}{d\Omega} d\Omega, \quad (18.38)$$

where $d\sigma/d\Omega$ is given by (18.26) and θ is the scattering angle. Show that

$$\sigma_{\text{Coul}} \simeq \pi (2s_\perp)^2 \ln(\lambda_D/s_\perp), \quad (18.39)$$

and justify the use of θ_m as the lower integration limit.

Kinetic energy is also exchanged between test and field particles during elastic collisions. Employing the same terminology as we used for momentum transfer, we can describe the change in kinetic energy of an incident test particle A by

$$\Delta \epsilon_A = \frac{1}{2} m_A (2\mathbf{v}_A \cdot \Delta \mathbf{v}_A + v_A^2).$$

For elastic collisions the magnitude of the relative velocity remains constant, which implies

$$\Delta u^2 + 2\mathbf{u} \cdot \Delta \mathbf{u} = 0.$$

Denoting the field-particle velocity distribution by $f_B(v_B)$ as before, and making use of the Coulomb scattering cross section (18.39), we find that the rate of change of a test particle's kinetic energy is given by

$$\frac{d\epsilon_A}{dt} = \frac{\mu^2}{m_A} \int f_B(v_B)\, d^3v_B\, \sigma_{\text{Coul}} |u|^3$$
$$- \mu \int f_B(v_B)\, d^3v_B\, \sigma_{\text{Coul}}\, u\, (\mathbf{u} \cdot \mathbf{v}_A). \quad (18.40)$$

We suppose that the test particles are distributed in velocity according to $f_A(v_A)$, and obtain for the time-rate of change of the kinetic energy E_A

$$\int f_A(v_A) \frac{d\epsilon_A}{dt} d^3v_A = \frac{dE_A}{dt}. \quad (18.41)$$

Expressions similar to (18.40) and (18.41) may be written for the change in ϵ_B and E_B. If the distributions are Maxwellian, then the kinetic temperatures may be defined by $E_A = 3kT_A/2$ and $E_B = 3kT_B/2$. An analysis like that leading to (18.35) yields

$$\frac{dT_A}{dt} = -\frac{T_A - T_B}{\tau^E_{AB}}, \quad (18.42)$$

where the relaxation time for energy exchange τ^E_{AB} is given by

$$\tau^E_{AB} = \frac{3}{8} \frac{(m_A T_B + m_B T_A)^{3/2}}{(m_A m_B)^{1/2} (2\pi)^{1/2} e^4 Z_A Z_B n_B \ln \Lambda}. \quad (18.43)$$

Finally, the rate of change in E_B or T_B can be obtained from energy conservation

$$n_A \frac{dT_A}{dt} + n_B \frac{dT_B}{dt} = 0. \quad (18.44)$$

For interstellar plasma, $n_e \approx n_i$ and $Z_i = 1$ and (18.43) give

$$\tau^E_{ee} = \left(\frac{m_e}{m_i}\right)^{1/2} \left(\frac{T_e}{T_i}\right)^{3/2} \tau^E_{ii}$$

$$= 2^{3/2} \frac{m_e}{m_i} \tau^E_{ei}. \qquad (18.45)$$

Problem 18.6. Suppose that hot e^- are introduced into a partially ionized hydrogen cloud. If $\tau^E_{ei} \equiv \tau_E$ is constant, find $T_e(t)$ and $T_i(t)$. What is the equilibrium state?

Generally this implies that $\tau^E_{ee} \lesssim \tau^E_{ii} \ll \tau^E_{ei}$. Comparing energy-exchange rates with momentum-exchange rates (deflection times), we see that if a hot distribution of electrons (photoelectrons, for example) is introduced into a cooler gas of hydrogen atoms and ions, the electrons will reach energy equilibrium first through e–e collisions. Then the ions will equilibrate to a different kinetic temperature through ion–ion collisions. Finally the e^- and ions will approach the same kinetic temperature through e^-–ion collisions. As expected, the deflection and energy exchange times are inversely proportional to the density of scattering centers.

18.3. INTERSTELLAR CLOUDS

The physical state of interstellar matter depends on its composition and the local temperature, pressure, and density. The stability of the state depends on the relative rates of energy gain and loss. The formation of interstellar HI clouds—more accurately, the inhomogeneity of the HI component of the interstellar medium—results because the gas is thermally unstable. These clouds are not gravitationally bound, as can be readily verified by simple order-of-magnitude estimates. Adopting a typical cloud mass and radius (as shown in Table 18.1), we find that the total thermal energy $U \sim (3MkT/2m_H) \sim 10^{46}$ ergs, but the gravitational binding energy $E_G \sim (3M^2G/5R) \sim 2 \times 10^{45}$ ergs.

Thermal stability of the interstellar gas is related to the rate of the heat transfer of the gas. If relatively dense clouds of HI are to form dynamically stable condensates in a dilute HI gas, then the pressure in the cloud must equal the pressure in the surrounding dilute gas. The rate of heat transfer to a unit volume of gas is, from the first law of thermodynamics, and $\rho = \mu m_H n$,

$$\rho T \frac{ds}{dt} = \rho \frac{d\epsilon}{dt} - \frac{P}{n}\frac{dn}{dt}, \qquad (18.46)$$

where s and ϵ are the entropy and internal energy per gram. For a monatomic ideal gas whose pressure is held constant, the right-hand side of (18.46) reduces to $(5nk/2)\, dT/dt$. The left-hand side will be written as the difference between the heating rate Γ and cooling rate Λ; so

$$\rho T \frac{ds}{dt} = \Gamma - \Lambda = \frac{5}{2} nk \frac{dT}{dt}. \qquad (18.47)$$

The heating and cooling rates are complicated functions of temperature, density, and composition. Before considering an example, we will develop (18.47) further.

The gas will be in thermal equilibrium if $\Gamma = \Lambda$ at a temperature T_E. We now consider the stability of the equilibrium state against small changes in temperature. First, define the rate of change in temperature when $T \neq T_E$ by

$$\frac{dT}{dt} = -\frac{T - T_E}{\tau}, \qquad (18.48)$$

which defines the relaxation time τ. Comparing the preceding two equations, and using the ideal-gas equation of state, we can express τ as

$$\tau = -\frac{5}{2}\frac{P}{T}\frac{T - T_E}{\Gamma - \Lambda}; \qquad (18.49)$$

as long as $\tau > 0$, (18.48) shows that if $T \neq T_E$ the gas will tend to return to its equilibrium temperature T_E. But if $\tau < 0$, then the difference $T - T_E$ will grow, leading to thermal instability. Define $Q \equiv \Gamma - \Lambda$, and expand it about $T = T_E$, remembering that P is to be held constant:

$$Q(T) \simeq Q(T_E) + (\partial Q/\partial T)_{T_E}(T - T_E). \qquad (18.50)$$

Since $Q(T_E) = 0$, we may rewrite (18.50) as

$$\tau = -\frac{(5P/2T)}{(\partial Q/\partial T)_p}. \qquad (18.51)$$

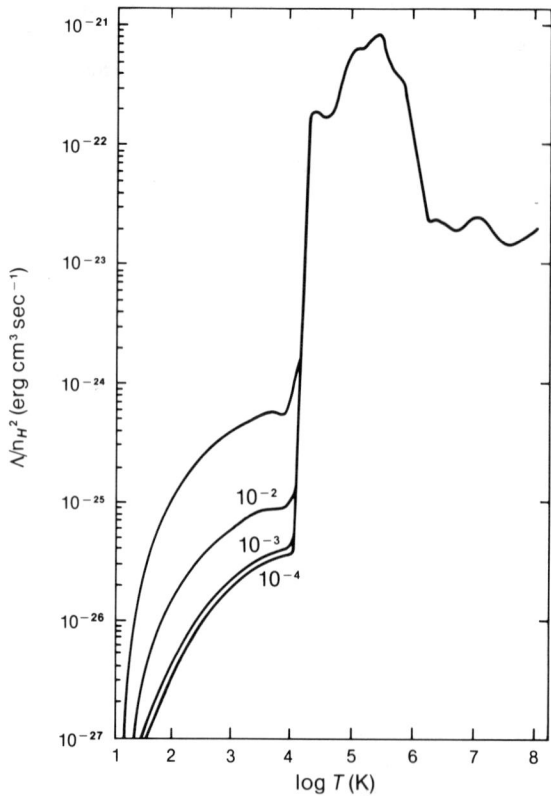

Figure 18.4. Cooling function for interstellar gas. For $T < 10^4$ K, different curves correspond to different values of n_e/n_H. For $T > 10^4$, collisional ionization is assumed for all elements.

The sign of τ depends on the change in Q with temperature at constant pressure. Since $Q = Q(\rho, T)$ the gas is unstable whenever

$$\left(\frac{\partial Q}{\partial T}\right)_P = \frac{\partial Q}{\partial T} + \left(\frac{\partial Q}{\partial \rho}\right)\left(\frac{\partial \rho}{\partial T}\right)_P > 0. \quad (18.52)$$

Therefore, given $Q = \Gamma - \Lambda$, one may evaluate (18.52) to find out if the gas is stable or not.

Problem 18.7. Hydrogen at temperatures of 10^7 K and low densities is thin to its own radiation ($\Gamma \approx 0$). The primary cooling mechanism is bremsstrahlung, for which

$$\Lambda_b \approx \Lambda_0 \rho^2 T^{1/2}. \quad (18.53)$$

Find out whether the gas is thermally stable.

If the gas is thermally stable, then $dT/dt = 0$, and the equilibrium temperature follows from (18.47) as the solution of

$$\Lambda(\rho, T_E) = \Gamma(\rho, T_E), \quad (18.54)$$

where $\rho \approx m_H n_H$ for HI regions where $n_H \gg n_e$. The cooling rate for HI regions arises from collisional excitation. The primary processes involved are believed to be excitation of CII (and to a lesser extent CI, OI, and FeII), excitation of rotational states in H_2 by H atoms, and excitation of H atoms by e^-. The latter process is important primarily at $T > 3,000$ K, and is negligible at lower temperatures. The region $T \lesssim 10^4$ K in Figure 18.4 shows a recent calculation of Λ as a function of T, excluding the effects of grains and H_2 molecules.

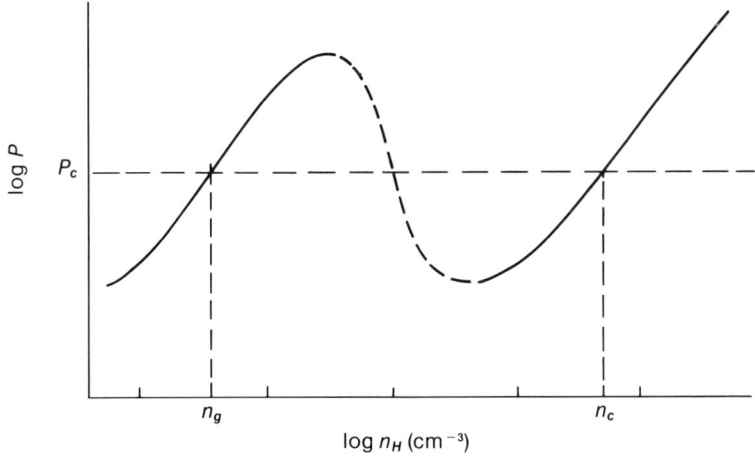

Figure 18.5. Pressure versus n_H for interstellar gas (HI regions). Two stable states develop corresponding to $n_H \lesssim n_g$ and $n_H \gtrsim n_c$.

Problem 18.8. A gas contains two types of atoms of number density n_a and n_b. Suppose that atoms of type b contain only two levels, the ground state l and an excited state u, and that collisions between atoms of types a and b excite b only. If the collisional-excitation cross section for b is σ_{lu}, and the gas is in local thermal equilibrium, show that the energy-loss rate upon deexcitation is

$$\Lambda = n_a n_b E_{ul} \langle v\sigma_{ul} \rangle \frac{g_u}{g_l} \exp^{(-E_{ul}/kT)}. \quad (18.55)$$

Here v is the relative velocity of a and b, $E_{ul} = E_u - E_l$, σ_{ul} is the deexcitation cross section, and $\langle v\sigma_{ul} \rangle_T$ denotes averaging over a Maxwellian distribution at temperature T. You may neglect collisional deexcitation.

Much less is known about the heating function for HI regions. Important processes include: photoionization of CII; cosmic-ray ionization of H atoms; energy release when molecular hydrogen forms on the surface of dust grains (2H → H_2 + 4.48 eV); and photoelectric emission by grains. Theoretical uncertainties arise with each of these processes, but the first three appear to be too weak to produce $T_E \approx 80$ K as is observed. Photoelectric emission by dust grains, which currently appears to be the primary heating process for HI regions, occurs when ultraviolet photons are absorbed, followed by the emission of an electron. Assuming that no other heating processes operate, the processes mentioned taken together are capable of predicting $T_E \approx 80$ K when $n_H \approx 30$ cm^{-3}.

Problem 18.9. Suppose that the heating and cooling rates in HI are given (in ergs/cm³ s) by

$$\Gamma = 2 \times 10^{-25} n_H,$$
$$\Lambda = 8 \times 10^{-27} n_H^2 e^{-T_0/T}, \quad (18.56)$$

where $T_0 = 92$ K. Find the equilibrium temperature and find out if the gas is thermally stable. Take $n_H = 100$ cm^{-3}

Suppose that $\Gamma - \Lambda$ is given, and that the gas is in a state of unstable thermal equilibrium about T_E. The interstellar medium is continually perturbed by such local phenomena as shock waves (from supernovae, novae, mass loss from stars), collisions between dense clouds, and local pockets of gravitational contraction. According to (18.48) and (18.51), these perturbations will cause the unstable gas to evolve away from the equilibrium value T_E. If the relaxation time is short enough, T will change [subject to the equilibrium constraint $\Lambda(\rho, T) = \Gamma(\rho, T)$] until a new stable equilibrium value T_E is reached. Since the gas can always be described by the ideal-gas equation of state, the thermal rearrangement traces out a curve $P(n_H)$, shown qualitatively in Figure 18.5. Equilibrium states for which $P(n_H)$ lies along the dashed curve connecting the local maximum and minimum pressure are unstable. If T_E and n_H correspond to a pressure P_c, then the gas separates into two phases, each of which is at the same pressure P_c. In the low-density phase, the gas is hot; in the high-density phase it is cool. Since the gas is ideal, the ratio of hot gas to cloud temperature is

$$\frac{T_g}{T_c} = \frac{n_c}{n_g}. \tag{18.57}$$

Typically $0.05 \lesssim n_g \lesssim 0.2 \text{ cm}^{-3}$, and $10 \lesssim n_c \lesssim 10^3 \text{ cm}^{-3}$.

18.4. INTERSTELLAR ELECTRON DENSITY

In the presence of free electrons, the interstellar gas behaves like a dielectric medium, through which electromagnetic waves travel at a speed that decreases with frequency. Since the rate of decrease depends on the free-electron number density n_e, a measurement of the difference in arrival time between waves at two different frequencies gives a measure of n_e along the line of sight.

Physically, the electromagnetic wave induces oscillations in the free electrons, which result in a polarization field. If we denote the electric field associated with the electromagnetic wave by **E**, then the force acting on an electron in its vicinity is

$$m_e \frac{d^2\mathbf{r}}{dt^2} = e\mathbf{E}(\mathbf{r},t) = e\mathbf{E}_0(r)e^{i\omega t}, \tag{18.58}$$

the displacement of the electron from equilibrium $\mathbf{r} = \mathbf{r}_0 e^{i\omega t}$ with $\mathbf{r}_0 = -e\mathbf{E}_0/m_e\omega^2$. The dipole moment that results is $e\mathbf{r}$, and the net polarization field is

$$\mathbf{P}(\mathbf{r}, t) = n_e e \mathbf{r} = -n_e e^2 \mathbf{E}/m_e \omega^2. \tag{18.59}$$

In a dielectric medium, the electric field satisfies the wave equation

$$\frac{\epsilon}{c^2} \frac{\partial^2 \mathbf{E}}{\partial t^2} = \nabla^2 \mathbf{E}, \tag{18.60}$$

where ϵ is the dielectric constant defined by the relation

$$\epsilon \mathbf{E} = \mathbf{E} + 4\pi \mathbf{P}. \tag{18.61}$$

It follows from the wave equation that the electromagnetic wave propagates through the medium with wave vector **k** and angular frequency ω, which satisfies the relation

$$\omega^2 \epsilon = k^2 c^2. \tag{18.62}$$

Problem 18.10. Show that the plane wave

$$\mathbf{E} = \mathbf{E}_0 \exp^{[i(\mathbf{k}\cdot\mathbf{r} - \omega t)]}$$

satisfies (18.60) and that the dielectric constant is

$$\epsilon = 1 - \omega_p^2/\omega^2, \tag{18.63}$$

where

$$\omega_p = (4\pi n_e e^2/m_e)^{1/2}$$
$$= 5.6 \times 10^4 \, n_e^{1/2} \text{ Hz}.$$

The dielectric constant is given by (18.63), and ω_p is the plasma frequency. For $\omega > \omega_p$, the wave vector $k = \omega \sqrt{\epsilon}/c$, and the wave can propagate without attenuation through the interstellar gas. The speed of propagation is

$$v = \frac{\partial \omega}{\partial k} = \frac{kc^2}{\omega} = c(1 - \omega_p^2/\omega^2)^{1/2}. \tag{18.64}$$

These results may be applied to the radio emission from pulsars. Consider the pulse profile of a typical pulsar (Figure 16.10b). The time required for a signal of angular frequency ω to travel a distance D from the pulsar to the observer is

$$t = \int \frac{dl}{v}. \tag{18.65}$$

Because v increases with frequency, the time-delay between two signals at ω_1 and ω_2 will be given by

$$\Delta t = t_2 - t_1 = \int_0^D \left(\frac{1}{v_2} - \frac{1}{v_1}\right) dl$$
$$\simeq \frac{\Delta \omega}{c\omega^3} \int_0^D \omega_p^2 \, dl = \frac{4\pi e^2 \Delta \omega}{m_e c \omega^3} \int_0^D n_e \, dl, \tag{18.66}$$

where we assume $\omega_p \ll \omega_1$, $\omega_1 \approx \omega_2 = \omega$, and define $\Delta \omega = \omega_2 - \omega_1$. Measurement of Δt determines the dispersion measure

$$DM \equiv D \langle n_e \rangle = \int_0^D n_e \, dl. \tag{18.67}$$

Observations of the dispersion measure for pulsars of known distance (Table 16.2) yield an average value for $\langle n_e \rangle \approx 0.03$ cm^{-3}. Alternatively, (18.67) may be used to approximate the distance to newly observed pulsars, assuming that $\langle n_e \rangle$ is known. The dispersion measure is generally expressed in pc cm^{-3}.

Problem 18.11. The time-delay Δt actually includes effects from the dense plasma surrounding the pulsar. How can these effects be removed observationally? Discuss your assumptions.

Problem 18.12. Radio waves of frequency much less than a megahertz can not be detected below the Earth's ionosphere. Explain why, and estimate the free-electron density in the ionosphere.

If the radio emission is linearly or partially linearly polarized, then its plane of polarization will be rotated as it passes through regions of nonzero n_e and nonzero magnetic field. This effect, known as Faraday rotation, can be calculated by replacing the electric force $e\mathbf{E}$ appearing in (18.58) by the complete Lorentz force

$$m_e \frac{d^2\mathbf{r}}{dt^2} = -e\mathbf{E} - \frac{e}{c}\mathbf{v} \times \mathbf{B}. \quad (18.68)$$

The electric field of the incident wave is again denoted by $\mathbf{E}$; the local (Galactic) magnetic field is denoted by $\mathbf{B}$, and the velocity of the electron by $\mathbf{v}$. It is assumed that the magnetic force due to the wave is small relative to that due to the Galactic field. For a uniform Galactic magnetic field, (18.68) may be solved for the dielectric constant

$$\epsilon_\pm = 1 - \frac{(\omega_p/\omega)^2}{1 \pm (\omega_c/\omega)}, \quad (18.69)$$

$$\omega_c \equiv eB/m_e c, \quad (18.70)$$

Problem 18.13. Find the polarization of a plasma subject to the force equation (18.68). Assume $\mathbf{B} = B_0 \hat{e}_z$, and denote the circularly polarized electric field by (plus for right, minus for left circular polarized)

$$\mathbf{E}_\pm = E_x \pm iE_y = E_0 e^{\pm i\omega t}. \quad (18.71)$$

The displacement of the electron due to E_+ can be written

$$r_\pm = x \pm iy = a_\pm e^{\pm i\omega t}. \quad (18.72)$$

---

where ω_c is the cyclotron frequency; ω_c is the angular velocity of an electron moving under the influence of a magnetic force veB/c. For typical Galactic fields ($B \simeq 10^{-6}$ gauss), $\omega_c \simeq 18$ sec. Note that if $B = 0$, then ϵ reduces to (18.63). The plus sign in ϵ corresponds to right-hand circular polarization, and the minus sign to left-hand circular polarization. The phase change of a wave that has traversed a path length L through the interstellar medium is given by $\int_0^L k\, dr$, where k is related to the frequency and dielectric constant by (18.62). Since k is different for right and left circularized waves, a net change in relative phase $\Delta \psi$ will result; assuming that $(\omega_p/\omega) \ll 1$, and that the magnetic field is uniform along the line of sight,

$$\Delta \psi \equiv \frac{1}{2} \int_0^L (k_+ - k_-)\, dr$$

$$= \frac{\omega}{2c} \int_0^L (\epsilon_+^{1/2} - \epsilon_-^{1/2})\, dr$$

$$\simeq \frac{2\pi e^3}{m_e^2 c^2 \omega^2} \int_0^L n_e B\, dr \equiv \text{RM } \lambda^2, \quad (18.73)$$

where

$$\epsilon_\pm^{1/2} \simeq 1 - \frac{1}{2}\left(\frac{\omega_p}{\omega}\right)^2 \left(1 \mp \frac{\omega_c}{\omega}\right).$$

In general the Galactic magnetic field varys from point to point, and B should be replaced by $B \cos \theta$, where θ is the angle between B and the line of sight. The quantity RM is the rotation measure, and it is conventionally given in rad cm^{-3}. Since $\Delta \psi$ and λ can be measured for many radio pulsars (Table 16.2), an estimate of the Galactic magnetic field can be made. In fact, as follows from (18.67) and (18.73),

$$\langle B \cos \theta \rangle \equiv \frac{\int_0^L n_e B \cos \theta\, dr}{\int_0^L n_e\, dr} \simeq 1.23 \frac{\text{RM}}{\text{DM}} \quad (18.74)$$

with B in μG. For example, the pulsars in Table 16.2 yield values of B in the range 0.6×10^{-6} to 3×10^{-6} gauss.

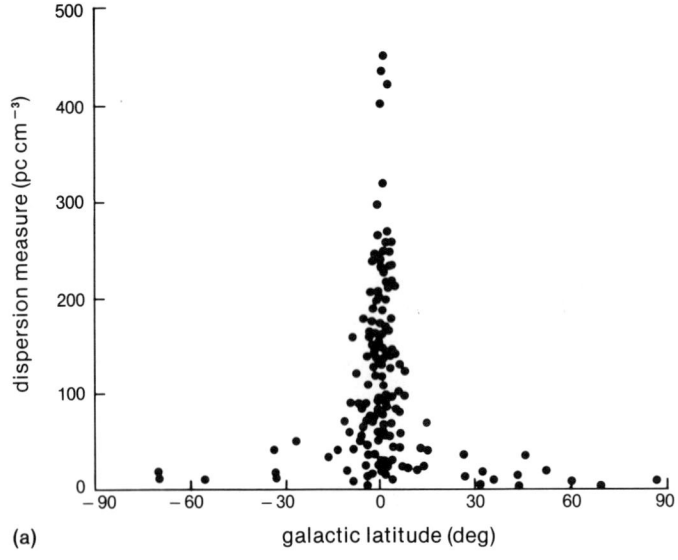

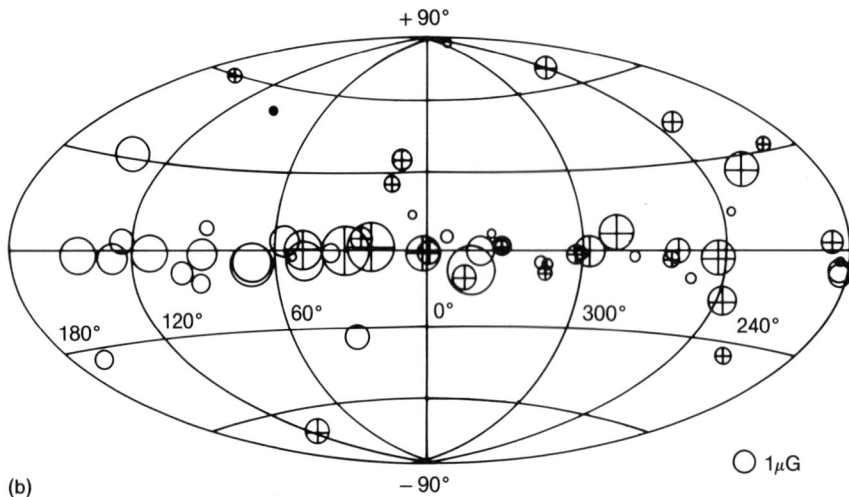

Figure 18.6. (a) Disperion measure versus Galactic latitude for 148 pulsars. (b) Mean magnetic-field components in direction of pulsars. For fields greater than 10^{-5} G, the area of the circle is proportional to the field strength. A cross within a circle means the rotation measure is positive.

The dispersion measure may also be used to estimate the distance of a new radio source. Independently, the distance to a pulsar can be estimated with reasonable accuracy if it is associated with a supernova remnant (as for the Crab or Vela pulsars), or if a portion of its signal is absorbed by interstellar HI (28 pulsars, whose Galactic latitude is less than 10°). Often $\langle n_e \rangle$ can be estimated in this way for sources of known distance in the general direction of a new source; then the average value can be used in (18.67) to estimate the new source's distance. This method is susceptible to error, but is nonetheless valuable when no other means exists for measuring distance. A value typically used for $\langle n_e \rangle$ is 0.03 cm^{-3}, and a scale height of 1 kpc for free electrons in the Galaxy. Figure 18.6 shows the dispersion measure as a function of Galactic

362 / INTERSTELLAR MATTER

latitude for 148 pulsars. Figure 18.6(b) shows the magnitude and direction of the Galactic magnetic field obtained from observed rotation measures for 61 pulsars. The Galactic center is toward longitude $l = 0°$, and the anticenter toward $l = 180°$. Generally the field points away from the Earth near $l = 90°$ and towards it near $l = 270°$, particularly if measurements of sources more distant than 2 kpc are removed.

18.5. Radio Emission and Absorption

Radio waves can excite transitions between molecular rotational levels, or coupled rotational-electronic levels (as in OH^-). Emission or absorption can occur, depending on the nature of the radio source and the intervening interstellar gas. Under special conditions, microwaves may be amplified by stimulated emission (maser) to produce sources with brightness temperatures of 10^{13} K or more. The following discussions make use of the basic elements of radiative transport developed in Chapter 5.

The primary source of radio emission and absorption by interstellar gas is the hyperfine transition at 1420 MHz (21.11-cm wavelength) that results when the electron spin changes its direction relative to the spin of the proton (nucleus) in atomic hydrogen (Section 6.1). The $S = 1$ state can be populated radiatively, or through collisions. Suppose that there are n_1 neutral H atoms per unit volume in the upper ($S = 1$) state, and denote the energy difference between this state and the $S = 0$ ground state by $h\nu_0 = E_1 - E_0$. Then the energy emitted by downward transitions per unit volume, per second, and per Hz is the product of: the energy emitted by the transition $h\nu$; the probability ϕ_ν that the photon will be emitted in the frequency interval ν to $\nu + d\nu$; the transition rate τ^{-1}; and the number of H atoms in the excited state $S = 1$. Collecting factors, and recalling the definition of the emission coefficient, we find that

$$4\pi j_\nu = n_1 \left(\frac{h\nu}{\tau}\right) \phi_\nu = n_1 h\nu A_{10} \phi_\nu, \quad (18.75)$$

where τ^{-1} is just the Einstein coefficient A_{10} (Section 6.3), and ϕ_ν is the line shape [defined after equation (6.57)]. In the absence of line broadening, $\phi_\nu = \delta(\nu - \nu_0)$. The natural line width for the transition is extremely small, about 10^{-16} sec. It can usually be assumed that the hydrogen is in LTE characterized by a spin temperature T_s, so that Kirchhoff's law applies, and the emission coefficient j_ν is related to the absorption coefficient by

$$j_\nu = B_\nu(T_s) k_\nu. \quad (18.76)$$

For radio waves at typical interstellar-matter temperatures, $kT \gg h\nu$, and the Planck function can be taken in the Rayleigh-Jeans limit (for the hyperfine transition in H, $h\nu_0/k = 0.07$ K):

$$B_\nu(T_s) \approx \frac{2\nu^2 kT_s}{c^2}. \quad (18.77)$$

The emission and absorption coefficients may then be written

$$\frac{j_\nu}{k_\nu} = \frac{2\nu^2 kT_s}{c^2}. \quad (18.78)$$

As is customary, the radio intensity observed is expressed in terms of brightness temperature T_b:

$$I_\nu(T_b) = \frac{2\nu^2 kT_b}{c^2}. \quad (18.79)$$

Since the hydrogen atoms are described by a local spin temperature T_s, the number density of atoms in the ground state ($S = 0$) is related to n_1 by

$$\frac{n_1}{n_0} = \frac{g_1}{g_0} e^{-h\nu_0/kT_s} = 3 e^{-h\nu_0/kT_s}. \quad (18.80)$$

Note that because $h\nu_0/kT_s$ is so small, the ratio $n_1/n_0 \approx 3$ for a wide range of spin temperatures, varying by less than a percent for T_s between 10 and 10^4 K.

The preceding results may be applied to the emission of 21.1-cm radiation by neutral H in the Galaxy. The transfer equation (5.38) describes the absorption and emission of radiation traversing a distance l that is related to the optical depth by

$$\tau_\nu = \int_0^l \kappa_\nu \rho \, ds. \quad (18.81)$$

Emission by Interstellar H

If the radiation is due entirely to emission from H atoms along the line of sight, then the transfer equation becomes

$$I_\nu = e^{-\tau_\nu} \int \frac{j_\nu}{\kappa\rho} e^{\tau_\nu'} d\tau_\nu'$$

$$= e^{-\tau_\nu} \frac{2k\nu^2}{c^2} T_s (1 - e^{-\tau_\nu})$$

$$= \frac{2k\nu^2 T_s}{c^2} (1 - e^{-\tau_\nu}), \qquad (18.82)$$

where the source function is (18.78), and T_s is assumed to be constant throughout the emission region. We would, for example, expect T_s within HI clouds to be different from HI in the intervening regions. The analysis may be modified to take into account variable T_s. The observed intensity may be replaced by the brightness temperature using (18.79) to obtain the relation

$$T_b = T_s (1 - e^{-\tau_\nu}). \qquad (18.83)$$

Problem 18.14. Modify (18.82) for a line of sight that traverses two HI clouds of thickness ℓ_c located a distance ℓ_1 and ℓ_2 from the observer. Suppose that the clouds' spin temperatures are $T_{s_0 c}$ and $T_{s_1 c}$.

Therefore, when there is no absorption, and the emission arises from atoms whose spin temperature is a constant, the brightness temperature is given by (18.83).

The column density N_H of neutral hydrogen along the line of sight may also be obtained from radio observations. Equations (18.76) and (18.78) may be combined to give

$$k_\nu = \frac{n_1 \phi_\nu}{8\pi\nu^2} \left(\frac{h\nu}{kT_s}\right) A_{10} c^2. \qquad (18.84)$$

Integrating the optical depth τ_ν over all frequencies, and using k_ν with $\phi_\nu = \delta(\nu - \nu_0)$, we find

$$\tau = \int_0^\infty \tau_\nu d\nu = \int_0^\infty d\nu \int ds\, k_\nu$$

$$= \frac{A_{10} c^2}{8\pi\nu_0^2} \left(\frac{h\nu_0}{kT_s}\right) \int n_1 ds. \qquad (18.85)$$

Defining the column density of atoms in $S = 1$ state

$$N_1 = \int n_1 ds \qquad (18.86)$$

and noting that $n_1 = 3n_0$, we can easily show that the column density of neutral hydrogen (including atoms in the $S = 0$ and $S = 1$ states) is given by

$$N_H = \frac{4}{3} N_1 = \frac{32\pi}{3} \frac{kT_s \tau}{hc\lambda A_{10}}$$

$$= 2.76 \times 10^{14} T_s \tau. \qquad (18.87)$$

For regions thin to radio waves ($\tau_\nu \ll 1$), the brightness temperature $T_b \approx T_s \tau_\nu$ as follows from (18.83). Therefore, N_H along the line of sight may be obtained from T_b. When the line of sight is optically thick, an estimate of T_s must be available. However, in this case absorption will occur as well as emission and the results above must be extended.

Absorption

Radio absorption may be included by retaining the term $I_\nu(0) e^{-\tau_\nu}$ in the transfer equation. This is appropriate when a radio source lies behind the emission region. This may be another neutral-hydrogen emitting cloud, or a distant radio galaxy. If the source temperature is T_0, and is related to $I_\nu(0)$ as in (18.74), then the transfer equation yields

$$T_{b\nu} = T_0 e^{-\tau_\nu} + T_s (1 - e^{-\tau_\nu}). \qquad (18.88)$$

The last term is the emission term obtained in (18.83), and the first term on the right represents the absorption of radiation from the source by intervening matter. It has been assumed that the source is a slowly varying function of frequency across the line. It is then possible to measure $T_{b\nu}$ at a frequency ν_0 corresponding to the line center, and outside the line. The difference is given by

$$T_{b\nu} - T_0 = (T_s - T_0)(1 - e^{-\tau_\nu}). \qquad (18.89)$$

If the source is brighter than the emitting region ($\tau_0 > \tau_s$), then the 21.1-cm line is seen in absorption. If $\tau_0 < \tau_s$, then the line appears in emission. Most of the emission and absorption occurs in HI clouds. When two adjacent points in a uniform cloud can be seen in emission and absorption, it is possible to estimate T_s and the optical depth separately.

Observed radio lines from neutral hydrogen are broadened by thermal motions in the gas, translational motion of HI clouds along the line of sight, Galactic differential rotation, and systematic departures from circular motion. Of these, differential rotation is usually the largest; so radio emission at 21.11 cm has been used to map out the spiral structure of the Galaxy (Chapter 30). Radio observations of emission from other molecules have also been used to identify structures near the Galactic center.

Once the location of absorbing HI clouds is known, the approximate distance of radio sources can be calculated, since the intensity will be absorbed only by clouds lying between the source and the observer. The extragalactic character of many sources can be established also. If a source is observed in absorption at all frequencies for which emission in that general direction is observed, then the source should lie outside the Galaxy. Cygnus A, a powerful radio galaxy, is one object whose extragalactic nature has been established in this way.

The absorption coefficient for HI clouds depends on the spin temperature, which has so far been treated as a parameter of the absorbing region. The neutral gas will have a characteristic kinetic temperature T_k, which may range from a few degrees to a few times 10^2 K. Depending on local conditions, T_s may be in equilibrium with the gas, or it may be in equilibrium with the background or source radiation at temperature T_R. Since collisions between H atoms can excite or deexcite the two hyperfine states, they may play a dominant role in establishing T_s. To analyze this possibility, define the transition rate from the $S = 1$ to $S = 0$ state per H atom by C_{10}, and the reverse rate by C_{01}. Imposing detailed balance, we find these rates to be related by

$$C_{01} = \frac{g_1}{g_0} C_{10} e^{-h\nu_0/kT_k}, \quad (18.90)$$

where $g_1/g_0 = 3$ for H, ν_0 is the hyperfine splitting as in the preceding, and T_k is the kinetic temperature of the gas. Finally, for radiation and gas in statistical equilibrium, the total number of upward transitions equals the total number of downward transitions, and

$$n_1(C_{10} + A_{10} + B_{10}U_{10}) = n_0(C_{01} + B_{01}U_{10}), \quad (18.91)$$

where the radiation energy density is assumed Planckian at radiation temperature T_R, and

$$U_{10} \equiv \frac{8\pi h\nu_0^3}{c^3}(e^{h\nu_0/kT_R} - 1)^{-1}. \quad (18.92)$$

The population ratio n_1/n_0 is given by (18.80), which introduces the spin temperature. The relations between the Einstein coefficients A_{ij} and B_{ij} (Section 6.3) may be used along with (18.91) and (18.92) to obtain n_1/n_0. If $h\nu_0/k$ is small compared with all the temperatures we have been discussing, then the result can be expanded to obtain the following relationship between T_s, T_R, and T_k:

$$T_s = \frac{\xi}{1+\xi}T_k + \frac{1}{1+\xi}T_R, \quad (18.93)$$

where

$$\xi = \frac{h\nu_0}{kT_k}\frac{C_{10}}{A_{10}}.$$

Collisions between neutral H atoms will deexcite the $S = 1$ hyperfine state at a rate $C_{10} \simeq C_0 n_H \times 10^{-11}$ sec^{-1}, where C_0 varies from about 0.2 at 10 K to 9.5 at 100 K. Thus $C_{10}/A_{10} \simeq 3 \times 10^4$ in HI clouds, and $h\nu_0/kT_k \approx 0.07/80$, so that $\xi \simeq 23$. The spin temperature in these clouds should therefore be comparable to T_k for the gas. This is, in fact, characteristic of radio observations whose line of sight lies within the Galactic disk. Values vary, but $T_s \approx T_k \approx 80$ K are typical.

Radio observations at 21.11 cm have also been used to investigate the distribution of neutral H in nearby galaxies. In directions away from the Galactic plane, absorption and emission by intergalactic neutral hydrogen should occur. If, as seems currently tenable, the density of HI between galaxies is low, then $\xi \ll 1$, and (18.93) implies that $T_s \approx T_R \approx 2.7$ K. The data are uncertain, but do suggest that an upper limit to n(HI) may be in the range 10^{-6} to 10^{-7} cm^{-3}.

Problem 18.15. Derive the column density N_H for neutral hydrogen atoms (18.87). Show that if nearby points in a single cloud give rise to emission and absorption separately, then N_H, T_s, and τ may each be found.

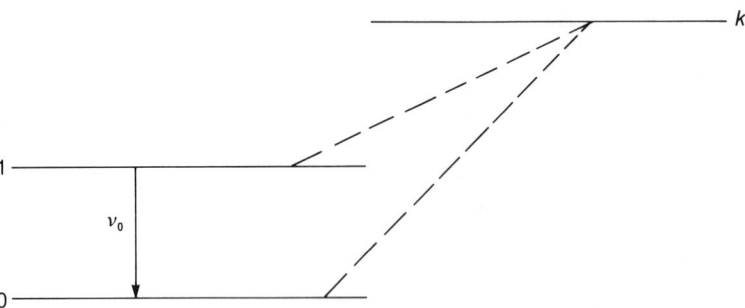

Figure 18.7. Three-level maser. The level k may represent more than one level. Absorption or collisions populate levels k, which then decay to level 1, overpopulating it relative to level 0.

Problem 18.16. What is the radio luminosity of a typical diffuse HI cloud?

Cosmic Masers

Brightness temperatures of more than 10^{13} K are sometimes observed at one or more isolated spots near compact HII clouds. The emission arises from regions whose angular extent is 10^{-6} to 10^{-4} times the extent of the cloud, and typical line widths are in the range 0.1 to 3 km/sec, compared with 2 to 100 km/sec found for emission from normal HI clouds. Clearly some process other than thermal emission must be operating. Many clouds containing these bright emission spots are coincident with infrared sources and HII regions, both characteristic of new star formation. These are referred to collectively as HII masers. A second class is associated with dust shells surrounding cool supergiants, and are designated IR masers. The cause of these unusual features is molecular masers. The most intense molecular masers use radio transitions in the OH^- or H_2O molecules with lines at 18 cm and 1.35 cm, respectively, but others have been detected using SiO (λ = 3.48 cm), CH_3OH, and H_2CO.

A maser is an object—in this case, simple molecules in interstellar gas clouds near an energy source—that amplifies a microwave signal by stimulating the emission of photons at the same wavelength. Three basic properties of cosmic masers need to be explained: the exponential gain in signal; narrowing of the emitted line; and eventual saturation as the signal traverses the cloud. These essential features can be obtained as an extension of the preceding treatment of radiative transport of radio waves. For simplicity, the molecular energy levels (designated 0 and 1) underlying the maser will be assumed to have statistical weight unity, and to be separated by energy $h\nu_0$ (Figure 18.7). The molecules are distributed among excited states thermally at temperature T_k. Additional levels k will exist above the ground and first excited states; and transition rates between these last two will be designated by the Einstein coefficients discussed in Section 6.3. The excitation rate from the state 0 to k and the deexcitation rate from k to 1 will be designated P_{0k} and P_{k1}, respectively. The process $0 \to k \to 1$ is the basic pump process for the maser. If the rate for the process $0 \to k \to 1$ is small relative to the deexcitation $1 \to 0$, then level 1 is metastable, and it will become overpopulated relative to the level 0. This is called *population inversion*.

Consider a cloud containing a molecular species capable of developing population inversion. The emission rate is given by (18.75), and the absorption coefficient is

$$k_\nu = \frac{h\nu}{c} B_{10} \phi_\nu (n_0 - n_1), \qquad (18.94)$$

and the populations of the two levels n_1 and n_0 are related by ($g_1 = g_0$)

$$n_1/n_0 = e^{-h\nu/kT_{ex}}. \qquad (18.95)$$

The line shape ϕ_ν is determined by molecular or gas thermal motion, but will be modified by the maser as the beam moves through the cloud. Initially it may be taken to be Maxwellian (Section 6.6). Population inversion implies $n_1 > n_0$, and the absorption coefficient is negative. Since $h\nu_0/k$ is positive, T_{ex} must be negative. The transfer equation for the emission coefficient (18.94) and absorption coefficient (18.75) is

$$\frac{dI}{ds} = \frac{h\nu}{4\pi} \phi_\nu \{(n_1 - n_0)B_{10}I_\nu + n_1 A_{10}\} \quad (18.96)$$

The population ratio n_1/n_0 is determined by spontaneous and induced radiative transitions, collisions, and pumping. In statistical equilibrium

$$n_1 \left(C_{10} + A_{10} + B_{10}I_\nu \frac{\Omega_b}{4\pi} + P_{10} \right)$$
$$= n_0 \left(C_{01} + B_{01}I_\nu \frac{\Omega_b}{4\pi} + P_{01} \right). \quad (18.97)$$

The coefficients have the same meaning as in (18.91), and P_{01} and P_{10} are the net rates of excitation and deexcitation of the metastable level 1 via states k. The coefficient of the induced transitions B_{10} and B_{01} require comment. Unlike transfer when the radiation is locally Planckian, as in (18.91), the coefficient contains I_ν, the unknown intensity. A further characteristic of maser radiation is that the photons are emitted in the direction of motion of those inducing the emission. Therefore I_ν is reduced by the ratio of the beam's solid angle Ω_b to 4π. The dependence on I_ν is important, since it results in saturation of the beam as we will show in the following.

A useful parameter characterizing the pumping efficiency of the molecules is the fractional population inversion when all processes except pumping are ignored; if we set $A_{10} = B_{10} = B_{01} = C_{10} = C_{01} = 0$, (18.97) becomes

$$\frac{\Delta n_0}{n} = \frac{(n_1 - n_0)_0}{n} = \frac{P_{01} - P_{10}}{P_{01} + P_{10}}. \quad (18.98)$$

The pump efficiency of typical cosmic masers is of order 10^{-2}. Under typical cloud conditions where maser emission is observed, it may be reasonably assumed that (recall $B_{10} = B_{01} = B$)

$$C_{10} = C_{01} = C,$$
$$A_{10} \ll BI_\nu \Omega_b/4\pi \text{ and } C. \quad (18.99)$$

Defining $\tilde{B} = B\Omega_b/4\pi$ and using (18.99), we find that the population ratio

$$\frac{n_1}{n_0} = \frac{C + \tilde{B}I_\nu + P_{01}}{C + \tilde{B}I_\nu + P_{10}}. \quad (18.100)$$

According to (18.99), collisions and stimulated processes dominate radiative decay. Combining (18.98) and (18.100), we find that the fractional population inversion is

$$\frac{\Delta n}{n} = \frac{\Delta n_0/n}{1 + (C + \tilde{B}I_\nu)P_{10}^{-1}}. \quad (18.101)$$

Collisions and stimulated emission reduce Δn as expected. The dependence on I_ν should be noted. In effect, the population inversion is unaffected for low intensities; but with increasing I_ν, the inversion decreases. This suggests that I_ν, which is produced in part by the overpopulation of level 1, will decrease. To verify this, consider the transfer equation with $\Delta n = n_1 - n_0$ given by (18.100):

$$\frac{dI_\nu}{ds} = \frac{h\nu_0}{4\pi} \phi_\nu [\Delta n \, BI_\nu + n_1 A_{10}]$$
$$= \frac{\alpha I_\nu}{1 + I_\nu/I_s} + \beta, \quad (18.102)$$

where

$$\alpha = \frac{h\nu_0}{4\pi} \phi_\nu \frac{B\Delta n_0}{1 + C/P_{10}},$$
$$\beta = \frac{h\nu_0}{4\pi} \phi_\nu n_1 A_{10},$$

and

$$I_s \equiv \frac{P_{10} + C}{B\Omega_b} 4\pi$$

is the saturation intensity of the maser. For simplicity, we will neglect β, since A_{10} is assumed to be small, and the solution of the transfer equation becomes

$$\ln \frac{I_\nu(l)}{I_\nu(0)} = \frac{I_\nu(0) - I_\nu(l)}{I_s} + \alpha l. \quad (18.103)$$

Consider the change in $I_\nu(l)$ as the beam traverses a cloud. Suppose the initial intensity $I_\nu(0) \ll I_s$; this

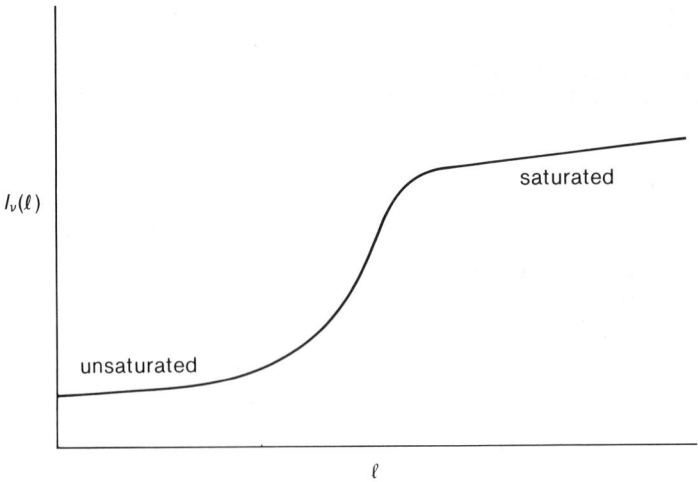

Figure 18.8. Intensity versus distance through interstellar gas, showing the unsaturated regime (exponential growth with l), saturation, and subsequent linear growth.

could arise as the emission from another cloud behind the one in question, or it could be the spontaneous emission of the cloud itself arising from the term $n_1 A_{10}$ neglected in (18.103); then (18.103) is approximately

$$I_\nu(l) = I_\nu(0) e^{\alpha l}. \qquad (18.104)$$

The product αl is the unsaturated gain. The initial signal grows exponentially with distance traversed through the cloud (see Figure 18.8), until $I_\nu(l)$ becomes comparable to I_s. For $I_\nu(l) \gg I_s$, the left-hand side of (18.103) is small and the intensity is approximately

$$I_\nu(l) \approx I_s \alpha l. \qquad (18.105)$$

In this regime $I_\nu(l)$ grows linearly with distance, and the maser is saturated. Figure 18.8 shows schematically the variation in $I_\nu(l)$ with distance l traversed through a cloud. The maximum brightness temperatures observed are as high as 10^{12} K for OH$^-$ masers, and 10^{15} K for H$_2$O masers. To achieve this amount of amplification from clouds whose kinetic temperatures are of order 10^2 K requires $\alpha l \approx 20$ to 30 at saturation.

The line shape of radiation arising from spontaneous emission in a cloud at kinetic temperature T_k will be thermally broadened, so that to good approximation

$$\phi_\nu = \frac{1}{\Delta \nu_D} \exp\left[-4 (\ln 2) (\nu - \nu_0)^2 / \Delta \nu_D^2\right], \quad (18.106)$$

where $\Delta \nu_D$ is the full width of the line at half maximum. Taking as input to the maser (18.104) with (18.105) in α, the line shape is Gaussian near ν_0, with width proportional to $\Delta \nu_D / (\alpha_0 l)^{1/2}$. Thus the line width decreases as the distance l that it travels through the cloud increases. At saturation, the width is about one fifth its original value. After the line saturates, its frequency dependence as given by (18.105) becomes Gaussian with the original width

$$\Delta \nu_D \approx \frac{\nu_0}{c} (kT_k / Am_H)^{1/2},$$

where A is the molecular weight of the molecules and T_k is the kinetic temperature of the gas.

Bright emission in H$_2$O masers is observed as spots whose typical linear dimensions are 10^{13} to 10^{14} cm. For a saturated gain $\alpha_0 l \approx 25$, theory predicts $\alpha_0 \approx 2 \times 10^{-14}$ cm^{-1}; so the beam must traverse about 10^{15} cm before it saturates, and must be close to saturation if it is to produce the observed brightness temperature. For a linear extent of 10^{14} cm, the gain would be $\alpha_0 l \approx 2.5$

and the amplification only a factor of 10^2, which is clearly inadequate. The apparent difficulty in obtaining a path length sufficient to produce the necessary amplification can be overcome by models in which the cloud contains filaments or an unsaturated core surrounded by a saturated shell. Detailed analyses of transport in these systems (which have linear dimensions of about 2×10^{15} cm) show that they can produce the necessary amplification, and appear to have radii in the range 4×10^{13} to 10^{14} cm.

Chapter 19

INTERSTELLAR DUST GRAINS

19.1. Interstellar Dust

About 1 percent by mass of the interstellar medium is in the form of solid dust grains whose characteristic dimensions are 10^{-5} to 10^{-6} cm. These grains are found throughout the interstellar gas, in HI and HII regions, in planetary nebulae, and in the dense condensations contained within these objects. Dust grains show up primarily by their effect on starlight, and as infrared emitters. They also play an important role in very dense clouds (gas-dust complexes), and as sites of molecular formation.

Dust grains scatter and (to a lesser extent) absorb electromagnetic radiation. The combined process is called *extinction,* and its effect on starlight was the original evidence for interstellar dust. We denote the intensity of light of wavelength λ near a source by $I_\lambda(0)$; after this light has traversed a region whose optical depth is τ_λ, its observed intensity will be reduced to

$$I_\lambda = I_\lambda(0) e^{-\tau_\lambda}. \tag{19.1}$$

Observations of stars of identical spectral type at different positions in the sky have shown that the extinction is roughly linear in $1/\lambda$,

$$\tau_\lambda = C/\lambda, \tag{19.2}$$

with C independent of λ (but dependent on stellar spectral type). This relation holds approximately for $3000 \text{ Å} \lesssim \lambda \lesssim 2 \times 10^4 \text{ Å}$, that is, from the near ultraviolet into the infrared. Scattering by atoms or molecules in the interstellar medium can not account for the observed extinction, since the corresponding wavelength dependence would go as λ^{-4} (Rayleigh scattering). Similarly, electron scattering would be independent of wavelength (Thomson scattering). The extinction implied by (19.1) is strongest toward the blue end of the spectrum; so the star's color appears to be reddened. Interstellar dust also explains the occurrence of blue reflection nebulae. These nebulae are rich in dust grains, and lie near bright stars. The nebula scatters most strongly toward the observer the shorter (blue) wavelengths from the star, giving the nebula its peculiar appearance.

A rough estimate of the grain size r_d can be obtained from the observed wavelength dependence of τ_λ. If r_d were much greater than the wavelength of starlight λ, then grains would absorb photons, and reradiate a continuous spectrum rather than scattering the light; if $r_d \ll \lambda$, the observed τ_λ should go as λ^{-4}. Thus grain sizes are expected to be comparable to the wavelength of visible light. As a typical grain dimension, we adopt $r_d = 2 \times 10^{-5}$ cm.

Problem 19.1. Dark globules contain enough dust to reduce starlight by a factor of 50 or more ($\tau \gtrsim 4$). Make a rough estimate of the number density of grains in a globule of radius 0.05 pc, if $r_d = 10^{-5}$ cm, and of the total mass of the globule.

As was noted in Section 18.1, the gas in dense nebulae is underabundant in C, N, and O, and nearly devoid of heavier elements, most of which must be locked up in grains. The extinction associated with grains suggests that several different grain compositions occur in the Galaxy, often coexisting in the same region. Grains consisting predominantly of graphite, SiC, magnesium and aluminum silicate, H_2O, NH_3, and iron are probably not uncommon. Some observations also require dirty ice (H_2O containing metal impurities), or refractory grains encased in mantles of H_2O, NH_3, or possibly graphite.

The extinction of starlight can be related to the grain number density n_d, and the geometric cross section for a spherical grain of radius r_d:

$$\sigma_d = \pi r_d^2. \quad (19.3)$$

The volume extinction coefficient (including the effects of scattering and absorption) is conveniently expressed as

$$k_{ex} = n_d Q_e(\lambda) \sigma_d. \quad (19.4)$$

The wavelength-dependent quantity $Q_e(\lambda)$ specifies the efficiency with which a grain of geometric cross section σ_d attenuates starlight. The optical depth τ_λ follows from (19.4), and may be written in terms of the grain column density n_d as

$$\tau_\lambda = \int k \, dr = Q_e \sigma_d n_d. \quad (19.5)$$

Extinction is usually expressed in magnitudes $A_\lambda = m - m_0$, where m_0 is the magnitude if no dust existed along the line of sight. It follows from the relation between magnitudes and luminosities that

$$A_\lambda = -2.5 \log \frac{I_\lambda}{I_\lambda(0)} = 1.086 \, n_d \sigma_d Q_e \quad (19.6)$$

using (19.1) and (19.5). Observations throughout much of the Galactic plane indicate that

$$n_d \sigma_d \approx 10^{-21} \, n_H;$$

therefore $A_\lambda \approx 3 n_H$ mag/kpc, where n_H represents an average number density of hydrogen along the line of sight, and Q_e has been set to unity. The extinction efficiency Q_e depends on the grain composition and shape. A value for $Q_e(\lambda)$ is established for a given grain composition and shape by evaluating the way electromagnetic waves scatter from the grain. The shape influences diffraction effects, and can result in polarization of starlight; the composition determines the grain's dielectric properties. Detailed theoretical evaluations of Q_e are limited to simple geometry, primarily spheroids and rods. For spherical grains, Q_e may be as large as 4; that is, the effective cross section exceeds the geometric cross section, because of diffraction effects.

The starlight that is scattered, or absorbed and then re-emitted by dust grains, produces a diffuse radiation background concentrated near the disk of spiral galaxies (including our Galaxy). Define the intensity of starlight traveling in the direction of Ω' by $I_\nu(\Omega')$; then the intensity scattered into solid angle Ω by dust grains is $I_\nu(\Omega') Q_s \sigma_d$, where Q_s, the scattering efficiency of the dust grains, is defined in analogy with Q_e. If the fraction $F(\Omega, \Omega')$ of the incident intensity $I_\nu(\Omega')$ is scattered into the Ω direction, then the emissivity is

$$j_{\nu,s} = Q_s \sigma_d n_d \int I_\nu(\Omega') F(\Omega', \Omega) \, d\Omega'. \quad (19.7)$$

To this must be added the thermal emissivity $j_{\nu,T}$ of the grains. Denote the absorption efficiency by $Q_a = Q_e - Q_s$, and the temperature of the grain by T_d; then if it radiates as a black body and is in LTE with the surrounding gas, Kirchhoff's law applies:

$$j_{\nu,T} = k_\nu^a B_\nu(T_d).$$

The absorption coefficient

$$k_\nu^a = n_d Q_a \sigma_d$$

may be used to obtain

$$j_{\nu,T} = n_d Q_a \sigma_d B_\nu(T_d). \quad (19.8)$$

For $\lambda > 2\pi r_d^2$, Q_a approaches unity; for $\lambda \lesssim 2\pi r_d$, its behavior depends on grain composition. For dirty ice grains, Q_a is less than unity, but for iron spheres it peaks at about 2, and then asymptotically approaches unity. The total emissivity for the background due to dust is given by the sum of $j_{\nu,s}$ and $j_{\nu,T}$ for each type of dust grain in the interstellar medium. Model results depend on grain shape and on composition, neither of which is well known. Although the results are not unique, they indicate that more than one grain type exists. For example, a reasonable fit to the data can be made by using graphite, with SiC and Mg and Al silicate abundances adjusted to reproduce irregularities.

The existence of interstellar grains is also supported by observed polarization of light from reddened stars. Nonmagnetic stars emit radiation whose electric field vector is randomly oriented relative to the observer; the light from these stars should show no polarization. The light from reddened stars, however, shows an amount of linear polarization that increases with the amount of reddening. For such stars, the intensity I_λ, when viewed through polarizing filters, will vary in magnitude with the angle of the wave's electric field. If the maximum and minimum intensity observed are denoted by $I_{\lambda,\text{max}}$ and $I_{\lambda,\text{min}}$, then the polarization $\mathcal{P}$ may be defined by

$$\mathcal{P} = \frac{I_{\lambda,\text{max}} - I_{\lambda,\text{min}}}{I_{\lambda,\text{max}} + I_{\lambda,\text{min}}}. \quad (19.9)$$

The polarization of starlight associated with interstellar dust varies up to about 0.07.

In order to polarize starlight, some of the interstellar dust grains must be nonspherical and aligned. Consider an elongated grain, which, for simplicity, is assumed to be axially symmetric, and an electromagnetic wave moving in the direction **k** whose wave front is parallel to the grain's major axis (Figure 19.1). The magnitude of the extinction coefficient $Q_e(\lambda)$ appearing in (19.4) will depend on whether the wave's electric-field vector lies parallel to or perpendicular to the grain's major axis, its value being conventionally denoted by $Q_{e,E}(\lambda)$ and $Q_{e,H}(\lambda)$, respectively (see Figure 19.1). The extinction coefficients for infinite rods (radius a) and spheroids (semiminor axis a) are similar; for radiation of wavelength $\lambda \lesssim a$, $Q_{e,E} > Q_{e,H}$, but for short wavelengths the two are essentially the same.

In any case the difference $Q_{e,E} - Q_{e,H} \lesssim 0.4$ (for spherical ice grains $Q_{e,\text{max}} \approx 4$). If enough grains are aligned, then components of the radiation with **E** parallel to the plane defined by **k** and the grain axis will suffer greater extinction than the component normal to this plane, and a small linear polarization will develop. The degree of polarization can be related to the column density of aligned grains along the line of sight, using (19.9), (19.1) for the intensity, and (19.5) for the optical depth. When a polarimeter is oriented to pass the maximum intensity, Q_e is replaced in τ_λ by the maximum value of $Q_{e,E}$, which we denote $Q_{e,\text{max}}$ and

$$I_{\lambda,\text{max}} = I_0 \exp(-n_d \sigma_d Q_{e,\text{max}})$$
$$\simeq I_0(1 - n_d \sigma_d Q_{e,\text{max}}). \quad (19.10)$$

The last step assumes that

$$n_d \sigma_d Q_{e,\text{max}} \ll 1.$$

$I_{\lambda,\text{min}}$ is similarly obtained as a function of $Q_{e,\text{min}}$, and the magnitude of the polarization becomes

$$\mathcal{P}_\lambda = \tfrac{1}{2} \xi_A n_d \sigma_d (Q_{e,\text{max}} - Q_{e,\text{min}}). \quad (19.11)$$

The factor ξ_A corrects for the fact that only the aligned, nonspherical grains included in the total grain column density n_d produce polarization ($0 \lesssim \xi_A \leq 1$). The extinction associated with grains will be given by (19.6) if Q_e is replaced by an average value, for example, $(Q_{e,\text{max}} + Q_{e,\text{min}})/2$. Both A_λ and $\mathcal{P}_\lambda$ can be measured, and their ratio gives

$$\frac{\mathcal{P}_\lambda}{A_\lambda} = 0.46 \xi_A \left(\frac{Q_{e,\text{max}} - Q_{e,\text{min}}}{Q_{e,\text{max}} + Q_{e,\text{min}}} \right). \quad (19.12)$$

The values of $Q_{e,\text{max}}$ and $Q_{e,\text{min}}$ obtained from theoretical models indicate that the observed polarization can be explained with $\xi_A < 1$, that is, if nonspherical grains are only partially aligned. Note that $\mathcal{P}_\lambda/A_\lambda$ is independent of n_d and σ_d. When we have a value for ξ_A, $\mathcal{P}_\lambda$ can be used to infer a grain radius a; for cylindrical grains, $a \approx 2 \times 10^{-5}$ cm is typical. Actual values vary, not only from cloud to cloud, but within clouds as well. The evidence indicates that somewhat larger grains are to be found within the denser regions of clouds, where, as shown in Section 19.3, the grain growth rate is greatest.

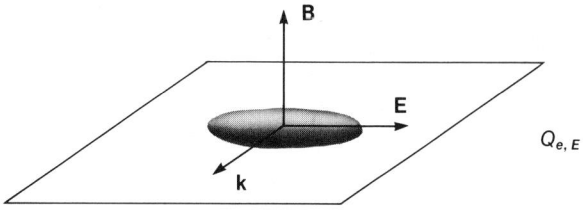
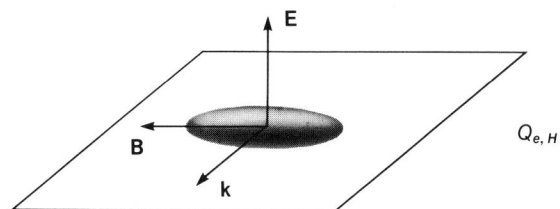

Figure 19.1. Interstellar grain alignment. The electromagnetic wave is moving along **k**.

19.2. Grain Properties

A complete description of the effects that interstellar grains have on starlight requires their shape and physical properties to be known. Many important characteristics of grains follow from their physical properties alone. These include their temperature and electric charge. In describing these parameters, we will assume the grains to be spherical.

The grain temperature T_d is determined by the steady state established between heat loss and gain. Grains lose energy essentially by radiation; if the radiation is black body, then the total emissivity of a grain is given by (19.8) for $n_d = 1$ integrated over solid angle. Therefore the loss rate from the grain is

$$\Lambda_d = 4\pi \int_0^\infty Q_a(\lambda) \, B_\lambda(T_d) \, d\lambda \text{ erg/sec}. \quad (19.13)$$

The energy-loss rate depends only on the grain's temperature T_d, and the dielectic properties of the grain through $Q_a(\lambda)$.

Three principal processes contribute to grain heating. The first of these is radiative heating from the stellar (and, to a lesser extent), diffuse photons* absorbed by the grain. The energy density in the local radiation field is given by (2.35), and the energy incident on the grain per second per unit area is cu_λ.

---
*Diffuse photons are photons scattered from dust grains in the gas cloud.

Multiplying by Q_a, the absorption efficiency, and integrating over wavelength gives

$$\Gamma_R = c \int_0^\infty u_\lambda Q_a(\lambda) \, d\lambda. \quad (19.14)$$

This is independent of T_d, but brings in the temperature of the local radiation field through u_λ.

A second heating mechanism is the transfer of atomic or molecular kinetic energy to a grain during inelastic collisions between grains and molecules. The change in kinetic energy of an atom or molecule of the kth species, whose velocity is v_k, is

$$E_k = \tfrac{1}{2} m_k v_k^2 - \overline{E}_k.$$

$\overline{E}_k$ is the average recoil kinetic energy of the atom. The energy-gain rate is obtained by averaging the energy flux $n_k v_k \Delta E_k$ over a Maxwellian velocity distribution $f(v)$, and summing over species:

$$\Gamma_c = \sum_k n_k \int f_k(v) \, dv_k \, \Delta E_k v_k. \quad (19.15)$$

If the grain is electrically charged (to be discussed shortly), then this result must be modified to include the electrostatic forces between the grain and the ions.

Molecular formation will be enhanced on grain surfaces. In fact, it is likely that most, if not all, complex molecules observed in space form on the surfaces of dust grains. When two atoms, or an atom and a simple molecule, bind, a significant fraction of their binding energy E_B will be deposited in the grain. The corresponding heating rate is

$$\Gamma_B = \sum_k \xi_k n_k \int f_k(v)\, dv_k \eta_k E_{B,k} v_k, \quad (19.16)$$

where ξ_k represents the fraction of atoms striking the grain that participate in molecular formation. Probably the most important process is $2H^0 \rightarrow H_2 + E_B$, where $E_B = 4.48$ eV. It is estimated that $\xi_H \simeq 1/3$, and that each molecule formed deposits about 1.5 eV in the grain as heat.

The net heating rate is the sum of Γ_R, Γ_C, and Γ_B, and the equilibrium temperature of the grain is such that

$$\Gamma_C + \Gamma_R + \Gamma_B = \Lambda_R. \quad (19.17)$$

This will be recognized as an energy-balance equation similar to the ones encountered in discussing the interstellar gas. The steady-state condition (19.17) simplifies considerably for an HI region that contains mostly hydrogen. Adopting $T = 80$ K for the gas, and $U = \int u_\lambda\, d\lambda \simeq 7 \times 10^{-13}$ erg/cm^3 for the local (diluted) energy density from starlight, it is straightforward to show that the reactive and collisional heating terms in (19.17) are negligible relative to Γ_R.

Problem 19.2. Verify this claim, using the representative values given in the preceding for T and U. Assume that reactive heating is due to formation of H$_2$ on grains, with an accompanying energy gain of 0.5 eV per H atom bound as H$_2$. Define

$$\int u_\lambda Q_a(\lambda)\, d\lambda = \overline{Q}_a U.$$

The steady-state condition is then $\Lambda_R = \Gamma_R$ or, using (19.13) and (19.14),

$$4\pi \int_0^\infty B_\lambda(T_d) Q_a(\lambda)\, d\lambda = c \int_0^\infty u_\lambda Q_a(\lambda)\, d\lambda, \quad (19.18)$$

where $B_\lambda(T)$ is the Planckian energy spectrum at grain temperature T_d, and u_λ is a diluted Planckian distribution at a temperature T_{eff},

$$u_\lambda = \frac{4\pi}{c} B_\lambda(T_{\text{eff}}) W. \quad (19.19)$$

Here T_{eff} is the effective temperature of a typical bright star that contributes most of the starlight, and W is a dilution factor. Combining (19.18) and (19.19) and canceling common factors yields

$$W \int_0^\infty \frac{\nu^3 Q_a(\nu)\, d\nu}{e^{h\nu/kT} - 1} = \int_0^\infty \frac{\nu^3 Q_a(\nu)\, d\nu}{e^{h\nu/kT_d} - 1}. \quad (19.20)$$

Over much of the visible spectrum $Q_a(\lambda) \sim \lambda^{-1}$; substituting this into (19.20), and rewriting (19.20) as functions of dimensionless variables $x = h\nu/kT$, we find

$$W T_{\text{eff}}^5 \int_0^\infty \frac{x_{\text{eff}}^4\, dx_{\text{eff}}}{e^{x_{\text{eff}}} - 1} = T_d^5 \int_0^\infty \frac{x_d^4\, dx_d}{e^{x_d} - 1}.$$

The two integrals cancel exactly, and we are left with the grain temperature

$$T_d = T_{\text{eff}} W^{1/5}. \quad (19.21)$$

Therefore, adopting $T_{\text{eff}} = 10^4$ K and $W = 10^{-14}$ the grain temperature $T_d \simeq 16$ K. This estimate is within a factor of two of results based on more detailed models. Interstellar grains are therefore several times hotter than the 3 K cosmic background-radiation field.

Dust grains are also found in HII regions and in planetary nebulae. In these regions the dominant heating mechanism is believed to be the absorption of Lyman α radiation from the central star, a process that is about 10^2 more effective than collisional heating. Unlike the case in HI regions typified by (19.18), the energy-gain rate per grain Γ_α due to Lyman α absorption depends on the local gas density n_H; so the grain temperature in these regions will also depend on n_H. Calculations indicate that $T_d \simeq 20 - 24$ K for $n_H = 10$ cm^{-3}, but is of order 50 K for $n_H = 10^3$ cm^{-3}. Finally, close in to the ionizing star, the grains may become hot enough to evaporate.

Problem 19.3. The number of Lyman α photons available for grain heating is fixed by requiring that their rate of absorption by grains per unit volume per second equal their rate of production by recombination of e$^-$ and H$^+$. Denote by ξ_α the fraction of all recombinations to excited states ($n \geq 2$) that eventu-

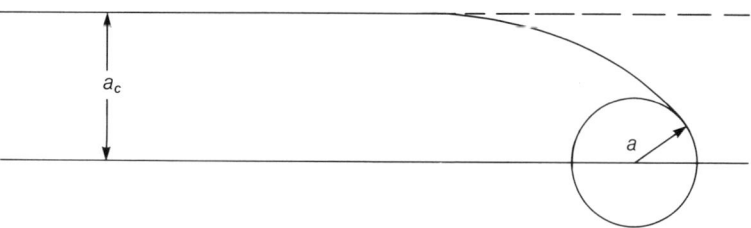

Figure 19.2. Greatest impact parameter a_c for charge captured by a spherical grain.

ally produce a Lyman α photon ($n = 2$ to $n = 1$ transition) of energy $h\nu_\alpha$, and show that the corresponding heating rate is

$$\Gamma_\alpha = \xi_\alpha n_H \left(\frac{n_H}{n_d \sigma_d}\right) \alpha' h\nu_\alpha$$
$$\simeq 5 \times 10^{-3} \xi_\alpha n_H \text{ erg/sec.}, \quad (19.22)$$

where α' is the recombination rate to all excited states. Assume that the gas is completely ionized at $T_e \simeq 10^4$ K.

Dust grains in HII regions and planetary nebulae will be continually bombarded by electrons and protons, and by high-energy photons capable of ejecting electrons from the grain's surface. These processes can lead to a substantial buildup of electric charge on the grain. If the grains are charged, their motion in the nebula will be constrained by local magnetic fields, and they will couple strongly to the ionic component of the gas. The combined effects may have important observational consequences (Section 19.4).

Photoejection of electrons from interstellar grains is analogous to the photoelectric effect in metals. For an electron to be removed from a neutral grain, the electron must acquire enough energy to overcome its binding to the surface. If we denote this energy (analogous to the work function in metals) by $h\nu_c$, only photons more energetic than $h\nu_c$ are capable of producing unbound electrons. If the grain is negatively charged ($Z_d < 0$), the threshold for ejection is also $h\nu_c$; however, for positively charged spherical grains, the electron must overcome the electrostatic potential energy $Z_d e^2 / ah$ as well. Therefore ejection requires that

$$h\nu \geq h\nu_z = \begin{cases} h\nu_c & \text{for } Z_d \leq 0, \\ h\nu_c + Z_d e^2/ah & \text{for } Z_d > 0. \end{cases} \quad (19.23)$$

Next, define ϕ_ν as the probability that a photon of energy $h\nu$ is absorbed by the grain and ejects an electron. Then the change in grain charge will be given by

$$\left(\frac{dZ_d}{dt}\right)_{\text{photo}} = \pi a^2 \int_{\nu_z}^{\infty} \frac{4\pi J_\nu}{h\nu} \phi_\nu \, d\nu. \quad (19.24)$$

For negatively charged grains this rate is independent of Z_d. Since photoejection is most likely near hot, bright stars, J_ν can be taken to represent the stellar component only [see (20.4)]; in which case it depends on the star's effective temperature.

Electron and ion collisions with grains also contribute to Z_d. The electron or ion capture cross section on grains will depend strongly on Z_d. Consider an individual charge $Z_i e$ incident on a spherical grain as shown in Figure 19.2. The initial kinetic energy of the charge is $m_i v_i^2 / 2$, and a_c is the maximum impact parameter for which capture can occur. Conservation of energy and angular momentum gives the relations

$$\tfrac{1}{2} m_i v_i^2 = \tfrac{1}{2} m_i v_i'^2 + Z_i eU, \quad (19.25)$$

$$a v_i' = a_c v_i, \quad (19.26)$$

where v_i' is the charge's velocity at impact with the grain, and $U = Z_d e / a$ is the electrostatic grain potential. Eliminating v_i', we solve for the effective cross section for capture of an ion of initial velocity v_i:

$$\pi a_c^2 = \pi a^2 \left(1 - \frac{2 Z_i eU}{m_i v_i^2}\right). \quad (19.27)$$

The increase in grain charge resulting from the capture of protons in HII regions or in planetary nebulae is given by the flux $n_p v$ times the cross section πa_c^2, integrated over the ion velocity distribution:

$$\left(\frac{dZ_d}{dt}\right)_p = \xi_p \int_{v_{\min}}^{\infty} n_p \pi a_c^2 v f(v) \, d\mathbf{v}, \quad (19.28)$$

where ξ_p is the probability that a colliding ion will actually stick to the grain. The lower limit of the integral depends on Z_d, which is given by $a_c^2 = 0$ in (19.27). If $Z_d \leq 0$, then the lower limit is zero; for $Z_d > 0$, it is

$$v_{\min} = (2Z_d e^2/m_i a)^{1/2}.$$

Suppose that $Z_d > 0$; then the integral becomes

$$4\pi \int_{v_{\min}}^{\infty} v^3 \left[1 - \frac{2Z_d e^2}{m_p v^2 a}\right] e^{-m_p v^2/2kT} \, dv \left(\frac{m_p}{2\pi kT}\right)^{3/2}$$

$$= 8\pi \left(\frac{m_p}{2\pi kT}\right)^{3/2} \left(\frac{kT}{m_p}\right)^2 \int_{x_{\min}}^{\infty} x \left(1 - \frac{Z_d e^2}{kTx}\right) e^{-x} \, dx,$$

where $x = m_p v^2/2kT$. The integral is elementary, and the rate of change of Z_d is readily shown to be ($Z_d > 0$):

$$\left(\frac{dZ_d}{dt}\right)_p = n_p \pi a^2 \xi_p \left(\frac{8kT}{\pi m_H}\right)^{1/2} e^{-Z_d e^2/akT}. \quad (19.29)$$

For grains with $Z_d \leq 0$, the result is the same except that the exponential is replaced by the factor $(1 - Z_d e^2/akT)$.

Problem 19.4. Evaluate the rate of change of Z_d for proton-grain collisions, assuming that $Z_d \leq 0$. Show that for $Z_d \leq 0$,

$$\left(\frac{dZ_d}{dt}\right)_p = n_p \pi a^2 \xi_p \left(\frac{8kT}{m_H \pi}\right)^{1/2} \left(1 - \frac{Z_d e^2}{akT}\right). \quad (19.30)$$

The charge-growth rate for electron-grain collisions is derived in a similar manner, and is

$$\left(\frac{dZ_d}{dt}\right)_e = -n_e \pi a^2 \xi_e \left(\frac{8kT}{\pi m_e}\right)^{1/2}$$

$$\times \begin{cases} 1 + \dfrac{Z_d e^2}{akT} & \text{for } Z_d > 0, \\ e^{Z_d e^2/akT} & \text{for } Z_d \leq 0, \end{cases} \quad (19.31)$$

where ξ_e is the probability that an electron striking a grain actually sticks to it. The charge Z_d is fixed by the steady-state solution of

$$\left(\frac{dZ_d}{dt}\right) = \left(\frac{dZ_d}{dt}\right)_e + \left(\frac{dZ_d}{dt}\right)_p + \left(\frac{dZ_d}{dt}\right)_{\text{photo}} = 0. \quad (19.32)$$

In the outer parts of hot nebulae, the photoejection term (19.24), which decreases with the square of the distance from the central star, becomes negligible, and Z_d is determined by the balance between electron-grain and proton-grain collisions. To order of magnitude,

$$\frac{(dZ_d/dt)_e}{(dZ_d/dt)_p} \approx \frac{n_e}{n_p} \left(\frac{m_p}{m_e}\right)^{1/2} \approx \left(\frac{m_p}{m_e}\right)^{1/2},$$

since $n_e \approx n_p$. Therefore, electron capture exceeds proton capture and the grain will be negatively charged. In fact, assuming $Z_d < 0$, and using (19.30) and (19.31) in (19.32) without the photoejection term, we find

$$\left(\frac{n_e}{n_p}\right) e^{Z_d e^2/akT} = \left(\frac{m_e}{m_p}\right)^{1/2} \left(1 - \frac{Z_d e^2}{akT}\right), \quad (19.33)$$

which may be solved iteratively to obtain (for $n_e \approx n_p$) $Z_d e^2/akT = 2.51$. For a spherical grain with $a = 2 \times 10^{-5}$ cm and a gas temperature $T = 10^4$ K, the grain's charge is $Z_d = 300$. In the central region of nebulae, photoejection becomes dominant and $Z_d > 0$. Numerical evaluation of (19.32) yields Z_d of order $+300$ in these regions. In general, Z_d will be positive near the ionizing star, but will decrease with increasing distance away from the star, becoming negative in the outer regions of the nebula.

19.3. Infrared Excess

Several classes of astronomical objects emit substantially more radiation in the infrared than their temperatures would justify. Notable among these objects are Mira variables, late spectral-type giants and supergiants, planetary nebulae, and novae. The infrared excess is observed to have a smooth spectrum, ranging from 3 μ to more than 700 μ. The temperatures obtained for the infrared emitting regions (assumed to be black bodies) vary from about 50 K to 150 K. The infrared excess can be explained if stellar radiation is absorbed by circumstellar grains, which then reradiate as black bodies with 50 K $\lesssim T_d \lesssim$ 150 K. It is believed that within the relatively cool distended atmospheres of these objects, grain formation is an ongoing process. The temperature of a grain depends in part on its distance from the central star; so a range of temperatures T_d will occur throughout the circumstellar material. A broad, smooth spectrum, consisting of a superposition of black-body spectra whose temperatures cover the range of T_d, will be emitted by the grain distribution. The correlation of infrared excess with interstellar reddening is generally cited as evidence that dust grains exist in the circumstellar gas shells associated with evolved stars.

Probable sites of grain formation are expanding gas shells ejected by novae. During the transition stage in the light curve of some novae (see Figure 17.18), the optical intensity decreases abruptly, and the expanding gas develops infrared excess. The observed spectrum is consistent with black-body radiation from grains. Presumably these grains condense in the relative dense expanding gas envelope which is being ejected by the nova. Once formed, grains would be driven by radiation pressure, or carried by escaping stellar gas into the surrounding medium. Infrared excess is also observed in HII regions. Evidently there are relatively few grains present, but they are very effective in absorbing stellar radiation. The cause of the apparent efficiency has to do with the nature of ionization-recombination processes. We will see in Chapter 20 that photons emitted in the interstellar gas by recombination to the ground state usually ionize another neutral hydrogen atom, but those emitted by recombination to states $n \geq 2$ generally cannot produce ionization or excitation and so escape from the nebulae. However, photons emitted by transitions from the state $n = 2$ to $n = 1$, or from any higher state to the ground state, have a reasonable chance of being absorbed by a neutral hydrogen atom. Consequently, these photons (the most important being the Lyman α photons from the $n = 2$ to $n = 1$ decay) are absorbed and re-emitted many times before escaping from the nebula. If there are even a few grains present, they will eventually absorb one of these photons, and reradiate the energy in the infrared. It can be shown that roughly half the luminosity from the central star in a planetary nebula or an HII region may eventually be absorbed by grains and reradiated in the infrared.

19.4. Grain Evolution

Interstellar grains contain primarily heavy elements, and must have formed after the more massive members of the first stellar generation completed their evolution. Theory must eventually account for their formation. Furthermore, given conditions representative of the interstellar medium today, the time required for a typical heavy atom to collide with and stick to an existing grain is of order $t_A \approx 4 \times 10^7/\xi_a$ years. This suggests that all heavy elements should by now be bound up in grains. Since we observe that they are not, some grains must be losing atoms and molecules to the interstellar medium.

Theory must also account for the evolution of dust grains. One of the most difficult problems about dust grains is how they form. A typical grain contains more than 10^9 atoms; as we shall see, growth from a condensation nucleus to this size can be explained. The formation theory of the original condensation nucleus, however, is only partially developed. Qualitatively, the initial formation stage probably proceeds as follows.

The atoms of a gas in thermal equilibrium are on average distributed uniformly throughout the gas volume. Nevertheless, at any instant the distribution of the atoms in the vicinity of an arbitrary point has a nonzero probability of undergoing statistical fluctuations, as a result of which an aggregate will temporarily form. If the gas is unsaturated, then these aggregates are unstable, and will rapidly disperse into the surrounding matter. Stable condensation nuclei do not form in an unsaturated gas phase. Statistical fluctuations also occur in saturated and supersaturated gases. The smaller aggregates that form are unable to release sufficient energy to condense into grains, and their constituent atoms disperse back into the gas. However, larger aggregates can condense into stable nuclei. If we denote by ΔE_{cond} the energy loss necessary for an aggregate to become a stable condensation nucleus, the probability for formation goes as

$$p \sim e^{-\Delta E_{\text{cond}}/kT}. \qquad (19.34)$$

In simple cases condensation nuclei form at a rate that increases from zero when $T = T_S$, the saturation temperature, by an amount that depends strongly on $(T_S - T)/T_S$. The latter quantity measures the degree of supersaturation of the vapor.

Consider an element of matter in the gradually expanding outer atmosphere of a cool supergiant (the argument could also be applied to planetary nebulae, or to the mass shells ejected by novae). Once the element cools by expansion below the saturation temperature, nucleation begins, and continues briefly as the temperature is reduced further by continued expansion. However, the combined effects of expansion (reducing the element's density) and of the removal of vapor atoms from the gas causes condensation to cease for T not too far below T_c. For still lower temperatures, no new nuclei form, but the existing nuclei will grow by the addition of atoms from the gas, as we will see shortly.

Now, consider the behavior of the various heavy elements as the gas expands. The condensation temperature T_S of heavy elements varies from nearly 2×10^3 K for Al and 1.5×10^3 K for Fe down to 10^2 K for C. Condensation begins first for the elements with the largest T_S. As the gas density decreases, the condensation rate decreases. Therefore, the fraction of an element that condenses out will be less for elements of low T_S than for those having a higher T_S. The depletion of elements relative to cosmic abundances shown in Figure 18.2 increases with T_S as expected. The process outlined in the preceding may also help to explain how grains might form that contain metallic cores surrounded by mantles of more volatile materials.

If we accept that condensation nuclei are somehow formed, then their growth to radii of order 2×10^{-5} cm in less than 10^9 years can result from accretion of interstellar atoms. For neutral grains, the increase in grain mass is given by

$$\left(\frac{dm_d}{dt}\right)_{\text{accretion}} = \sum_k n_k v_k \pi a^2 \xi_{a,k} m_k, \quad (19.35)$$

where the sum is over all atomic species in the gas, $m_k = A_k m_H$ is the atomic weight, and $\xi_{a,k}$ is the sticking factor as defined in Section 19.2. It is assumed that atoms are absorbed onto the grain surface. For neutral atoms $\xi_{a,k} \approx 1$. Defining an average sticking factor ξ_a for all atoms, integrating (19.35) over a Maxwellian velocity distribution for each species, and using $m_d = 4\pi a^3/3\rho_S$ where ρ_S is the grain's density, we find

$$\frac{da}{dt} = \frac{\xi_a}{4\rho_s}\left(\frac{8kT}{\pi m_H}\right)^{1/2} \sum_k \frac{\rho_k}{A_k^{1/2}}$$
$$\simeq 2 \times 10^{-13} \xi_a \text{ cm/yr}. \quad (19.36)$$

The last form follows if $T = 80$ K and $n_H = 20$ cm^{-3}, as is typical of HI regions, and the atomic abundances are similar to those in Figure 18.2. Even if $\xi_a \simeq 0.1$, growth to 2×10^{-5} cm would be accomplished within 10^9 yrs.

Under most interstellar conditions, grain growth as described by (19.36) will be countered by several disruptive processes that tend to reduce their size, and return heavy elements to the medium. The most important processes are sputtering, vaporization, and grain-grain collisions. Sputtering occurs when either a high-energy ion or a photon impacts a grain, knocking off atoms. The sputtering efficiency for particles is small if the particle's energy is only a few eV (as in HII regions), but can produce a significant reduction rate at energies of order 10^2 eV or more. Photon sputtering may also be an important process countering grain growth; although less is known about this process, it may destroy ice grains containing impurities in $10^2 - 10^3$ years. If the grain temperature becomes large enough, atoms will evaporate from its surface. The evaporation rate will exceed the growth rate if the material's vapor pressure exceeds the ambient pressure of the atoms in the surrounding medium. Under typical conditions, the evaporation temperature T_v is of order 20 K for CH_4, 60 K for NH_3, and 100 K for H_2O. Vaporization is most important in the central region of nebulae. Finally, grain-grain collisions with a relative velocity of several km/sec or more will completely vaporize both grains. This is likely to occur in shock waves moving through interstellar gas. However, vaporization in shock fronts probably results in the destruction of only a small fraction of the total grain population in the interstellar medium.

19.5. Dust Dynamics

Dust grains can have a strong influence on the dynamics of nebulae because of radiation acceleration, and because they are collisionally coupled to the gas. Consider a grain of radius a near a bright star. If the stellar flux at r is $\mathcal{F}_\nu(r)$, then the radiation force (Section 5.2) on the grain is

$$F_{\text{rad}} = \int_0^\infty \frac{\pi \mathcal{F}_\nu}{c} \pi a^2 \, d\nu$$
$$= \frac{\pi a^2}{c} \int_0^\infty \frac{L_\nu}{4\pi r^2} d\nu$$
$$= \frac{1}{4}\left(\frac{a}{r}\right)^2 \frac{L}{c}, \quad (19.37)$$

where L is the star's luminosity. Clearly the radiation force is greatest near the star, and falls off as $1/r^2$. In the absence of the other effects, an accelerated grain would eventually be driven out of the nebula.

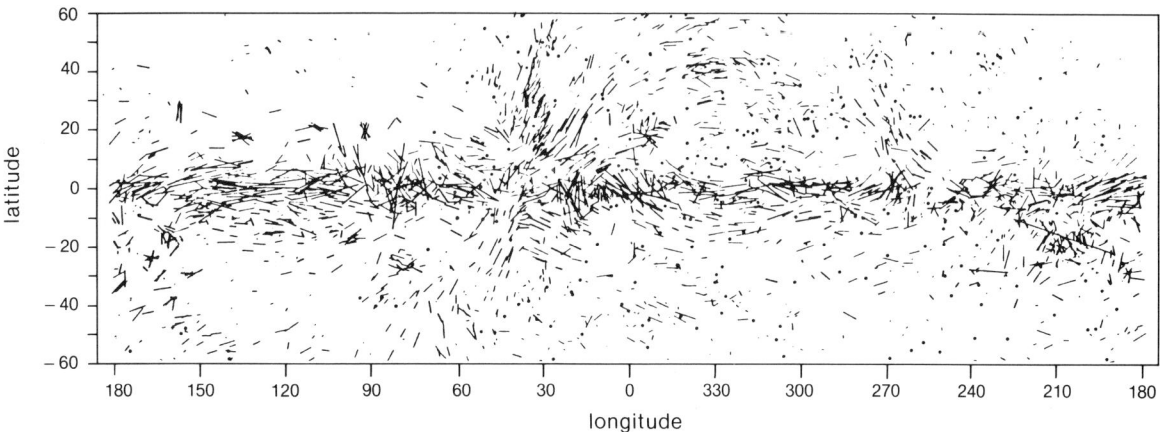

Figure 19.3. Polarization of starlight in our Galaxy due to interstellar dust grains.

Problem 19.5. How long would it take for a typical dust grain to be driven out of a planetary nebula (of radius 2×10^{17} cm) surrounding a star of $10^3 \, L_\odot$? Assume that only radiation acceleration acts on the grain.

An accelerated grain develops velocity u relative to the gas atoms, which results in a collisional drag force on the grain. The methods discussed in Section 18.2 can be used to show that the drag force is

$$F_{\text{drag}} = m_H n_H \pi a^2 \langle v \rangle u$$

$$= m_H n_H \pi a^2 \left(\frac{8kT}{\pi m_H} \right)^{1/2} u. \qquad (19.38)$$

The constant of proportionality varies from unity (when the average thermal velocity of the atoms $\langle v \rangle$ is small compared with u) to 4/3 (when $\langle v \rangle \gg u$). The net force on a grain is $F_{\text{rad}} - F_{\text{drag}}$. Near the star, F_{rad} dominates, and the velocity of the grain relative to the gas increases. Since n_H and $T^{1/2}$ are nearly constant throughout the nebula, F_{drag} will increase as u increases, until eventually $F_{\text{drag}} = F_{\text{rad}}$, and the net force is zero. If the grains are charged, then the drag force can be substantially larger than (19.38).

The motion of dust grains can be transmitted to the nebular gas by collisional drag forces. If the radiation force is great enough, the dust will be driven out of the central regions of the nebula, sweeping most of the gas with it. The central core of the Rosetta nebula, an HII region in Monoceros, may have become devoid of dust and gas in this manner.

Problem 19.6. What effect would gravitational forces have on the analysis of Problem 19.5? Discuss the gravitational force due to the central star, and due to the mass of nebular gas.

Interstellar grains must be at least partially aligned if they are to polarize starlight. Figure 19.3 shows the observed polarization in the Galaxy (the plane of the disk corresponds to zero latitude; the Galactic center to zero longitude). Each line represents the angle of polarization (plane of the electric-field vector), and its length is proportional to the percent of polarization [see (19.11)]. Apart from local irregularities, $\mathcal{P}_\lambda$ is greatest in the plane, and becomes small at higher latitudes. Faraday rotation also indicates that the Galactic magnetic field lies primarily in the disk. These observations suggest that the local magnetic field may play an important role in grain alignment, if the grains can couple to it. Any such tendency will, however, be countered by several processes that tend to randomize a grain's orientation. These include continual bombardment by photons and gas atoms, and recoil accompanying molecular formation on the grain surface. The degree of orientation, and of polarization, will depend on how strong grain-field coupling is relative to the randomizing processes.

Each grain will acquire rotational energy if it is in

LTE with the interstellar gas as a result of continual, random bombardment by gas atoms. The principle of equipartition states that for each particle (atom or grain) in thermodynamic equilibrium at temperature T, an average energy $\tfrac{1}{2}kT$ is associated with each of the particle's degrees of freedom. For monatomic atoms, this reduces to the well-known result $(3/2)kT$. For rotating grains,

$$\tfrac{1}{2} I \langle \omega^2 \rangle \approx \tfrac{1}{2} kT, \quad (19.39)$$

where I is the grain's moment of inertia, and $\langle \omega^2 \rangle$ is its root mean square angular velocity. It must be emphasized that the processes resulting in (19.39) are statistical, and no alignment results from them. In fact, if a grain were aligned, then thermal bombardment by gas atoms would tend to randomize the grain's angular momentum. For typical grain parameters, and $T = 80$ K for HI regions,

$$\langle \omega^2 \rangle^{1/2} = \left(\frac{5}{2} \frac{3}{4\pi} \frac{kT}{\rho_s a^5} \right)^{1/2} \simeq 5 \times 10^4 \text{ sec}^{-1}, \quad (19.40)$$

where, for a spherical grain, $I = 2 m_d a^2/5$, with $m_d = 4\pi a^3/3\, \rho_S$ the mass of the dust grain. Evidently, random collisions are effective in establishing a state of rapid spin for grains.

Problem 19.7. The processes leading to equipartition (19.40) are statistical (random walk). The average angular momentum due to random collisions $\langle J \rangle = 0$, but the root mean square $\langle J^2 \rangle^{1/2}$ is nonzero even for spherical grains. After N random collisions with gas atoms,

$$\langle J^2 \rangle \approx N (\Delta J)^2,$$

where ΔJ is the change in grain angular momentum per collision. The equilibrium angular momentum results after roughly

$$N \approx \frac{m_d}{m_H}, \quad (19.41)$$

where m_d is the grain mass. Use simple dimensional arguments to show that $\langle \omega^2 \rangle^{1/2}$ is of order $(kT/\rho_s a^5)^{1/2}$ as in (19.40). Justify the equilibrium condition $N \approx m_d/m_H$.

The time required for a grain to equilibrate should be comparable to the time it takes to collide with N gas atoms, that is, collide with a mass of gas equal to m_d. If τ_{eq} is the equipartition time, then N/τ_{eq} is equal to the flux of gas atoms $n_H v$ times the grain's cross section, or

$$\tau_{eq} \approx \frac{m_d}{m_H v \pi a^2} \approx \frac{\rho_s a}{n_H (m_H kT)^{1/2}}$$
$$= 2 \times 10^{11} a/n_H \text{ yrs}, \quad (19.42)$$

which is roughly 2×10^5 years for typical grains in HI regions.

The preceding discussion is applicable when the collisions are elastic, and no external torques act on the grains. As noted in Section 19.2, dust grains are expected to be electrically charged. A spinning charged grain sets up a magnetic moment μ parallel to the angular momentum of the grain. For a spherical grain whose charge is $Z_d e$, and whose angular velocity is ω,

$$\mu = \frac{Z_d e}{c} \frac{\omega}{2\pi} a_{\text{eff}}^2 \pi \frac{\mathbf{J}}{J}. \quad (19.43)$$

This corresponds to a current $Z_d e \omega / 2\pi$ in a loop of effective radius a_{eff}. In the presence of the Galactic magnetic field, a torque τ results that acts on the grain, changing the direction of its angular momentum:

$$\boldsymbol{\tau} = \boldsymbol{\mu} \times \mathbf{B} = \frac{d\mathbf{J}}{dt}. \quad (19.44)$$

The change $\Delta \mathbf{J}$ is along $\boldsymbol{\tau}$, and normal to both $\mathbf{B}$ and $\mathbf{J}$, and causes the grain's spin to precess about $\mathbf{B}$. The average Galactic magnetic field is of order 10^{-6} gauss; taking $Z_d \gtrsim 20$, $\rho_S = 1$ g cm^{-3}, and $\theta = 45°$ (see Problem 19.8), the precession rate $\omega \approx 10^{-4}$ yrs. Although precession about $\mathbf{B}$ results in weak orientation of the grains, it is probably not enough to account for the observed polarization. Furthermore, the precession time is comparable to the time τ_{eq} required for thermal randomization.

Problem 19.8. Show that the average charge radius of a spherical grain is given by $a_{\text{eff}}^2 = 2a^2/3$, and that the angular velocity of precession implied by

(19.44) is

$$\omega = \frac{5}{8\pi} \frac{Z_d e}{c} \frac{B \sin\theta}{a^3 \rho_s}$$

$$= 3.2 \times 10^{-21} \left(\frac{B \sin\theta}{a^3 \rho_s} Z_d \right) \sec^{-1}, \quad (19.45)$$

where θ is the angle between μ and $\mathbf{B}$.

A second alignment mechanism involves paramagnetic grains. These are grains containing atoms that have an intrinsic magnetic moment. The average moment of a grain will be zero, since individual atomic moments are randomly oriented (we exclude the possibility that the grain formed in the presence of a strong magnetic field; such grains could be ferromagnetic). However, in an external applied field, a paramagnetic grain will develop a magnetic moment $\mathbf{M} = \chi_m \mathbf{B}$, where χ_m is the magnetic susceptibility. When the relation between $\mathbf{M}$ and $\mathbf{B}$ is static, χ_m is real. In the presence of grain rotation, the induced magnetic moment continually changes orientation within the grain. This continual change in $\mathbf{M}$ is accompanied by energy dissipation; in effect, rotational energy is absorbed by the grain. The loss of rotational energy can be related to the rate of magnetic work performed, $\mathbf{B} \cdot d\mathbf{M}/dt$:

$$-\frac{dE_{\rm rot}}{dt} = -\frac{1}{2} I \frac{d\omega^2}{dt} = \frac{V}{P} \int_0^P \mathbf{B} \cdot \frac{d\mathbf{M}}{dt} dt, \quad (19.46)$$

where the rotation period $P = 2\pi/\omega$, and V is the grain's volume. The right-hand side represents the magnetic work done on the grain in a single cycle, divided by the period. When magnetic energy is absorbed, χ_m is complex, and the integrand above can be shown to be $\chi_m'' B \, dB/dt$, where χ_m'' is the imaginary part of χ_m. The loss of rotational energy described by (19.46) is accompanied by a reduction in the component of the grain's angular momentum perpendicular to the rotation axis. Dimensionally, (19.46) can be written as

$$\frac{I\omega^2}{\tau_p} \approx \frac{V \chi_m'' B^2}{P} = \frac{V \chi_m'' B^2 \omega}{2\pi},$$

where τ_p is the paramagnetic relaxation time. The suceptibility χ_m'' for interstellar grains is uncertain; recent models suggest that $\chi_m'' \sim \omega/T_d$, and that

$$\tau_p \simeq \frac{2\pi I \omega}{V B^2 \chi_m''} = 2 \times 10^{11} \frac{a^2 \rho_s T_d}{B^2 \sin\theta} \approx 10^7 \text{ yrs}.$$

Paramagnetic relaxation, according to this model, would be expected to align a typical grain in about 10^7 yrs. Thus $\tau_p \gg \tau_{\rm eq}$, suggesting that grain alignment will be extremely weak if paramagnetic relaxation is its primary cause.

Chapter 20

GASEOUS NEBULAE

Chapter 18 concentrated on neutral hydrogen distributions in the Galaxy, which show up visually as absorption features against brighter objects. In contrast, ionized hydrogen (HII) regions and planetary nebulae stand out primarily in emission.

20.1. Gaseous Nebulae

Although HII regions in spiral arms and planetary nebulae are associated with opposite extremes of the evolutionary sequence, they share a common feature: both are produced by strong sources of ultraviolet radiation. HII regions in spiral arms mark the birthplaces of stars, usually bright O-B types, whose spectra peak in the ultraviolet corresponding to $T_e \approx 20 - 50 \times 10^3$ K. They are often found in association with dense HI clouds or condensations, and infrared sources. Several contain molecular masses (Section 18.5). Planetary nebulae, on the other hand, are associated with the instabilities arising during the final evolutionary stages of low-mass ($M_\odot \lesssim M \lesssim 4\ M_\odot$) stars (Section 11.7). The central star of a planetary nebula is a hot, ultraviolet dwarf whose surface temperature ranges up to a few times 10^5 K, and which is a strong source of ultraviolet radiation. The expanding shells of matter that encompass the nebulae are believed to have been ejected from the diffuse, low-temperature envelope of the star in its distended red-giant stage; they are rich in dust grains, and often exhibit strong infrared emission.

The qualitative characteristics of nebulae, and their relation to stars emitting in the ultraviolet, are typified by the effect of a young O-B star's radiation on the surrounding interstellar gas. The stellar spectrum contains a substantial amount of radiation at energies in excess of 13.6 eV, which is the ionization energy of neutral hydrogen in its ground state. In steady state, the photons create a cavity of ionized hydrogen in the neutral medium [the formation of an HII cavity (Strömgren sphere) will be discussed in Section 24.4]. The size of the HII region is determined primarily by the competing processes of photoionization and recombination of H and, to a lesser extent, of He. The photoelectrons rapidly thermalize at a temperature comparable to that of the stellar atmosphere. Energy is transferred from the electron gas to the protons by e-p collisions, with the result that the matter (initially at a temperature of about 100 K) heats up.

The thermal electrons collisionally excite intermediate-mass ions (primarily O^+, O^{+2}, and N^+), which eventually radiatively decay by forbidden transitions (see Section 6.1), emitting photons in the visible spectrum. Because of the low density and the small absorption cross section for forbidden lines of these ions, and because the energy of the emitted radiation is incapable of exciting H or He, the radiation escapes from the region. These and other cooling processes act as thermostats to regulate the steady-state temperature of the gas.

The simplest model of a nebula assumes that ionization results from a single star, and occurs in an initially uniform and homogeneous background. Stars appear to be born in associations, and are often surrounded by, or lie near, irregular distributions of neutral gas, with the result that nebulae of highly nonspherical appearance are the rule rather than the exception.

20.2. Ionization and Recombination

Ionization represents the primary means whereby stellar electromagnetic energy is converted into thermal energy of the surrounding gas. Ionization is accompanied by a certain amount of recombination, and the balance between these two processes determines (as a function of the ionizing star's luminosity) the ionization state of the gas. Theory and observation show that the chemical abundances of gaseous nebulae are more or less cosmic, with about 90 percent of the mass in hydrogen, 10 percent in helium, and less than 1 percent in heavier elements (see Figure 18.2). Generally speaking, photoionization of elements other than H and He is small, and can be ignored. In fact, most of the heating is due to photoionization of H.

For H, or hydrogenic ions of charge Z, photoionization will not occur for photon energies $h\nu < h\nu_1 \equiv \chi_1$, where the ionization energy is χ_1 (for H, $\chi_1 = 13.6$ eV and for He^0 $\chi_1 = 24.5$ eV). For $\nu > \nu_1$, the photoionization cross section from the ground state is given approximately by (Figure 20.1)

$$a_\nu(k) = \frac{\sigma_0}{Z_k^2}\left(\frac{\nu_k}{\nu}\right)^3 \qquad (20.1)$$

where k is an atomic index, ν_k is the threshold frequency, and

$$\sigma_0 \equiv 6.3 \times 10^{-18} \text{ cm}^2. \qquad (20.2)$$

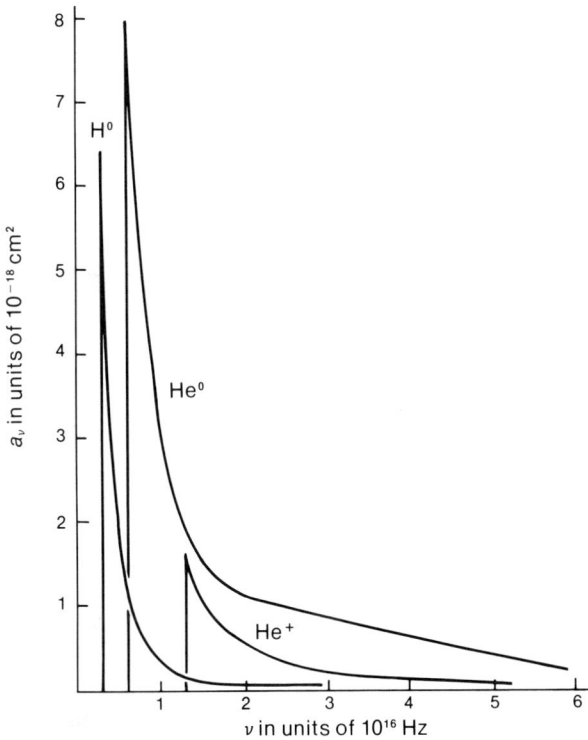

Figure 20.1. Photoionization absorption cross sections for HI, HeI, and HeII.

For $k = H$, $Z_1 = 1$ and $h\nu_k = 13.6$ eV. The approximation in (20.1) for a_ν is best near threshold. We note that a_ν is independent of the thermal state of the gas (in particular the electron kinetic energy T_e). The ionization rate (sec^{-1}) is given by the product of the photon number intensity and the ionization cross section a_ν, integrated over photon energies above threshold:

$$\int_{\nu_1}^{\infty} \frac{4\pi J_\nu(r)}{h\nu} a_\nu(H) \, d\nu = \frac{1}{\tau_i}, \quad (20.3)$$

where $J_\nu(r)$ is related to the ionizing star's luminosity at r by

$$4\pi J_\nu(r) = \frac{L_\nu(r)}{4\pi r^2}. \quad (20.4)$$

We note that $L_\nu(r)$ does not equal the star's luminosity $L_\nu(R)$, where R is the stellar radius, because of absorption (see Section 20.1) in the circumstellar gas.

Problem 20.1. Estimate the photoionization rate for a star whose effective temperature $T_{\text{eff}} = 4 \times 10^4$ K at a distance from the star of 5 pc.

If the temperature of the medium is not too great, then there is a significant probability that a free proton will capture an electron. The resulting neutral atom may be in its ground state or in an excited state. The recombination rate τ_R^{-1} depends on the temperature and density of the medium. Photoionization produces a nonthermal electron distribution that is proportional to the photon flux $J_\nu/h\nu$ and the capture cross section a_ν. Since $J_\nu \sim L_\nu$ and $a_\nu \sim \nu^{-3}$, the electron distribution is proportional to L_ν/ν^4. Elastic e^--e^- scattering produces a Maxwellian velocity distribution on a timescale much shorter than τ_R; so the distribution of electrons that recombine can be characterized by a local kinetic termperature T_e. One goal of the theory of nebulae is to predict T_e (Section 20.5); for now we simply need to know that one exists.

Define the cross section to capture an electron of energy $m_e v^2/2$ into the atomic state n ($n = 1$ is the ground state) by $\sigma_n(v)$. Then the recombination rate to state n is given by the product of the flux of electrons of velocity v, $n_e v f(v)$, and the cross section $\sigma_n(v)$, integrated over velocity. The total recombination rate to all states is then

$$\tau_R^{-1} = \sum_{n=1}^{\infty} \int n_e \sigma_n(v) v f(v) \, dv$$
$$\equiv \alpha n_e, \quad (20.5)$$

where $\alpha = \alpha(T_e)$ is the recombination coefficient, which depends on the electron temperature through the electron's Maxwellian velocity distribution. The recombination cross section $\sigma_n(v) \approx v^{-2}$. Under typical conditions prevailing in nebulae, $\sigma_n \approx 10^{-20}$ to 10^{-21} cm^2, which is at least 10^{-3} times smaller than the ionization cross section. Figure 20.2 shows α as a function of the electron temperature T_e. The curve marked α' is the recombination coefficient to all states except the ground state, and is given by (20.5) if the sum starts with the first excited state $n = 2$. Recombination to the ground state $n = 1$ will produce a photon capable of ionizing another neutral atom. In relatively dense nebulae, the emission and ionization will occur at essentially the same point. When this occurs, recombination to the ground state has no net effect on the thermal state of the gas, and the process can be excluded from the sum (20.5), yielding the coefficient α'.

Problem 20.2. Show that
$$\tau_R^{-1} \sim \rho T_e^{-1/2}. \quad (20.6)$$

The photoionization and recombination processes occur with equal rates when the medium is in a steady state. For $H^0 + \gamma \rightleftharpoons H^+ + e^-$, the steady state obtains if

$$n(H^0) \int_{\nu_1}^{\infty} \frac{4\pi J_\nu(r)}{h\nu} a_\nu(H^0) \, d\nu$$
$$= n_e n(H^+) \int dv \sum_{n=1}^{\infty} v f(v) \sigma_n(v)$$
$$= n_e n(H^+) \alpha(H^0), \quad (20.7)$$

which states that the number of ionizations per second, $n(H^0)/\tau_i$ equals the number of recombinations per second, $n(H^0)/\tau_R$. Similar equations can be con-

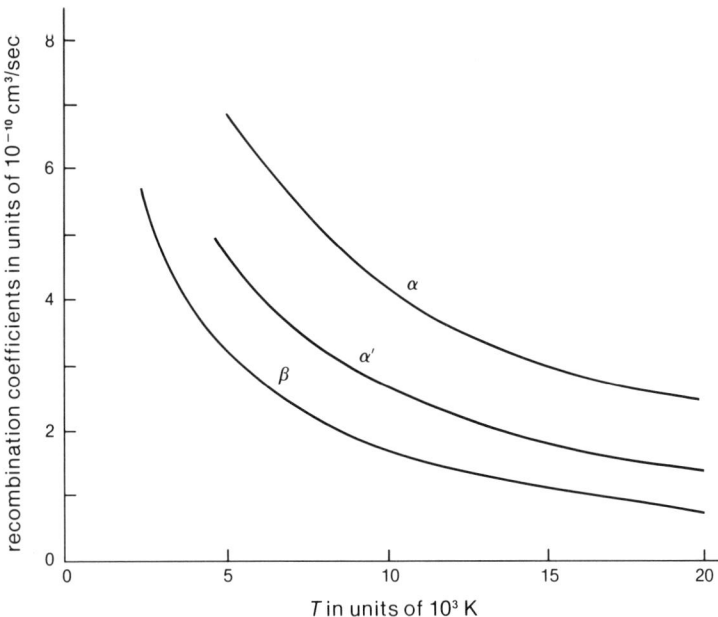

Figure 20.2. Recombination coefficients for hydrogen versus temperature: α includes recombination to all levels; α' excludes recombination to the ground state. The coefficient β is defined by (20.11).

structed for other atoms or ions; however, care must be taken when more than one atomic or ionic species can be ionized by the star's radiation. For example, in typical nebulae the temperature will be high enough to produce He^+ as well as H^+. Recombination of H^+ to form H^0 will produce few photons capable of ionizing He^0, but all photons emitted by recombination of He^+ to excited states of He^0 can ionize H^0.

In HII nebulae and planetary nebulae, heating is due to the photoelectrons, which are emitted with an energy

$$\epsilon_e = \tfrac{1}{2} m_e v^2 = h\nu - \chi_k = h(\nu - \nu_k), \quad (20.8)$$

where χ_k is the ionization energy from the ground state of the kth species. The rate at which the electron energy tends to heat the medium is given by the product of the flux of ionizing photons, the ionization cross section, the number density of neutral atoms, and the electron energy, integrated over frequency. For hydrogen

$$\Gamma = n(H^0) \int_{\nu_1}^{\infty} \frac{4\pi J_\nu(r)}{h\nu}$$
$$\times \epsilon_e a_\nu(H^0)\, d\nu \; \text{erg cm}^{-3}\,\text{sec}^{-1}. \quad (20.9)$$

Using (20.8) and (20.7) to eliminate $n(H^0)$, we can rewrite the heating rate in the form

$$\Gamma = n(H^+) n_e \alpha(H^0)$$
$$\times \frac{\int_{\nu_1}^{\infty} \dfrac{4\pi J_\nu(r)}{h\nu} h(\nu - \nu_1) a_\nu(H^0)\, d\nu}{\int_{\nu_1}^{\infty} \dfrac{4\pi J_\nu(r)}{h\nu} a_\nu(H^0)\, d\nu}$$
$$\equiv n(H^+) n_e \alpha(H^0) h\bar{\nu}, \quad (20.10)$$

where $h\bar{\nu}$ represents an average energy of the ionizing photons. For most nebulae, $h\nu_1 \gg kT_{\text{eff}}$, where T_{eff} is the effective temperature of the ionizing star. If the local spectrum $J_\nu(r)$ is nearly Planckian, and there is little absorption, then $h\bar{\nu} \approx kT_{\text{eff}}$. Thus T_{eff} can be thought of as the initial temperature of the medium before recombination and cooling are considered. As stellar photons are absorbed, the average energy $h\bar{\nu}$ increases because the ionization cross section a_ν decreases rapidly with energy (20.1). Therefore $h\bar{\nu}$ shows a tendency to increase with distance away from the ionizing star. Finally, we note that $h\bar{\nu}$ can be written in terms of the star's luminosity at r using (20.4).

Problem 20.3. Show that for $h\nu_1 \gg kT_{\text{eff}}$, the average ionizing photon energy $h\bar{\nu} \approx kT_{\text{eff}}$, where the luminosity at r is Planckian. Note that

$$\int_x^\infty e^{-t}\frac{dt}{t} \simeq \frac{e^{-x}}{x}\left(1 - \frac{1}{x} + \frac{2!}{x^2} + \cdots\right) \qquad x \gg 1.$$

Problem 20.4. Set an upper limit to the number of photoionizations from photons produced by recombination of He^+ to excited states of He^0. The recombination coefficients $\alpha'(H^0) \approx \alpha'(He^0)$ to about 5 percent.

The principal heating mechanism for gaseous nebulae is of the form (20.9), and should in principle include a contribution for $\gamma + He^0 \to He^+ + e^-$. In the hottest planetary nebulae, an additional contribution may be needed to describe the photoionization $\gamma + He^+ \to He^{+2} + e^-$.

20.3. Energy Loss Mechanisms

Effective cooling of gaseous nebulae depends on the production of photons whose energy is too low to lead to further ionization of H or He. Once produced, these photons will leave the nebula. Three primary mechanisms can be identified: recombination of H^+ and He^+; collisional excitation by electrons of the atoms or ions having excited states lying only a few eV or less above their ground states; and free-free transitions (*Bremsstrahlung*) of electrons in the field of positive ions. Each process can be expressed as an energy-loss rate per unit volume, which counteracts the photoionization heating rate discussed in Section 20.2.

Radiative Recombination

The radiative-recombination loss rate can be obtained directly from the discussion of (20.5). Each recombination removes from the gas an energy equal to that of the captured electron $\epsilon_e = m_e v^2/2$. The energy-loss rate Λ_R is obtained if the electron flux in (20.5) is replaced by the electron energy flux $\epsilon_e n_e v f(v)$, and the result multiplied by the density of target particles $n(H^+)$:

$$\Lambda_R = n(H^+) \sum_{n=2}^\infty \int_0^\infty n_e v \frac{1}{2} m_e v^2 \sigma_n(v) f(v)\, dv$$

$$= n(H^+) n_e \beta'(H^0) kT_e. \qquad (20.11)$$

The second line defines β'; recombination to the ground state $n = 1$ has been excluded, since the emitted photon will reionize another H^0 rather than escape from the nebula. Recombination is assumed to occur from the thermalized electron distribution characterized by the kinetic temperature T_e, as outlined in the discussion preceeding (20.5). Because $\sigma_n(v) \approx v^{-2}$, low-energy electrons are preferentially captured, and their average energy is slightly below the average of the initial distribution. Consequently, if the equilibrium state of the gas were defined by $\Gamma = \Lambda_R$, then its temperature T_e would exceed T_{eff}. An expression similar to (20.11) could also be written for the energy loss due to recombination of He^+, or other ions present. The contribution of these processes is small, since they are proportional to the ion density. In fact, generalizing (20.11) to other ions, we find that the net recombination cooling rate is

$$\Lambda_R = \Lambda_R(H^+) + \Sigma \Lambda_R(X_k^i)$$

$$= \Lambda_R(H^+)\left[1 + \sum \frac{n(X_x^i)}{n(H^+)} \frac{\beta'(X_k^i)}{\beta'(H^0)}\right], \qquad (20.12)$$

where the sum covers all atomic species k and all ionization states i of the gas. Evidently the contribution of He^+ is small (less than 10 percent), and that of the heavier elements is, in total, less than about 1 percent. In practice He^+ recombination is included when calculating Λ_R, but the recombination of heavier elements (such as O^+, O^{+2}, and N^+) is omitted.

Collisional Excitation

The most effective cooling mechanism in gaseous nebulae is the emission of forbidden-transition radiation by intermediate-mass ions that have been excited by collisions with electrons to states whose energies are roughly kT_e above the ground state. Figure 20.3 shows the low-lying energy levels of three of the most important coolants in gaseous nebulae, and the principal transitions in the visible spectrum. The lowest-energy transitions in O^+ produce lines in the violet. Transitions in O^{+2} produce lines in the green, and the transitions in N^+ lie in the red. Spontaneous emission rates A_{mn} vary from about one per sec to 4×10^{-5} sec^{-1} for these forbidden processes. Although the abundances of intermediate-mass ions are extremely low (see Figure 18.2), the radiation they emit cannot excite or ionize other elements and therefore escapes from the nebula.

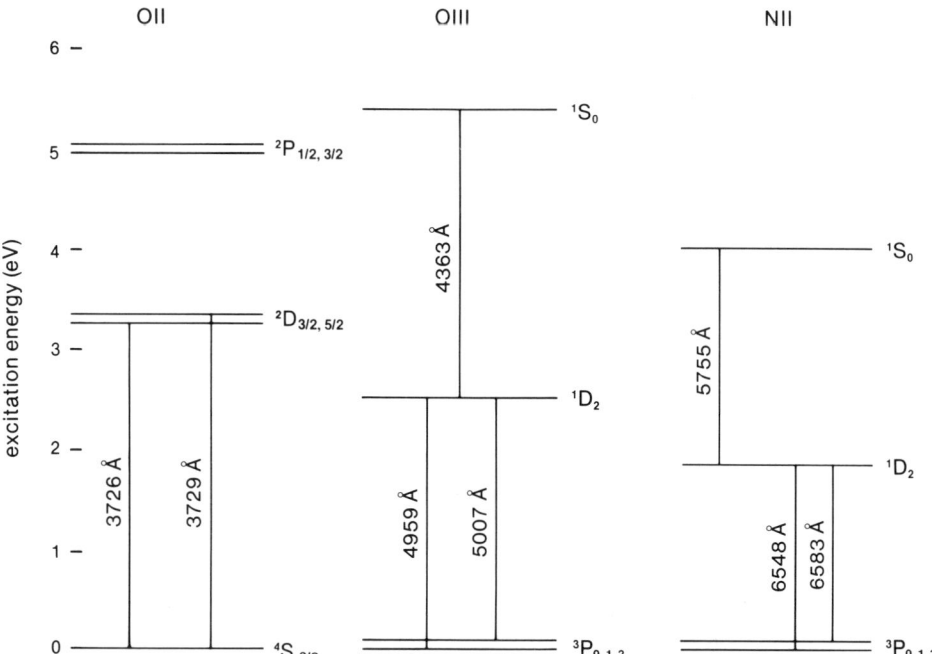

Figure 20.3. Low-lying energy levels of OII, OIII, and NII, the primary coolant ions in interstellar HI regions. Excitation energy relative to the ground state of the ion is shown in eV at left. Wavelength of transitions in Å is given next to each transition.

Electron-ion collisions populate excited states in the ions, whereas radiative (spontaneous) emissions to the ground state depopulate them. Collisions can also depopulate excited states, in which case the excitation energy goes into kinetic energy of the electron rather than into radiation. Consider a typical electron-ion collision that excites an ion from state i to j (i is usually the ground state in a nebula). The excitation cross section for this process is given by

$$\sigma_{ij} = \frac{\pi \hbar^2}{m_e^2 v^2} \frac{\Omega_{ij}}{g_i} = \frac{4.21}{v^2} \frac{\Omega_{ij}}{g_i} \text{ cm}^2, \quad (20.13)$$

where v is the electron velocity, and g_i the statistical weight of the initial state of the ion; Ω_{ij}, which depends weakly on velocity, is called the *collision strength*. It is dimensionless, and is usually of order unity under nebular conditions. For $T_e \approx 7{,}000$ K, $v \approx 6 \times 10^7$ cm/sec and $\sigma_{ij} \approx 10^{-15} (\Omega_{ij}/g_i)$ cm^2. If $m_e v^2/2 < h\nu_{ij}$, the excitation energy, then the cross section is zero.

Problem 20.5. Use the principle of detailed balance to show that the de-excitation cross section (used to calculate the rate of collision-induced transition from level j to level i) is

$$\sigma_{ji} = (g_i/g_j)(v_i/v_j)^2 \sigma_{ij}. \quad (20.14)$$

The rate of collisional de-excitation from level j to level i caused by electrons of velocity v is described by the product of the electron flux, the number density of target ions in the jth excited state $n(X_k^j)$, and the de-excitation cross section; integrating the result over electron velocities, we find the rate per unit volume to be

$$n_e n(X_k^j) r_{ij} = n(X_k^j) \int_0^\infty n_e v \sigma_{ji}(v) f(v) \, dv. \quad (20.15)$$

The electron-velocity distribution $f(v)$ is Maxwellian. Assuming that Ω_{ij} is a constant, we find that the de-excitation rate becomes

$$n_e n(X_k^j) r_{ij} = n_e n(X_k^j) \left(\frac{2\pi}{kT_e}\right)^{1/2} \frac{\hbar^2}{m_e^{3/2}} \frac{\Omega_{ji}}{g_j} \quad (20.16)$$

$$= 8.6 \times 10^{-6} \frac{n_e n(X_k^j)}{g_j T_e^{1/2}} \Omega_{ij} \text{ cm}^{-3} \text{ sec}^{-1}.$$

For typical nebular temperatures and abundances, the rate r_{ji} for the important coolants is roughly 10^{-7} cm^3/sec.

Problem 20.6. Show that the collisional excitation rate is given by

$$r_{ij} = (g_j/g_i) r_{ji} e^{-\chi_{ji}/kT_e}, \quad (20.17)$$

where

$$\chi_{ji} = m_e (v_j^2 - v_i^2)/2,$$

v_j is the velocity of the electron exciting the ion, and v_i is the velocity inducing de-excitation. Recall that no photon is emitted in collisional de-excitation.

Expressions (20.16) and (20.17) for r_{ij} and r_{ji} may be used to obtain the collisional cooling rate. For most nebular coolants (such as those in Figure 20.3), the initial or final levels are multiple, and transitions among all states must be considered. For example consider a simple two-level ion (such as the transition between the $^1S_0$ and $^1D_2$ states of O^{+2} or N^+ in Figure 20.3). The energy loss per unit volume per second due to the spontaneous emission of a photon from a collisionally excited state 2 to the lower level 1 is

$$\Lambda_0 = n(X_k^2) A_{21} h\nu_{21}, \quad (20.18)$$

where $n(X_k^2)$ is the number density of coolant ions in the excited state. In statistical equilibrium the number of upward transitions (due to collisions) equals the number of downward transitions (due to collisions and spontaneous decay):

$$n(X_k^2)(n_e r_{21} + A_{21}) = n_e n(X_k^1) r_{12}. \quad (20.19)$$

The rates r_{21} and r_{12} are given by (20.16) and (20.17).

Solving (20.19) for the ratio of ions in the excited and ground states, and substituting into (20.18) yields

$$\Lambda_{c,21} = \frac{n_e n(X_k^1) r_{12} h\nu_{21}}{1 + n_e r_{21}/A_{21}} \text{ erg cm}^{-3} \text{ sec}^{-1}. \quad (20.20)$$

In nebulae the electron density n_e is usually low enough that spontaneous emission dominates collisional de-excitations. Taking the $n_e \to 0$ limit of $\Lambda_{c,21}$, we find

$$\Lambda_{c,21} \simeq n_e n(X_k^1) h\nu_{21} r_{12} \quad (20.21)$$

$$= n_e n(X_k^1) h\nu_{21} \frac{g_2}{g_1} e^{-\chi/kT_e} r_{21}$$

$$= n_e n(X_k^1) h\nu_{21} \frac{e^{-\chi/kT_e}}{g_1} \frac{8.6 \times 10^{-6}}{T_e^{1/2}} \Omega(1,2),$$

using (20.16), and $\chi = h\nu_{21}$. At temperatures well below χ/k, the exponential in Λ_c dominates, and the cooling rate is low; but at T_e of order $0.2\chi/k$, it increases rapidly, reaching a maximum at $T_e \approx 2\chi/k$. Thereafter Λ_c decreases approximately as $T^{-1/2}$ for fixed $n(X_k^i)$. For the primary coolants (O^+, O^{+2}, and N^+) at $T_e \approx 7{,}000$ K, the ratio $\Lambda_c/n_e n(H^0)$ is of order 10^{-24} erg cm^3/sec.

Problem 20.7. What is Λ_c in the limit $n_e \to \infty$? Explain why the form of this result corresponds to thermal equilibrium.

When forbidden transitions involve more than one level (g_i or $g_j > 1$), Λ_c may be constructed as follows. First, the net cooling rate for the ion k due to transitions from the multiple levels j to the state i is

$$\Lambda_{c,k} = \sum_i n(X_k^i) \sum_{j>i} A_{ji} h\nu_{ji}. \quad (20.22)$$

The relation between the population of various levels j and i is given by the obvious extension of (20.19), resulting in one equation for each possible transition (collisional or spontaneous). Finally, for each pair of levels between which transitions can occur, the rates $r_{ij}^{(k)}$ analogous to (20.16) may be obtained. Finally, the total number of ions in the ground or excited states must equal the number of ions in the gas. This set of equations can be used to construct the net cooling rate (20.22) as a function of n_e, $n(X_k^1)$, and the electron temperature. Once the fractional abundance of the ionized elements has been decided (see Section 20.5), the cooling rate Λ_c becomes a function of T_e.

Bremsstrahlung

A continuous spectrum is emitted by thermal electrons that scatter off ions in the gas. Most of the energy is emitted in the radio and in the infrared, and can readily escape the nebula. The loss rate for ions of number density n_i is

$$\Lambda_{ff} \simeq \frac{2^5 e^6 \pi Z^2}{3^{3/2} h m_e c^3} \left(\frac{2\pi k T_e}{m_e}\right)^{1/2} n_e n_i$$

$$= 1.4 \times 10^{-27} Z^2 T_e^{1/2} n_e n_i \text{ erg cm}^{-3} \text{ sec}^{-1}. \quad (20.23)$$

Usually energy loss via Λ_{ff} is small compared with that from recombination and collisional processes, unless the ion density is extremely small. Note that (20.23) would produce losses from protons in a pure hydrogen nebula in which Λ_c would be zero.

The net cooling rate for a gaseous nebula is the sum of radiative recombination losses (20.12), collisional losses (20.22) for all coolant ions, and *Bremsstrahlung* (20.23).

20.4. STRUCTURE EQUATIONS

In principle the structure of an HII region or a planetary nebula could be obtained from the spectrum L_ν of the ionizing star, as well as the density distribution and abundances of elements in the surrounding medium. From these could be obtained the ionization states $X_k^i(\mathbf{r})$ and the plasma temperature $T_e(\mathbf{r})$ at each point in the nebula. These would be used to construct the spectrum of emitted radiation, including lines and continuum from the radiative transfer equation. Unfortunately, many processes produce lines in the ultraviolet or infrared that are not easily observed through the atmosphere. Furthermore, it is difficult to resolve the internal structure in nebulae with enough accuracy to pinpoint the spatial origin of individual spectral components. Because of these limitations, most models of nebulae assume spherical symmetry.

All the processes to be considered in this section operate on time-scales that are short relative to the time required for large-scale changes in the structure of nebulae. The structure equations are limited to a description of the steady state.

The input for model nebulae consists of those quantities specifying the energy input of the ionizing star: $L_\nu(R)$ in ergs cm^{-2} sec^{-1} Hz^{-1} ster^{-1}; the stellar radius R and inner radius of the nebula r_0; as well as the density distribution $\rho(\mathbf{r})$ and relative abundances $n(X_k)/n(\text{H})$ of all elements X_k. The radiation intensity I_ν at point r in the nebula consists of two parts:

$$I_\nu = I_\nu^s + I_\nu^d; \quad (20.24)$$

the first is the stellar component, and the second is due to emission from atoms in the nebula itself. Defining the mean intensity as (5.2), we find the stellar component to be

$$4\pi J_\nu^s = L_\nu(r)/4\pi r^2. \quad (20.25)$$

The transport equations may be written in terms of each component. For the stellar component,

$$\frac{dL_\nu(r)}{dr} = -k_\nu L_\nu(r), \quad (20.26)$$

$$\frac{dI_\nu^s}{ds} = -k_\nu I_\nu^d + j_\nu. \quad (20.27)$$

The emission coefficient j_ν describes the photons emitted from the gas, and does not include direct radiation from the ionizing star. The absorption coefficient will be defined as

$$k_\nu = \sum_i n_i a_\nu(i)$$

$$\simeq n(\text{H}^0) a_\nu(\text{H}^0) + n(\text{He}^0) a_\nu(\text{He}^0). \quad (20.28)$$

In general the sum is over all ions or atoms that can be further ionized by the radiation source (20.25).

Equations describing the ionization state of the gas as a function of the ionizing star's luminosity are of the form (20.7) for H and He, but require modification for the intermediate-mass ions. For these, the source $J_\nu(r)$

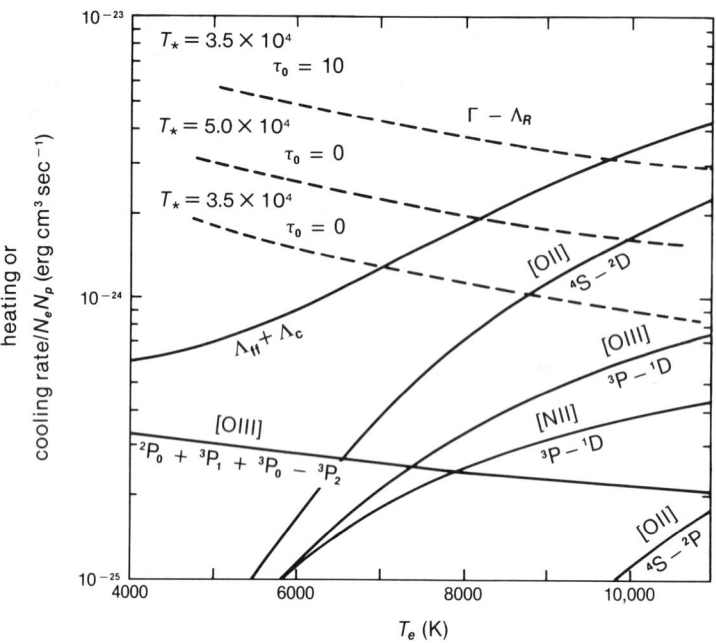

Figure 20.4. Heating and cooling rates versus temperature. Dashed curves give $\Gamma - \Lambda_R$ for three stellar input spectra (T_* denotes the spectral temperature), and τ_o is the optical depth of the medium at the ionization limit. The radiative cooling curve is labeled $\Lambda_{ff} + \Lambda_c$. dominant contributions to the radiative cooling are shown by the light solid curves.

includes the diffuse emission from the gas as well as the stellar emission (as we will see). Thus for each ionic state in the gas, an equation of the general form (20.7) results. Together these equations describe the relative concentrations of ions in the nebula.

Finally, the thermal state of gas is described by the energy equilibrium equation

$$\Lambda_c + \Lambda_{ff} + \Lambda_R = \Gamma, \quad (20.29)$$

which states that the net energy loss equals the energy gain per unit volume per second. The quantities in (19.29), which are defined in Sections 20.2 and 20.3, are functions of the density (assumed given), the ionic concentrations (derived from ionization equilibrium), and T. Therefore (20.29) describes T_e as a function of these variables and $L_\nu(r)$. The heating function Γ does not depend on T_e, but the cooling rates Λ_R (20.11), Λ_c (20.21), and Λ_{ff} (20.23) do contain T_e. In practice the procedure outlined in the preceding must be solved iteratively, until the value of T_e used to evaluate the ionization state and the terms in (20.29) and the derived T_e converge.

Several general conclusions about the equilibrium state of a nebula follow from (20.29) and the expressions for Γ and the loss rates developed in earlier sections. First, because each term in (20.29) varies linearly with n_e (in the low-density limit), the equilibrium temperature T_e is independent of n_e. Each term is also proportional to the density of an ion in the gas. Therefore, the equilibrium state is independent of the density, though it does depend through Λ_c on the abundances of coolant ions relative to $n(\mathrm{H}^+)$. Finally, at high densities, Λ_c is independent of n_e; therefore, as n_e increases, so must T_e.

Problem 20.8. Show that the preceding conclusions follow for the simple case of a nebula containing only hydrogen and a single coolant ion.

Figure 20.4 shows the heating and cooling rates as a function of T_e for a simple model of an HII region in

390 / GASEOUS NEBULAE

which the elements have abundances comparable to those shown by the open circles in Figure 18.1. Here 80 percent of the coolant ions are assumed to be singly ionized, and the rest doubly ionized, and

$$n(H^+)/n(H^+ + H^0) = 0.9.$$

The dashed curves represent three possible results for the net photoionization heating $\Gamma - \Lambda_R$. The lower two curves result for stellar effective temperatures T_{eff} of 35,000 K (spectral type O7) and 50,000 K (spectral type O5) at points near the star. The upper dashed curve is at an optical depth = 10; the net heating is greater there because the average photon energy $h\bar{\nu}$ (20.10) increases with distance. The thin solid lines labeled by an ionic state and the electronic transition in the ion (see Figure 20.3) are the contributions to Λ_c. At low electron temperatures, transitions between the triplet P states in O^{+2} are the dominant cooling mechanism, but at higher values O^+ becomes the more important. Notice the rapid rise in the contributions of O^+, N^+, and O^{+2} that results when T_e almost doubles. The heavy solid curve gives the net cooling rate $\Lambda_{ff} + \Lambda_c$ including all ions, and its intersection with the dashed curve defines the electron temperature of the nebulae. Depending on the distance from the ionizing star (optical depth) and assumed surface brightness, T_e varies from 7,000 K to 9,000 K. The density n_e is low enough in this example for collisional de-excitation to be negligible. If n_e approaches 10^4 cm^{-3}, collisional de-excitation becomes important, and the relative importances of various ionic cooling terms change. However, the equilibrium T_e increases by only about 20 percent.

Problem 20.9. Fluctuations in T_e can occur at arbitrary points in a nebula. Show that the equilibrium state defined by (19.29) and shown in Figure 20.4 is stable against small variations in T_e.

20.5. Model Nebulae

The results developed in Section 20.4 can be applied to a model of a planetary nebula composed of H, He, and trace ions. The initial data required to solve the structure equations, discussed at the beginning of Section 20.4, will be summarized here for convenience.

They are:

$$\begin{pmatrix} \text{stellar luminosity} \\ \text{at stellar surface } R \end{pmatrix} = L_\nu(R) \, \text{erg Hz}^{-1} \text{sec}^{-1}; \quad (20.30)$$

$$\text{inner radius of nebula} = r_0 \quad (20.31)$$

$$n(\text{He})/n(\text{H}) \quad (20.32)$$

$$n(X_k)/n(\text{H}) \quad (20.33)$$

$$\rho(r) = m_H[n(\text{H}) + 4n(\text{He})]. \quad (20.34)$$

The number density of a specific ionic state (i) of an element X_k will be denoted by $n(X_k^i)$. Thus in (20.32)

$$n(\text{H}) = n(\text{H}^+) + n(\text{H}^0),$$

where H^0 denotes neutral hydrogen. First, the stellar energy input can be obtained at $r > R$ from (20.26):

$$L_\nu(r) = L_\nu(R)e^{-\tau_\nu(r)}, \quad (20.35)$$

where

$$\tau_\nu(r) = \int_{r_0}^r k_\nu(r')dr' \quad (20.36)$$

and the absorption coefficient is given by the second line of (20.28). In general r_0 will not equal R, but $L_\nu(r_0) = L_\nu(R)$. Evidently, $\tau_\nu(r)$ is the optical depth of the stellar radiation measured into the nebula, and $L_\nu(r)$ gives the stellar energy spectrum (including effects of absorption) carried across a sphere of radius r in the nebula. Although (20.35) is a formal solution for $L_\nu(r)$, it can not be evaluated until the abundances of absorbing atoms (H^0 and He^0) are specified.

Ionization State (H and He)

We will assume at the outset that the nebula contains no He^{+2}, that $n(\text{He})/n(\text{H}) \approx 0.1$, and that the total abundance of intermediate-mass ions

$$\sum_k n(X_k)/n(\text{H}) = 10^{-3}.$$

The first step in constructing a model nebula is to find a value for the ionization state of H and He, assuming a local value for T_e; the latter must be checked against the value derived from the energy-balance equation (20.29) for consistency. It is difficult to solve the structure equations in terms of the total intensity (stellar plus diffuse components). In most cases one of two approximations may be made.

In the first, the diffuse radiation (that emitted by atoms in the gas) escapes directly from the nebula upon emission, and all ionization is due to the stellar component. With these assumptions, the absorption term in the transfer equation (20.27) is absent, and

$$I_\nu^d = \int j_\nu \, ds. \qquad (20.37)$$

We call this the *thin limit;* it is applicable to low-density or small nebulae, and to the radiation emitted by coolant ions.

The second approximation is that the nebula is optically thick. Diffuse radiation is absorbed where it is created (*on-the-spot* approximation); the change in I_ν^d with path length is then small, and (20.27) becomes

$$I_\nu^d = j_\nu / k_\nu. \qquad (20.38)$$

This limit applies to the photoionization processes that determine the ionization state of H and He, which we now consider.

Hydrogen recombination to all states except the ground state produces photons that can not ionize H^0 or He^0. We therefore write (20.7), excluding recombination to the ground state, and replacing $J_\nu(r)$ by $L_\nu(r)$, using (20.4), in the form

$$n(H^0) \int_{\nu_1}^{\infty} \frac{L_\nu(r)}{h\nu} \alpha_\nu(H^0) \, d\nu$$
$$= 4\pi r^2 n_e n(H^+) \alpha'(H^0), \qquad (20.39)$$

where $\nu_1 = 13.6$ eV. Recombination of He^+ to excited states of He^0 is more difficult; it can not ionize He^0, but can ionize H^0. It is customary to assume that all photons produced by He^+ recombination ionize H^0 but not He^0 (see Problem 20.4); the ionization state of He is then described, in analogy with (20.39) by

$$n(He^0) \int_{\nu_2}^{\infty} \frac{L_\nu(r)}{h\nu} a_\nu(He^0) d\nu$$
$$= 4\pi r^2 n_e n(He^+) \alpha'(He^0) \qquad (20.40)$$

with $\nu_2 = 24.5$ eV. Charge neutrality requires that the local abundances of He^+ and H^+ satisfy

$$n_e = n(H^+) + n(He^+). \qquad (20.41)$$

The ionization state of the nebula is now specified; the stellar energy input given by (20.35), (20.36), and (20.28), together with the ionization equation (20.39) and (20.40), mass density (20.34), and charge neutrality (20.41), represent seven equations in eleven unknowns. Given the initial data $L_\nu(r)$, r_0, ρ, and $n(He)/n(H)$, they may be solved for $n(H^+)$, $n(H^0)$, $n(He^+)$, $n(He^0)$, and n_e. Figure 20.5 shows results for a model nebula as a function of distance from the ionizing star whose effective temperature $T = 3.5 \times 10^4$ K. In this model, $r_0 = 8 \times 10^{16}$ cm; the nebula consists of an inner sphere ($r \lesssim 1.75 \times 10^{17}$ cm) in which essentially all the H and He are singly ionized, surrounded by a shell 5.5×10^{16} cm thick containing He^0 and H^+. The transition between two ionization states is quite sharp. The outer boundary of the nebula, where $n(H^+)$ falls essentially to zero, is about 2.3×10^{17} cm from the ionizing star. Planetary nebulae are observed to range in diameter from 0.06 to 0.4 pc, and the effective temperature of the central star usually lies in the range 3 to 30×10^4 K.

The ionization structure established thus may then be used to obtain the intensity of the diffuse radiation emitted by the gas. Using the approximation (19.38) for an optically thick nebula, and remembering that j_ν and k_ν result almost entirely from H and He and are dependent only on the assumed T_e and on known ionization abundances, we can find I_ν^d. The stellar component (recall that the mean intensity is used) J_ν^s follows from the solution of (20.35) and (20.4).

Ionization State (Coolants)

The coolant ions occur with very low abundances relative to H or He, and essentially all their emitted radiation escapes from the nebula. This justifies treating them as if they were decoupled from the nebular gas (H and He) as a whole. Nevertheless, these trace elements play a dominant role in the energy budget of the nebula, and in setting the electron temperature. Given the results summarized in Figure 20.5, we can obtain the ionization state of the coolant ions as follows.

We assume first that all radiation from these ions escapes. Therefore, ionization is due to stellar photons

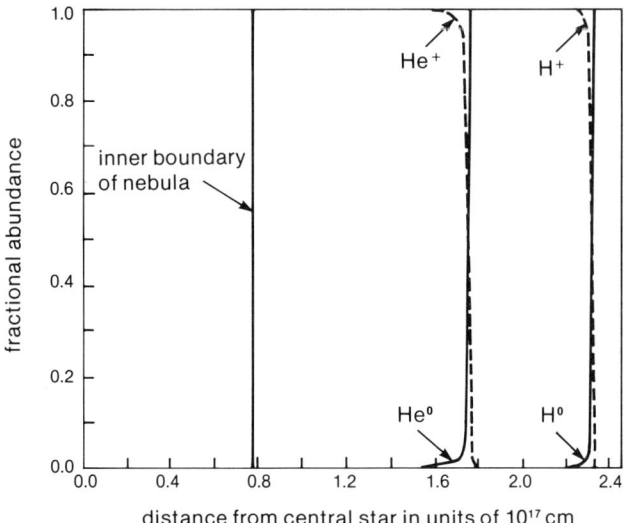

Figure 20.5. Fractional abundances of He^0 and H^0 (solid), and of He^+ and H^+ (dashed) in model nebula. (See Figures 20.6 and 20.7 also.)

and to photons in the diffuse component emitted by H and He. In statistical equilibrium the photoionization and recombination rates are equal, and an equation of the form

$$n(X_k^i) \int_{\nu_k}^{\infty} \frac{J_\nu}{h\nu} a_\nu(X_k^i) \, d\nu = n_e n(X_k^{i+1}) \alpha(X_k^{i+1}) \quad (20.42)$$

results for each ion k, and for each possible ionization state. Two comments about (20.42) are in order. First, the recombination coefficient includes recombination to the ground state, because the emitted photon is assumed to escape from the nebula. Second, the source J_ν is

$$4\pi J_\nu(r) = \int (I_\nu^d + I_\nu^s) \, d\Omega = J_\nu^s(r) + \int I_\nu^d \, d\Omega. \quad (20.43)$$

Finally, for each atomic species

$$n(X_k) = \sum_i n(X_k^i), \quad (20.44)$$

with the sum extending over all ionization states for which an equation of the form (20.42) is assumed. The set of equations (20.42) to (20.44) can be solved for the abundance $n(X_k^i)/n(H)$ for each ionic state, assuming that the total relative abundance $n(X_k)/n(H)$ is known. The final model may be used to predict the relative strength of emission lines as a function of the assumed abundances. In practice the latter are chosen so that the relative line strengths agree with these observations.

Figure 20.6 shows the most important ionic abundances $n(X_k^i)$ for the model. Oxygen, nitrogen, and neon occur in the singly and doubly ionized states within the nebula, but recombine to the neutral state across the outer boundary. Carbon, which is singly, doubly, and triply ionized within the nebula, exists as C^+ outside it. Notice that the transition between ionization states is gradual except near the outer boundary.

Thermal Equilibrium

The abundances shown in Figures 20.5 and 20.6 are now used to evaluate (20.29) for the kinetic temperature T_e of the electron gas. It will be recalled that T_e appears in the recombination coefficient (its dependence there is weak, $\approx T_e^{-1/2}$) and in the cooling rates Λ_c, and that one must assume an average value

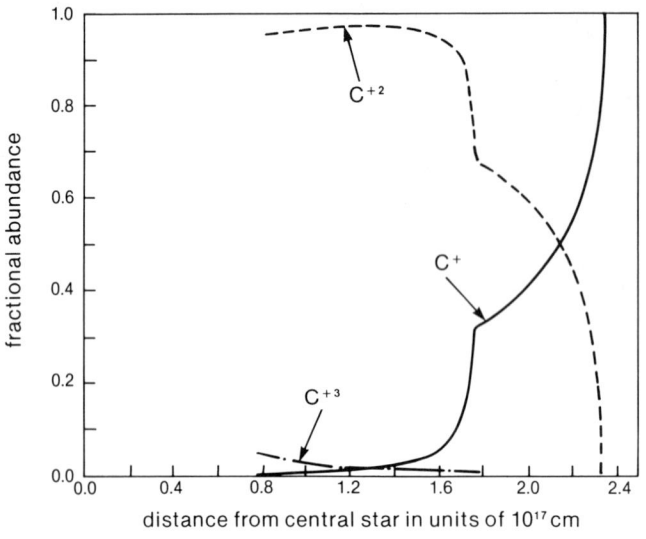

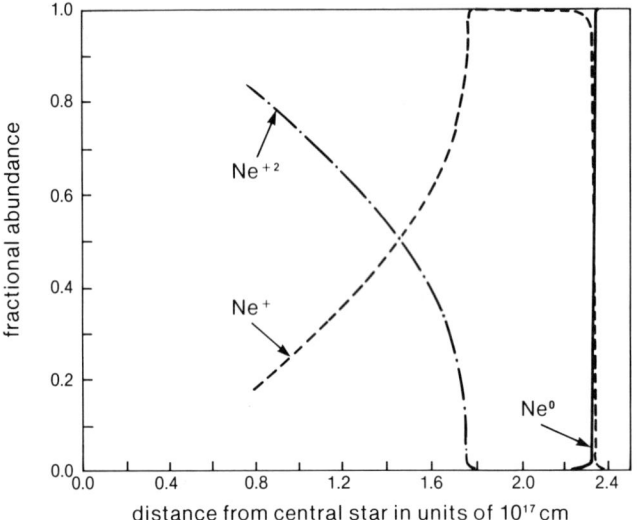

Figure 20.6. Fractional abundances of coolant ions in the model nebula shown in Figure 20.5.

initially in order to evaluate the ionic abundances and cooling rates. Therefore, if precise results are desired, the entire procedure may need to be repeated, using the T_e derived in the following as input for a new iteration of the structure equations. The following discussion applies in this case as well.

Let us assume initially that only H and the coolant ions are present in the nebula, so that the equation of thermal equilibrium becomes

$$h\bar{\nu} \equiv \frac{\beta'(H^0) + \beta_{ff}(H^0)}{\alpha(H^0)} kT_e$$
$$+ \frac{\Lambda_c}{n_e n(H^+)\alpha(H^0)} \quad (20.45)$$

with $h\bar{\nu}$ defined as in (20.10) and $\beta'(H^0)$ as in (20.11). The *Bremsstrahlung* contribution $\beta_{ff}(H^0)$ follows immediately on comparison with (20.23); but this

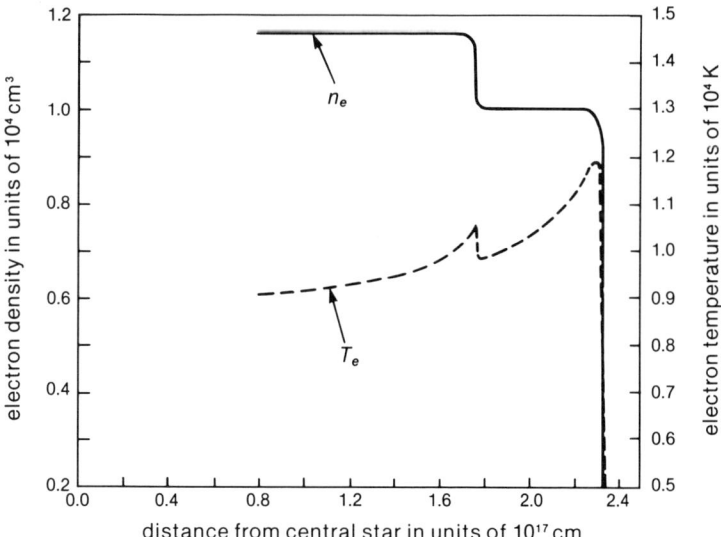

Figure 20.7. Temperature and electron number-density profiles in model nebula (see Figures 20.5 and 20.6).

effect is extremely small, and can be ignored here. Therefore, recalling (20.21) for the low-density regime ($n_e \ll 10^4$ cm^{-3}), we have

$$h\bar{\nu} \simeq \frac{\beta'(\mathrm{H}^0)}{\alpha(\mathrm{H}^0)} kT_e + \sum_{i,k} A_k \frac{n(X_k^i)}{n(\mathrm{A}^+)} \frac{e^{-\chi_i/kT_e}}{T_e^{1/2}}, \quad (20.46)$$

where A_k is a constant, and the summation in the last term is over all coolant ionic species present in the nebula. The energy-balance equation is transcendental in T_e, but may be solved numerically.

Several features of the solution can be seen in (20.46). First, we recall that the left-hand side $h\bar{\nu}$, which contains an integral over the photoionization cross section $a_\nu \sim (\nu_1/\nu)^3$, increases with distance into the nebula; see the discussion following (20.10). Consider the right-hand side of (20.46). For $T_e \lesssim$ several times χ_i, the collisional cooling terms increase or remain essentially constant with increasing T_e. Thus, since $h\bar{\nu}$ increases radially outward, so will T_e, until we reach the outer boundary of the nebula. At the boundary, $h\bar{\nu}$ continues to change slowly, but $n(\mathrm{H}^+)$ goes rapidly to zero (see Figure 20.5). The abundance of coolants changes across the outer boundary also, but the net effect is to replace one ionic coolant (O^+, for example) by another coolant (O^0). As a result, the collisional cooling terms in (20.46) dominate, and T_e must decrease abruptly. The equilibrium electron temperature therefore increases gradually with distance from the ionizing star, reaches its maximum value just behind the $\mathrm{H}^+ \rightarrow \mathrm{H}^0$ transition zone at the nebula's boundary, and then drops rapidly to values typical of the interstellar medium. Figure 20.7 shows T_e and n_e obtained numerically for the model (which includes heating from photoionization of He) corresponding to Figures 20.5 and 20.6. The reduction in n_e at $r \simeq 1.7 \times 10^{17}$ cm corresponds to He$^+$ recombination; since $n(\mathrm{He}) \simeq 0.1 \, n(\mathrm{H})$, it is about a 10 percent effect, as expected. At the He$^+ \rightarrow$ He^0 transition, the ratios $n(\mathrm{O}^+)/n(\mathrm{O})$ and $n(\mathrm{N}^+)/n(\mathrm{N})$ increase from relatively low values to essentially unity. Since these elements are extremely efficient coolants (Figure 20.4), the local electron temperature drops abruptly, as shown in Figure 20.7, but increases gradually until the outer boundary is reached.

In general, the value of T_e obtained from (20.45) will differ from the value assumed at the start of the calculation. If the difference is significant, then the entire process described above may be repeated using the new value of T_e as input for the next iteration. In practice the procedure usually converges rapidly. For planetary nebulae, such as the one described here, $T_e \approx 10^4$ K, though it rises to slightly higher values just behind the $He^+ - He^0$ and $H^+ - H^0$ transition boundaries.

20.6. RELATIVE LINE STRENGTHS

Much of the line emission observed in nebulae arises from the coolant ions, and their relative strength is one of the most useful probes of T_e, n_e, and the coolant abundances. Figure 20.3 shows the energy levels most frequently involved in line emission from nebulae; at $T_e \approx 10^4$ K, only the first two excited states are populated, and the ions may be considered to be relatively simple three-level systems.

For simplicity, consider a representative three-level atom whose ground state is labeled 1, and whose two upper (excited) states are 2 and 3 (Figure 20.8). The relative population of the two excited states is straightforward to obtain, but can involve a large number of coupled equations. Fortunately, transitions between all levels in Figure 20.3 are not equally likely. Because of the relatively weak radiation intensity in nebulae, induced transition rates in the coolant ions can usually be neglected. In fact, we may assume that only collisional excitation and de-excitation and spontaneous decay occur between the ground state and either excited state. Furthermore, we neglect transitions between levels 2 and 3 (this will be justified in examples below). Figure 20.8a shows these transitions. Dashed lines indicate collisional processes, and wavy lines indicate radiative processes. Under these assumptions, the number of ions per unit volume in each state is given by

$$n_3(C_{31} + A_{31}) = n_1 C_{13} = n_1 \frac{g_3}{g_1} C_{31} e^{-E_{13}/kT_e}, \quad (20.47)$$

$$n_2(C_{21} + A_{21}) = n_1 C_{12} = n_1 \frac{g_2}{g_1} C_{21} e^{-E_{12}/kT_e}, \quad (20.48)$$

where $E_{ij} = E_i - E_j$. The line emissivity from the transition $i \to j$ is just the product of the number density n_i of ions in the ith excited state, the energy difference $E_{ij} = h\nu_{ij}$, and the spontaneous emission rate A_{ij}. Therefore the ratio of intensity emitted by the $3 \to 1$ transition to that emitted by the $2 \to 1$ transition is

$$\frac{j(3 \to 1)}{j(2 \to 1)} = \frac{n_3 h\nu_{31} A_{31}}{n_2 h\nu_{21} A_{21}}$$

$$= \frac{g_3 A_{31} \nu_{31}}{g_2 A_{21} \nu_{21}} \left(\frac{1 + A_{21}/C_{21}}{1 + A_{31}/C_{31}} \right) e^{-E_{23}/kT_e}, \quad (20.49)$$

using (20.47) and (20.48) to eliminate n_3 and n_2. Before proceeding to specific processes, we should note two important general features of the relative line strength (20.49). First, if the two excited states are such that E_{23} is comparable to E_{12}, then the relative

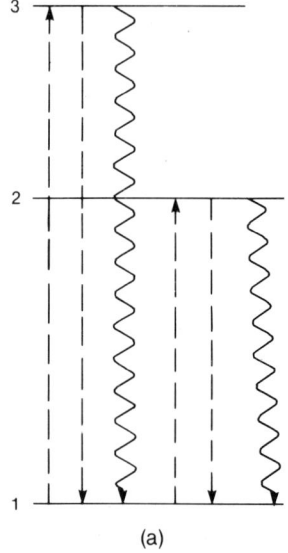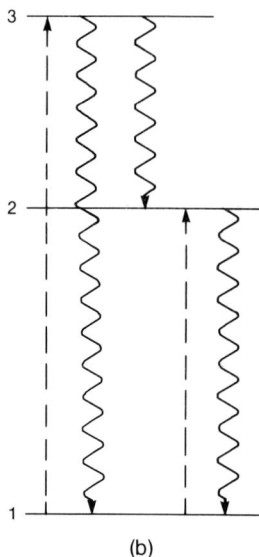

Figure 20.8. Three-level atom showing principal radiative transitions (wavy lines) and collisionally induced transitions (dashed lines): (a) transitions giving rise to (20.47) and (20.48); (b) transitions used to obtain (20.50) and (20.51).

strength will be very sensitive to T_e. If, on the other hand, $E_{23} \ll kT_e$, then the relative strength will vary most strongly with electron density (recall $C_{ij} \approx n_e/T_e^{1/2}$). Consequently, measurements of (20.49) for certain ions can be used to fix T_e, n_e, and the coolant abundances.

Most of the emission lines in HII regions and planetary nebulae that are generated by these collisional excitation–radiative de-excitation processes are optically forbidden. They occur with low transition probabilities ($A \sim 1$ sec^{-1}). Gaseous nebulae are the most common instances of these lines in astronomy. Other instances are: (1) the solar corona; (2) the spectrum of the night sky; and (3) the aurora borealis.

Electron temperature measurements utilize the transitions $^1S_0 \to {}^1D_2$ ($\lambda = 4363$ Å), $^1D_2 \to {}^3P_2$ ($\lambda = 5007$ Å), and $^1D_2 \to {}^3P_1$ ($\lambda = 4959$ Å), which occur in [OIII] (Figure 20.3). All of these lie in the visible spectrum, the first with an energy of 2.84 eV, the other two with energies of nearly 2.51 eV, and all may be observed in most nebulae. Both excited states will be populated, and the relative populations will be a good indicator of T_e. We may consider as a reasonable approximation to [OIII] in nebulae a three-level system (as described in the preceding) in which excitation of the ^1S or ^1D levels is due to collisions, but the decay is radiative (no collisional de-excitation), an assumption that is reasonable as long as n_e is not too great. Furthermore, we shall allow the radiation decay $^1S \to {}^1D$. The possible transitions are shown in Figure 20.8. Setting up level densities as in (20.47) and (20.48), we can find immediately that the ratio of line intensity at $\lambda = 4363$ Å to that at $\lambda = 4959$ Å and $\lambda = 5007$ Å combined is

$$\frac{j(4363)}{j(5007) + j(4959)} \simeq \frac{n_3 h\nu_{32} A_{32}}{n_2 h\nu_{21} A_{21}} \quad (20.50)$$

$$= \frac{g_3 \nu_{32} C_{31}}{g_2 \nu_{21} C_{21}} \frac{A_{32}}{A_{32} + A_{31}} e^{-E_{32}/kT_e}.$$

For [OIII], the energy difference $E_{32} = 2.84$ eV and the coefficient of the exponential is 0.12. Thus

$$\frac{j(4363)}{j(5007) + j(4959)} = 0.12 \, e^{-32{,}900/T_e}, \quad (20.51)$$

where T_e is in K. This simple relation fits quite well for $n_e \lesssim 10^5$ cm^{-3}. A similar result can be found for [NII] ($E_{32} = 2.16$ eV, and the coefficient is 0.13). It is evident that small changes in T_e produce large shifts in the relative line strength. In fact, the ratio varies by nearly a factor of 10^2 when T_e varies from 6000 K to 2×10^4 K. Observations of the [OIII] ratio in HII regions give T_e in the range 8 to 9×10^3 K; for planetary nebulae, T_e is found to vary from 10 to 18×10^3 K. Measurements of the relative strength for the analogous transitions in [NII] yield $7{,}000 \lesssim T_e \lesssim 1.1 \times 10^4$ K in HII regions, and $10^4 \lesssim T_e \lesssim 1.5 \times 10^4$ K in planetary nebulae. Presumably the difference between [NII] and [OIII] measurements reflects the fact that the emission arises from different portions of the nebulae (see Figure 20.6).

The $^2D_{3/2} \to {}^4S_{3/2}$ ($\lambda = 3726$) and $^2D_{5/2} \to {}^4S_{3/2}$ ($\lambda = 3729$) transitions in [OII] are used to measure n_e in nebulae. If only collisional processes and radiative decay between the ^2D and ^4S states are considered, then we again have a simple three-level model as in Figure 20.8(a), except that $E_{23} \ll kT_e$, or $\nu_{21} \simeq \nu_{31}$. Proceeding as above, we easily find the relative line strength of the $\lambda = 3729$ Å and $\lambda = 3726$ Å transitions to be

$$\frac{j(3729)}{j(3726)} = \frac{g_3 A_{31}}{g_2 A_{21}} \left(\frac{1 + A_{21}/C_{21}}{1 + A_{31}/C_{31}} \right), \quad (20.52)$$

which depends weakly on T_e through C_{ij}. In fact, since $C_{ij} \approx n_e/T_e^{1/2}$, the ratio is sensitive to n_e. In relatively dilute regions $C_{ij} \to 0$, and the ratio reduces to

$$g_3 C_{31}/g_2 C_{21} = \Omega(3,1)/\Omega(2,1),$$

which for [OII] is 1.5. In the high-density limit, the ratio (20.52) approaches $g_3 A_{31}/g_2 A_{21} = 0.32$ for [OII]. At temperatures of order 10^3 to 10^4 K, the change in relative line strength is most sensitive to n_e. A similar set of transitions in the red occurs in [SII], and these are also used to measure n_e.

Electron number densities obtained in this way for HII regions vary from 10^2 to several times 10^3 cm^{-3} within individual nebulae. The denser regions are believed to be condensations. A gradual reduction in n_e from the center outward is also observed.

Line intensities from coolant ions in planetary nebulae are of limited use in measuring n_e, since at temperatures of 10^4 K or more, few ions are found in their lower excited states. The emission that is observed is believed to arise from condensations, or from the outer regions of the nebula. Observed ratios for [OII] (20.39) imply that $n_e/T_e^{1/2}$ varies from roughly unity to about 400, with typical values around 40.

Problem 20.10. Derive the relative line strength (20.52) for the processes shown in Figure 20.8(a), and plot it as a function of $n_e/T_e^{1/2}$. For the [OII] transition rates, take $A_{31} = 4.2 \times 10^{-5}$ sec^{-1}, $A_{21} = 1.8 \times 10^{-4}$ sec^{-1}, $\Omega(3,1) = 0.88$ and $\Omega(2,1) = 0.59$.

The relative abundance of some intermediate-mass elements can also be found from line-strength measurements. For example, the $\lambda = 4959$ Å and $\lambda = 5007$ Å transitions in [OIII] are the dominant cooling mechanisms in some nebulae. To the extent that other cooling processes are small, the energy-balance equation (20.46) becomes

$$h\bar{\nu} = \frac{n(O^{+2})}{n(H^+)} e^{-\chi/kT_e}. \quad (20.53)$$

Examination of Figure 20.6 indicates that when [OIII] is likely to dominate the cooling rates, $n(O^{+2}) \approx n(O)$, the total abundance of oxygen, and similarly, $n(H^+) \approx n(H)$. Therefore, if T_e is known,

$$n(O)/n(H) \approx h\bar{\nu} \, e^{\chi/kT_e}. \quad (20.54)$$

Furthermore, if the strength of emission from other ions can be measured relative to the [OIII] lines, then some estimate of their abundances may be made as well.

20.7. Thermal Radio Emission

Diffuse nebulae and some types of galaxies are known to be radio sources. The simplest sources (in terms of their radio emission) have thermal spectra. One source of radio emission is a hot, tenuous plasma, such as is found in HII regions or in planetary nebulae (supernova remnants will be discussed in Chapter 25). The analysis of emission lines implies temperatures of order 10^4 K; so for wavelengths in the radio, $B_\nu(T) \approx 2kT/\lambda^2$. The observed intensity will be given by the solution to the transfer equation (5.38). Assuming that the plasma is in thermal equilibrium and that the temperature is uniform, I_ν will be given by (18.82), with T_s replaced by T. The primary contribution to the opacity of the gas κ_ν is from absorption caused by electron free-free transitions, or *Bremsstrahlung*, which was discussed in Section 6.4; so we obtain the opacity by the following argument.

In thermal equilibrium $j_\nu = I_\nu k_\nu = B_\nu(T) k_\nu$, and k_ν is given by (6.82) and (6.83) integrated over the electron velocity distribution. Denoting the emission rate per gram in the frequency interval ν to $\nu + d\nu$ by $j_\nu d\nu$, we find that, to within constant factors,

$$j_\nu \, d\nu \approx \frac{n_e n_i}{T^{1/2}} d\nu. \quad (20.55)$$

The absorption rate is given by

$$k_\nu B_\nu(T) d\nu \approx \rho \kappa_\nu T \nu^2 \, d\nu. \quad (20.56)$$

In equilibrium, these two rates must be equal, which gives the desired result

$$\rho \kappa_\nu \approx \frac{n_e^2}{T^{3/2} \nu^2}, \quad (20.57)$$

where we set $n_e = n_i$. The preceding arguments are not exact, and the actual opacity contains an additional weak dependence on T and ν of the form $\log(T^{3/2}/\nu)$, which is ignored in the following discussion.

According to (20.57), the plasma will be transparent at high frequencies, but becomes opaque at lower frequencies (longer wavelengths). Therefore, at long wavelengths, $\tau_\nu \ll 1$, and the intensity becomes

$$I_\nu \simeq \frac{2kT\nu^2}{c^2} \tau_\nu. \quad (20.58)$$

If the temperature is uniform throughout the emission region, then the optical depth becomes

$$\tau_\nu = \int_0^l \kappa_\nu \rho \, dr \approx \frac{\int_0^l n_e^2 \, dr}{T^{3/2} \nu^2}. \quad (20.59)$$

The quantity $\int n_e^2 \, dr$ is called the *emission measure*, and l is the length along the line of sight of the emitting region. Combining (20.58) and (20.59) yields

$$I_\nu \approx \frac{\int_0^l n_e^2 \, dr}{T^{1/2}} \quad \text{for} \quad \tau_\nu \ll 1, \quad (20.60)$$

which is independent of frequency. At low frequencies the plasma is transparent ($\tau_\nu \gg 1$), and the intensity is simply

$$I_\nu \approx B_\nu(T) \sim \frac{2kT\nu^2}{c^2} \quad \text{for} \quad \tau_\nu \gg 1, \quad (20.61)$$

which is independent of density. Figure 20.9 shows I_ν from the opaque to the transparent region.

In planetary nebulae and HII regions, the radio spectrum is as shown in Figure 20.9. This means that two radio-intensity measurements (one where $\tau_\nu \gg 1$, and one where $\tau_\nu \ll 1$), and an estimate of the system's dimension l appearing in the emission measure, will allow one to calculate the plasma temperature and n_e, and to estimate the total mass of the emitting region. Emission from [OIII], [OII], and [NII] typically imply temperatures of order 10^4 K in planetary nebulae. Expressions similar to (20.61) can also be applied to the gas shells surrounding novae to obtain the gas temperature and an estimate of the amount of ejected mass (see Section 17.6).

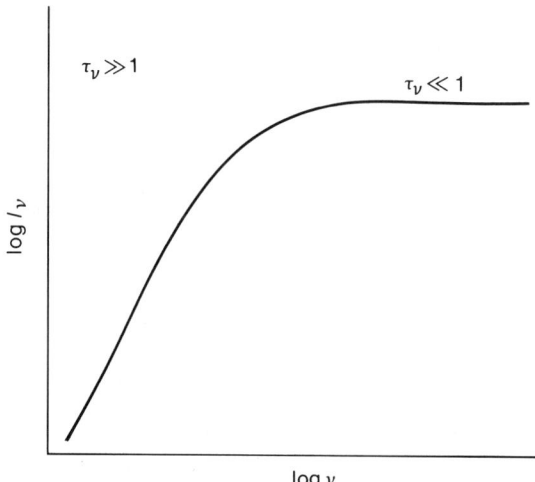

Figure 20.9. Thermal radio intensity versus frequency, showing effect of *Bremsstrahlung*.

Problem 20.11. Observations of the planetary nebula NGC 7027 give $T = 1.1 \times 10^4$ K for the plasma, and an emission measure 5.4×10^7 cm^{-6} pc. Find n_e and the mass of gas in the nebula, assuming n_e is a constant. The nebula is spherical with diameter $l \simeq 0.1$ pc. Discuss the assumptions made in obtaining your results.

Problem 20.12. What do the solar corona, the night sky, and the aurora borealis have in common with HII regions that could explain the presence of forbidden lines of highly ionized elements in their spectra?

Chapter 21

HYDRODYNAMICS

Many aspects of stellar and nonstellar astrophysics involve hydrodynamic phenomena. Some matter coexists with (or may even be the source of) magnetic fields extending over large regions of space. Often the material is in the form of ions and electrically charged particles, and so we must incorporate the laws of classical electrodynamics. We then deal with magnetohydrodynamics. Usually charge neutrality will hold, at least down to the level of volume elements that, though macroscopically small, are nonetheless microscopically large. As a result, the local electric fields must be small or zero. Finally, when all particle velocities and usually all volume element velocities are small compared to the speed of light, the underlying physics is to sufficient accuracy described by Newtonian mechanics.

In principle, a system containing many particles in motion could be described by writing Newton's equations for each particle, including interparticle collisions, and using each charged constituent as a source in Maxwell's equations. The result, augmented by appropriate boundary conditions, would completely describe the behavior of the material. Such a program is obviously impractical. Instead, we develop a description that includes the necessary aspects of local microscopic physics, but focuses on macroscopic portions of the system. Basic to this approach is the concept of a fluid element of volume ΔV and density $\rho(\mathbf{x}, t)$ having velocity $\mathbf{v}(\mathbf{x}, t)$. We consider ΔV to be macroscopically small (as compared to the dimensions of regions over which external fields vary, for example) but microscopically large (so that the number ΔN of particles in ΔV is large). The properties of the material in ΔV may then be described by the laws of thermodynamics. In general we may assume that the limit $\Delta r \to 0$ may be taken in such a way that quantities like $\Delta m / \Delta v$ remain finite.

21.1. Reference Frames

The two most frequently encountered reference frames in hydrodynamics are the Lagrangian and Eulerian frames. Lagrangian coordinates are fixed with each fluid element Δm and move with it (comoving coordinates). This system is especially convenient, because the fluid elements retain their identity during flow, and because it is the most natural extension of particle mechanics, where the equations of motion follow each particle in the system. In the Lagrangian description each fluid element Δm is characterized by its initial coordinates $\mathbf{x}_0$ at time t_0, and its current position $\mathbf{x}$ at

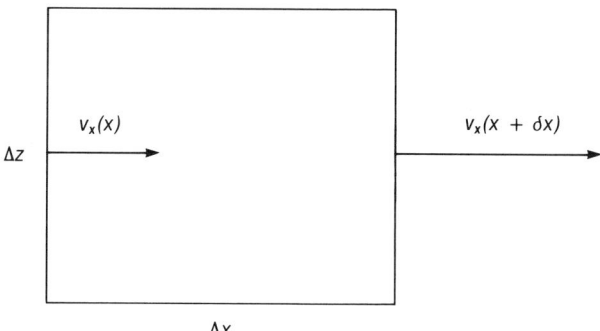

Figure 21.1. Time-rate of change of rectangular volume element result when opposite faces move with different velocities.

time t, which we denote by $\mathbf{x}(\mathbf{x}_0, t)$. The initial position serves as a label identifying the fluid element. The Lagrangian system is useful only where the geometry of the volume element occupied by Δm remains relatively simple during its motion, as it does in simple flow or when shocks form, but not, for example, when the flow becomes turbulent.

In the Eulerian approach, attention is directed to specific points $\mathbf{x}$ fixed in space through which the fluid flows. To these are assigned values of ρ and $\mathbf{v}$. The solution then reduces to finding their time-dependence. It is important to note that the values of these quantities are not carried by fluid elements as in the Lagrangian framework. In the Eulerian approach, the time-rate of change of a quantity $Q(\mathbf{x}, t)$ is just $(\partial Q/\partial t)_\mathbf{x}$ evaluated for fixed $\mathbf{x}$.

When we employ the Lagrangian method, where the frame of reference is attached to the fluid element ΔV, both $\mathbf{x}$ and t change as the fluid element moves. In this case the rate of change of $Q(\mathbf{x}, t)$ is

$$\frac{dQ}{dt} = \frac{\partial Q}{\partial t} + \frac{\partial Q}{\partial x^k}\frac{dx^k}{dt} = \frac{\partial Q}{\partial t} + \mathbf{v} \cdot \nabla Q, \quad (21.1)$$

where $v^k = dx^k/dt$ are the components of a fluid element's velocity. The first term on the right is the Eulerian time-derivative, and it gives the change in Q at fixed $\mathbf{x}$ due to its explicit time-dependence. The second term gives the change from one point to a nearby point that results from the implicit time-dependence of the space coordinates. Note that $k = 1$, 2, or 3, and that we sum over repeated indices ($a^k b^k \equiv \Sigma_k a^k b^k$). The last term in (21.1) is often called the *advection term*, since it may be thought of as the change in Q due to the net flow of material into a volume ΔV.

A useful identity may be obtained by examining the way in which a volume element $\Delta V = \Delta x \Delta y \Delta z$ changes in time. Consider what happens along the $\mathbf{x}$ direction (see Figure 21.1). The velocity of the left and right edges of the element are $\mathbf{v}(\mathbf{x})$ and $\mathbf{v}(\mathbf{x} + \Delta \mathbf{x})$, respectively, with $\mathbf{x}$ components $v_x(\mathbf{x})$ and $v_x(\mathbf{x} + \Delta \mathbf{x})$. If these two velocity components are not equal, there will be a change in ΔV given by

$$\frac{d}{dt}\Delta V = \Delta y \Delta z \frac{d\Delta x}{dt} + \Delta y \Delta x \frac{d\Delta z}{dt} + \Delta x \Delta z \frac{d\Delta y}{dt}.$$

The second and third terms arise when the velocity components on opposite edges along the y and z directions are unequal. The first term is readily seen to be

$$dt \frac{d\Delta x}{dt} = v_x(\mathbf{x} + \Delta \mathbf{x})\, dt - v_x(\mathbf{x})\, dt$$

$$= v_x(\mathbf{x})\, dt + \frac{\partial v_x}{\partial x}\Delta x\, dt + \ldots - v_x(\mathbf{x})\, dt$$

$$= \frac{\partial v_x}{\partial t}\Delta x\, dt + O_2,$$

where O_2 denotes second- and higher-order terms in the Taylor-series expansion of $v_x(\mathbf{x} + \Delta \mathbf{x})$ for small $|\Delta \mathbf{x}|$. Continuing in this way, we can show that the volume element changes by

$$\frac{d}{dt}\Delta V = (\nabla \cdot \mathbf{v})\Delta V = \frac{\partial v^k}{\partial x^k}\Delta V. \quad (21.2)$$

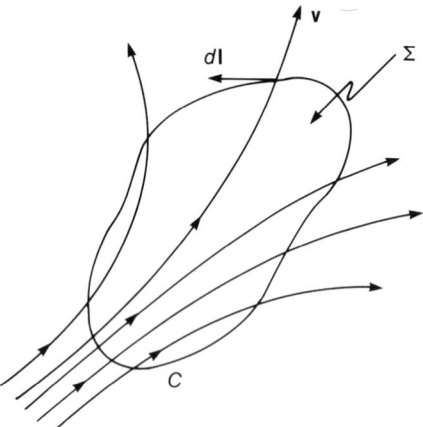

Figure 21.2. Contour enclosing fluid flow lines. The area enclosed by C is Σ.

Equation (21.2) demonstrates that an incompressible fluid, that is, one whose volume elements remain constant in magnitude, is defined by the relation $\nabla \cdot \mathbf{v} = 0$ on the velocity field.

An additional constraint on fluid motion is contained in the quantity $\nabla \times \mathbf{v}$. Imagine a path in the fluid whose infinitesimal length is $d\mathbf{l}$ (see Figure 21.2). At $\mathbf{x}$ evaluate $\mathbf{v} \cdot d\mathbf{l}$ and integrate around the closed loop C:

$$\oint_C \mathbf{v} \cdot d\mathbf{l} = \int_\Sigma \nabla \times \mathbf{v} \cdot d\mathbf{s}. \quad (21.3)$$

Here Σ is the arbitrary surface whose boundary is C, and $d\mathbf{S}$ is an infinitesimal area $\hat{n}\,dA$ of Σ with $\hat{n}$ normal to Σ at $\mathbf{x}$. If the fluid flow lines (denoted by $\mathbf{v}$) do not change direction, then the fluid flows essentially without rotation; the contribution to the integral in (21.3) for $\mathbf{v} \cdot d\mathbf{l} > 0$ will equal that for $\mathbf{v} \cdot d\mathbf{l} < 0$; it will vanish for any contour C. Thus the right side (obtained by using Stokes' theorem) will vanish. Since Σ is arbitrary, the integral must vanish. We therefore find $\nabla \times \mathbf{v} = 0$ for a fluid flow that is irrotational.

21.2. Equations of Motion

The equations of motion are Newton's second law applied to fluid elements, such as the one shown in Figure 21.3. Consider two classes of forces: those due to the pressure $P(\mathbf{x})$ times an infinitesimal area dA at every point on the surface S (directed opposite to the normal $\hat{n}$ to the surface); all others (gravitational, electromagnetic, etc.) are given by the product $\mathbf{f}\,\Delta V$, where $\mathbf{f}$ is called the *body-force density*. The net force on the entire surface due to $P(\mathbf{x})$ is given by

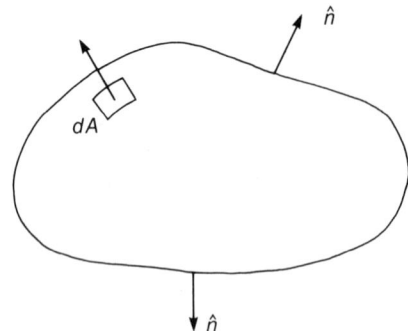

Figure 21.3. Unit vectors and infinitesimal area of arbitrary volume element.

$$-\int_\Sigma P\hat{n}\,dA = -\int_v \nabla P\,dV. \quad (21.4)$$

The minus sign is necessary since $\hat{n}$ points outward; the right-hand side of (21.4) results if we apply the generalized form of Gauss's theorem (Problem 21.1). The net force on the element is

$$\int_{\Delta V} \rho \frac{d\mathbf{v}}{dt} dV = \int_{\Delta V} \mathbf{f}\,dV - \int_{\Delta V} \nabla P\,dV. \quad (21.5)$$

The quantity on the left is the mass-averaged acceleration of the fluid element; the terms on the right represent the sum of all forces acting on it. All integrals are over the fluid element volume ΔV, which may be chosen arbitrarily. Therefore we conclude that

$$\rho \frac{d\mathbf{v}}{dt} = \mathbf{f} - \nabla P. \quad (21.6)$$

This is the equation of motion in Lagrangian form. Note that (21.6) applies to fluid elements of any shape, but assumes that no shearing stress exists (the pressure P is a scalar). This assumption also holds for fluids that are nonviscous. In hydrostatic equilibrium the acceleration of the volume element is zero; so the left side of (21.6) vanishes and $\mathbf{f} = \nabla P$. If the matter distribution is spherically symmetric as well, and the only body force is due to self-gravitation, we recover the usual result,

$$\frac{dP}{dr} = -\frac{m(r)G\rho(r)}{r^2}, \qquad (21.7)$$

where $\rho(r)$ is the density at point r, and

$$m(r) = \int_0^r 4\pi r^2 \rho\, dr$$

is the mass contained within a surface of radius r about the center. Equation (21.7) will be recognized as one of the stellar structure equations. For dynamical (nonrelativistic) stages of stellar evolution, equation (21.6) replaces (21.7).

Problem 21.1. Consider a rectangular volume element $\Delta V = \Delta x\, \Delta y\, \Delta z$ as shown in Figure 21.1. By adding up all pressure contributions to the force on the element, show that they are equal to the expression (21.4). Then show that the surface integral may be replaced by the volume integral as claimed.

If the dynamical motion of the fluid element is governed in part by electromagnetic effects, these may be introduced through the term $\mathbf{f}$ in (21.6), as is shown in Section 21.3.

Conservation Laws

In a classical, nonrelativistic fluid, mass is conserved. This means that

$$\frac{d(\rho\, \Delta V)}{dt} = 0$$

for all ρ and ΔV. Recalling (21.2), we can rewrite this as

$$\frac{d\rho}{dt} + \rho \nabla \cdot \mathbf{v} = 0. \qquad (21.8)$$

This is the mass-continuity equation in Lagrangian form. It is sometimes useful to employ the integral form $\rho\, \Delta V = $ constant. Using (21.2) and (21.1), we also have

$$0 = \frac{d}{dt}(\rho\, \Delta V) = \Delta V \frac{d\rho}{dt} + \rho \frac{d\Delta V}{dt}$$

$$= \Delta V \frac{\partial \rho}{\partial t} + \Delta V \mathbf{v} \cdot \nabla \rho + \rho(\nabla \cdot \mathbf{v})\, \Delta V.$$

Since ΔV is arbitrary,

$$\frac{\partial \rho}{\partial t} + \mathbf{v} \cdot \nabla \rho + \rho \nabla \cdot \mathbf{v} = \frac{\partial \rho}{\partial t} + \nabla \cdot (\rho \mathbf{v}) = 0, \qquad (21.9)$$

which is the equation of continuity in Eulerian form. Equation (21.9) states that the mass of a volume element changes only when mass crosses its surface. The final form of (21.9) uses the identity

$$\nabla \cdot (\rho \mathbf{v}) = \mathbf{v} \cdot \nabla \rho + \rho \nabla \cdot \mathbf{v}. \qquad (21.10)$$

Problem 21.2. Show that an incompressible fluid flows at right angles to its density gradient. Recall that incompressibility is defined by the condition $\nabla \cdot \mathbf{v} = 0$.

Problem 21.3. Start with the hydrostatic equilibrium condition $\nabla P = \mathbf{f}$, where $\mathbf{f}$ is the gravitational force on an element of matter in a spherical mass distribution, and derive (21.7).

The momentum of a fluid element is given by $\rho \mathbf{v}\, \Delta V$. From (21.6) we see that

$$\mathbf{f}\, \Delta V - \nabla P\, \Delta V = \rho\, \Delta V \frac{d\mathbf{v}}{dt}$$

$$= \frac{d}{dt}(\mathbf{v}\, \Delta V \rho) - \mathbf{v} \frac{d}{dt}(\rho\, \Delta V)$$

$$= \frac{d}{dt}(\rho \mathbf{v} \Delta V). \qquad (21.11)$$

By mass conservation, $\rho\, \Delta V$ is constant, and the element's time-rate of change is given by $(\mathbf{f} - \nabla P)\, \Delta V$. If this vanishes (i.e., $\mathbf{f} = \nabla P$), hydrostatic equilibrium obtains and momentum is conserved:

$$\frac{d}{dt}(\rho \mathbf{v}\, \Delta V) = 0 \qquad \mathbf{f} = \nabla P. \qquad (21.12)$$

We deduce from (21.11) that a state of hydrostatic equilibrium is equivalent to constant linear momentum of the fluid elements.

Forming the dot product of $\mathbf{v}$ and (21.11) and using the continuity equation yields

$$\rho \Delta V \mathbf{v} \cdot \frac{d\mathbf{v}}{dt} = \frac{d}{dt}\left(\frac{1}{2}\rho v^2 \Delta V\right)$$
$$= (\mathbf{f} \cdot \mathbf{v} - \mathbf{v} \cdot \nabla P) \Delta V. \quad (21.13)$$

The left-hand side is the rate of change of the element's kinetic energy. The term $\mathbf{v} \cdot \nabla P \Delta V$ may be rewritten using (21.1) with P in place of Q:

$$\mathbf{v} \cdot \nabla P \Delta V = \Delta V \left(\frac{dP}{dt} - \frac{\partial P}{\partial t}\right)$$
$$= \frac{d}{dt}(P\Delta V) - \frac{\partial P}{\partial t}\Delta V - P\nabla \cdot \mathbf{v} \Delta V.$$

The last step uses (21.2) for $d\Delta V/dt$. Substituting this in (21.13) gives

$$\frac{d}{dt}\left(\frac{1}{2}\rho v^2 + P\right)\Delta V$$
$$= \left(\mathbf{f} \cdot \mathbf{v} + \frac{\partial P}{\partial t} + P\nabla \cdot \mathbf{v}\right)\Delta V. \quad (21.14)$$

The term $(\mathbf{f} \cdot \mathbf{v}) \Delta V$ represents the rate at which work is done by all body forces $\mathbf{f}\Delta V$ on the fluid element. The last term vanishes ($\nabla \cdot \mathbf{v} = 0$) for an incompressible fluid.

Problem 21.4. Show that the last term in (21.14) corresponds to work done per second because of contraction or expansion of the fluid element. Use $\Delta m = \rho \Delta V$ to show that

$$\frac{d}{dt}\left(\frac{1}{2}\rho v^2 + P + \rho\epsilon\right)\Delta V = \mathbf{f} \cdot \mathbf{v} \Delta V, \quad (21.15)$$

where the specific energy of compression or expansion is given by

$$\epsilon = \epsilon_0 + \int_{P_0}^{P} \frac{P}{\rho^2} d\rho \quad (21.16)$$

per unit mass. The energy density is $\epsilon\rho$ in this case.

Two special cases of (21.15) are of interest. First, when the body forces $\mathbf{f}$ do no work on the fluid elements, $\mathbf{f} \cdot \mathbf{v} = 0$. Equation (21.15) becomes

$$\frac{d}{dt}\left(\frac{1}{2}v^2 + P/\rho + \epsilon\right)\rho\Delta V$$
$$= \rho\Delta V \frac{d}{dt}\left(\frac{1}{2}v^2 + P/\rho + \epsilon\right) = 0,$$

since $\rho \Delta V = \Delta m$ is constant. The sum in parentheses is thus constant:

$$\tfrac{1}{2} v^2 + P/\rho + \epsilon = \text{constant}. \quad (21.17)$$

This is one familiar form of Bernoulli's law. A more common form includes gravitational potential-energy effects. If $\mathbf{f}$ is just the gravitational force density, and may be expressed in the form $\mathbf{f} = -\rho\nabla\phi$, then (21.14) becomes

$$\frac{d}{dt}\left(\frac{1}{2}\rho v^2 + P + \rho\phi\right)\Delta V = \left(\frac{\partial P}{\partial t} - \rho\frac{\partial \phi}{\partial t}\right)\Delta V$$
$$+ P(\nabla \cdot \mathbf{v}) \Delta V. \quad (21.18)$$

For incompressible fluids in static gravitational fields, $\partial\phi/\partial t = 0$, and arguments similar to those used in arriving at (21.17) show that

$$\tfrac{1}{2}v^2 + P/\rho + \phi = \text{constant}. \quad (21.19)$$

Specifically, (21.17) and (21.19) may be evaluated at two different points. Bernoulli's law, as given by (21.17) or (21.19), is a statement of energy conservation for fluids.

Energy Equation

The equation of motion (21.6) and the continuity equation (21.9) must be supplemented by an equation of state $P = P(U,V)$, where U is the internal energy of the fluid element ΔV, and is given by the first law of thermodynamics (2.16):

$$\frac{dU}{dt} = \frac{dQ}{dt} - P\frac{dV}{dt}. \quad (21.20)$$

A more convenient form is obtained by setting $dQ = T\,dS$, substituting $\Delta V = \Delta Nm/\rho$, where the mass of the particles making up the fluid element is m, and expressing the result in terms of the internal energy per gram, $\epsilon = U/m\,\Delta N$ and the entropy per gram $s = S/m\,\Delta N$:

$$\frac{d\epsilon}{dt} = T\frac{ds}{dt} + \frac{P}{\rho^2}\frac{d\rho}{dt}. \tag{21.21}$$

The heat-transfer rate $dQ/dt = T\,dS/dt$ may depend on local processes, such as thermonuclear fusion, or ionization and recombination. In general, it will depend on the local thermodynamic state of the fluid, in which case the additional equation of state $T = T(\epsilon, \rho)$ will be needed. The change in entropy may also be due to irreversible processes, such as the passage of shock waves.

The complete set of hydrodynamic equations in Lagrange form are

$$\rho\frac{d\mathbf{v}}{dt} = -\nabla P + \mathbf{f}, \tag{21.22}$$

$$\frac{d\rho}{dt} + \rho\nabla\cdot\mathbf{v} = 0, \tag{21.23}$$

and

$$\frac{d\epsilon}{dt} = T\frac{ds}{dt} + \frac{P}{\rho^2}\frac{d\rho}{dt}. \tag{21.24}$$

These must be supplemented by the equations of state

$$P = P(\rho, \epsilon), \tag{21.25}$$

$$T = T(\rho, \epsilon), \tag{21.26}$$

and by expressions for the nonadiabatic changes in energy $T\,ds/dt$.

Adiabatic Perturbations: Sound Waves

Special cases of great importance are adiabatic processes for which $ds/dt = 0$, and in which the change in ϵ is caused by expansion or compression of the fluid element Δm. Here the energy equation may be replaced by the relation

$$P = K\rho^\gamma; \tag{21.27}$$

for a perfect fluid, the complete set of relations (2.15) applies. For example, consider a uniform fluid in which the density ρ_0 and pressure P_0 are independent of position, and imagine that a small localized disturbance occurs, so that $\rho = \rho_0 + \rho_1$ and $P = P_0 + P_1$. If ρ_1 is small relative to ρ_0, then the resulting fluid motion will be characterized by a velocity $\mathbf{v}$ that is also small. Substituting ρ and P as defined here into the hydrodynamic equations in Eulerian form yields

$$(\rho_0 + \rho_1)\frac{\partial \mathbf{v}}{\partial t} + (\rho_0 + \rho_1)\mathbf{v}\cdot\nabla\mathbf{v} = -\nabla P_1,$$

$$\frac{\partial \rho_1}{\partial t} + \mathbf{v}\cdot\nabla\rho_1 + (\rho_0 + \rho_1)\nabla\cdot\mathbf{v} = 0,$$

where we have used the fact that ρ_0 and P_0 are constants.

An approximate solution of these equations may be found by retaining only lowest-order terms. Remembering that ρ_1, P_1, and v are small, we find

$$\rho_0\frac{\partial \mathbf{v}}{\partial t} = -\nabla P_1 = -(\partial P/\partial \rho)_s\nabla\rho_1, \tag{21.28}$$

$$\frac{\partial \rho_1}{\partial t} + \rho_0\nabla\cdot\mathbf{v} = 0. \tag{21.29}$$

Note that the equation of state (21.27) has the functional dependence $P = P(\rho)$, which permits ∇P_1 to be replaced by $\nabla\rho_1$ in (21.28). For small disturbances, $(\partial P/\partial \rho)_s$ will be a constant. Taking the divergence of (21.28) and subtracting it from the time-derivative of (21.29) yields

$$\frac{\partial^2 \rho_1}{\partial t^2} = \left(\frac{\partial P}{\partial \rho}\right)_s \nabla^2 \rho_1. \tag{21.30}$$

Since this is a wave equation for ρ_1, the wave velocity must be given by (Figure 21.4)

$$c_s = (\partial P/\partial \rho)_s^{1/2}. \tag{21.31}$$

This is the adiabatic sound speed, and (21.30) states that small adiabatic disturbances propagate with speed c_s.

Problem 21.5. Show that the small change in pressure P_1 also propagates with speed c_s.

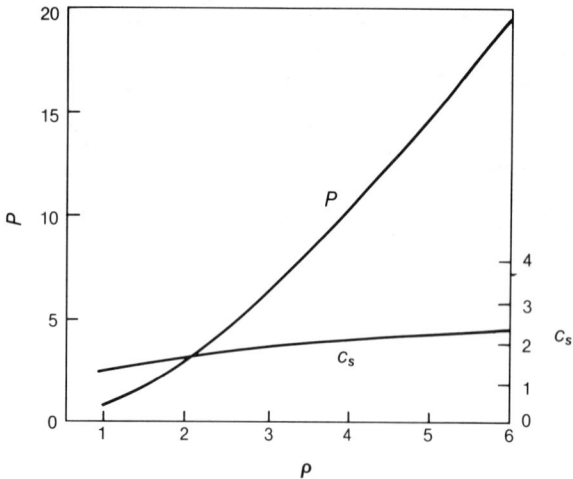

Figure 21.4. Pressure and sound speed for an ideal gas corresponding to the polytropic equation of state $P = \rho^{5/3}$.

Problem 21.6. Evaluate c_s for an ideal gas, expressing the result in terms of the gas temperature.

Problem 21.7. Compare the time required for a sound wave to cross a uniform-density sphere of radius R (sound transit time) to the sphere's free-fall time. Why is the relationship physically reasonable?

If the amplitude of the disturbance is not small, the preceding linear analysis breaks down, and nonlinear terms become important. An extreme example, to be discussed in Chapter 24, is given by shock waves.

21.3. MAGNETOHYDRODYNAMIC EFFECTS

Magnetic fields, if present, are introduced through **f** in equation (21.6). Suppose a fluid element moving with velocity **v** has associated with it a current density **j**. The force exerted on this element by an external magnetic field **B** is

$$\mathbf{f}_M = \mathbf{j} \times \mathbf{B}. \tag{21.32}$$

(The "external" field **B** may be set up by the combined motions of the other fluid elements surrounding the one of interest.) From Maxwell's equations without displacement currents (and setting the magnetic permeability $\mu = 1$),

$$\nabla \times \mathbf{B} = \frac{4\pi}{c} \mathbf{j}, \tag{21.33}$$

we obtain

$$\mathbf{f}_M = \mathbf{j} \times \mathbf{B} = \frac{(\nabla \times \mathbf{B}) \times \mathbf{B}}{4\pi c}$$

$$= \frac{1}{4\pi c}\left[(\mathbf{B} \cdot \nabla)\mathbf{B} - \frac{1}{2}\nabla B^2\right], \tag{21.34}$$

where $B = |\mathbf{B}|$. Substituting (21.34) into (21.6), we obtain, with the aid of $\mathbf{f}_G = -\rho\nabla\phi$, the magnetohydrodynamic equation of motion

$$\rho\frac{d\mathbf{v}}{dt} = -\nabla P - \rho\nabla\phi$$

$$- \frac{1}{8\pi c}\nabla B^2 + \frac{1}{4\pi c}(\mathbf{B} \cdot \nabla)\mathbf{B}. \tag{21.35}$$

The gravitational potential ϕ is found by solving Poisson's equation

$$\nabla^2 \phi = 4\pi G \rho. \tag{21.36}$$

Finally the magnetic field satisfies the equation

$$\nabla \times (\mathbf{v} \times \mathbf{B}) = \frac{\partial \mathbf{B}}{\partial t} - \frac{1}{4\pi\sigma}\nabla^2 \mathbf{B}, \tag{21.37}$$

which follows from the expression for the current density **j** of a moving element (σ is the electrical conductivity)

$$\mathbf{j} = \sigma(\mathbf{E} + \mathbf{v} \times \mathbf{B}), \tag{21.38}$$

and the density is given by the equation of continuity (21.9),

$$\frac{\partial \rho}{\partial t} + \nabla \cdot (\rho\mathbf{v}) = 0. \tag{21.39}$$

Equations (21.35) to (21.39) describe the magnetohydrodynamic behavior of fluid elements in magnetic and gravitational fields **B** and ϕ under general conditions. A direct solution of these equations is impractical because of their complexity. Notice that in general the quantities $\mathbf{v}$, ϕ, ρ, and $\mathbf{B}$ (an assumed relation between ρ and P is also needed) must be self-consistently determined in the following sense. The motion of a fluid element may be responsible for the fields $\mathbf{B}$ on other elements. The fields alter the other element's velocity, which changes the field acting on the original element. Because of the complexity of these equations it is desirable to explore other, more approximate, methods of describing the fluid properties of a plasma. Two methods, the virial theorem and linearized theory, will be considered in subsequent chapters.

21.4. Cylindrical Coordinates

The hydrodynamic equations may be written in cylindrical coordinates for problems relating to galactic disk structure. The transformation from Cartesian coordinates (x, y, z) to (r, θ, z) is straightforward, but tedious. We summarize the steps here.

Start with the continuity equation (21.39) and assume that $\rho = \sigma(r, \theta)$ and $\mathbf{v} = u\hat{e}_r + (v + r\Omega)\hat{e}_\theta$. We assume that the mass is confined to the x,y-plane, so that $\sigma(r, \theta)$ is a mass per unit area, and write the equations for $z = 0$. The extension to a volume density and to motion out of the plane is straightforward. Now evaluate $\sigma \nabla \cdot \mathbf{v}$ and $\mathbf{v} \cdot \nabla \sigma$ in cylindrical coordinates:

$$\mathbf{v} \cdot \nabla \sigma = u \frac{\partial \sigma}{\partial r} + (v + r\Omega) \frac{1}{r} \frac{\partial \sigma}{\partial \theta}, \quad (21.40)$$

using

$$\nabla = \hat{e}_r \partial/\partial r + (\hat{e}_\theta/r)(\partial/\partial \theta).$$

Since σ and $\mathbf{v}$ are assumed to be independent of z, we work in the r, θ-plane. Next we note that

$$\sigma \nabla \cdot \mathbf{v} = \sigma \frac{\partial u}{\partial r} + \frac{\sigma}{r} \frac{\partial}{\partial \theta} (v + r\Omega) + \frac{u\sigma}{r}. \quad (21.41)$$

The last term arises from the change in $\hat{e}_r$ with θ. Combining (21.40) and (21.41) in the first expression for the continuity equation (21.39) yields

$$\frac{\partial \sigma}{\partial t} + \frac{1}{r} \frac{\partial}{\partial r} (ru\sigma) + \frac{1}{r} \frac{\partial}{\partial \theta} \sigma(v + r\Omega) = 0. \quad (21.42)$$

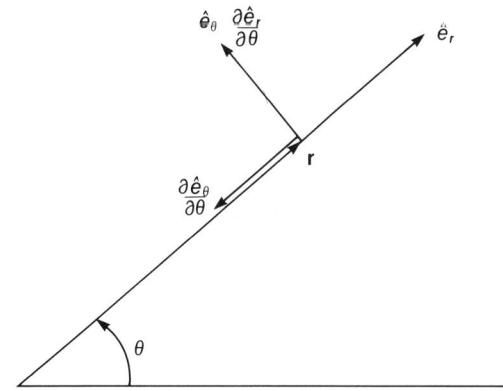

Figure 21.5. Unit vectors and their derivatives in polar coordinates.

The equations of motion are obtained from (21.6), the left side of which is proportional to

$$\frac{d\mathbf{v}}{dt} = \frac{\partial \mathbf{v}}{\partial t} + \mathbf{v} \cdot \nabla \mathbf{v}. \quad (21.43)$$

To evaluate $\nabla \mathbf{v}$, note that

$$\partial \hat{e}_r / \partial \theta = \hat{e}_\theta \text{ and } \partial \hat{e}_\theta / \partial \theta = - \hat{e}_r$$

(Figure 21.5). Therefore

$$\begin{aligned}
\mathbf{v} \cdot \nabla \mathbf{v} &= u \frac{\partial \mathbf{v}}{\partial r} + \frac{(v + r\Omega)}{r} \frac{\partial \mathbf{v}}{\partial \theta} \\
&= u \frac{\partial u}{\partial r} \hat{e}_r + u \frac{\partial}{\partial r} (v + r\Omega) \hat{e}_\theta \\
&+ \frac{v + r\Omega}{r} \frac{\partial u}{\partial \theta} \hat{e}_r + \frac{u(v + r\Omega)}{r} \hat{e}_\theta \\
&- \frac{(v + r\Omega)^2}{r} \hat{e}_r + \hat{e}_\theta \frac{(v + r\Omega)}{r} \frac{\partial}{\partial \theta} (v + r\Omega).
\end{aligned} \quad (21.44)$$

The r component is thus

$$\frac{\partial u}{\partial t} + u \frac{\partial u}{\partial r} + \frac{(v + r\Omega)}{r} \frac{\partial u}{\partial \theta}$$

$$- \frac{(v + r\Omega)^2}{r} = f_r - \frac{\partial P}{\partial r} \quad (21.45)$$

and the θ component is

$$\frac{\partial v}{\partial t} + u\frac{\partial}{\partial r}(v + r\Omega) + u\frac{(v + \Omega r)}{r}$$
$$+ \frac{(r\Omega + v)}{r}\frac{\partial v}{\partial \theta} = f_\theta - \frac{1}{r}\frac{\partial P}{\partial \theta}, \quad (21.46)$$

since $r\Omega$ is independent of time and θ.

Problem 21.8. Carry out the step leading to (21.44).

Equations (21.45) to (21.46) are the equations of motion in the plane $z = 0$ for a distribution of matter that is axially symmetric, and whose fluid elements rotate about the axis with angular velocity $\Omega(r)$. They enable discussion of density waves in the disks of spiral galaxies (Chapter 27).

Chapter 22

THE VIRIAL THEOREM

We have already discussed a simplified version of the virial theorem (Section 4.2). If we are to apply it to the interstellar medium, we must include magnetic fields and rotation. Therefore the virial theorem will be derived here in general form. Although the virial theorem contains some dynamic information, it focuses attention on macroscopic parameters of the entire system, such as total energy.

22.1. GENERAL FORM OF THE VIRIAL THEOREM

Starting with the equation of motion (21.6), form the dot product with **r**, and integrate over the system mass:

$$\int dm\, \mathbf{r} \cdot \frac{d^2\mathbf{r}}{dt^2} = \int \mathbf{r} \cdot \mathbf{F}\, dm - \int \mathbf{r} \cdot \nabla P \frac{dm}{\rho}. \quad (22.1)$$

Here **r** is the position of the element and $\mathbf{v} = d\mathbf{r}/dt$, and the force per unit mass $\mathbf{F} = \mathbf{f}/\rho$. The first term is rewritten, using the identity

$$\mathbf{r} \cdot \frac{d^2\mathbf{r}}{dt^2} = \frac{d}{dt}\left(\mathbf{r} \cdot \frac{d\mathbf{r}}{dt}\right) - \frac{d\mathbf{r}}{dt} \cdot \frac{d\mathbf{r}}{dt} = \frac{d^2}{dt^2}\frac{r^2}{2} - \frac{d\mathbf{r}}{dt} \cdot \frac{d\mathbf{r}}{dt},$$

to obtain

$$\int dm\, \mathbf{r} \cdot \frac{d^2\mathbf{r}}{dt^2} = \frac{1}{2}\frac{d^2 I}{dt^2} - 2\langle T \rangle$$

$$= \frac{1}{2}\frac{d^2}{dt^2}\int r^2\, dm - \int \left|\frac{d\mathbf{r}}{dt}\right|^2 dm. \quad (22.2)$$

The first term is a generalized moment of inertia (note r is not the perpendicular distance from a rotation axis), and the second is twice the kinetic energy of mass motion within the fluid.

Now consider the last term in (22.1): since $dm = \rho\, dV$, this is

$$\int \mathbf{r} \cdot \nabla P\, dV = \int \nabla \cdot (\mathbf{r}P)\, dV - 3\int P\, dV$$

$$= \oint_\Sigma P\mathbf{r} \cdot d\mathbf{S} - 3\int P\, dV. \quad (22.3)$$

The identity

$$\nabla \cdot (P\mathbf{r}) = (\nabla \cdot \mathbf{r})P + \mathbf{r} \cdot \nabla P$$

409

with $\nabla \cdot \mathbf{r} = 3$ has been used, and the volume integral of the divergence converted to a surface integral by Gauss' theorem. The surface Σ is assumed to include the entire mass M of the system. Two special cases of (22.3) are useful. These follow if $P(\Sigma)$ is constant, or if the system is spherically symmetric, so that (22.3) becomes

$$\int \mathbf{r} \cdot \nabla P \, dV = P_s \oint_\Sigma \mathbf{r} \cdot d\mathbf{S} - 3 \int P \, dV$$
$$= P_s \int \nabla \cdot \mathbf{r} \, dV - 3 \int P \, dV$$
$$= 3 P_s V - 3 \int P \, dV$$
$$= 4\pi R^3 P_s - 3 \int P \, dV. \quad (22.4)$$

The last expression is valid if the distribution is spherically symmetric (radius R), or if its volume is defined to be equal to $4\pi R^3/3$ (which defines R). Combining these equations we find

$$\frac{1}{2}\frac{d^2 I}{dt^2} = 2\langle T \rangle + 3 \int P \, dV$$
$$+ \int \mathbf{r} \cdot \mathbf{F} \, dm - \oint_\Sigma P\mathbf{r} \cdot d\mathbf{S}, \quad (22.5)$$

where

$$\oint_\Sigma P\mathbf{r} \cdot d\mathbf{S} = 4\pi R^3 P_s, \quad (22.6)$$

when $P(\Sigma)$ is constant or the distribution is spherically symmetric. All forces not associated with mechanical pressure are contained in the virial term $\mathbf{r} \cdot \mathbf{F}$. Two specific cases will be encountered in what follows.

Gravitational Forces

The force per unit mass for an arbitrary matter distribution is given by

$$\mathbf{F}_G = -\int dV' \rho(\mathbf{r}') \frac{\mathbf{r} - \mathbf{r}'}{|\mathbf{r} - \mathbf{r}'|^3} G. \quad (22.7)$$

Taking the dot product of $\mathbf{F}_G$ with $\mathbf{r}$, and taking half the sum of the result and the $\mathbf{r} \to \mathbf{r}', \mathbf{r}' \to \mathbf{r}$ term, we easily find

$$\int \mathbf{F}_G \cdot \mathbf{r} \, dm = -\int \frac{Gm(r) \, dm(r)}{r} \equiv \Omega, \quad (22.8)$$

which is the total gravitational potential energy of the system.

Problem 22.1. Prove that the gravitational potential energy of a spherically symmetric mass distribution of constant density is

$$\Omega = -\frac{3}{5}\frac{M^2 G}{R}. \quad (22.9)$$

The total mass and radius are M and R. It is useful first to show that $m(r)/M = (r/R)^3$.

Problem 22.2. A disk of radius a and thickness $2h$ consists of matter of uniform density ρ_0. Show that its gravitational potential energy is given by

$$\Omega = -\frac{1}{3}\frac{M^2 Gh}{a^2}, \quad (22.10)$$

where M is the mass of the disk. Check your integrals before doing the integration to see if some vanish by symmetry!

Magnetic Forces

Magnetic fields couple to charged particles (electrons or ions) in the interstellar medium. At low temperatures their numbers may be quite small, but the effect of the magnetic force may be transmitted to the neutral gas through collisions. The magnetic force is the $\mathbf{E} = 0$ component of the Lorentz force

$$e(\mathbf{E} + \mathbf{v} \times \mathbf{B}/c),$$

and the corresponding force per unit mass is

$$\mathbf{F}_M = \frac{\mathbf{j} \times \mathbf{B}}{\rho c}. \quad (22.11)$$

The electric current density $\mathbf{j} = ne\mathbf{v}$ for a plasma with positive and negative charges of equal magnitude. The magnetic virial is

$$\int dm \, \mathbf{F}_M \cdot \mathbf{r} = \int dm \, \frac{\mathbf{r} \cdot (\mathbf{j} \times \mathbf{B})}{\rho c}$$
$$= \int dV \, \mathbf{r} \cdot (\mathbf{j} \times \mathbf{B}). \quad (22.12)$$

This may be rewritten using $\nabla \times \mathbf{B} = (4\pi/c)\mathbf{j}$ to eliminate the current:

$$\int dm\, \mathbf{F}_M \cdot \mathbf{r} = \frac{1}{4\pi} \int dV\, \mathbf{r} \cdot [(\nabla \times \mathbf{B}) \times \mathbf{B}]$$

$$= \int \frac{dV}{4\pi} \left\{ \mathbf{r} \cdot (\mathbf{B} \cdot \nabla)\mathbf{B} - \frac{1}{2} \mathbf{r} \cdot \nabla B^2 \right\}. \quad (22.13)$$

We now write the integral as a divergence and use Gauss' theorem to express as much of the magnetic virial as possible in terms of surface integrals. The integrand is (summation over repeated indices is implied)

$$\{\} = x^i B_k \frac{\partial B_i}{\partial x^k} - \frac{1}{2} x^k \frac{\partial B^2}{\partial x^k} - \frac{B^2}{2} \frac{\partial x^k}{\partial x^k} + \frac{3}{2} B^2$$

$$= \frac{3}{2} B^2 - \frac{1}{2} \frac{\partial}{\partial x^k} (B^2 x^k) + x^i B_k \frac{\partial B_i}{\partial x^k}$$

$$+ x^i B_i \frac{\partial B_k}{\partial x^k}$$

$$= \frac{3}{2} B^2 - \frac{1}{2} \frac{\partial}{\partial x^k} (x^k B^2) + x^i \frac{\partial}{\partial x^k} (B_i B_k)$$

$$+ \left(B_i B_k \frac{\partial x^i}{\partial x^k} - B^2 \right)$$

$$= \frac{B^2}{2} + \frac{\partial}{\partial x^k} (x^i B_i B_k - \frac{1}{2} x^k B^2)$$

$$= \frac{B^2}{2} + \nabla \cdot (\mathbf{B}(\mathbf{B} \cdot \mathbf{r}) - \frac{1}{2} \mathbf{r} B^2). \quad (22.14)$$

Substituting (22.14) into the magnetic virial, we may use Gauss' theorem to rewrite the divergence as a surface integral, the first term being just the magnetic-energy density of the field

$$U_M = \frac{1}{8\pi} \int B^2\, dV. \quad (22.15)$$

The magnetic virial is thus

$$\int dm\, \mathbf{F}_M \cdot \mathbf{r} = U_M + \int_\Sigma (\mathbf{B} \cdot \mathbf{r}) \frac{\mathbf{B} \cdot d\mathbf{S}}{4\pi}$$

$$- \int_\Sigma B^2 \frac{\mathbf{r} \cdot d\mathbf{S}}{8\pi}. \quad (22.16)$$

The last term in (22.16) may be combined with the last term in (22.5) containing the mechanical pressure at the surface. This shows that $B^2/8\pi$ behaves like a magnetic pressure. In the special case where B is generated by the motion of the plasma itself, and where no mass loss from the distribution occurs, we expect $\mathbf{B} \cdot \hat{n} = 0$ on the surface, in which case only the term U_M survives. In more general cases the last term in (22.6) must be retained.

Internal Energy

The second term on the right-hand side of (22.5) may be related to the internal energy of the fluid; if it is assumed to be ideal and nondegenerate, then thermodynamics tells us that the pressure P and the internal energy density u are related at the point r by

$$\left(\frac{\partial P}{\partial u} \right)_V = -\left(\frac{\partial \ln T}{\partial \ln V} \right)_s = \Gamma_3 - 1. \quad (22.17)$$

The total internal energy of the matter distribution is then obtained by integrating (22.17) over a local region to arrive at

$$\int (\Gamma_3 - 1)\, du = u(\Gamma_3 - 1)$$

$$= \int_{\Delta V} \left(\frac{\partial P}{\partial u} \right)_V du = P \quad (22.18)$$

evaluated at the point r. Note that Γ_3 depends on r in general. The total energy is then the volume integral

$$\int_V P\, dV = \int_V (\Gamma_3 - 1) u\, dV$$

$$\equiv (\gamma - 1) \int_V u\, dV = (\gamma - 1) U, \quad (22.19)$$

with γ defined by the average over the entire mass distribution of the locally defined adiabatic index Γ_3:

$$\gamma \equiv \int_V u \Gamma_3\, dV \Big/ \int_V u\, dV. \quad (22.20)$$

Recall (from Chapter 2) that only under specialized circumstances will γ be equal to the ratio of specific heats c_P/c_V. Nevertheless, we will find (22.19) useful in the following.

The final form of the virial theorem, including gravitational and magnetic fields, is

$$\frac{1}{2}\frac{d^2 I}{dt^2} = 2\langle T \rangle + 3(\gamma - 1)U + \Omega + U_M$$

$$+ \frac{1}{4\pi}\int_\Sigma (\mathbf{B} \cdot \mathbf{r})\mathbf{B} \cdot d\mathbf{S}$$

$$- \oint_\Sigma \left(P + \frac{B^2}{8\pi}\right)\mathbf{r} \cdot d\mathbf{S}. \qquad (22.21)$$

The first four terms are given by

$$\langle T \rangle = \frac{1}{2}\int \rho \left|\frac{d\mathbf{r}}{dt}\right|^2 dV \qquad (22.22)$$

and (22.19), (22.8), and (22.15), respectively. The generalized moment of inertia I is

$$I \equiv \int r^2 dm = \int r^2 \rho \, dV. \qquad (22.23)$$

It is also possible to include turbulence, rotation, and other effects by means of appropriately defined virials $\int \mathbf{f}_\alpha \cdot \mathbf{r} \, dm$, where $\mathbf{f}_\alpha$ is the appropriate force density acting on a fluid element.

The left-hand side of (22.21) vanishes ($\ddot{I} = 0$) for matter in hydrostatic equilibrium. Collapse results when $\ddot{I}$ is negative, and expansion occurs when $\ddot{I}$ is positive.

22.2. Stability (Macroscopic)

The macroscopic stability conditions may be obtained from the virial theorem. In addition, we find constraints on the internal energy and gravitational energy changes, which can occur if an evolving distribution of matter is to remain in, or nearly in, hydrostatic equilibrium.

Problem 22.3. A spherically symmetric cloud is in hydrostatic equilibrium. There are no magnetic fields and no macroscopic mass motion ($\langle T \rangle = 0$). Denoting the system's total energy by $E = U + \Omega$, show that the internal energy U and E satisfy

$$\Delta U = -\frac{\Delta \Omega}{3(\gamma - 1)}, \quad \Delta E = \left(\frac{\gamma - 4/3}{\gamma - 1}\right)\Delta \Omega. \qquad (22.24)$$

Suppose that $\gamma > 4/3$, and discuss the energetics of a gas cloud during contraction ($\Delta R < 0$), and during expansion ($\Delta R > 0$). Assume that $\Omega = -\alpha M^2 G/R$ when α is positive, and that γ is constant.

The total energy of an equilibrium configuration is given by

$$E = \frac{\gamma - 4/3}{\gamma - 1}\Omega \quad < 0, \text{ if } \gamma > 4/3 \quad (22.25)$$

since $\Omega < 0$. Thus the distribution will be self-bound only if $\gamma > 4/3$. A careful analysis of the frequency of small disturbances away from equilibrium shows that they will decay in time (the system returning to its original state) only if $\gamma > 4/3$, where γ is also given by (22.20).

The case $\gamma = 4/3$ is also of interest, but must be handled with more care. From (22.25) we see that $E = 0$ and remains zero for all changes in U and Ω when $\gamma = 4/3$. The latter are constrained by (22.24) to satisfy $\Delta U = -\Delta \Omega$. This case is not applicable to a system that was at some previous time self-bound, such as a collapsing cloud (with $E < 0$), or a star. We must treat these situations by returning to the more general form of the virial theorem (22.21).

Suppose that $E < 0$ initially, with $\gamma > 4/3$, and that the system was bound and stable. Then at a subsequent stage in its evolution, $\gamma = 4/3$ because of internal processes. Such situations arise, for example, when H_2 dissociates in regions of a contracting protostellar cloud, or when ionization zones develop in stellar envelopes. We also encountered this state in stellar cores following silicon burning, where the subsequent collapse may trigger a supernova. If we assume that the mass distribution is spherically symmetric, and that $\mathbf{B} = 0$, the internal energy is

$$E = U + \Omega, \qquad (22.26)$$

and (22.21) reduces to

$$E = \frac{\gamma - 4/3}{\gamma - 1}\Omega + \frac{d^2 I/dt^2}{6(\gamma - 1)}. \qquad (22.27)$$

When $\gamma > 4/3$ and $E < 0$, the last term can vanish. However, if $\gamma = 4/3$ and E is still negative, the term d^2I/dt^2 cannot vanish, but is given by $2E$ (the first term still vanishes). If $E < 0$ before $\gamma = 4/3$ occurs, then

$$d^2I/dt^2 < 0, \qquad (22.28)$$

and the configuration must begin to collapse. The time-scale for the collapse is $t_{ff} \approx (R^3/MG)^{1/2}$, which corresponds to free fall; it is thus dynamic. Collapse continues until internal processes cause $\gamma > 4/3$, at which stage collapse is halted and a new phase of hydrostatic equilibrium results. This will occur, for example, when ionization or dissociation has been completed. In many situations γ may be less than $4/3$, such as in local ionization zones or if pair formation occurs. When this happens the general analysis is the same as in the preceding, except that $\Omega(\gamma - 4/3)/(\gamma - 1)$ is now positive, and I must be of greater magnitude than when $\gamma = 4/3$. The collapse is thus more violent.

Problem 22.4. A cloud of interstellar gas and dust is spherically symmetric and in hydrostatic equilibrium. Suppose that there is no rotation or mass motion, and that the surface pressure is zero. Take $\mathbf{B} \cdot \hat{n} = 0$ on the surface, but do not set $\mathbf{B}$ equal to zero, so that magnetic fields exist within the cloud. The total energy $E = U + \Omega + U_M$ shows that

$$E = -\frac{(\gamma - 4/3)}{\gamma - 1}(|\Omega| - U_M). \qquad (22.29)$$

Define an average magnetic field by $\langle B^2 \rangle V = \int B^2 dV$, and show that if the magnetic cloud is to remain bound by its own gravitational field, its mass must be given by

$$M \geq \frac{(5/2G)^{3/2} \langle B^2 \rangle^{3/2}}{48\pi^2 \rho^2} \equiv M_c. \qquad (22.30)$$

Values typical for interstellar matter are $\langle B^2 \rangle^{1/2} \approx 3 \times 10^{-6}$ gauss, $\rho \simeq 20\, m_H\, \text{cm}^{-3}$ with $m_H = 1.67 \times 10^{-24}$ g. Under these conditions, what is the minimum value of the mass M for a self-bound cloud? Find the radius of such a cloud.

Problem 22.5. From the preceding problem, it is found that if a cloud of interstellar matter is to form, then the self-bound system will have a mass of 10^3 to $10^4\, M_\odot$ confined to a region whose linear dimension is ~ 100 pc. Assuming that this matter eventually ends up in the form of stars normally found in the Galactic disk, discuss whether the model is in reasonable agreement with what is currently known about young disk stars.

Chapter 23

STAR FORMATION

A fundamental problem associated with interstellar matter is that of star formation. Observations strongly support the idea that star formation is an ongoing process in spiral galaxies, and thus demand that any complete theory of astrophysical phenomena include current star formation. We will review three of the strongest arguments supporting this view.

23.1. Matter Condensations and Star Formation

The first is the existence of massive, luminous main-sequence stars in our Galaxy. Stellar evolution theory predicts that the more massive a star, the shorter its main-sequence lifetime. In fact, reasonable estimates give for the main-sequence lifetime

$$\tau \simeq (M/M_\odot)(L_\odot/L) \times 10^{10} \text{ yrs} \qquad (23.1)$$

and a mass-luminosity relation $L \sim M^\alpha$, where α is between 3 and 4.5. For an OB-type star, with mass $M \approx 10\, M_\odot$ and $\alpha \approx 4.5$, the lifetime is $\tau \simeq 3 \times 10^6$ yrs, which is much less than the age of the Galaxy itself. Such stars therefore could not have formed when the Galaxy was formed about 5×10^9 years ago.

Further direct support comes from the existence of unbound stellar associations (such as the Orion association), whose members are separated by 30 to 200 pc. Since each of these stars partakes in the differential rotation of the disk, two stars at r and $r + dr$ will rotate at slightly different rates. An approximate expression for the rotation speed of a star at a distance r from the Galactic center, and in the disk, which holds for $3 \leq r \leq 13$ kpc, is

$$v_R = 67.76 + 50.06r - 4.0448r^2 + 0.0861\, r^3 \text{ km/sec.} \qquad (23.2)$$

Here r is in kpc $= 10^3$ pc. Taking 100 pc as an initial separation between two stars in an association (23.2), we can show that after 10^6 yrs the interstellar separation will be about one pc, whereas after 10^{10} yrs (current age of the Galaxy) it will be at least 10 kpc, which exceeds the association's diameter by about fifty times. If the associations were formed from older disk stars, they would no longer be recognizable as associations, so they must be formed from younger stars.

Finally, the existence of hot, young stars in clouds of interstellar dust and gas, from which they presumably condensed, offers further indirect support. Specifically it is found that matter densities of 10^{-22} g/cm^3 are typical in these regions, but that values as high as 10^{-18} g/cm^3 are not uncommon in smaller portions of individual clouds (Bok globules).

Ideally, a complete theory of star formation would permit calculation of the stellar birthrate function $\psi(M)$, giving the number of stars of mass M formed per year per cubic pc (Chapter 2). This function may be empirically calculated for stars in the solar neighborhood, and would serve as one check on the theory.

We shall adopt as a working hypothesis that star formation is an on-going process in the disk of spiral galaxies and in irregular galaxies, and that it is associated with regions where interstellar matter has greater than average density. The next section describes an approximate set of hydrodynamic equations for dilute gases, and then develops simplified models for the formation of gravitational instabilities in an otherwise uniform gas. The subsequent collapse and fragmentation of the initial cloud into smaller collapsing regions is then reviewed. The importance of magnetic fields and the effects of rotation greatly complicate the analysis, but they are likely to lead to qualitative changes in the conclusions obtained for highly simplified models. In a few instances these complications will be included very approximately. Our ultimate goal will be to show that the formation of stars out of interstellar matter as it is known to exist today is not inconsistent with simple models of magnetohydrodynamics.

23.2. LINEARIZED HYDRODYNAMIC EQUATIONS

The full set of hydrodynamic equations needed to describe star formation must generally be solved numerically, or recourse to approximations must be made. Approximations are often limited to situations in which deviations from equilibrium are small. When the nonequilibrium effects depend on time, the results generally are applicable only over time-scales small compared with the time necessary for the effects to build significantly. And it must always be remembered that some solutions to the nonlinear equations may not be obtained from simple approximation schemes built on perturbation theory (see Section 11.1).

One approach assumes that the dynamic system deviates only slightly from some (known) equilibrium state (whose parameters ρ_0, P_0, v_0, ϕ_0, and $\mathbf{B}_0$ are denoted by subscript zero). Any physical quantity Q is then expanded about its equilibrium value

$$Q = \sum_{n=0}^{\infty} \lambda^n Q_n, \qquad (23.3)$$

where the λ is introduced to count orders of smallness, and will ultimately be set equal to unity. For (23.3) to be useful, one assumes that $Q_{n+1}/Q_0 \ll 1$, and one hopes that $Q_{n+1}/Q_n \ll 1$ also.

We may use (23.3) to expand ρ, P, v, ϕ, and $\mathbf{B}$, and substitute the results into the hydrodynamic equations

$$\rho \frac{d\mathbf{v}}{dt} = \mathbf{f} - \nabla P, \qquad (23.4)$$

$$\frac{\partial \rho}{\partial t} + \nabla \cdot (\rho \mathbf{v}) = 0, \qquad (23.5)$$

$$\nabla^2 \phi = 4\pi G \rho, \qquad (23.6)$$

$$\nabla \times (\mathbf{v} \times \mathbf{B}) = \frac{\partial \mathbf{B}}{\partial t} - \frac{1}{4\pi\sigma} \nabla^2 \mathbf{B}, \qquad (23.7)$$

where, in the presence of gravitation and magnetic fields, $\mathbf{f}$ is given by

$$\mathbf{f} = -\rho \nabla \phi - \frac{1}{8\pi} \nabla B^2 + \frac{1}{4\pi} (\mathbf{B} \cdot \nabla) \mathbf{B}. \qquad (23.8)$$

If all terms corresponding to order λ^2 and higher in the expansions are dropped, then we arrive at a set of equations that are linear in the unknowns Q_n for $n = 1$. In this approximation, the equations must hold in each order. We consider the linear case next. Since the equations hold to each order, the Q_0s will themselves satisfy (23.4) to (23.8), a result that will be used below.

Linearized Equations

To order λ, (23.3) reduces to $Q = Q_0 + \lambda Q_1$; thus we substitute the following into (23.4) to (23.8):

$$\rho = \rho_0 + \lambda \rho_1, \qquad P = P_0 + \lambda P_1,$$
$$\phi = \phi_0 + \lambda \phi_1,$$
$$\mathbf{v} = \mathbf{v}_0 + \lambda \mathbf{v}_1, \qquad (23.9)$$
$$\mathbf{B} = \mathbf{B}_0 + \lambda \mathbf{B}_1.$$

The equation of continuity (23.5) and the Poisson equation for the gravitational potential (23.6) are the easiest to work out. Poisson's equation (23.6) becomes

$$\nabla^2 \phi_0 + \lambda \nabla^2 \phi_1 = 4\pi G \rho_0 + 4\pi G \lambda \rho_1.$$

But $\nabla^2 \phi_0 = 4\pi G \rho_0$, since unperturbed quantities are assumed to obey the equilibrium equations, and therefore

$$\nabla^2 \phi_1 = 4\pi G \lambda \rho_1. \quad (23.10)$$

Similarly, the unperturbed continuity equation is assumed to be obeyed and (23.5) becomes,

$$\frac{\partial \rho_1}{\partial t} + \nabla \cdot (\rho_0 \mathbf{v}_1 + \rho_1 \mathbf{v}_0) = 0. \quad (23.11)$$

Equation (23.7) reduces in first order to

$$\nabla \times (\mathbf{v}_1 \times \mathbf{B}_0) + \nabla \times (\mathbf{v}_0 \times \mathbf{B}_1)$$
$$= \frac{\partial \mathbf{B}_1}{\partial t} - \frac{1}{4\pi\sigma} \nabla^2 \mathbf{B}_1. \quad (23.12)$$

Actually, the conductivity σ appearing in (23.12) could depend strongly on density or temperature, in which case it would also have to be expanded, by analogy with (23.9). However, for our purposes we assume that σ is a constant. Finally, if (23.9) is substituted into (23.4), we obtain the linearized equation of motion

$$\frac{\partial \mathbf{v}_1}{\partial t} + \mathbf{v}_0 \cdot \nabla \mathbf{v}_1 + \mathbf{v}_1 \cdot \nabla \mathbf{v}_0$$
$$= -\nabla(\phi_1 + P_1/\rho_0)$$
$$- \frac{\nabla B_1^2}{8\pi \rho_0} + \frac{(\mathbf{B} \cdot \nabla)\mathbf{B}_0}{4\pi \rho_0}. \quad (23.13)$$

Jeans Model

Usually the linearized equations are not easily solved. However, an interesting, though limited, model does admit an exact solution. If in the unperturbed state

$$\rho_0 = \text{constant}, \quad \mathbf{v}_0 = 0, \quad \text{and} \quad \mathbf{B} = 0, \quad (23.14)$$

then the hydrodynamic equations (23.10) to (23.13) reduce to

$$\rho_0 \nabla \cdot \mathbf{v}_1 = -\partial \rho_1/\partial t, \quad (23.15)$$

$$\nabla^2 \phi_1 = 4\pi G \rho_1, \quad (23.16)$$

$$\frac{\partial \mathbf{v}_1}{\partial t} = -\nabla(\phi_1 + P_1/\rho_0). \quad (23.17)$$

Since $\mathbf{B}_0$ and $\mathbf{B}_1$ vanish, there are no magnetic terms in (23.15) to (23.17). These equations may be solved when the motion is adiabatic, since then $P_1 = c_s^2 \rho_1$ (with c_s^2 constant) may be used to eliminate P_1 in (23.17). Taking the divergence of (23.17) gives

$$\nabla \cdot \frac{\partial \mathbf{v}_1}{\partial t} = \frac{\partial}{\partial t}(\nabla \cdot \mathbf{v}_1)$$
$$= -\nabla^2 \phi_1 - (c_s^2/\rho_0)\nabla^2 \rho_1, \quad (23.18)$$

since ρ_0 is constant. Next, the partial time-derivative of (23.15) yields

$$\frac{\partial}{\partial t}(\nabla \cdot \mathbf{v}_1) = -\frac{1}{\rho_0} \frac{\partial^2 \rho_1}{\partial t^2}. \quad (23.19)$$

Now, if (23.16) is used to eliminate $\nabla^2 \phi_1$ in (23.18) and (23.19) to eliminate the term $\partial \nabla \cdot \mathbf{v}_1/\partial t$, we rewrite the result in the form

$$\left(\nabla^2 - \frac{1}{c_s^2}\frac{\partial^2}{\partial t^2} + \frac{4\pi G \rho_0}{c_s^2}\right)\rho_1 = 0. \quad (23.20)$$

This is a wave equation with a "mass" term: $4\pi G \rho_0/c_s^2$. A solution is easily seen to be

$$\rho_1 = A e^{i(\mathbf{k} \cdot \mathbf{r} - \omega t)}, \quad (23.21)$$

which is a plane-wave density fluctuation with wave vector $\mathbf{k}$, and frequency ω, satisfying the dispersion relation

$$\omega^2 = k^2 c_s^2 - 4\pi G \rho_0. \quad (23.22)$$

This relation shows that disturbances will not grow if k is greater than the critical value k_J given by $\omega^2 = 0$:

$$k_J^2 = \frac{4\pi G \rho_0}{c_s^2} = \frac{4\pi G \mu m_H \rho_0}{kT}. \quad (23.23)$$

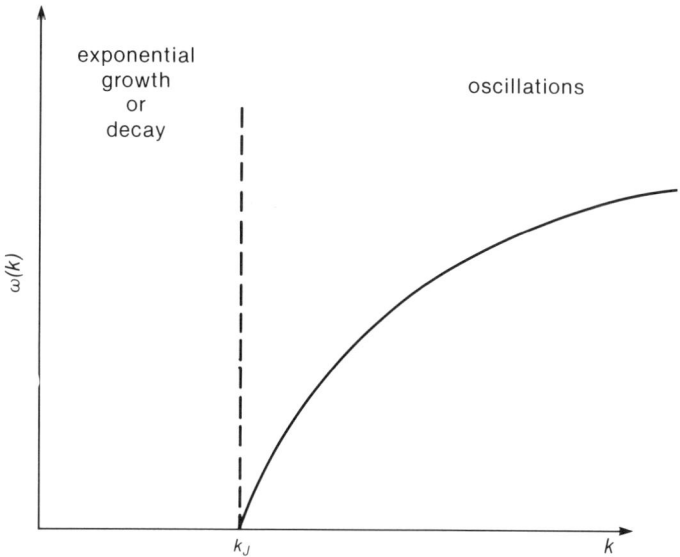

Figure 23.1. Frequency of density fluctuations in the Jeans model. Disturbances with $k < k_J$ grow or decay with time. Disturbances with $k > k_J$ are stable oscillations.

If $k < k_J$, then ω is imaginary, $\omega = \pm i\eta$ and ρ_1 will grow exponentially with time (Figure 23.1). This corresponds to the onset of an instability and results in a local increase in the system's density over a region whose characteristic linear dimension $l_J = 2\pi/k_J$. The stable ($\omega^2 > 0$) and unstable regions ($\omega^2 < 0$) are separated by the case $\omega = 0$, called marginal stability.

There is a serious difficulty with the preceding model, as can be seen from the unperturbed equations. Since ρ_0 and P_0 are constant, and $\mathbf{v}_0 = 0$, (23.17) shows that $\nabla \phi_1 = 0$, or that $\rho_0 = 0$. If $\rho_0 \neq 0$, then $\nabla \phi_1 = 0 \rightarrow \nabla^2 \phi_1 = 0$ and (23.15) then forces $\rho_0 = 0$. Thus (23.20) to (23.23) is consistent only with the trivial solution, and $k_J = 0$. We will see in the following that the general form of (23.23) is still applicable in many cases, although with a different numerical coefficient that depends on the specifics of the problem.

Infinite Disk

A model that is slightly more complicated, but that does not suffer from the limitations of the Jeans model, is an infinite disk in which $\mathbf{v}_0 = 0$, and where $\rho_0 = \rho_0(z)$ and $\phi_0 = \phi_0(z)$ depend only on the height above or below the central plane. The distribution extends to $\pm\infty$ in the x and y directions. A solution to the linearized equations may be obtained as above. However, a less ambitious analysis that yields the necessary results uses the fact that for this system the $\omega^2 = 0$ mode of small disturbances away from equilibrium separates stable and unstable behavior. We thus limit attention to the $\omega^2 = 0$ mode. The analysis is outlined in Problems 23.1 to 23.3.

Problem 23.1. Consider an infinite disk in the x, y-plane (in which $\mathbf{B} = 0$ everywhere), subject only to self-gravitational forces. The equilibrium state is described by

$$\mathbf{v}_0 = 0, \qquad \rho_0 = \rho_0(z), \qquad \phi_0 = \phi_0(z). \quad (23.24)$$

(a) Following the analysis of the Jeans model, use equations (23.16) and (23.17), and the constraints (23.24) to show that the density of the equilibrum state $\rho_0(z)$ is a solution of

$$\frac{d}{dz}\left(\frac{1}{\rho_0}\frac{d}{dz}\rho_0\right) = -\frac{4\pi G}{c_s^2}\rho_0. \quad (23.25)$$

Assume $P_0 = c_s^2 \rho_0$ with c_s^2 constant, and note that ρ_0 is time-independent.

(b) Solve the second-order differential equation (23.25) by changing variables to $y = \ln \rho_0$, and noting that multiplication of the resulting differential equation by dy/dz places all terms in the form of perfect differentials. The resulting equation is integrable. Of the two integration constants, one is proportional to $\rho_0(0)$ at $Z = 0$, and the second vanishes by requiring that $\rho_0(+Z) = \rho_0(-Z)$ by symmetry. The answer is

$$\rho_0(z) = \rho_0(0) \operatorname{sech}^2 \frac{z}{l_J}, \qquad (23.26)$$

$$l_J^2 = \frac{kT}{2\pi G \rho_0 \mu m_H}. \qquad (23.27)$$

Problem 23.2. *Disk mass* (continuation of Problem 23.1). Evaluate the mass of a cylindrical section of the disk perpendicular to the x, y-plane and having a unit cross section that extends a distance $\pm z$ from the midpoint $z = 0$ (Figure 23.2). The result may be expressed in the form

$$m(z) = M \tanh \frac{|z|}{l_J}, \qquad (23.28)$$

with $M \equiv M(\infty) = 2\rho_0(0)l_J$, the total mass of a column of unit cross section. Plot the mass distribution $M(z)$ and the density $\rho_0(z)$.

Problem 23.3. *Marginal stability* (continuation of Problem 23.2). Assuming a general time-dependence for small deviations from equilibrium of the form $e^{i\omega t}$, consider the mode $\omega = 0$ corresponding to marginal stability. Assume that $\rho_0(z)$ is given by (23.26), and set $\rho_1(\zeta)$ equal to

$$\rho_1(\zeta) = \rho_0(\zeta)\theta(\zeta)e^{ik_c x}, \text{ where } \zeta = \tanh \frac{z}{l_J} \quad (23.29)$$

for the $\omega = 0$ mode. The critical wave vector is k_c, and $\rho_1(\zeta)$ corresponds to a density variation in the x direction only (no y dependence). Show that if (23.29) is substituted into (23.16) and (23.17), the following differential equation for $\theta(\zeta)$ results:

$$\nabla^2 \theta(\zeta)e^{ik_c x}$$
$$= -\frac{4\pi G \rho_0(0)}{c_s^2}\rho_0(0)(1-\zeta^2)\theta(\zeta)e^{ik_c x}. \quad (23.30)$$

Use the second of (23.29) to show that the final differential equation takes the form

$$\frac{d^2\theta}{d\zeta^2} - \frac{2\zeta}{1-\zeta^2}\frac{d\theta}{d\zeta}$$
$$+ \left\{ \frac{2}{1-\zeta^2} - \frac{(k_c l_J)^2}{(1-\zeta^2)^2} \right\} \theta = 0. \quad (23.31)$$

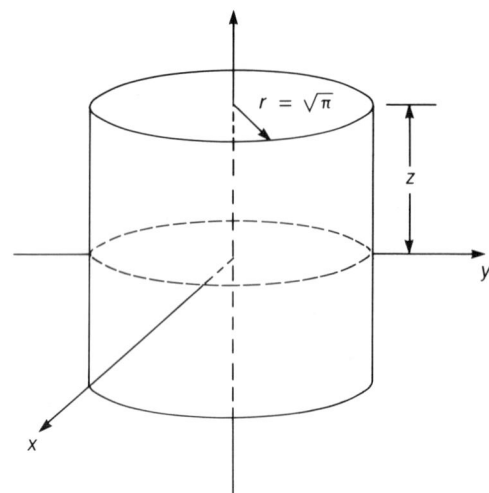

Figure 23.2. Cylindrical section of the Galactic disk.

The general solution of (23.31) may be written in the form

$$\theta(\zeta) = A \left(\frac{1+\zeta}{1-\zeta} \right)^{k_c l_J/2} (k_c l_J - \zeta)$$
$$+ B \left(\frac{1-\zeta}{1+\zeta} \right)^{k_c l_J/2} (k_c l_J + \zeta). \quad (23.32)$$

Explain why $k_c l_J$ must equal 1 in order for (23.32) to be an acceptable solution to the infinite-disk problem, thereby justifying the relation

$$k_c^2 = \frac{4\pi^2}{l_J^2} = \frac{2\pi G \mu m_H \rho_0(0)}{kT} \quad (23.33)$$

for the critical wave vector corresponding to the $\omega^2 = 0$ mode.

The critical wave vector for the infinite disk (23.33) has the same form, apart from a factor of 2, as in the Jeans model (23.23). However, the model and the intermediate results leading to (23.33) enjoy the advantage of internal consistency. A more complete analysis shows that $\omega^2 > 0$ for $k > k_J$, and that if $k < k_J$ the modes are imaginary, leading to an exponential growth with time, as suggested by (23.21) and (23.22) of the Jeans model. This signals the development of an actual instability in the system, setting the stage for the contraction of all the matter within a region of characteristic dimension $l_J \sim 2\pi/k_c$. Numerically, this corresponds to

$$l_J = 8 \times 10^7 (T/\rho_0)^{1/2} \text{ cm}, \qquad (23.34)$$

where ρ_0 and T are the $z = 0$ matter density and temperature, respectively. Order-of-magnitude values for interstellar matter in the disk of the Galaxy are $\rho \sim 2 \times 10^{-24}$ g/cm^3 and $T \sim 10^2$ K, giving $l_J \sim 200$ pc. The total amount of matter contained within the instability is roughly $\rho_0 l_J^3$, which corresponds to $M \sim 2$ to $10 \times 10^5 \, M_\odot$. This is certainly far too much matter to be the protostellar stage of an individual star (except for supermassive stars), but it agrees well with the estimated masses of larger stellar associations. We will return to this facet of star formation in Section 23.6.

23.3. Effects of Rotation

If the material in the cloud or disk is rotating, the situation can become extremely complex. The simplest effect may be thought of as a weakening of the gravitational field. In the rotating frame of the material, this weakening may be thought of as an outward centrifugal force that counters gravity. If rotation is too great, collapse may be halted. The condition on the rotational velocity $v = \omega r$ at a distance r from the axis, obtained by setting the gravitational force equal to the centrifugal force, is

$$\omega = (MG/R^3)^{1/2}. \qquad (23.35)$$

This leads to a restriction on the density ρ of a cloud, which, if it is spherical (no rotational distortion), is

$$\rho > \frac{3\omega^2}{4\pi G} = \rho_{\text{crit}} = 3.6 \times 10^6 \, \omega^2 \text{ g cm}^{-3} \qquad (23.36)$$

with ω in sec^{-1}. If ρ exceeds ρ_{crit}, then the rotating system may still collapse as a result of an instability. If the system is a disk rather than a sphere, then the numerical coefficient in (23.36) is increased somewhat (an increase by a factor of 4 may not be unreasonable).

Differential Galactic rotation, with linear velocity $v_R(r)$ at a distance r from the Galactic center, is well described by (23.2). Since $v_R(R + r) < v_R(R) < v_R(R - r)$, the matter at $R + r$ will tend to drift backward, and matter at $R - r$ will move forward as seen by an observer moving with the gas at R (Figure

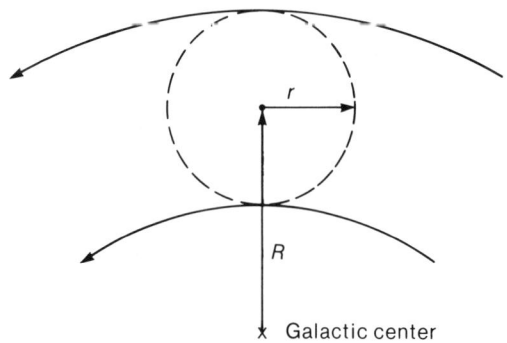

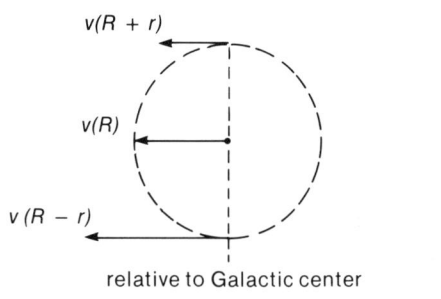

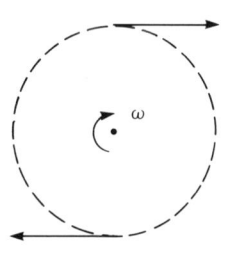

Figure 23.3. Rotation induced by differential motion in the galactic disk.

23.3). As a result, regions centered at r begin to rotate with angular velocity

$$\omega = \frac{v_R(R - r) - v_R(R + r)}{r}. \qquad (23.37)$$

Using (23.2), and taking $r \simeq 150$ pc (a reasonable minimum for the radius of a condensing region in the Galactic plane), we obtain from (23.37) $\omega \simeq 2 \times 10^{-15}$ sec^{-1}. When substituted into (23.36), this gives a critical density $\rho_{\text{crit}} \simeq 10^{-23}$ g/cm^3. This is about five times greater than the average density of 2×10^{-24} g/cm^3 observed for interstellar matter, but this observation is not certain; furthermore we could anticipate densities this high in the leading edges of spiral density waves in the Galaxy. According to this view, protostar formation is triggered in the regions of increased density along the spiral arms, and would be less likely elsewhere. Such a view is consistent with current observational evidence. The passage of a shock wave from a supernova would also increase the interstellar gas density locally, and trigger star formation.

23.4. Collapse of Isolated Clouds

The preceding discussion emphasized the onset of gravitational instabilities in a system assumed initially to be in equilibrium. We now consider the system, or a portion of it, as collapse begins. Again only the simplest model will be considered in detail. Imagine a large cloud of density ρ' and temperature T', in which a spherical cloud of radius R, density ρ, and temperature T is embedded (Figure 23.4). The cloud is isothermal, nonrotating, and nonmagnetic ($\mathbf{B} = 0$). We assume further that

$$\rho' \ll \rho, \qquad T' \gg T; \qquad (23.38)$$

the surrounding matter is hot and dilute, but the spherical cloud is cool and "dense." The hot surrounding gas will exert a pressure P_{ext} on the cloud's surface (Figure 23.4). The virial theorem (22.21), with the surface pressure term given by (22.6), may be applied to this example:

$$3(\gamma - 1)U + \Omega - 4\pi R^2 P_{\text{ext}} = 0 \qquad (23.39)$$

(we set $P_s = P_{\text{ext}}$). Furthermore, Ω will be given by (22.9) if we assume uniform density $\rho_c = 3M_c/4\pi R^3$, where M_c is the cloud's total mass. For the initial state of the cloud, we will assume an ideal-gas equation of state with $\gamma = 5/3$, so that

$$U = \frac{3}{2} \frac{MkT}{\mu m_H}. \qquad (23.40)$$

This corresponds to an neutral monatomic gas at a temperature of 50 to 100 K (an HI region). With these assumptions, the virial theorem yields

$$P_{\text{ext}} = \frac{3kTM}{4\pi \mu m_H R^3} - \frac{3}{20\pi} \frac{M^2 G}{R^4}. \qquad (23.41)$$

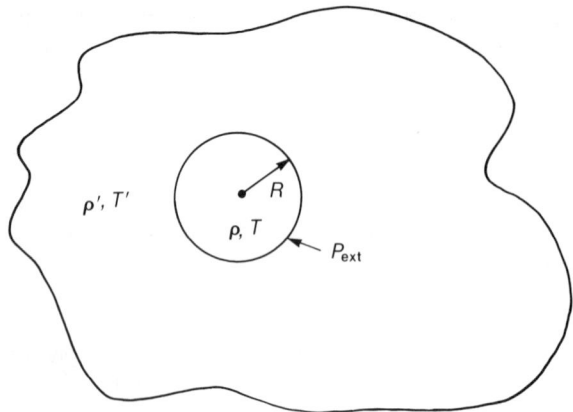

Figure 23.4. Dense, cool gas cloud imbedded in uniform medium. The pressure on the cloud due to the surrounding medium is P_{ext}.

The first term on the right is the internal energy; the second is the gravitational energy. For large R the internal energy dominates, but P_{ext} may be quite small. For a fixed cloud mass M and isothermal contraction, P_{ext} reaches a maximum when the radius reaches

$$R_J \equiv \left(\frac{3}{\pi} \frac{15}{16} \frac{kT}{\mu m_H \rho G}\right)^{1/2}. \qquad (23.42)$$

This may be compared with the Jeans length l_J given by (23.7) for an infinite disk, or to $1/k_J$ for the Jeans model (23.23).

Examination of Figure 23.5 indicates that for $R > R_J$, an increase in pressure leads to contraction until $R = R_J$. However, as the cloud continues to contract beyond R_J, the pressure decreases, and the system becomes bound. The total energy

$$E = U + \Omega = \frac{3MkT}{\mu m_H} - \frac{3}{5} \frac{M^2 G}{R} \qquad (23.43)$$

becomes negative when $R_J < R_c = (2/\sqrt{3})R_J$. Thus the cloud is self-bound while it is still being compressed by the surrounding matter.

Problem 23.4. Find the radius of a spherical, isothermal cloud for which the total energy (23.43) is zero. Show that the radius R_c is given by

$$R_c^2 = (4/3)R_J^2. \qquad (23.44)$$

420 / STAR FORMATION

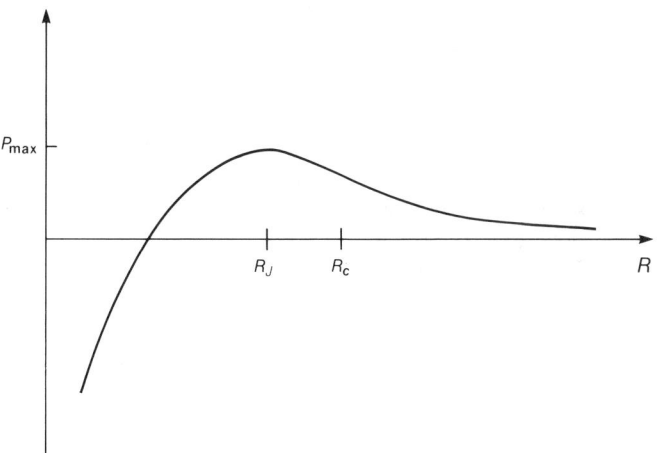

Figure 23.5. External pressure on cloud in Figure 23.4 versus radius of the cloud. R_J is the cloud's Jeans radius; R_c is the radius of the cloud whose total energy vanished.

Finally, we estimate the maximum pressure P_{max} for the cloud if it is to be in equilibrium by substituting (23.42) into (23.41), obtaining

$$P_{max} \simeq 3.14(kT/\mu m_H)^4 (M^2 G^3)^{-1}. \quad (23.45)$$

If the surrounding pressure is greater than P_{max}, then the cloud will not be in equilibrium, and the left-hand side of (23.39) will not vanish. But the left-hand side equals $(1/2)\, d^2I/dt^2$ [see (22.21)], and (23.39) becomes

$$\frac{1}{2}\frac{d^2I}{dt^2} = 3(\gamma - 1)U + \Omega - 4\pi R^2 P_{ext},$$
$$P_{ext} > P_{max}. \quad (23.46)$$

When $4\pi R^3 P_{ext}$ exceeds its equilibrium value, d^2I/dt^2 must be negative, and the cloud must collapse.

23.5. Effects of Magnetic Fields

The presence of magnetic fields, even of small magnitude ($B \sim 3 \times 10^{-6}$ gauss), may have a dramatic effect on the evolution of interstellar clouds, because the magnetic energy stored in a cloud of dimension ~ 100 pc, even though it is due to weak fields $B \simeq 10^{-6}$ gauss, can be comparable to the system's gravitational potential energy. Furthermore, the magnetic field tends to resist compression of the material, and thus works against gravity.

The energy density associated with a field **B** in a region whose permeability is unity is

$$u_M = \frac{B^2}{8\pi}. \quad (23.47)$$

Dimensionally, and thermodynamically, we expect a pressure to be associated with u_M, as shown in Problem 23.5. Unlike ordinary gas pressure, the pressure exerted by magnetic fields is not isotropic, and may be positive or negative (tension) depending on local conditions. For example, consider an infinitesimal cube throughout which **B** is nearly constant. If the field lines are tied to the matter,* then the magnetic field resists compression transverse to **B**, but assists compression parallel to **B**.

Problem 23.5. A cubic element of plasma having sides Δx, Δy, and Δz is small enough that **B** is essentially uniform throughout its volume. Suppose the cube is compressed parallel to one of its faces. The initial energy E_0 of the cube due to **B** is

$$E_0 = (B_0^2/8\pi)\, \Delta x\, \Delta y\, \Delta z.$$

---

*This occurs when the matter is a good conductor.

If the volume changes, because of compression or expansion, by an amount $\delta(\Delta V)$, then the pressure exerted by the field **B** will be given by

$$P = -\frac{\delta E}{\delta \Delta V}. \qquad (23.48)$$

(a) Assume that the magnetic flux remains constant, consider a small change in volume that results if $\Delta z \rightarrow \Delta z + \delta z$, where $\delta z \ll \Delta z$, and show that the resulting pressure is

$$P_\parallel = -\frac{B^2}{8\pi}. \qquad (23.49)$$

(b) Suppose the volume change is due to $\Delta x \rightarrow \Delta x + \delta x$, with $\delta x \ll \Delta x$, and show that

$$P_\perp = +\frac{B^2}{8\pi}. \qquad (23.50)$$

Note that $P_\parallel$ is a tension, and that (23.50) applies also to the case $\Delta y \rightarrow \Delta y + \delta y$.

An important application of magnetic pressure may be made to a cloud of magnetic interstellar matter. We suppose that gas in equilibrium is permeated by a magnetic field, as shown in Figure 23.6. The gas may, for example, lie in a spiral arm through which the field runs. The matter within the curve C may be of less than critical mass, and normally would not condense as a result of small perturbations. But suppose a slight compression is induced, so that the central region contracts slightly to C', as shown at the right. The pressure (23.50) resists continued contraction transverse to the field, but the tension (23.49) along the field lines causes matter initially outside C and C' to flow into the contracted region. This increases the mass within C', which then contracts further until the transverse pressure (23.50) brings it back into equilibrium. Additional matter then flows into C' because of (23.49), and the process is repeated until the thermal properties of the gas have changed enough to lead to a stable equilibrium state.

Problem 23.6. Carefully analyze the response of the gas outside C (in the preceding example) to a compression of the central region, assuming that the field lines outside C' are expanded slightly during the process.

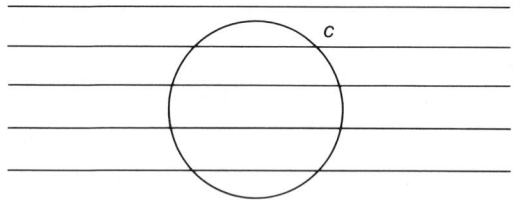

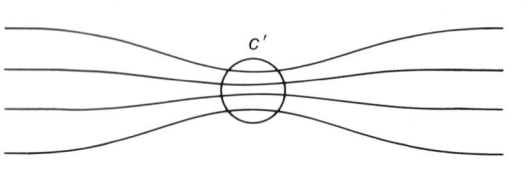

Figure 23.6. Cloud collapse in galactic magnetic field.

The example above assumed that the magnetic field was trapped with the moving plasma, and was carried with it. The behavior of the magnetic field can be obtained by solving Maxwell's equations

$$\nabla \times \mathbf{E} = -\frac{1}{c}\frac{\partial \mathbf{B}}{\partial t}, \qquad \nabla \times \mathbf{B} = \frac{4\pi}{c}\mathbf{j} + \frac{\partial \mathbf{D}}{\partial t}. \qquad (23.51)$$

Introducing the conductivity σ and dielectric constant ϵ of the plasma through the relations $\mathbf{j} = \sigma \mathbf{E}$ and $\mathbf{D} = \epsilon \mathbf{E}$, we can reduce (23.51) to the second-order equation

$$\nabla^2 \mathbf{B} = \epsilon \frac{\partial^2 \mathbf{B}}{\partial t^2} + \frac{4\pi\sigma}{c}\frac{\partial \mathbf{B}}{\partial t}. \qquad (23.52)$$

A time-scale may be associated dimensionally with each term on the right: the relaxation time $\tau_R \sim \sqrt{\epsilon} l/c$; and the diffusion time $\tau_D \sim (4\pi\sigma/c^2)l^2$, where l is the typical length scale of the system. When $\tau_R \gg \tau_D$, (23.52) reduces to the wave equation whose solution $\mathbf{B}(\mathbf{r}, t) \simeq \mathbf{B}(r)e^{i\omega t}$ describes the propagation of radiation through a medium. In the opposite extreme, $\tau_D \gg \tau_R$, and (23.52) becomes a diffusion equation whose solution is $\mathbf{B}(\mathbf{r}, t) \simeq \mathbf{B}(r)e^{-\alpha t}$. In this extreme the field lines diffuse through the matter with a time-scale of order τ_D. The diffusion time, which is normally small for terrestrial materials, may be large for an astrophysical process because it depends on the linear size of the system. In fact, $\tau_D \gg \tau_R$ is typical for interstellar gas, and the field behaves as if it were frozen into the matter. It is possible to show that this is equivalent to the flux constraint

$$\Phi \equiv \oint_s \mathbf{B} \cdot d\mathbf{s} = \text{constant}, \quad (23.53)$$

where the surface S is closed and moves with the matter. Dimensionally, this implies that $Bl^2 \simeq$ constant, where l is a characteristic dimension of the region permeated by the field. Magnetic fields carried by moving matter are observed in the filaments of nebulae formed by supernova explosions, and in the filamentary structures believed to have been ejected from active galactic nuclei.

The analysis of stellar formation becomes exceedingly complex when magnetic fields are present. Nevertheless, relatively simple application of the virial theorem reveals some of the important effects that can result. Since these are straightforward extensions of the preceding discussions, they have been given in Problems 23.7 and 23.8. Perhaps the most important of these is the probability that when magnetic fields permeate a cold cloud of matter, there will be a critical mass (or, for given mass, a critical density), such as (23.55) that must be exceeded if the cloud is to collapse, even under the influence of an external pressure. This situation is similar to what was encountered in the presence of rotation.

Problem 23.7. Consider a spherical cloud embedded in a hot, dilute medium as shown in Figure 23.4. The cloud's radius, density, and temperature are R, ρ, and $T = $ constant. It is subject to the external pressure P_ext, and is permeated by a uniform magnetic field B_0 in magnitude. If the field outside the cloud falls off rapidly enough $[B_\text{out}^2 = B_0^2(R/r)^6]$, it can be shown that all terms in the virial theorem containing B (surface and volume terms) reduce to an effective U_mag given by

$$U_\text{mag} = \frac{1}{3} B_0^2 R^3. \quad (23.54)$$

Use this expression in the virial theorem with U given by (23.40) and $\gamma = 5/3$, and analyze the conditions for collapse of a cloud. Show that there is a maximum mass, below which collapse is not possible:

$$M_c' = 1.37 \times 10^9 B_0^3/\rho^2. \quad (23.55)$$

Note that this is $2^{3/2}$ times M_c as given by (22.30). Show that if the magnetic flux $\Phi \sim B_0 R^2$ is constant as the cloud contracts, then magnetic fields can not halt collapse once it starts. Will magnetic fields alter the Jeans radius? Finally, show that the maximum pressure $P_\text{max}(B)$ is given by

$$P_\text{max}(B) = 3.14 \left(\frac{kT}{\mu m_H}\right)^4 \frac{1}{\mu^2 G^3 [1 - (M_c/M)^{2/3}]^3}. \quad (23.56)$$

Problem 23.8. The total masses $M/M_\odot$ of typical visible dark nebulae are given in Table 23.1. Their density $\rho = m_p n$, where $m_p = 1.67 \times 10^{-24}$ g. Find the maximum magnetic field of the region from which they condensed, assuming that the critical mass is given by (23.55) and that the initial density in the cloud was 4×10^{-23} g/cm^3. (For an intermediate cloud, $B_0 = 1.2 \times 10^{-6}$ gauss).

A plasma permeated by a magnetic field may sustain material wave motion because of its high conductivity. If the field is frozen into a rotating mass, angular momentum may be carried away by this wave motion, and the rotating mass will slow down. For example, consider a small element of fluid whose velocity $\mathbf{v}$ is along the negative x direction (into the page) in a region of space permeated by a magnetic field $\mathbf{B}$ in the z direction (Figure 23.7). The motion of the fluid element will result in a Lorentz force proportional to $(e\mathbf{v} \times \mathbf{B})$ on the ions and on the electrons in the element, while producing a slight charge separation, as shown in Figure 23.7(a). If the element were moving in a nonconducting medium, not much would happen. However, when the medium itself is a plasma of high conductivity, the charge built up will tend to

Table 23.1
Total masses of some typical, visible dark nebulae.

	Globules		Clouds	
Characteristic	Small	Large	Intermediate	Large
$M/M_\odot$	>0.1	3	8×10^2	1.8×10^4
R(pc)	0.03	0.25	4	20
n(H/cm^3)	$>4 \times 10^4$	1.6×10^3	100	20

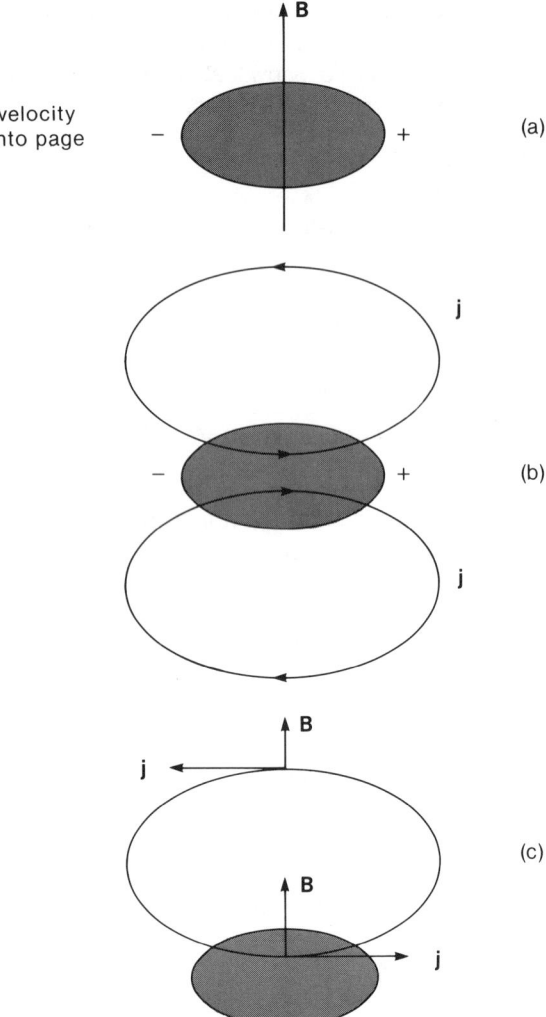

Figure 23.7. Alfvén wave in a plasma: (a) volume element (shaded) moving normal to **B** into the page sets up a charge separation. Induced currents (b) interact with field, producing a Lorentz force that slows the volume element and accelerates one above it (c). The result is a plasma pulse traveling normal to **B**.

dissipate as currents flow [Figure 23.7(b)]. But a magnetic field exerts a force $\mathbf{j} \times \mathbf{B}$ on these currents, [Figure 23.7(c)]. The force exerted on the plasma outside the fluid element will tend to accelerate it in the negative x direction, but the force on the element itself is opposite to its initial velocity **v**.

This may be applied to a rotating, magnetic cloud embedded in interstellar plasma (or to a neutron star embedded in a plasma, a pulsar) to show that the cloud's angular momentum will gradually be radiated away. The disturbances that carry off angular momentum are Alfvén waves, and travel with velocity $\mathbf{v}_A = v_A(\mathbf{B}/B)$, where

$$v_A = (B^2/4\pi\rho)^{1/2}, \tag{23.57}$$

and ρ is the density of the surrounding matter. For interstellar matter ($B \simeq 3 \times 10^{-6}$ gauss and $\rho \simeq 4 \times 10^{-23}$ g/cm³), v_A is of order 10^5 cm/sec. Note that this exceeds the local speed of sound, c_s. For a typical HI region with $T = 50$ K, $c_s = 6 \times 10^4$ cm/s.

The rate of angular momentum loss from a spherically symmetric rotating cloud with an axially symmetric magnetic field **B** is discussed in Problem 23.9.

Problem 23.9. A spherically symmetric cloud of radius R rotates with angular velocity $\boldsymbol{\omega}$. A uniform, axial symmetric magnetic field **B** fills space and is parallel to $\boldsymbol{\omega}$. Surrounding the cloud is interstellar matter ($\rho = 4 \times 10^{-23}$ g/cm³, $T = 50$ K) assumed to be of high conductivity. The magnetohydrodynamic equations (with $\sigma \to \infty$) are given by (21.35), (21.37), and (21.38), and $\nabla \cdot \mathbf{B} = 0$. Assume initially that the surrounding plasma is at rest, and that its velocity remains zero in the radial and z directions (polar coordinates r, θ, z are most convenient):

$$\mathbf{v}(0) = 0; \quad v_r(t) = v_z(t) = 0. \tag{23.58}$$

Assume that radial components of **B** vanish for all times, and that pressure and gravitational force terms are negligible. Then show that the hydrodynamic equations reduce to the set

$$\frac{\partial B_\theta}{\partial t} = B_z \frac{\partial v_\theta}{\partial z}, \quad \frac{\partial v_\theta}{\partial t} = \frac{B_z}{4\pi\rho} \frac{\partial B_\theta}{\partial z}. \tag{23.59}$$

You may find it helpful to use the vector identities

$$\nabla B^2 = 2\mathbf{B} \times (\nabla \times \mathbf{B}) - 2(\mathbf{B} \cdot \nabla)\mathbf{B}$$

$$\nabla \times (\mathbf{v} \times \mathbf{B}) = \mathbf{v}(\nabla \cdot \mathbf{B}) - \mathbf{B}(\nabla \cdot \mathbf{v}) + (\mathbf{B} \cdot \nabla)\mathbf{v} - (\mathbf{v} \cdot \nabla)\mathbf{B}.$$

Assume that ρ is constant in the plasma, and that B_z is a function only of r, and show that B_θ and v_θ satisfy wave equations with velocity (23.57):

$$\frac{\partial^2 B_\theta}{\partial t^2} = v_A^2 \frac{\partial^2 B_\theta}{\partial z^2}, \tag{23.60}$$

$$\frac{\partial^2 v_\theta}{\partial t^2} = v_A^2 \frac{\partial^2 v_\theta}{\partial z^2}. \tag{23.61}$$

Show that the solutions of (23.60) and (23.61) are given by

$$v_\theta = f(t - z/v_A, r) + g(t + z/v_A, r), \quad (23.62)$$

$$B_\theta = -(4\pi\rho)^{1/2}$$
$$\times [f(t - z/v_A, r) - g(t + z/v_A, r)], \quad (23.63)$$

where f and g are arbitrary functions.

The solutions to the wave equations contain advanced (g) and retarded (f) parts. Physically, if the disturbance starts at the cloud's surface at time $t = 0$, it should propagate away from the cloud at a speed v_A. In time it will have traveled a distance $v_A t = d$ in the $\pm z$ directions. For distances $>d$, the medium will be undisturbed. Thus for propagation along $+z$ we set $g \equiv 0$ and along $-z$ we set $f \equiv 0$.

The rate of transfer of angular momentum from the cloud to the interstellar medium by Alfvén waves may be found from (23.62). As the waves advance into the medium, they set it into rotation about **B** with velocity v_θ. The angular momentum of an element of interstellar matter will thus be $\rho r v(t - z/v_A, r)$ per unit volume at the point (r, z) at time t. Furthermore, all matter along z from the origin $z = 0$ to the point $d = v_A t$ will have been set into motion, so that the total angular momentum in the medium (assuming $d \gg R$) will be

$$2 \int_0^d \rho r v(t - z/v_A, r) \, v_A t \, 2\pi r \, dr = J. \quad (23.64)$$

The extra factor of 2 accounts for matter set into motion along $z < 0$ as well. Dividing by t gives the rate of transfer of angular momentum from the cloud to the interstellar medium, which we equate to the time-rate of change of the cloud's angular momentum J_{cloud}:

$$\frac{dJ_{\text{cloud}}}{dt} = -4\pi v_A \rho \int_0^{d = v_A t} v(t - z/v_A, r) r^2 \, dr. \quad (23.65)$$

In principle, the depth into the medium along $\pm z$ to which angular momentum may be transferred is unlimited. However, it is unlikely that much will reach further from the rotation axis than $d \simeq R$, the cloud's radius. In other words, Alfvén waves in this case tend to set up a rotating column of diameter $\sim R$ and length $2d = 2v_A t$, particularly if effects of viscosity may be neglected. Setting $d = R$ in (23.65), and assuming uniform rotation of the medium, we have, for $z \lesssim v_A t$,

$$v(t - z/v_A, r) = \omega(t)r, \quad (23.66)$$

where $\omega(t)$ is the cloud's angular velocity.

Problem 23.10. If the rotating cloud has angular momentum $J = I\omega(t)$, where $I = 2MR^2/5$, use (23.65) and (23.66) with $d = R$ to show that $\omega(t)$ is given by

$$\omega(t) = \omega(0)e^{-t/\tau}, \quad (23.67)$$

$$\tau = \frac{4M}{5\sqrt{\rho\pi}BR^2} = \left(\frac{M}{M_\odot}\right)\left(\frac{B_0}{B}\right)\frac{10^{-6}}{R^2\rho^{1/2}} \text{ yrs}, \quad (23.68)$$

where $B_0 = 3 \times 10^{-6}$ gauss, and R is in pc.

Problem 23.11. A 10 $M_\odot$ subcondensation of interstellar matter reaches an average density $\rho \simeq 4 \times 10^{-17}$ g/cm^3 when fragmentation stops because of increasing internal temperature. The initial angular velocity associated with the condensed cloud was 2×10^{-15} sec^{-1}. Suppose that the collapsing cloud loses angular momentum according to (23.67) and (23.68). How long will it take to increase its angular velocity to 2×10^{-14} sec^{-1} (typical of observed stars when R reaches $R_\odot$)? How does it compare to the free-fall time, assuming an initial radius $R = 0.1$ pc?

Alfvén waves may carry off most of the excess angular momentum of protostellar clouds before star formation is complete.

Available data, as summarized in Table 23.1, indicates that the collapse of magnetic clouds is possible. However, many details remain to be worked out. Before proceeding to some of the difficulties, we consider two more stages of star formation: the dynamic collapse of a cloud and cloud fragmentation.

23.6. FRAGMENTATION OF COLLAPSING CLOUDS

A major difficulty with the model for protostar formation outlined in the preceding is the tendency toward development of massive condensations (typically $\sim 10^4$ $M_\odot$ or more, particularly when $\mathbf{B} \neq 0$). It remains to be explained how individual stars having observed masses in the range $0.05 \lesssim M/M_\odot \lesssim 60$, as observed, may be formed. The fact that the critical masses are in good agreement with the total stellar mass of associations (at least for $|\mathbf{B}| \lesssim 2 \times 10^{-5}$ gauss) is encouraging, and suggests that we pursue this line of attack further. Therefore, we return to the Jeans length discussed in Section 23.2. All the models predicted that instabilities, due either to exponential growth of perturbations or to external pressure on the cloud's surface, set in over regions whose typical length is of order $(T/\rho)^{1/2}$. Detailed collapse models suggest that initially the cloud remains isothermal, with T constant, partly as a result of extremely low opacities. Eventually, when the density has increased significantly, the opacity will rise, and so will the temperature. However, before this happens (i.e., while $T \simeq$ constant), ρ increases as the cloud contracts, the Jeans length decreases, and it is possible for smaller and smaller regions to become themselves unstable, leading to fragmentation of the original mass into subcondensates. If this process can continue, perhaps in several stages (until individual cloud masses reach stellar sizes) before T begins to increase, then we would have a reasonable mechanism for protostar formation. Of course, this requires that when the subcondensate masses reach stellar magnitude, T must increase rapidly enough to balance the increase in ρ and keep the Jeans length greater than the protostellar radius. In the remainder of this section we consider a simple collapse model illustrating these points.

Zero-Pressure Model

In this highly idealized model, the collapsing cloud's pressure is taken to be zero, and the infalling matter's acceleration is determined entirely by gravitation. Despite its simplicity, it is not a bad approximation to real collapse, at least during intermediate stages. The gravitational force per unit volume of collapsing matter goes as $F_G/V \sim M^2/R^2V$, which, for M constant, gives $F_G/V \sim R^{-5}$. The pressure gradient, assuming T constant and $P \sim \rho$, is $dP/dr \sim P/R \sim \rho/R \sim R^{-4}$, and resists collapse. Initially the pressure gradient wins out unless instabilities result; but if the matter begins to collapse, the term F_G/V will take over, and the pressure gradient will become negligible, at least until the material heats up. During this intermediate stage we may approximately treat collapse by setting ∇P in (21.6) to zero, and for nonmagnetic cases use the gravitational force density $f = m(r)\rho G \mathbf{r}/r^3$. Thus

$$\rho \frac{d\mathbf{v}}{dt} = -\frac{m(r)G\rho(r)}{r^3}\mathbf{r}. \qquad (23.69)$$

As a specific model, assume spherical symmetry, that the initial velocity of each element is identically zero, and that the density depends only on time:

$$\mathbf{v}_0(0) = 0 \qquad \rho_0(\mathbf{r}, t) = \rho_0(t). \qquad (23.70)$$

Further assume that no two mass shells [of radius $r(0) = a$ at $t = 0$] cross during collapse. Thus the mass within any shell remains constant, and is given by

$$m(r, t) = m(a) = \int_0^a 4\pi r^2 \rho_0(0) \, dr$$
$$= \frac{4\pi a^3}{3} \rho_0(0). \qquad (23.71)$$

We proceed to calculate the unperturbed infall [all quantities with subscript (0) as in (23.70)] and then investigate the growth of initially small perturbations during collapse.

To find the unperturbed motion, substitute (23.71) into (23.69), multiply both sides by $\dot{r} = dr/dt$, and note that both sides are perfect differentials:

$$\dot{r}\ddot{r} = \frac{1}{2}\frac{d\dot{r}^2}{dt} = -\frac{4\pi a^3}{3}\rho_0(0)G\frac{\dot{r}}{r^2} = \frac{4\pi}{3}a^3\rho_0(0)G\frac{d}{dt}\frac{1}{r}.$$

Integrating and using (23.70), we get

$$\dot{r}^2 = \left(\frac{dr}{dt}\right)^2 = \frac{8\pi a^2}{3}\rho(0)\left(\frac{a}{r} - 1\right)G. \qquad (23.72)$$

This may be solved easily by setting

$$x = r/a = \cos^2\beta \qquad (23.73)$$

to get

$$\left(\frac{8\pi}{3}G\rho(0)\right)^{1/2} t = \beta + \frac{1}{2}\sin 2\beta. \qquad (23.74)$$

Collapse starts for $\beta = 0$ and $t = 0$ [as seen from (23.74) and (23.73)], terminating when $\beta = \pi/2$, or $t = [3\pi/32G\rho(0)]^{1/2}$, which is just the free-fall time. Equation (23.74) applies to each mass shell, showing that each reaches the origin at the same instant. The density within a given mass shell of initial radius a is, at time t, given by

$$\frac{4\pi}{3}r^3\rho_0(t) = \frac{4\pi}{3}a^3\rho_0(0),$$

or

$$\rho_0(t) = \rho_0(0)\sec^6\beta. \qquad (23.75)$$

The first equality is a statement of mass conservation; the second follows from this and (23.73); β is given by (23.74) for each instant. The instantaneous velocity $v_0(t) = dr/dt$ is obtained from (23.72), by taking the negative root, and (23.73), and is radial:

$$v_0(t) = -a \tan \beta \, [8\pi\rho(0)G/3]^{1/2}$$
$$= -r \tan \beta \sec^2 \beta \, [8\pi\rho(0)G/3]^{1/2} \quad (23.76)$$
$$= -2r \tan \beta \, d\beta/dt.$$

The last form follows if $[8\pi\rho(0)G/3]^{1/2}$ is eliminated for $d\beta/dt$ as calculated from (23.74). An obvious limitation to (23.76) is that v_0 becomes infinite when $\beta = \pi/2$. Presumably the opacity becomes great enough to lead to a temperature (and pressure) increase long before this point is reached.

We may now consider the growth of perturbations (assumed to be present initially) during the collapse. The preceding model will be taken as the unperturbed case, and the linearized theory of Section 23.2 used to solve for ρ_1 and v_1 as functions of time. The perturbations ρ_1 and v_1 are assumed to be small initially, but may grow as the cloud collapses. Specifically, if $\rho = \rho_0 + \rho_1$ grows by very much, then the resulting Jeans length will decrease, and the cloud could fragment, with the smaller condensed regions continuing their collapse. In this way ultimately, it is hoped, protostars with the observed stellar masses would result.

Our starting points are the linearized momentum and continuity equations (23.13) and (23.11), with $\mathbf{B} = 0$ and $P_1 = 0$. The continuity equation is [recall that $\rho_0(t)$ is independent of r at fixed t]

$$\frac{\partial \rho_1}{\partial t} + \rho_0 \nabla \cdot \mathbf{v}_1 + \rho_1 \nabla \cdot \mathbf{v}_0 + \mathbf{v}_0 \cdot \nabla \rho_1$$
$$= \frac{\partial \rho_1}{\partial t} + \rho_0(\nabla \cdot \mathbf{v}_1) + \mathbf{v}_0 \cdot \nabla \rho_1 - 2\rho_1 \nabla \cdot \left(\mathbf{r} \tan \beta \frac{d\beta}{dt}\right)$$
$$= \frac{\partial \rho_1}{\partial t} + \rho_0(\nabla \cdot \mathbf{v}_1) + \mathbf{v}_0 \cdot \nabla \rho_1 - 6\rho_1 \tan \beta \frac{d\beta}{dt}$$
$$= 0. \quad (23.77)$$

The last form of (23.76) was used for $\mathbf{v}_0$. Next, the momentum equation is

$$\frac{\partial \mathbf{v}_1}{\partial t} + \mathbf{v}_0 \cdot \nabla \mathbf{v}_1 + \mathbf{v}_1 \cdot \nabla \mathbf{v}_0 = -\nabla \phi_1.$$

Take the divergence of this equation to obtain

$$\frac{\partial}{\partial t}(\nabla \cdot \mathbf{v}_1) + \nabla \cdot (\mathbf{v}_0 \cdot \nabla \mathbf{v}_1)$$
$$+ \nabla \cdot (\mathbf{v}_1 \cdot \nabla \mathbf{v}_0) = -\nabla^2 \phi_1. \quad (23.78)$$

The middle two terms may be rewritten in more convenient form. The first is, using the last form of (23.76) for $\mathbf{v}_0$ and the fact that β depends only on t,

$$\nabla \cdot (\mathbf{v}_0 \cdot \nabla \mathbf{v}_1) = -\nabla \cdot (\mathbf{r} \cdot \nabla \mathbf{v}_1) \, 2 \tan \beta \frac{d\beta}{dt} \quad (23.79)$$

$$\nabla \cdot (\mathbf{r} \cdot \nabla \mathbf{v}_1) = \nabla \cdot \mathbf{v}_1 + \mathbf{r} \cdot \nabla(\nabla \cdot \mathbf{v}_1). \quad (23.80)$$

Problem 23.12. Following the preceding method if necessary, show that the third term in (23.78) is

$$\nabla \cdot (\mathbf{v}_1 \cdot \nabla \mathbf{v}_0) = -2 \tan \beta \frac{d\beta}{dt} (\nabla \cdot \mathbf{v}_1). \quad (23.81)$$

Combining (23.79) to (23.81), we obtain, for the divergence of the linearized momentum equation,

$$\frac{\partial}{\partial t}(\nabla \cdot \mathbf{v}_1) + \mathbf{v}_0 \cdot \nabla(\nabla \cdot \mathbf{v}_1)$$
$$- 4 \tan \beta \frac{d\beta}{dt}(\nabla \cdot \mathbf{v}_1) = -\nabla^2 \phi_1. \quad (23.82)$$

Equations (23.76), (23.82), and $\nabla^2 \phi_1 = 4\pi G \rho_1$ completely determine the behavior of the perturbation. As they stand they are too difficult to solve; so we make further approximations. One approximation might be to set $\nabla \cdot \mathbf{v}_1 = 0$, in which case the total rate of change of the volume of a fluid element would be determined solely by $\nabla \cdot \mathbf{v}_0$, and ρ_1 would vanish. Instead we assume

$$\nabla \cdot \mathbf{v}_1 = H(\beta) \sec^4 \beta f(a). \quad (23.83)$$

The first two factors contain all the time-dependence; the last, which we assume to be arbitrary except that it vanishes on the cloud's surface, depends only on the original radius a of the mass shell under consideration. Equation (23.83) implicitly assumes that $\rho_1(\mathbf{r}, t) = \rho_1(t)$, which is at least consistent with (23.80) for $\rho_0(t)$. Finally, the form $H \sec^4 \beta$ is taken for convenience (by hindsight!).

We must solve for $H(\beta)$. Start by noting that $\nabla \cdot \mathbf{v}_1$ is independent of $\mathbf{r}$ explicitly, so that the middle term of the left-hand side of (23.82) vanishes, as does $\nabla \rho_1$ in (23.77). Thus

$$0 = \frac{\partial \rho_1}{\partial t} + \rho_0(\nabla \cdot \mathbf{v}_1) - 6\rho_1 \tan\beta \frac{d\beta}{dt}$$

$$= \frac{\partial \rho_1}{\partial t} - \rho_0 H \sec^4\beta f(a) - 6\rho_1 \tan\beta \frac{d\beta}{dt}; \quad (23.84)$$

$$4\pi G \rho_1 = \nabla^2 \phi_1 = -\frac{\partial}{\partial t}(\nabla \cdot \mathbf{v}_1) + 4\tan\beta \frac{d\beta}{dt}(\nabla \cdot \mathbf{v}_1)$$

$$= 4\tan\beta \frac{d\beta}{dt} H \sec^4\beta f(a) \quad (23.85)$$

$$- f(a)\left(\sec^4\beta \frac{dH}{d\beta} + 4H\sec^4\beta \tan\beta\right)\frac{d\beta}{dt}.$$

The last equation gives

$$\rho_1 = -\frac{f(a)\sec^4\beta}{4\pi G}\frac{dH}{d\beta}\frac{d\beta}{dt}\left(\frac{G\rho_0(0)}{24\pi G}\right)^{1/2}\frac{dH}{d\beta}$$

$$= -f(a)\sec^6\beta \quad (23.86)$$

Taking the time-derivative of ρ_1 in (23.86), substituting the result into (23.84), and using (23.74) to evaluate the time-derivative $d\beta/dt$ leads to the final expression

$$\frac{d^2H}{d\beta^2} = 6H\sec^2\beta. \quad (23.87)$$

Problem 23.13. Carry out the steps leading to (23.87). Solve this equation by any method, or by the following procedure. Set $z = \cos\beta$, and show that (23.87) becomes

$$(1-z^2)\frac{d^2H}{dz^2} - z\frac{dH}{dz} - \frac{6H}{z^2} = 0. \quad (23.88)$$

Expand $H(z)$ in a power series $H(z) = \sum_{n=0}^{\infty} c_n z^{n+s}$ and substitute this into (23.88) to obtain the coefficient c_n. Solve the indicated equation for s. Of the two solutions, choose the one that increases with time.

The solution to this equation, as shown in Problem 23.13, is

$$H(\beta) = C(3\sec^2\beta - 2). \quad (23.89)$$

Therefore the density ρ_1 as given by (23.86) is

$$\rho_1(t) = -f(a)\sec^6\beta \left(\frac{G\rho_0(0)}{24\pi G}\right)^{1/2}$$

$$\times 6C\sec^2\beta \tan\beta. \quad (23.90)$$

For C negative, $\rho_1(t)$ is positive and the ratio

$$\frac{\rho_1(t)}{\rho_0(t)} \sim \sec^2\beta \tan\beta f(a) = f(a)\sin\beta \sec^3\beta. \quad (23.91)$$

As $\beta \to \pi/2$ (late collapse stages), $\sin\beta \to 1$ and $\sec^3\beta \to 0$. Although the linearized theory may not be trusted for such large values of β, equation (23.91) nevertheless indicates a definite, initial growth in $\rho_1(t)$ relative to $\rho_0(t)$. Thus if an initial perturbation $\rho_1(0)$ is present, then as the cloud collapses it may begin to grow significantly before the temperature changes because of increased opacity. This would lead to an accelerated decrease in the Jeans length, the onset of smaller instabilities within the collapsing cloud, and just possibly the formation of protostars with masses in the correct range.

Problem 23.14. Evaluate the collapse time $t_{\text{ff}} = [3\pi/32G\rho_0(0)]^{1/2}$ for initial matter densities 2×10^{-24} g/cm^3 and 4×20^{-23} g/cm^3. Discuss at least two effects that have been left out of the preceding simplified analysis, and try to estimate what effect they would have on the results.

Problem 23.15. A cloud's initial density $\rho_0(0) = 4 \times 10^{-23}$ g/cm^3 and its temperature (which remains constant during collapse) is 50 K. Using the results of this section as a guide, find the mass of the cloud, assuming that it is just large enough to become Jeans unstable. As it begins to collapse $\rho_1(0)/\rho_0(0) \simeq 0.1$, but grows as in (23.91) above. Show that this is equivalent to

$$\rho_1/\rho_0 \simeq \sqrt{\rho_0}.$$

If the effects of pressure are included in the analysis, then it can be shown that the ratio $\rho_1/\rho_0 \sim \rho_0^{3/2}$. Show that the critical mass $M_c \sim \sqrt{\rho_0}$ at any instant. Use these relations ($\rho_1/\rho_0 \sim \rho_0^{3/2}$ and $M_c \sim \rho_0^{1/2}$) to find out how many times the cloud must fragment in order to develop condensations of proper mass to be candidates for stars. What kind of association would this probably be? Estimate the total time for the process up to the point where individual protostars begin to collapse. Is the result reasonable? Assume that fragmentation occurs when $\rho_1 \approx \rho_0$.

23.7. DIFFICULTIES

The last sections indicate that rotation sets a lower limit to the density of a cloud that can develop an instability, and magnetic fields set a lower limit to the mass. Realistic models are likely to be much more involved than the simple models discussed in preceding sections, and the relative influence of neglected effects on the conclusions is not yet fully understood. Therefore we cannot say with certainty that star formation occurs as we have outlined. The exact formation mechanism (assuming only one exists!) probably involves many complicated effects. Some of these are outlined in the rest of this section.

Rotation

If the material out of which the initial cloud forms is rotating with the Galaxy, then the latter's differential rotation will endow the cloud with a nonzero angular momentum **J**, perpendicular to the disk, that must be conserved during collapse, unless an external torque is exerted on the cloud during its collapse. Consider the collapse, after fragmentation to subcondensates of about one $M_\odot$, of matter contained within a cylinder of radius $r = 0.1$ pc and height $h = 40$ pc parallel to **J** (Figure 23.8). The initial density $\rho_0 = 4 \times 10^{-23}$ g/cm³, as is characteristic of dense, dark nebulae. This configuration minimizes the angular momentum that must be conserved. Furthermore, condensation along the rotation axis will not be altered by the rotation itself. As shown in Problem 23.15, if the initial angular velocity of the cloud is $\omega(0) = 2 \times 10^{-15}$ sec^{-1}, then by the time it has contracted to a star of radius $R_\odot$ its period will be several minutes, which not only disagrees with known stellar rotation periods, but can be ruled out because the centrifugal force on matter at the equator would exceed gravity. It becomes evident, then, that some mechanism must be found by which the contracting cloud may greatly reduce its angular momentum. One way to achieve this would be to

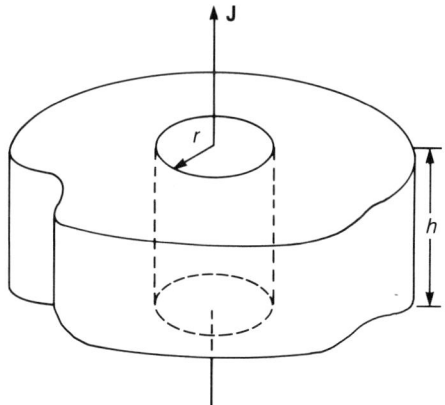

Figure 23.8. Rotating mass column (angular momentum **J** in the Galactic disk.

assume that an external torque is applied to the cloud; this could occur magnetically (Section 23.5). The presence of magnetic fields in the protostellar material, however, leads to new complications.

Problem 23.16. Find the final rotation period of a star formed from collapse of the column shown in Figure 23.8. Its radius $r = 0.1$ pc, height $h = 40$ pc, and $\rho = 4 \times 10^{-23}$ g/cm³ initially. If its initial angular velocity $\omega(0) = 2 \times 10^{-15}$ sec^{-1}, and its final radius is $R_\odot$, show that it cannot be in equilibrium in its assumed final state. By how much would the angular momentum have to be reduced if the star's equatorial rotation velocity were to be comparable to the observed values shown in Table 23.2?

If angular momentum transfer by magnetic fields does not occur, then star formation with observed **J** may still result from condensation. We know that most stars near the Sun are members of multiple-star sys-

Table 23.2
Observed equatorial velocities for main-sequence stars.

Type	Equatorial velocity	Average radius
O, BV	200 km/s	10 $R_\odot$
AV	<170 km/s	2 $R_\odot$
F5V	<30 km/s	1 $R_\odot$

tems (about 50 percent by number are binary, and 40 percent belong to systems of higher multiplicity). In addition to the stars' intrinsic angular momentum, these bound systems have orbital angular momentum. It is conceivable that the excess angular momentum initially present in the original cloud (before fragmentation to stellar masses) could go into orbital motion of multiple-star systems, with just enough left over to account for individual star rotation. This is still, however, an open question.

Problem 23.17. Two stars contract from cylinders of matter, as in Problem 23.16, and end up in the form of a binary system of semimajor axis a and period P. Assuming that each star's mass is $M_\odot$, find a relation between P and a, and discuss whether the resulting binary system could correspond to observed cases. Assume that the final, intrinsic angular momentum of each star $J = (2/5) M_\odot R_\odot^2 \omega$, with $\omega = 8.7 \times 10^{-13}$ sec^{-1}, corresponding to an equatorial velocity of about 100 km/sec.

Magnetic Fields

We have already discussed the effects of magnetic fields on the initial collapse and fragmentation. If the material was magnetic in its initial state, then it will be so in the final state as well, unless a loss mechanism is operative. The properties of interstellar matter are such that, as a cloud contracts, the magnetic flux is expected to remain constant. Roughly speaking, then, the magnitude of the field B, and the linear extent of the cloud R at any instant, must satisfy $R^2 B =$ constant. Therefore the final field strength of a star of radius R will be approximately

$$B \simeq B_0 (R_0/R)^2. \qquad (23.92)$$

Typical observed values for B_0 are of order 10^{-6} gauss. Thus, with $R_0 \simeq 0.1$ pc and $R = R_\odot$, we have $B \sim 10^7$ gauss. Magnetic stars are observed, the best known being Ap type, but in these cases $B_{\text{ave}} \sim 10^3$ gauss, with extreme values as high as 10^4 gauss or so. The only known stars having fields as high as, or greater than, 10^7 gauss are some white dwarfs, and probably neutron stars. It appears likely that some method of magnetic-flux loss is needed if stars are to condense with observed properties.

Problem 23.18. Show that formation of a one-solar-mass star of radius $R_\odot$ and average $B \lesssim 10^8$ gauss is possible, in the sense that it may be gravitationally bound.

Mechanisms have been proposed that would allow a collapsing cloud to shed magnetic flux. For example, a model applicable to HI regions (neutral hydrogen with a small admixture of ions, say, $n_{\text{ions}}/n_{\text{neutral}} \simeq 10^{-3}$) suggests that in the presence of a pressure gradient, the charged components (which carry with them the field lines) separate from the neutral matter. The difficulties with these models tend to be that the diffusion time exceeds the free-fall time, and thus would not be completed before the protostar formed.

23.8. Summary

The models we have discussed are the most elementary ones that illustrate the basic features of collapse, fragmentation, rotation, and magnetic fields, usually taken one at a time. Much progress is needed before any convincing arguments can be made about protostar formation out of interstellar matter. Even simple models with all these effects included have not been developed. With this proviso, we summarize what the simple models tell us about protostar formation, assuming the following parameters to characterize interstellar matter:

$\bar{\rho} =$ (average density of matter in clouds) $\sim 2 \times 10^{-24}$ g/cm^3;
$B =$ (magnitude of Galactic magnetic field in clouds) $\sim 3 \times 10^{-6}$ gauss;
$\omega =$ (angular velocity of cloud due to galactic differential rotation) 8×10^{-16} to $\sim 2 \times 10^{-15}$ sec^{-1}.

The following conclusions hold for clouds in the solar neighborhood.

(1) Gravitational instabilities may develop, but require that the cloud density exceed some lower limit, estimated to be 5.4×10^{-24} g/cm^3.

(2) The condensing cloud's mass probably exceeds $1.2 \times 10^4 \, M_\odot$ for $B = 3 \times 10^{-6}$ gauss. In any case, if $\mathbf{B} \neq 0$, then a lower limit to the mass results.

(3) The initial cloud will probably have an angular momentum due to differential Galactic rotation.

(4) Fragmentation of the originally unstable cloud

into subcondensates appears likely, but must occur at several stages before it can result in gravitationally bound masses that could be considered candidates for protostars. Rotation has little effect on fragmentation along the axis, but may halt it perpendicular to the axis. Fragmentation along magnetic field lines is somewhat enhanced, but is opposed in transverse directions.

(5) It appears likely that some mechanisms exist that effectively carry off angular momentum and magnetic flux before the protostar forms, since the initial **B** and ω given above would result in stars whose rotation rates and magnetic fields are in excess of observed values.

A complete theory of how stars form out of interstellar matter will probably require extensive two-dimensional (or even three-dimensional) numerical computations, but efforts to carry out such computations have only recently begun, and are far from producing definitive results.

Chapter 24

SUPERSONIC FLOW AND SHOCK WAVES

In this chapter we continue our discussion of hydrodynamics, and consider what happens when a mass of moving fluid encounters another object, or when disturbances are set up that propagate through the fluid. Situations in astrophysics where these processes are important include: cloud collisions; passage of a stellar wind past a compact star (binary companion, such as a neutron star, or black hole); expansion of ionized hydrogen (HII) into surrounding neutral hydrogen (HI); shock waves, formed in the late evolutionary stages of stellar evolution, triggering a supernova, or formed by the leading edge of a spiral density wave propagating in a galactic disk; and the rapid infall of matter onto a recently formed high-density core in protostars. In all these situations the fluid under consideration flows past (or into) an obstacle (which may be stationary fluid) with some velocity v relative to the obstacle. Now, small-amplitude disturbances in a fluid will usually propagate adiabatically with the local speed of sound, defined by (Chapter 21)

$$c_s^2 \equiv (\partial P/\partial \rho)_s. \qquad (24.1)$$

The derivative is at constant entropy S, since the time-scale for heat transfer from one region (of higher density corresponding to a local compression) to another is much longer than the dynamic time-scales on which the disturbances act. If $v < c_s$, then disturbances reach all parts of the fluid, and the flow is subsonic. When $v > c_s$, however, some regions in the flowing matter will be unaffected by the perturbation (see Figure 24.1). This type of flow for which $v > c_s$ is called *supersonic flow*. When the motion is subsonic, all quantities of physical interest will vary continuously throughout the fluid, and with time. The analysis leading to (24.1) as the sound speed is valid only for small-amplitude disturbances. When the amplitude of the disturbance becomes large, terms such as $\rho v \cdot \nabla v$ cannot validly be neglected, and the full set of hydrodynamic equations must be solved. If the disturbances are strong enough, they will result in a shock wave rather than a sound wave. We will consider shock waves after we mention several preliminary concepts.

There are two interesting illustrations of the way shock waves can be established. The first involves the propagation of a pressure wave of large amplitude through an otherwise uniform fluid. The pressure wave can be thought of as a region of rarefaction ($\rho < \rho_0$) followed by a region of compression ($\rho > \rho_0$). The initial profile of the density variations is shown in Figure 24.2(a), where ρ_0 is the unperturbed density

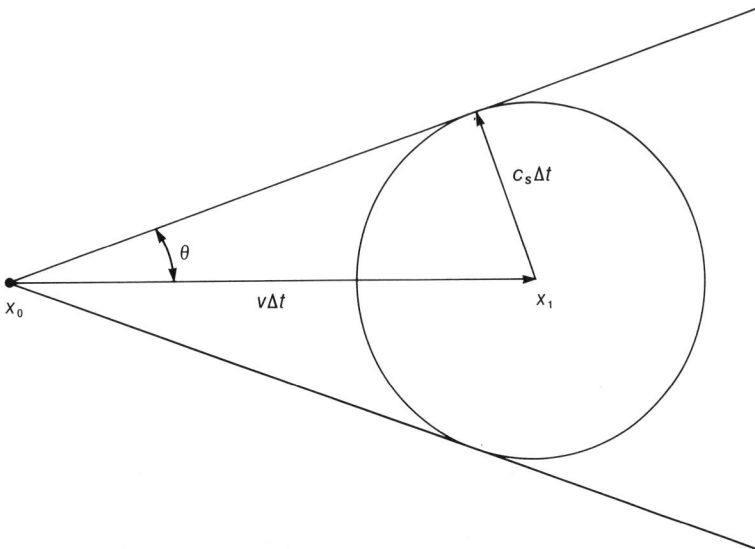

Figure 24.1. Disturbance in a supersonic flow. Location of initial disturbance x_0 in a fluid moving with $v > c_s$ moves to x_1 in time Δt. The effect of the disturbance moves spherically outward from the point of origin a distance $c_s \Delta t$ in the same interval. Fluid outside the Mach cone (apex angle $2\theta = 2\sin^{-1} c_s/v$) is unaffected.

(the wave moves to the right). As the pulse moves through the medium, the speed of sound (24.1) increases as the density increases; the compressions move with increased speed, but the rarefactions tend to slow down, as shown at time t_1 (the wave profile if c_s were constant is shown in the dashed curve). This separation of the sections of the pulse continues until a discontinuity develops (at time t_s). The discontinuity in ρ and the corresponding one in P and the temperature T represent the formation of a shock front. If the maximum pressure is not much greater than P_0, the shock front travels through the fluid with the speed of sound $(\partial P_0/\partial \rho_0)_s$ to first approximation. Figure 24.2 explains, for example, how shock waves are set up by spiral density waves propagating into a galactic disk.

The second example involves the propagation of a pulse, whose initial amplitude need not be large, in a medium whose density decreases in the direction of propagation. In this situation a shock wave will eventually form as shown in Problem 24.1. These conditions arise when a compression wave is set up in a stellar core or envelope, and then propagates down the density gradient, that is, outward through the star.

Problem 24.1. A pressure pulse, having the profile shown in Figure 24.3, propagates into a medium of decreasing density. Use (24.1) and the adiabatic relation $P = K\rho^\gamma$, where K and γ are constants, to show that a shock front will be formed. Assume that

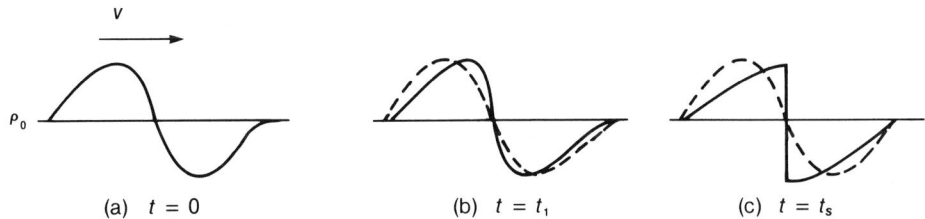

Figure 24.2. Formation of a shock wave. Density (and pressure) perturbation moving to right at $t = 0$. At $t = t_1$ the crest has begun to overtake the trough. Shock forms at $t = t_s$.

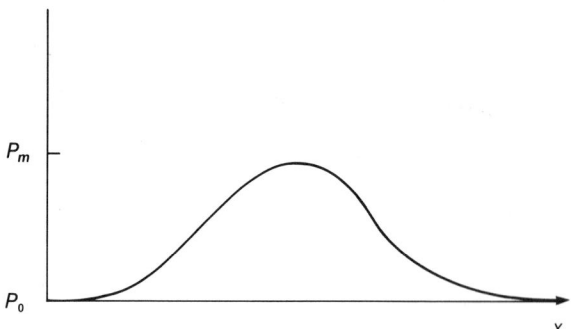

Figure 24.3. Pressure pulse (P_0 is unperturbed pressure).

the pulse amplitude P_m is small enough that all parts of the wave travel at the local unperturbed speed of sound independently of the degree of compression of the wave.

The following sections deal with shock waves and related phenomena in ideal gases. It is often convenient to work with the enthalpy $H = U + PV$, whose independent variables are the entropy S and pressure P. Denoting the enthalpy density by $h = H/V$ and recalling (2.6), we have for an ideal gas

$$h = u + P = \frac{\gamma P}{\gamma - 1}. \qquad (24.2)$$

Since the speed of sound (24.1) is defined adiabatically, we may use the polytropic equation of state to express

$$c_s^2 = \gamma \frac{P}{\rho} = \frac{\gamma k}{\mu m_H} T. \qquad (24.3)$$

This shows that, for an ideal gas, c_s depends only on the temperature, so that in regions of varying density but constant T, the adiabatic speed of sound is a constant.

24.1. Shock Waves

Near-discontinuities in ρ, P, and T (other than at boundary regions) in fluids constitute shock fronts. In practice, the region of rapid change in these variables may be replaced conceptually by a true discontinuity, or shock front. For example, in Figure 24.4, the fluid in region 2 is flowing to the left, into region 1. Across the plane AB, the thermodynamic variables are discontinuous. The discontinuity, or shock front, moves to the left with velocity $v_1 > c_1$, where c_1 is the local speed of sound in region 1. The flow is thus supersonic. Since the shock propagates with $v_1 > c_1$ into region 1, the fluid in front of the shock will not be affected by the discontinuity until the shock reaches it.

A more convenient reference frame is one at rest relative to the shock front, as in Figure 24.4(b). In this frame the fluid in region 1 flows into the shock with velocity v_1 and emerges with velocity v_2 in region 2. The effect of the shock is to compress and heat the gas. Sometimes the gas may be ionized in passing through the shock. We will apply the usual rules of thermody-

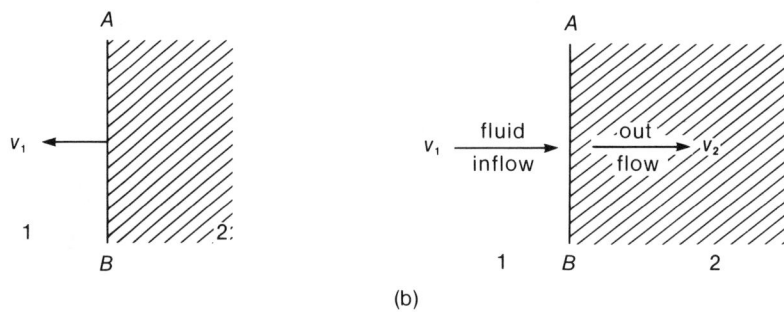

Figure 24.4. Shock front: (a) frame of reference at rest relative to unshocked fluid (region 1); (b) frame at rest relative to shock front AB. The unshocked fluid moves into the front, and emerges from the back side with velocity v_2.

namics to regions 1 and 2 separately, but conditions must be obtained from nonthermodynamic arguments in order to relate such properties as ρ, P, and T across the shock front. These are called *junction conditions*.

Junction Conditions

Quantities in region 1 or 2 will be denoted by subscripts. We now consider how these are related, by noting that the following three quantities must be continuous across a shock front: (1) mass flux; (2) momentum flux; and (3) energy flux. These are sufficient to relate the physics of region 1 to that of region 2. When ionization occurs, or if the shock triggers local energy release, as in a detonation, additional input will be needed, as discussed in the following.

Mass conservation. Mass is conserved across the shock. This follows from the continuity equation (21.9), which if integrated over the shock front yields

$$v_1 \rho_1 = v_2 \rho_2 \equiv J. \quad (24.4)$$

This equation states simply that mass neither appears nor disappears in the shock front. Clearly the mass flux J is constant.

Momentum conservation. Although we may not apply thermodynamics to the shock, we can apply the laws of mechanics in their hydrodynamic form. If we combine the continuity equation (21.9) and the equation of motion (23.4), and integrate over the volume, we may show that the momentum flux out of a fluid element is given by the tensor $P\delta_{ij} + \rho v_i v_j$, which for flow normal to the shock becomes

$$P_1 + \rho_1 v_1^2 = P_2 + \rho_2 v_2^2. \quad (24.5)$$

The term ρv^2 is the momentum flux due to the motion of the fluid element; the term P represents momentum flux associated with the thermal motion of particles. Physically, equation (24.5) states that all acceleration must vanish in the shock's frame of reference.

Energy conservation. In the absence of ionization or detonation, the energy flux into the shock must equal the flux out of it. This is equivalent to Bernouilli's principle, which, for a compressible fluid, is given by (21.17), and leads to the equality

$$\frac{1}{2} v_1^2 + (P_1/\rho_1) + (u_1/\rho_1)$$

$$= \frac{1}{2} v_2^2 + (P_2/\rho_2) + (u_2/\rho_2). \quad (24.6)$$

If energy is transferred into the fluid as it passes through the shock front, then this amount per unit mass must be added to the left-hand side of (24.6).

The conditions to be satisfied by the fluid as it passes through a shock front are given by (24.1) to (24.3), which we will use frequently in what follows. Finally, we observe that only the component of the fluid velocity normal to the front suffers a discontinuity; the tangential components are continuous, as is implicit in our discussion of the reference frame at rest with respect to the shock wave. If magnetic fields permeate the gas, then changes in the tangential component may result.

We now obtain a relation between the pressures and specific volumes, $V_1 = 1/\rho_1$ and $V_2 = 1/\rho_2$, on either side of the shock by rewriting (24.6) and using (24.2) to obtain

$$\frac{1}{2} v_1^2 + h_1/\rho_1 = \frac{1}{2} v_2^2 + h_2/\rho_2. \quad (24.7)$$

The enthalpy h is defined by (24.2). Next use (24.4) to express $v_2 = v_1(\rho_1/\rho_2)$, and eliminate v_2 from (24.7) and (24.5). Combining the results to eliminate v_1, we find that

$$\frac{h_1}{\rho_1} - \frac{h_2}{\rho_2} + \frac{1}{2} \frac{\rho_1 + \rho_2}{\rho_1 \rho_2} (P_2 - P_1) = 0. \quad (24.8)$$

Substituting specific volumes for densities gives the final form

$$H_1 - H_2 = \frac{1}{2}(P_1 - P_2)(V_1 + V_2), \quad (24.9)$$

which defines the shock adiabat or *Hugoniot*. It should be distinguished from the usual Poisson adiabat (2.11), which is of the form $F(P, V) =$ constant. The Hugoniot depends on the initial and final thermodynamic state of the gas, but is independent of dynamic variables such as velocity. It therefore relates static properties of the gas in much the same way as does an equation of state.

Problem 24.2. For an ideal gas $H_i = h_i V_i = h_i/\rho_i$ is given by (24.2). Show that the shock adiabat becomes

$$\frac{V_2}{V_1} = \frac{\rho_1}{\rho_2} = \frac{(\gamma + 1)P_1 + (\gamma - 1)P_2}{(\gamma - 1)P_1 + (\gamma + 1)P_2}. \quad (24.10)$$

A strong shock wave is defined by a large pressure discontinuity $P_2 - P_1$. Suppose $P_2 \gg P_1$, and find the maximum density increase behind the shock wave for a monatomic gas with $\gamma = 5/3$. Use (24.10) and the perfect-gas equation of state to show that

$$\frac{T_2}{T_1} = \frac{P_2}{P_1} \frac{(\gamma + 1)P_1 + (\gamma - 1)P_2}{(\gamma - 1)P_1 + (\gamma + 1)P_2}. \quad (24.11)$$

Next use (24.5) and (24.10) to express the increasing velocity v_1 as

$$v_1^2 = \frac{(\gamma - 1)P_1 + (\gamma + 1)P_2}{2\rho_1}. \quad (24.12)$$

Then show that

$$v_2^2 = \frac{[(\gamma + 1)P_1 + (\gamma - 1)P_2]^2}{2\rho_1[(\gamma - 1)P_1 + (\gamma + 1)P_2]}. \quad (24.13)$$

Show that the strong shock limits of v_1 and v_2 are

$$v_1 \to \left(\frac{\gamma + 1}{2} \frac{P_2}{\rho_1}\right)^{1/2}$$

and

$$v_2 \to v_1 \left(\frac{\gamma - 1}{\gamma + 1}\right), \quad (24.14)$$

respectively.

As Problem 24.2 shows, the pressure, temperature, and density behind the shock front are greater than in front of it. Even if $P_2/P_1 \to \infty$, the density increases only by a finite amount. Equation (24.14) shows that $v_2 < v_1$. The shock wave moves into the unperturbed fluid with speed v_1 [see Figure 24.4(a)]. Combining (24.5) and (24.4), we readily see that the speed of the shock wave relative to region 1 is

$$v_s^2 = v_1^2 = \frac{\rho_2}{\rho_1} \frac{P_2 - P_1}{\rho_2 - \rho_1}. \quad (24.15)$$

For an ideal gas we may combine (24.12) and (24.3) evaluated in region 1 to obtain

$$\frac{v_1^2}{c_1^2} = \frac{1}{2}\left(\frac{\gamma - 1}{\gamma} + \frac{\gamma + 1}{\gamma} \frac{P_2}{P_1}\right). \quad (24.16)$$

Since $P_2 > P_1$, if a shock front exists between regions 1 and 2, then (24.16) leads to the inequality $v_1 > c_1$. The shock front thus moves into region 1 supersonically, as expected.

The Hugoniot and adiabat starting from an initial density $\rho_1 = 1/V_1$ are shown schematically in Figure 24.5. Equation (24.15), which may be rewritten as a linear relation

$$P_2 - P_1 = \rho_1^2 v_1^2 (V_1 - V_2)$$

connecting the initial state (P_1, V_1) and the final state (P_2, V_2), is also shown. This is called the Rayleigh line. The area under it is easily shown to be

$$\frac{1}{2}(P_2 + P_1)(V_1 - V_2).$$

But this is just $\epsilon_2 - \epsilon_1 = \Delta\epsilon$ or the change in internal energy of the gas due to the shock. The area under the isentrope (curve S) is the change in internal energy $\Delta\epsilon_s$ that would result if the gas were compressed at constant entropy from V_1 to V_2. The difference $\Delta\epsilon - \Delta\epsilon_s$ represents shock heating of the gas, and is accompanied by an increase in entropy as well.

Problem 24.3. Use the fact that $v_1 > v_2$, and show that if $P_2 > P_1$, the motion behind the shock front is subsonic ($v_2 < c_2$).

Problem 24.4. Make a graph of the shock adiabat for an ideal gas with $\gamma = 5/3$, and on the same figure plot the Poisson adiabat

$$P_2 V_2^\gamma = P_1 V_1^\gamma.$$

Start at point P_1, V_1, and consider values of $P_2 > P_1$. In this way show that the pressure behind a shock front that results from a given compression $\Delta V = V_1 - V_2$ is greater than the pressure that results from the same amount of compression obtained by a sequence of equilibrium states.

Problem 24.5. The entropy of the ideal fluid on either side of the shock front is given by (2.13). Use

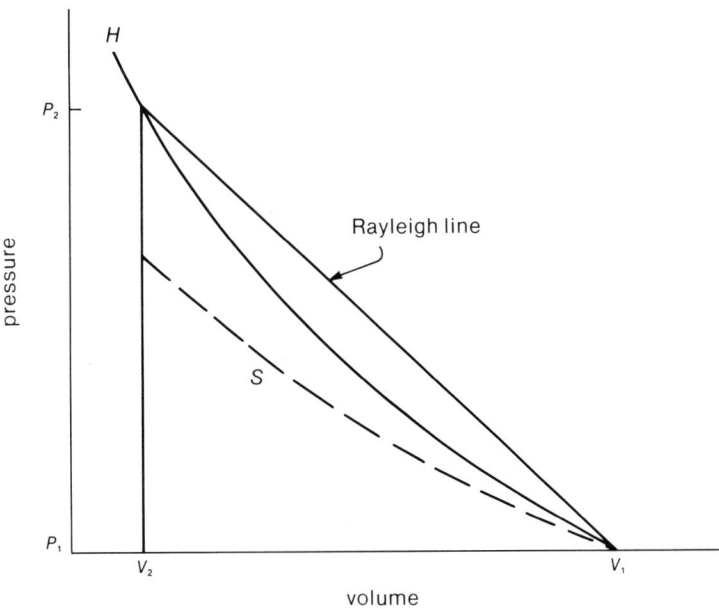

Figure 24.5. Pressure-volume diagram showing an adiabat (S), the Hugoniot (H), and the Rayleigh line connecting the initial unshocked state (P_1, V_1) and the final state (P_2, V_2). The area under the Rayleigh line gives the change in internal energy due to the shock.

the identity for an ideal gas

$$c_V = Nk/\mu(\gamma - 1),$$

and show that if $P_2 > P_1$, $\Delta s \equiv s_2 - s_1 > 0$. This shows that entropy is increased by passage of a shock wave.

The shock fronts we have been discussing are of zero width (i.e., are discontinuities) because the parameters of an ideal gas do not define a length scale. Real gases are to some extent viscous, and have nonzero thermal conductivity. The coefficient of viscosity, for example, has units of g cm^{-1} sec^{-1}, and does define a characteristic length. Therefore shock fronts in nonideal gases will be of finite width, and interesting processes may occur within this region. It can be shown that the width of the shock wave is inversely proportional to the pressure difference across the shock. Approximating a strong shock wave by a discontinuity is therefore reasonable under most circumstances.

A detailed analysis of processes within a shock wave requires the kinetic theory of gases. The width will be proportional to the mean free path, which is given by

$$\lambda \sim (n_a \sigma_a)^{-1}$$

for a neutral gas. Shock waves are observed in interstellar gas, where n_a may be small. For example, the atomic cross section $\sigma_a \approx 10^{-15}$ cm^2, and the atom number density in front of the shock wave $n_a \approx 10$ cm^{-3}, so that $\lambda \approx 10^{-5}$ pc. Detailed calculations indicate that $\lambda \sim 6 \times 10^{-6}$ pc is a better estimate.

24.2. Luminous Shock Waves

The shock waves discussed in Section 24.1 involve adiabatic compression. There are some astrophysically interesting situations in which shock waves upset local energy balance, so that the region immediately behind the shock front becomes a source of radiation that is luminous enough to be detectable. We refer to shock waves that accompany light emission as *luminous shocks*. The Veil (or Loop) Nebula in Cygnus may be such an object. It is believed that the Veil Nebula is the result of a supernova that occurred about 70,000 years ago. Observation indicates that the bright filaments are moving outward with speeds of order 100 km/sec (the more rapid expansion being furthest from the center).

Consider a region of interstellar gas in thermal equilibrium with radiation; the specific heat-loss rate Λ (ergs/g sec) of the gas must equal the specific heat gain Γ (Chapter 20):

$$\Gamma = \Lambda. \quad (24.17)$$

As the gas enters a strong shock wave, it is compressed, and the electron collision rate with ions and atoms increases. This leads to an increase in atomic excited states in the hydrogen gas, and in ionic excited states in trace elements such as O, N, S, Ca, and Na. Sometimes immediately after the formation or arrival of the shock wave, the electrons and ions will not be in equilibrium with one another, the e^- having a lower effective temperature. However, once equilibrium between particles has been reached, the net effect of collisions is to temporarily store kinetic energy as excitation energy. When excited states decay, the radiation escapes, and the gas is cooled. Furthermore, cooling leads to an increase in density in addition to that caused by the shock wave itself.

We must now include the possibility that energy balance (24.17) will not hold during the passage of a shock front; we do so by relating the imbalance to the heat flow. Using the first law of thermodynamics, we can write the heat loss per second as

$$\frac{dQ}{dt} = \frac{dU}{dt} + P\frac{dV}{dt}$$

$$= \frac{3}{2}\frac{Nk}{\mu m_H}\frac{dT}{dt} + \frac{\rho kT}{\mu m_H}\frac{dV}{dt}. \quad (24.18)$$

The second form uses $dU = c_V\,dT$ with

$$c_V = \frac{3Nk}{2\mu m_H}$$

for an ideal gas, and the ideal-gas equation of state. Dividing (24.18) by the volume, we obtain the total energy loss rate per cm³, which we rewrite as

$$\frac{1}{V}\frac{dQ}{dt} = \frac{NkT}{\mu m_H V}\left(\frac{3}{2}\frac{1}{T}\frac{dT}{dt} + \frac{1}{V}\frac{dV}{dt}\right)$$

$$= P\frac{d}{dt}\ln(T^{3/2}V).$$

Dividing by the density gives the difference

$$\Gamma - \Lambda \equiv (1/\rho V)\,dQ/dt$$

$$= \frac{kT}{\mu m_H}\frac{d}{dt}\ln(T^{3/2}V). \quad (24.19)$$

If the total mass of the gas $M = N\mu m_H$ is constant, we may replace $1/V$ in (24.19) by the density ρ. Applications of (24.19) require expressions for Γ and Λ, like those discussed in Chapter 20. Usually numerical methods must be used to study energy-loss rates.

Two of the most common features of interstellar matter, HI and HII regions, were discussed in Chapter 20. Consider a typical HII region. Observed temperatures are of order 10^4 K; so much of the hydrogen gas will be ionized. HII regions are common around young, hot OB-type stars, whose ultraviolet radiation is responsible for the ionization. The liberated electrons form a gas whose temperature is denoted by T_e. When an electron is initially ionized, it acquires an energy of order kT, where T is the temperature of the star. After many collisions with other particles, the electron's energy is reduced to kT_e, whereupon it may recombine with a proton. In this way the electrons serve as catalysts to transfer ultraviolet radiation from a star to thermal energy of the gas. The energy gained by the gas as a result is proportional to the energy difference $k(T - T_e)$, the number of protons and electrons per unit volume, n_p and n_e, respectively, and the recombination rate α:

$$k(T - T_e)\frac{n_p n_e}{\rho}\alpha(T_e). \quad (24.20)$$

The recombination rate α must be quantum-mechanically calculated, and may depend on the electron temperature. A more detailed analysis would include the exact energy distribution of the electrons, and the frequency dependence of the coefficient α.

The principal sources of energy loss are the deexcitation of atomic states, and electron-ion *Bremsstrahlung*. The energy-loss rate per unit volume because of deexcitation is proportional to the line frequency $\nu_i = (E_f - E_i)/h$, where E_i and E_f are the initial and final-state energies; to the number density of electrons, n_e; and to the transition rate $n_i \sigma_i v_e$, where n_i is the number density of ions in initial-energy states E_i, σ_i is the excitation cross section of the ith level, and v_e is the electron velocity. The complete relationship is

$$\rho\Lambda = \sum_i h\nu_i n_e (n_i \sigma_i v_e) + n_e n_i kT_e. \quad (24.21)$$

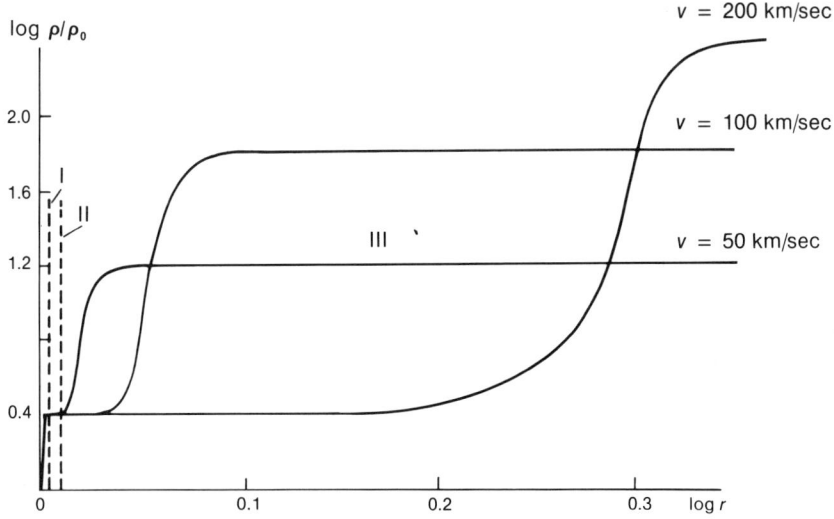

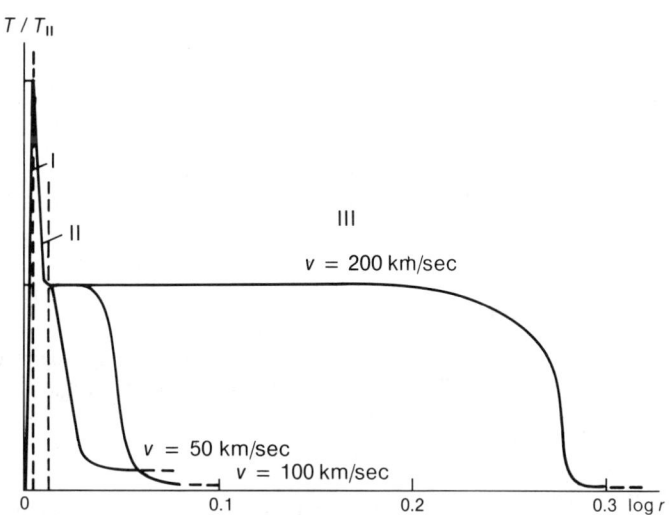

Figure 24.6. Density and temperature behind a luminous shock front moving into interstellar HI with velocities 50 km/sec, 100 km/sec, and 200 km/sec. The density ahead of the shock is $\rho_0 = 10^{-24}$ g cm^{-3}.

The last term is the free-free emission rate. It should be evident that Λ and Γ depend not only on the detailed thermodynamic state of the gas (recall that n_e and n_i will be interrelated by the Saha equation), but on the microscopic coefficients α and σ_i as well.

The situation in a neutral hydrogen (HI) region is more complex. Not only will processes such as those just described occur, but the presence of molecules (principally H_2) must also be considered, since temperatures are often as low as 50 K. Ionization by cosmic rays may also have an important influence on the energy balance.

The preceding results may be used to modify the theory of shock waves to include light emission. The junction conditions (24.4) to (24.6) still apply, since the emission of light does not alter mass or momentum conservation. They may be rewritten in the form

$$J = \rho v = \text{constant}, \quad (24.22)$$

$$\rho v^2 + P = \rho \left(v^2 + \frac{kT}{\mu m_H} \right) = \text{constant}. \quad (24.23)$$

These may be used to eliminate two of the variables appearing in (24.19), which is then solved by means of (24.20) and (24.21) to obtain the energy-loss rate immediately behind the shock front. The density variation through the emission region may then be used to calculate the radiant flux $\mathcal{F}$ (ergs cm^{-2}sec^{-1}) from the shock front:

$$\mathcal{F} = \int_{\text{front}} \rho(\Lambda - \Gamma) \, dl. \quad (24.24)$$

The integration is in the direction of mass flow, which we assume is normal to the shock front. Studies indicate that observable effects could be associated with shock fronts that propagate into interstellar matter at speeds in excess of 80 km/sec (recall that the observed velocity of filaments in the Veil Nebula is of order 100 km/sec).

The density and temperature profiles are shown in Figure 24.6 for a simple model of a luminous shock. The distance behind the shock is denoted by r. The three regions, I, II, and III are characterized by the following properties.

I. The thickness of the actual shock front is proportional to the mean free path of the heavy particles in the gas. The ions and atoms increase in temperature dramatically, but the electron temperature increases slowly. The density increases according to (24.10).

II. In this somewhat broader region, the electron and ion temperatures reach equilibrium. The temperature drops, partly because of ionization, and the density remains nearly constant.

III. Finally the radiative region is reached. The gas exists in a high state of ionization. The temperature levels off, and the density increases again. For moderate velocities the density increase may be as much as 100 times the initial gas value.

The material on either side of a luminous shock is usually in equilibrium with the interstellar radiation field. As a result, the gas in front of the shock and that behind the light-emitting region are at the same temperature. According to (24.13) the speed of sound will be the same in both regions, which we denote by subscripts 1 and 2. From these observations we obtain a relation between the densities ρ_1 and ρ_2, before and after the shock. Using $P = \rho c_s^2$ in (24.5), we obtain

$$P_1 - P_2 = (\rho_1 - \rho_2) c_s^2 = \rho_2 v_2^2 - \rho_1 v_1^2$$
$$= (\rho_1 - \rho_2) v_1^2 (\rho_1/\rho_2).$$

We have used (24.4) in the last step. If $\rho_1 \neq \rho_2$, this leads to the expression

$$\rho_2/\rho_1 = v_1^2/c_s^2. \quad (24.25)$$

This relation shows that large density increases are possible following a luminous shock. Recall that the maximum density increase following a shock in an ideal gas is $\rho_2 = 4\rho_1$. As shown in Figure 24.6, an increase in ρ by as much as a factor of 10^2 is possible.

24.3. Ionization Fronts and Strömgren Spheres

The junction conditions across a shock may be extended to include ionization of gas flowing through the front. A specific example is the boundary separating a region of ionized hydrogen (HII) from neutral hydrogen (HI), called an *ionization front*. The bulk of interstellar matter is nonionized hydrogen gas. However, the immediate vicinity of hot young stars is often filled with ionized hydrogen. We will next consider the properties of such regions, and one scenario for their establishment and maintenance.

Light passing through hydrogen gas scatters primarily off electrons. In HI regions the electrons are bound to protons, and the effective cross section is, to order of magnitude, proportional to the size of the atom, $\sigma_I \sim r_b^2 \sim 3 \times 10^{-17}$ cm^2. If, on the other hand, the gas is ionized, scattering occurs primarily from free electrons (Thomson scattering), and the cross section $\sigma_{II} \sim (e^2/mc^2)^2 \sim 10^{-24}$ cm^2.

If radiation emitted (from a hot young star, for example) into hydrogen gas is less energetic than the binding energy of hydrogen ($E_B = 13.6 \, eV$), scattering is from atoms, and the mean free path of the radiation is of order

$$l_1 \sim (n_A \sigma_1)^{-1} \sim (3/n_A) \times 10^{16} \text{ cm}$$
$$\sim n_A^{-1} \times 10^{-2} \text{ pc}. \quad (24.26)$$

Here n_A is the number density of atoms. For radiation more energetic than E_B (wavelengths $\lambda \lesssim 912$ Å), the gas will be ionized, and the mean free path is

$$l_{II} \sim (n_e \sigma_{II})^{-1} \sim n_e^{-1} \times 10^{24} \text{ cm}$$
$$\sim n_e^{-1} \, 3 \times 10^5 \text{ pc}, \quad (24.27)$$

where n_e is the electron number density. It follows that $l_I \ll l_{II}$, and ultraviolet radiation is immediately absorbed, ionizing the gas. Subsequent ultraviolet radiation may then move freely through the HII region until it reaches the neutral gas, where it is absorbed within a layer of thickness l_I. Figure 24.7 summarizes the regions surrounding a hot, young star that is emitting a significant amount of light with $\lambda < 912$ Å. The spherical zone of volume $4\pi R_s^3 / 3$ is called the Strömgren sphere.

Evolution of the Strömgren Sphere

Stars radiate approximately like black bodies: the hotter the star's surface, the more energy it emits at shorter wavelengths. The energy distribution function for a black body is $B_\nu(T)$ and is given by (5.47) and the photon flux (number/cm² sec) is $B_\nu(T)c/h\nu$. The flux of photons more energetic than $E_B = 13.6$ eV is

$$\mathcal{F}_* = \int_{\nu_{max}}^{\infty} \frac{cB_\nu(T)\,d\nu}{h\nu} = \frac{2}{c^2} \int_{\nu_{max}}^{\infty} \frac{\nu^2\,d\nu}{e^{h\nu/kT}-1}, \quad (24.28)$$

where $\nu_{max} = E_B/h$. Equation (24.28) gives the number of photons emitted per cm² per second from the star's surface. Denoting the stellar radius by R_*, we find the emission rate to be

$$\frac{dN_i}{dt} = 4\pi R_*^2 \mathcal{F}_* = \frac{4\pi R_*^2}{c^2} \int_{\nu_{max}}^{\infty} \frac{2\nu^2\,d\nu}{e^{h\nu/kt}-1}. \quad (24.29)$$

The preceding analysis is not in exact agreement with observations. The difficulties involve blanketing in the stellar atmosphere, which results in fewer high-energy photons being emitted in the ultraviolet than is predicted by $B_\nu(T)$.

Problem 24.6. The maximum of the black-body distribution $B_\nu(T)$ occurs for $\lambda_{max} T \simeq 0.29$ (with λ_{max} in cm). Assume that $\lambda_{max} > 912$ Å and that $T \ll 0.3/\lambda_{max}$, and estimate the emission rate (24.29). Retain lowest-order terms in $h\nu_{max}/kT$.

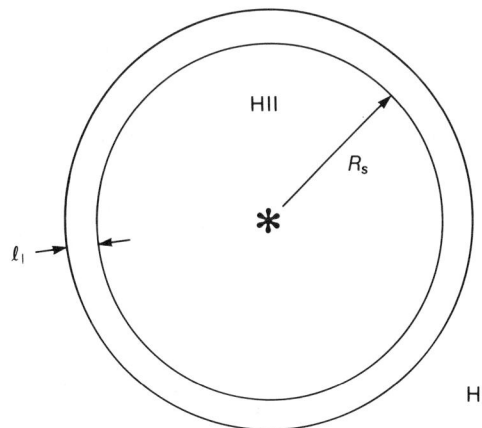

Figure 24.7. Strömgren sphere separating HI and HII regions.

Equation (24.29) and the results of Problem 24.6 indicate that the emission rate depends on only the stellar parameters. Typical values for $\mathcal{F}_*$ as a function of spectral type are given in Table 24.1. The radii are those expected for a main-sequence star.

Now consider what happens if a young star suddenly brightens, increasing its ultraviolet emission. At first all photons in the ultraviolet will be absorbed as the surrounding gas is ionized. For typical values of n in the variety of young stars ($n_H \sim 10^3$ cm^{-3}), $l_I \sim 10^{-5}$ pc ~ 1 A.U. Once the gas is ionized, $l_{II} \sim 3 \times 10^2$ pc; the radiation emitted into the HII region grows, expanding into the neutral gas. The rate of growth may be written as

$$\begin{pmatrix} \text{rate of change} \\ \text{of volume of} \\ \text{HII region} \end{pmatrix} \times \begin{pmatrix} \text{number of} \\ \text{hydrogen atoms} \\ \text{per unit vol.} \end{pmatrix}$$

$$= 4\pi r^2 \frac{dr}{dt} n_H, \quad (24.30)$$

assuming that each ultraviolet photon is absorbed, ionizing an atom, and that no recombination of ions

Table 24.1
Stellar flux and radius for main-sequence stars in HII regions.

Spectral type	O5	O7	O9	B0
$\mathcal{F}_*$ ($\times 10^{-23}$ cm^{-2} sec^{-1})	8.7	5.5	2.3	0.36
Radius R_* ($R_\odot$)	17.8	13	11	7.4

and electrons follows. This is probably a reasonable first approximation during the early growth. However, during later stages recombination will occur. As a result, some of the emitted photons will be used up in reionizing atoms in the HII region. The number of photons used up in this way per second is equal to the recombination rate, which is proportional to the ion and electron number densities n_i and n_e, respectively, and to the volume of the HII region,

$$\text{(rate of recombination)} = \frac{4\pi r^3}{3} n_i n_e \alpha(T), \quad (24.31)$$

where $\alpha(T)$ is the recombination rate factor for the ion. For typical HII regions, $T \lesssim 10^4$ K, and $\alpha(T)$ at this temperature is roughly

$$\alpha(T) \sim 4 \times 10^{-13} \text{ cm}^3 \text{ sec}^{-1}. \quad (24.32)$$

This value excludes recombination to the ground state, since the re-emitted ionizing photon would simply ionize another atom, as discussed in chapter 20.

The total ionizing-photon emission rate must be responsible for HII growth and must also compensate for recombination within the HII region. Thus we write dn_i/dt as

$$\frac{dn_i}{dt} = 4\pi n_H r^2 \frac{dr}{dt} + \frac{4\pi\alpha}{3} r^3 n_e n_i. \quad (24.33)$$

Recall that n_H, n_e, and n_i will also depend on density and temperature. Since the gas is primarily hydrogen, we set $n_e = n_i$. Equation (24.33) governs the evolution of the HII region.

Problem 24.7. The growth rate of the HII region will be large initially. Show that during initial stages, the last term in (24.33) is small. Then express the growth rate dr/dt in terms of the ionizing-photon flux rate of the star.

As the HII region grows, more and more of the star's ionizing radiation is needed to balance the effects of recombination. Eventually recombination ($\sim r^3$) will dominate ionization ($\sim r^2$), and a steady state is reached in which the zone separating the HI and HII regions is no longer advancing into the neutral gas. This happens when the radius of the HII region is equal to R_s:

$$R_s n_e^{2/3} \simeq (3 R_*^2 \mathcal{F}_*/\alpha)^{1/3}. \quad (24.34)$$

This depends on the characteristics of the ionizing star, and on the density of matter in the Strömgren sphere.

Problem 24.8. Show that (24.34) follows from (24.33) and (24.29). Assuming that α is given by (24.32) and that $n_e = 10^3$ cm^{-3}, show that the ionized hydrogen region is of radius $R_s \simeq 0.72$ pc for an O5 star (see Table 24.1 for parameters), and $R_s = 0.14$ pc for a B0 star.

A closer look at (24.30) and the discussion preceding it shows that R_s as given by (24.34) is an underestimate of the size of the HII region, because of the photons re-emitted as a result of recombination. According to the simplified model we have been using, these re-emitted photons no longer contribute to the growth or maintenance of the HII region. However, the re-emitted photon's energy will often exceed the ionization energy E_B, and subsequent ionization may be possible. This effect means that the actual radius of the HII region exceeds R_s. Analysis shows that the increase may be as much as a factor of 10.

Another complication needs to be included in the analysis of HII regions. For each hydrogen atom present in an HI region, two particles (proton and electron) will be present once the region has been ionized. Thus, even if the temperatures of the HI and HII regions were the same (they are not), the pressures would be different, with $P_{II} > P_I$. In fact, characteristic parameters for both regions are given in Table 24.2. We see that in actuality $P_{II} \simeq 200\, P_I$. This suggests that the pressure will drive the front separating the two regions into the neutral gas. This dynamic effect may be handled by the methods of shock-wave motion, as we will see.

Table 24.2
Temperature and density in HI and HII regions.

Region	Temperature	Number density
HI	10^2 K	0.1–10^2
HII	10^4 K	10^3–10^5

Problem 24.9. Suppose that the ionization front is driven into the surrounding HI region by the pressure difference across it. If the HII region is to expand in this manner, the entire mass M of the HI region must be accelerated. Take $P = 2 n_i k T$ (with T the temperature in the HII region) and show that the acceleration of the front is given by

$$\frac{d^2 R}{dt^2} = \left(\frac{8\pi n_i k T}{M}\right) R^2. \quad (24.35)$$

From this show that the expansion velocity $\dot{R} = DR/dt$ is equal to 3.4×10^5 cm/sec if $n_i = 10^4$ cm^{-3}, $T = 10^4$ K, $M = M_\odot$, and $R = 10^{17}$ cm. Assume spherical symmetry.

Problem 24.10. Find the speed of sound in an HI region in which $\mu \simeq 1$ and $\gamma = 5/3$. Show that

$$c_s \simeq 1.2 \times 10^5 \text{ cm/sec}.$$

Problem 24.11. Suppose that the HI region surrounds the HII region of Problem 24.9. Noting that the effects of overpressure in the HII region propagate normally only at the local speed of sound to distant portions of the HI region, discuss carefully the modifications that must be made to explain the expansion of the HII region. Comment specifically on the actual mass accelerated during the expansion, and on the expected expansion velocity.

Shock Waves with Ionization

The implication of the last three problems—specifically, of the fact that $v_{II} > c_s$—is that the narrow zones between HI and HII regions move into interstellar gas at supersonic speed. Furthermore, the large pressure differences existing across relatively narrow regions, and the large temperature differences (factor of 100), suggest that the transition region is a shock front. We therefore modify our earlier discussion of shock waves to include the effects of ionization across the front. Besides the relative difference in ionization states across the front, another feature must be taken into account. The rate at which the ionization shock front (ionization front for short) moves into the HI region is determined not by gas velocity (as is typical of regular shocks), but by the flux of ionizing photons emitted by the central star or stars. Although an ionization front may be set up around groups of stars, we will assume, in the discussion to follow, that it is due to a single star, and that the front is spherically symmetric. In actuality both extremes are observed.

Now consider how the junction conditions across a shock front are to be modified when ionization is caused by an external photon flux. We denote fluid variables in the neutral gas by subscript 1, and those in the ionized gas by subscript 2. (Actually, the ionization front moves into the neutral gas, but it is more convenient to fix our reference frame in the front. Then the neutral gas is observed to flow into the front and ionized gas out of it, as shown in Figure 24.4.)

We now modify the junction conditions to include ionization effects across a shock front. The mass-conservation relation (24.4) to (24.6) remains unchanged; however, the mass flux J is no longer arbitrary. Instead, it is related to the flux of ionizing radiation from the central star. This follows since the emission of each ionizing photon ultimately converts a hydrogen atom into a photon plus an electron. Assuming conservation of flux, we see that $\mathcal{F}_*$ and the corresponding photon flux at an arbitrary distance r from the star, $\mathcal{F}(r)$, are related by

$$4\pi R_*^2 \mathcal{F}_* = 4\pi r^2 \mathcal{F}(r). \quad (24.36)$$

Denoting the mass flux across a front at r due to ionization by $J = m_H \mathcal{F}(r)$, we see that the mass-conservation condition becomes

$$J = \frac{m_H R_*^2 \mathcal{F}_*}{r^2} = \rho v. \quad (24.37)$$

No subscripts appear on the right, since J is constant.

The momentum and energy conditions (24.5) and (24.6) must also be modified to include the momentum and energy flux across an arbitrary surface due to the radiation. Denoting an average photon energy by $h\bar{\nu}$, we find that the total momentum associated with the ionizing radiation in the HII region is

$$\frac{1}{c}\left(\begin{array}{c}\text{average photon}\\ \text{energy}\end{array}\right)\left(\begin{array}{c}\text{photon}\\ \text{flux}\end{array}\right) = \frac{h\bar{\nu}}{c}\frac{R_*^2 \mathcal{F}_*}{r^2}, \quad (24.38)$$

which must be added to $P + \rho v^2$ in the ionized-gas region (HII). Thus we obtain

$$P_1 + \rho_1 v_1^2 = P_2 + \rho_2 v_2^2 + \frac{h\bar{\nu}}{c^2}\frac{R_*^2 \mathcal{F}_*}{r^2}. \quad (24.39)$$

The last term in (24.39) is determined (for fixed r) by

the nature of the ionizing star, and the binding energy of the atoms in the surrounding gas.

Shock waves are usually driven by some external agent. For example, a shock set up by a piston is driven by the piston. Shock waves set up by novae or supernovae, which appear as shells of expanding gas, move partly as a result of the initial momentum imparted to them. This is not so for ionization fronts, which move as a result of the increased pressure within the front. The resulting expansion often corresponds to free expansion of the ionized gas, occurring at the local speed of sound of the gas itself. Therefore

$$c_{s,2}^2 = \frac{\gamma k T_2}{\mu_2 m_H} = v_2^2. \quad (24.40)$$

We thus take, as the fluid velocity in the HII region, the speed of sound $c_{s,2}$. This is known as the *Jouquet point condition*. Substituting (24.40) into (24.37), evaluated in the HII region, we find its density to be

$$\rho_2 = \frac{J}{v_2} = \frac{m_H R_*^2 \mathcal{F}_* \mu_2^{1/2}}{r^2 (\gamma k T_2 / m_H)^{1/2}}. \quad (24.41)$$

Substituting (24.40) into the momentum condition (24.39), we can easily obtain an equation for the velocity v_1:

$$v_1^2 - v_1 \left[\frac{h\bar{\nu}}{m_H c} + (\gamma + 1) \left(\frac{kT_2}{\mu_2 m_H} \right)^{1/2} \right]$$
$$+ \frac{kT_1}{\mu_1 m_H} = 0. \quad (24.42)$$

This may be solved for v_1. However, for conditions typical of many HII regions, the ratio $h\bar{\nu}/m_H c$ is a small correction to the second term in the brackets, and we may neglect it, finding

$$v_1 = (\gamma + 1) \left(\frac{kT_2}{4\mu_2 m_H} \right)^{1/2}$$
$$\pm \left[\frac{(\gamma + 1)^2}{4} \left(\frac{kT_2}{\mu_2 m_H} \right) - \frac{kT_1}{\mu_1 m_H} \right]^{1/2}. \quad (24.43)$$

Since HII regions are typically a factor of 100 hotter than neighboring HI regions, $T_2 \gg T_1$, and we may

expand (24.43), obtaining the two velocities $v_1^{(+)}$ and $v_1^{(-)}$,

$$v_1^{(+)} \simeq (\gamma + 1) \left(\frac{kT_2}{\mu_2 m_H} \right)^{1/2} \left[1 - \frac{1}{(\gamma + 1)^2} \frac{\mu_2 T_1}{\mu_1 T_1} \right],$$
$$v_1^{(-)} \simeq \frac{1}{\gamma + 1} \left(\frac{kT_1}{\mu_1 m_H} \right) \left(\frac{\mu_2 m_H}{kT_2} \right)^{1/2}, \quad (24.44)$$

with $v_1^{(+)} \gg v_1^{(-)}$. Examination of the density ratio ρ_1/ρ_2 using each of (24.44) shows that

$$\rho_2 \gtrsim \rho_1, \quad v_1 = v_1^{(+)} \quad (24.45)$$

$$\rho_2 \ll \rho_1, \quad v_1 = v_1^{(-)}. \quad (24.46)$$

Therefore, as the ionization front moves into the neutral gas with speed $v_1^{(+)}$ (which is easily shown to exceed the speed of sound c_1 in region 1), the front corresponds to a compressing wave, and its motion is supersonic. If the front moves into the HI region with $v_1 = v_1^{(-)}$, then the wave is a rarefaction wave, since $\rho_2 \ll \rho_1$.

Problem 24.12. Show that if $v_1 = v_1^{(+)}$ the resulting front moves supersonically, and is a compression wave (i.e., $\rho_2 > \rho_1$). Taking as typical values $T_2 = 10^2 T_1 = 10^4$ K, $\mu_1 = 1$, and $\mu_2 = 1/2$, find $v_1^{(+)}$.

Given the temperatures T_1 and T_2, we may use (24.41) and (24.42) to find values for the densities ρ_1 and ρ_2. However, we have not yet made use of the energy-conservation equation. First we must modify it to include the photon-energy flux in the HII region. If the photon energy is $h\nu$, then the excess energy after ionization is taken by the electron. However, the ionization itself uses up energy, which must be subtracted from the flux in region 2. Denoting the energy given to the electron in ionizing it by ϵ_0, we find that the energy condition becomes

$$\rho_1 v_1 \left(\frac{\gamma}{\gamma - 1} \frac{P_1}{\rho_1} + \frac{v_1^2}{2} \right)$$
$$= \rho_2 v_2 \left(\frac{\gamma}{\gamma - 1} \frac{P_2}{\rho_2} + \frac{v_2^2}{2} \right) - \epsilon_0 \frac{R_*^2 \mathcal{F}_*}{r^2}. \quad (24.47)$$

The last term is just the photon flux times the energy

loss due to ionization. Using expression (24.40) for $v_2 = c_{s,2}$ and (24.37) for mass conservation, we may solve (24.47) for the velocity v_1, obtaining

$$v_1^2 = \frac{\gamma(\gamma+1)}{\gamma-1}\frac{kT_2}{\mu_2 m_H} - \frac{2\gamma}{\gamma-1}\frac{kT_1}{\mu_1 m_H} - \frac{2\epsilon_0}{m_H}. \quad (24.48)$$

When $T_1 \ll T_2$ we may expand v_1. The result denoted by $v_1^{(\pm)}$ may then be equated to previous results for $v_1^{(\pm)}$ given by (24.44). When this is done, we find the expression for T_2 given in (24.49), which is then substituted into the expressions for the velocity v_1. Next we use v_1 thus obtained to find an expression for the density ρ_1. The results are, for $v_1^{(+)}$,

$$T_2 = \frac{\gamma(\gamma-1)}{\gamma+1}\frac{2\epsilon_0 \mu_2}{k},$$
$$v_1^{(+)} = \sqrt{(\gamma^2-1)2\epsilon_0/m_H},$$
$$\rho_2 = \frac{\gamma+1}{\gamma}\rho_1, \quad (24.49)$$
$$\rho_1 = \frac{R_*^2 \mathcal{F}_*}{r^2}\sqrt{\frac{m_H^3}{2\epsilon_0(\gamma^2-1)}},$$

and, for $v_1^{(-)}$,

$$T_2 = \frac{2\mu_2 \epsilon_0}{k}\frac{\gamma-1}{\gamma(\gamma+1)},$$
$$v_1^{(-)} = \frac{\mu_2}{\mu_1}\frac{T_1}{T_2}\sqrt{\frac{\gamma-1}{(\gamma+1)^3}\frac{2\epsilon_0}{m_H}}, \quad (24.50)$$
$$\rho_2 = \frac{\mu_2}{\mu_1}\frac{T_1}{T_2}\frac{\rho_1}{\gamma+1},$$
$$\rho_1 = \frac{R_*^2 \mathcal{F}_*}{r^2}\frac{\mu_1}{\mu_2}\frac{T_2}{T_1}\sqrt{\frac{(\gamma+1)^3}{\gamma-1}\frac{m_H^3}{2\epsilon_0}}.$$

Since $\mu_1 \sim \mu_2$, and $T_2 \gg T_1$, the first case (24.49) corresponds to a compression wave. We will return to aspects of the two solutions later.

Motion of Ionization Fronts

An analysis of the motion of ionization fronts can be quite complicated; so we will restrict our discussion to the qualitative aspects of their motion. We again label the HI region by subscript 1 and the HII region by subscript 2, and refer all motion to a frame of reference fixed in the ionization front, so that the neutral gas flows into the front, and the ionized gas flows out of it (recall that this is equivalent to motion of the front into the neutral gas with speed v_1).

Define the density ratio ψ as follows:

$$\psi \equiv \frac{\rho_2}{\rho_1}$$
$$= \left(\frac{\text{density of gas flowing out of front}}{\text{density of gas flowing into front}}\right). \quad (24.51)$$

We may then combine the mass, momentum, and energy-conservation conditions across the front to obtain a quadratic equation for ψ. Neglecting the last term in (24.39) and using $\rho_1 v_1 = \rho_2 v_2$, we can show that

$$P_2 = P_1 - \rho_1 v_1^2 \frac{1-\psi}{\psi}. \quad (24.52)$$

Using this to eliminate P_2 from the energy-conservation equation, (24.47) and noting that the speed of sound is given by $\gamma P_1/\rho_1$ in region 1, we find

$$\left(\frac{c_1^2}{\gamma-1} + \frac{v_1^2}{2} + \frac{\epsilon_0}{m_H}\right)\psi^2 - \left(\frac{c_1^2}{\gamma-1} + \frac{v_1^2}{\gamma-1}\gamma\right)\psi$$
$$+ \frac{1}{2}\frac{\gamma+1}{\gamma-1}v_1^2 = 0. \quad (24.53)$$

In arriving at (24.52), we find it convenient to replace v_2 by $v_1\rho_1/\rho_2$. The solutions ψ to (24.53) are easily found. It is important to observe that in general there will be two solutions, and that those of physical interest must be real. This restricts the discriminant in the quadratic formula for ψ to nonnegative values; that is,

$$\left(\frac{c_1^2}{\gamma-1} + \frac{\gamma}{\gamma-1}v_1^2\right)^2$$
$$\geq 2v_1^2 \frac{\gamma+1}{\gamma-1}\left(\frac{c_1^2}{\gamma-1} + \frac{v_1^2}{2} + \frac{\epsilon_0}{m_H}\right). \quad (24.54)$$

If this is expanded, we find the relatively simple equivalent condition

$$(c_1^2 - v_1^2)^2 \geq 2(\gamma^2-1)v_1^2 \epsilon_0/m_H. \quad (24.55)$$

Two cases of interest are: (1) $v_1 < c_1$, corresponding to subsonic motion in region 1; (2) $v_1 > c_1$, corresponding to supersonic flow. In either case the critical velocity results if the equality is used in (24.55).

Case 1: $v_1 < c_1$. The solution of (24.55) is, in this case, obtained by equating the two sides and taking the positive root. Then

$$v_D = -\sqrt{\frac{\epsilon_0(\gamma^2 - 1)}{2m_H}} + \sqrt{\frac{\epsilon_0(\gamma^2 - 1)}{2m_H} + c_1^2}. \quad (24.56)$$

This represents the critical speed of the fluid inflow (or of the front into the neutral gas). It is not difficult to show that (24.55) is satisfied in case 1 as long as $v_1 \leq v_D$.

Problem 24.13. Typically $\epsilon_0 \simeq 5 \times 10^{-12}$ ergs and $\gamma = 5/3$ for HI regions. Using representative values for the temperature in these regions, show that

$$c_1^2 \ll \frac{\epsilon_0(\gamma^2 - 1)}{2m_H}, \quad (24.57)$$

where $m_H = 1.67 \times 10^{-24}$ g. Use this fact to show that the critical speed v_D is approximately

$$v_D = c_1^2 \sqrt{\frac{m_H}{2\epsilon_0(\gamma^2 - 1)}}. \quad (24.58)$$

Assuming $v_1 = v_D$, show that the density ratio $\psi < 1$; that is,

$$\rho_2 < \rho_1 \quad (D\text{ condition}). \quad (24.59)$$

As is shown in Problem 24.13, when $v_1 \leq v_D$, the wave is a rarefaction wave with $\rho_2 < \rho_1$. This is known as the D condition ("Dense," since the matter into which the wave propagates is dense relative to the fluid behind the front). The mass-conservation condition, (24.37) may be used to show that $\rho_D v_D = \rho_1 v_1$ and, using $v_1 \leq v_D$, that

$$\rho_1 \geq \rho_D, \quad v_1 \leq v_D \quad (D\text{ condition}). \quad (24.60)$$

Case 1 corresponds to an ionization front moving with speed $v_1 \leq v_D$ into neutral gas of density $\rho_1 \geq \rho_D$.

Case 2: $v_1 > c_1$. In this case we take the negative root of (24.54), and obtain for the critical speed

$$v_R = \sqrt{\frac{\epsilon_0(\gamma^2 - 1)}{2m_H}} + \sqrt{\frac{\epsilon_0(\gamma^2 - 1)}{2m_H} + c_1^2}. \quad (24.61)$$

We can show that in general $v_1 \geq v_R$. Following the procedure used in the last problem, we can easily show that v_R is given approximately by

$$v_R \simeq \sqrt{\frac{2\epsilon_0(\gamma^2 - 1)}{m_H}}, \quad (24.62)$$

and that the density ratio $\psi > 1$, so that

$$\rho_2 > \rho_1, \quad v_1 \geq v_R \quad (R\text{ condition}). \quad (24.63)$$

The R condition corresponds to an ionization front moving supersonically into the relatively less-dense neutral gas, and is a compression wave. If the inequalities (24.60) or (24.63) are not satisfied, then the solutions of (24.52) for ψ will be complex, which is clearly not a physical possibility.

Problem 24.14. Show that $\rho_1 \leq \rho_R$.

A possible scenario for the formation and motion of an ionization front may be constructed from the preceding analysis. The photon flux at the ionization front and J fall off as r^{-2} (24.37), since m_H, R_*, and $\mathcal{F}_*$ are fixed (or change slowly compared to the time-scale governing the front's motion). Initially the front is near the star ($r \gtrsim R_*$), and the flux is high. The expanding front will thus move with a velocity $v_1 > v_R$ as a compression wave into the neutral gas. As the front expands, the flux decreases, as does the velocity v_1. Eventually v_1 decreases until $v_1 = v_R$, and the density has increased until $\rho_1 = \rho_R$. As the flux continues to drop, so does the velocity v_1. However, when $v_1 < v_R$, no real solutions ψ exist to (24.53). Mathematically this means that the boundary conditions across the ionization shock front can not be satisfied. Physically this means that the shock front and the ionization front are not in contact. In order for the system to continue moving into the neutral gas, the shock wave moves ahead of the ionization front, compressing the neutral gas. The critical speed and density at which the ionization and shock fronts separate are, for an ideal, monatomic gas,

$$v_R = \frac{4}{3}\left(\frac{2\epsilon_0}{m_H}\right)^{1/2},$$

$$\rho_R = \frac{3R_*^2 \mathcal{F}_*}{4r^2}\left(\frac{m_H^3}{2\epsilon_0}\right)^{1/2}. \quad (24.64)$$

Further expansion leads to continued reduction in v_1, until the D condition is reached, $v_1 = v_D$ and $\rho_1 = \rho_D$. When this happens, the solutions ψ of (24.53) are again real, and the boundary conditions may be satisfied. The critical speed at which this occurs for a monatomic gas ($\gamma = 5/3$) is

$$v_D = c_1^2 \, [m_H/2\epsilon_0(\gamma^2 - 1)]^{1/2}. \qquad (24.65)$$

When this point is reached, the ionization and shock fronts will again propagate together. For densities ρ_1 in the range

$$\rho_R < \rho_1 < \rho_D \qquad (24.66)$$

the two fronts will not be in contact. The final motion of the ionization front will be subsonic.

24.4. Accretion onto Compact Objects

The infall of matter onto compact objects (white dwarfs, neutron stars, and black holes) may lead to electromagnetic radiation, and possibly cosmic-ray production. Simply speaking, a significant fraction η of the gravitational potential energy of the infalling particle is converted to radiation:

$$\frac{dE}{dt} = \eta \frac{mGM}{R} \mathcal{F} = \eta \mathcal{F} mc^2 \, (r_g/R). \qquad (24.67)$$

Here $r_g = MG/c^2$ is the gravitational radius of the compact object, whose mass is M; R is the distance from the mass at which most of the infall energy is emitted, and $\mathcal{F}$ is the particle flux times the surface area. Typically $R \simeq 0.01 \, R_\odot$ for white dwarfs, $R \simeq 10^6$ cm for neutron stars, and $R \simeq 3 \, r_g$ for black holes. The efficiency factor η may be as large as 0.1, and the energy release per infalling baryon will be large (see Table 24.3); far more energy is released near neutron stars and black holes than can be realized even from nuclear reactions (maximum energy released per reaction $\lesssim 0.007 \, m_H c^2$).

Here the analysis of accretion processes is complicated by several factors. First, although the radiation emitted near the surface of all but white dwarfs has energies in the hard x-ray and gamma-ray range, this radiation is probably absorbed and re-emitted by the surrounding matter. Much of it may escape only after having been degraded to ultraviolet or even visible wavelengths. Therefore the final emergent spectrum of the object may depend critically on the density and temperature of the infall material.

Table 24.3
Infall energy per particle near compact objects.

Compact object	Energy per particle
White dwarf	0.4 MeV
Neutron star	280 MeV
Black hole	280 MeV

For neutron stars and black holes, the accretion process involves speeds that approach the speed of light, and light emission occurs near enough to the mass M that non-Newtonian aspects of the gravitational field must be taken into account. One must use a relativistic formulation of hydrodynamics, and treat the gravitational field relativistically whenever $2 \, MG/c^2 \approx R$, where R is the distance to the mass. Only when $R \gg 2 \, MG/c^2$ will the Newtonian approach developed earlier be strictly applicable.

Finally, a complication that must be overcome in either the nonrelativistic or the relativistic regime is the effect of outgoing radiation on the inflowing gas. As the rate of accretion increases, so does the amount of radiation leaving the central regions surrounding M. The radiation exerts an outward force on the infalling matter, tending to slow the accretion process. Roughly speaking, the maximum luminosity is expected to be the Eddington limit

$$L \simeq 1.3 \times 10^{38} \, (M/M_\odot) \text{ erg/sec.} \qquad (24.68)$$

In arriving at (24.68), we assume that Compton scattering is the principal source of opacity, since the surrounding matter is ionized completely. The electron-photon cross section is thus, to order of magnitude, $\sigma_e \simeq 7 \times 10^{-25}$ cm^2.

The infalling matter may be interstellar matter or supernova ejecta that do not escape the remnant, or may originate from a binary companion as the result of mass exchange or stellar winds.

In the following analysis we neglect relativistic effects and assume spherical symmetry for the compact mass M, as well as for the inflowing gas. The mass continuity and momentum equations for spherical symmetry are

$$\frac{\partial v}{\partial t} + v \frac{\partial v}{\partial r} + \frac{1}{\rho} \frac{\partial P}{\partial r} = -\frac{\partial \phi}{\partial r}, \qquad (24.69)$$

$$\frac{\partial \rho}{\partial t} + v\frac{\partial \rho}{\partial r} + \rho \frac{\partial v}{\partial r} + \frac{2\rho v}{r} = \frac{\partial \rho}{\partial t} + \nabla \cdot \rho \mathbf{v} = 0, \quad (24.70)$$

where we have denoted the radial velocity by v.

We now consider two approximations that are often valid in accretion problems, at least for periods of time that are not too long: (1) the inflow is stationary; (2) the inflow does not significantly alter the total mass of the compact object. The first condition is equivalent to the requirement

$$\frac{\partial \rho}{\partial t} = \frac{\partial v}{\partial t} = 0; \quad (24.71)$$

the second implies that M is constant. Thus we may write

$$\frac{\partial \phi}{\partial r} = \frac{MG}{r^2}, \quad (24.72)$$

where M is approximately constant; so (24.69) and (24.70) become independent of time. The mass-continuity equation (24.70) may be integrated immediately to give

$$\rho v r^2 = \text{constant}. \quad (24.73)$$

Finally we obtain the energy equation. Denoting the pressure, considered a function of the density ρ and entropy S, by $P = P(\rho, S)$, we can show that

$$dP = c_s^2 d\rho - \frac{\gamma - 1}{v}(\Gamma - \Lambda)\rho dr, \quad (24.74)$$

where the last term represents the heat loss of the infalling matter (if $\Gamma > \Lambda$), and Γ and Λ are as defined in Section 24.2. Equations (24.69), (24.70), and (24.74) completely determine the inflow behavior. We now use (24.74) to eliminate $\partial P/\partial r$ in (24.69), and then use (24.70) to eliminate $\partial \rho/\partial r$ to obtain the result

$$\frac{1}{v}\left(\frac{v^2}{c_s^2} - 1\right)\frac{dv}{dr}$$
$$= -\frac{MG}{r^2 c_s^2} + \frac{2}{r} - \frac{(\gamma - 1)(\Lambda - \Gamma)}{vc_s^2}, \quad (24.75)$$

where the speed of sound c_s is defined by $(\partial P/\partial \rho)_s^{1/2}$

and γ is the usual adiabatic index. Usually the analysis must be carried further by numerical means.

Adiabatic Inflow or Outflow

Equation (24.75) is applicable not only to inflow problems (accretion), but to outflow as well. If the gas motion is adiabatic ($\Gamma = \Lambda$), then we may find some simple approximate results indicative of the behavior in more general situations. Suppose that the first term on the right side of (24.75) dominates; that is, assume that the matter begins to move inward or outward from a distance $r \ll MG/2c_s^2$ from the center. This holds near large masses as long as the speed of sound is small compared to the speed of light. Then (24.75) becomes

$$\left(\frac{v^2}{c_s^2} - 1\right)\frac{dv}{dr} = -\frac{MGv}{r^2 c_s^2} < 0. \quad (24.76)$$

The right-hand side is always negative. Suppose the motion is supersonic ($v > c_s$). It then follows that dv/dr is negative, and that dv and dr have opposite signs. This means that supersonic mass inflow ($dr < 0$) is accelerated, since then $dv > 0$. Conversely, supersonic outflow ($dr > 0$) will be decelerated, because $dv < 0$. If the motion is initially subsonic ($v < c_s$), then $dv/dr > 0$, and dv and dr must have the same sign. In this case, subsonic inflow will be decelerated, and subsonic outflow will be accelerated.

For distances $r \gg MG/2c_s^2$ the second term on the right side of (24.75) will dominate (for $\Lambda = \Gamma$), and the preceding conclusions will be reversed. Far from the "critical" radius

$$r_s = MG/2c_s^2, \quad (24.77)$$

we have approximately

$$\left(\frac{v^2}{c_s^2} - 1\right)\frac{dv}{dr} = \frac{2v}{r} > 0. \quad (24.78)$$

Thus, for example, subsonic inflow ($v < c_s$ and $dr > 0$) will lead to $dv > 0$, and the matter will accelerate. At the critical radius r_s, the first two terms on the right of (24.78) cancel, and we have $dv/dr = 0$. Material initially falling inward subsonically will become supersonic beyond this point. Equation (24.77) may be written in terms of the speed of sound in the surrounding gas far from the compact mass M [note that c_s in

(24.77) is the speed of sound at r_s, which depends on the rate of increase of density or temperature of the accreting gas]. Noting that at large distances the gas is essentially at rest ($v_\infty = 0$), and denoting the speed of sound there by c_∞, we use Bernouilli's equation, in the form

$$\frac{v^2}{2} + \frac{c_s^2}{\gamma - 1} = \frac{c_\infty^2}{\gamma - 1} + \frac{MG}{r}, \qquad (24.79)$$

and set $r = r_s$, at which point $v = c_s$. Then we find from (24.78) and (24.79) that

$$c_s^2 = \frac{2}{5 - 3\gamma} c_\infty^2 \qquad (24.80)$$

and therefore that

$$r_s = \frac{5 - 3\gamma}{4} \frac{MG}{c_\infty^2}. \qquad (24.81)$$

Notice that this depends only on the properties of the gas far from the accreting mass M (in practice, $r_\infty \gtrsim 2r_s$ is usually distant enough to be considered "at infinity").

Finally, we observe that the density of the infalling gas remains nearly constant down to a radius r_{critical}, which is estimated to be about $10r_s$. Between r_{critical} and r_s the gas is compressed, and beyond r_s its motion is supersonic. Typically the temperature of the matter "far" from the mass M (at r_∞) will be $\sim 10^4$ K, and the matter, as observed earlier, will be ionized. Roughly speaking

$$\begin{aligned} r_s &\simeq 10^{12} - 10^{13} \text{ cm} \\ r_{\text{critical}} &\sim 10^{13} - 10^{14} \text{ cm} \end{aligned} \qquad (24.82)$$

for $M \approx M_\odot$. Notice that the value for r_s is at least $10 R_\odot$. Recalling that $R = 10^{-2} R_\odot$ for white dwarfs, and $R \simeq 10^{-4} R_\odot$ for neutron stars or black holes, we can expect the matter near compact objects to be in supersonic motion.

The rate of mass infall may be obtained from the expression (mass-continuity equation in spherical coordinates)

$$\frac{dM}{dt} = 4\pi r^2 \rho \frac{dr}{dt} = 4\pi r_s^2 \rho_s v_s, \qquad (24.83)$$

evaluated at the point r_s when the infall becomes supersonic. It is possible to rewrite this expression in terms of the density and temperature of the gas at r_{critical}, as

$$\frac{dM}{dt} = \alpha \rho c r_g^2 \left(\frac{mc^2}{kT}\right)^{3/2}. \qquad (24.84)$$

We denote $2MG/c^2 \equiv r_g$, and all quantities are evaluated at $r = 10 r_s$. The coefficient α ranges from about 0.3 for monatomic gases to 1.5 for polytropic index $n \to \infty$ ($\gamma \to 1$). It will be noted that dM/dt does not depend on the radius of the configuration.

Problem 24.15. Accretion rate for noninteracting particles. Matter at "large" distances from a compact mass M of radius R falls without collisions onto a central object. The initial speed of the particles is $v_\infty \neq 0$, and their impact parameter is denoted by s (see Figure 24.8). Note that angular momentum is conserved during infall, and that the particles will be captured only if their speed at the moment of impact is $v \leq (2MG/R)^{1/2}$, corresponding to the escape velocity. Show that the accretion rate is given by

$$\frac{dM}{dt} = \pi r_g^2 \rho c \left(\frac{c}{v_\infty}\right)\left(\frac{R}{r_g}\right). \qquad (24.85)$$

This follows if $\rho \equiv m n_\infty$, with m the particle mass, and n_∞ the number density at "large" distances from M. Taking $R = 3 r_g$, $r_g = 3$ km (corresponding to 1 $M_\odot$), $\rho = 2 \times 10^{-23}$ g/cm^3, and $v_\infty = 10^6$ cm/sec, find the accretion rate in solar masses per year. How does this compare with (24.84) under the same conditions? [Estimate T from $(1/2)mv^2 \approx 3/2 \, kT$.]

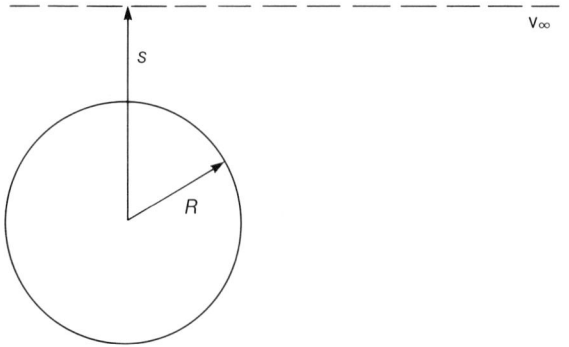

Figure 24.8. Impact parameter for accretion onto star of radius R.

The luminosity due to accretion (24.67) may be obtained by noting that $m\mathcal{F} = dM/dt$. Combining (24.67) and (24.84), we can show the luminosity to be

$$\frac{dE}{dt} = \eta \frac{n_c}{T_c^{3/2}} \left(\frac{M}{M_\odot}\right)^2 \left(\frac{r_g}{R}\right) \times 10^{38} \text{ ergs/sec.} \quad (24.86)$$

Both n_c and T_c are evaluated at r_{critical}, and η is the efficiency factor. Setting $\eta = 0.1$ and assuming $M = M_\odot$, $R = 3 r_g$, we find that the predicted luminosity due to accretion is

$$\frac{dE}{dt} = 3 \times 10^{37} \, n_c / T_c^{3/2} \text{ ergs/sec.} \quad (24.87)$$

Finding a value for T_c is difficult, since it depends on the rate of radiation caused by the infalling mass.

Chapter 25

DIFFUSE SUPERNOVA REMNANTS

A supernova explosion (Chapter 15) produces, in addition to a possible neutron-star or black-hole remnant, a rapidly expanding shell of hot gas. Initially the expanding shell contains the matter from the star (presumably the envelope and perhaps some of the mantle), ejected at velocities of 2 to 10×10^3 km/sec, and has a total mass of 0.1 to 0.2 $M_\odot$ for Type I supernovae, or more than several $M_\odot$ for Type II supernovae. Within a few years, the shell's expansion decelerates as more and more interstellar gas is swept up. Since typical remnants may persist for up to 10^5 years, much additional mass may have been swept up by the system. We can observe also bright filamentary regions whose light shows strong polarization effects associated with local magnetic fields. Even in relatively old remnants (e.g., the Veil nebula, about 70,000 years old), the leading edge of the bright regions is expanding at more than 100 km/sec. These remnants supply at least two additional pieces of data about the original explosion: when it occurred, which is important in estimating supernova rates; and the amount of energy released by the explosion. We consider the second of these now.

25.1. Expanding Nebulae

We begin the discussion of diffuse supernova remnants with a simple blast-wave model of an expanding nebula. We then discuss filamentary structure; the nonthermal radio component; and finally the structure of the Crab nebula.

We will suppose that the supernova in exploding releases an amount of energy E_T, and that the subsequent shock wave travels into the interstellar medium. Furthermore, assume that (1) the energy release is instantaneous; (2) the shock wave is spherically symmetric; and (3) the interstellar medium is uniform and homogeneous in the vicinity of the supernova. The first assumption is quite good, since the time-scale for core collapse and envelope ejection is typically a few tenths of a second. The second assumption is reasonable for nonrotating and nonmagnetic stars. Assumption (3) probably does not apply, at least within the disk of spiral galaxies. For example, the expansion of some remnants is highly irregular, suggesting that the supernova was located near dense clouds of dust and gas. Others, like the Crab nebula, show only slight signs of asphericity, in which case assumption (3) may not be too extreme.

The motion of gas on either side of the shock front is

described by the hydrodynamic equations discussed in Section 21.1, and the entropy conservation equation

$$\frac{d}{dt}\log(P/\rho^\gamma) = 0. \qquad (25.1)$$

We denote variables ahead of the shock in the unperturbed gas by subscript 0, and those behind it by subscript 1, and assume that the shock is strong; then the junction conditions yield (Section 24.2)

$$\rho_1 = \rho_0 \frac{\gamma+1}{\gamma-1},$$
$$P_1 = \frac{2}{\gamma+1} v_s^2 \rho_0, \qquad (25.2)$$
$$u_1 = \frac{2v_s}{\gamma+1}.$$

The gas on either side of the shock has the same γ, and v_s is the expansion velocity of the front (Figure 25.1); $v_s = dR/dt$. Recall that u_1 in (25.2) is the radial velocity of a fluid element in the gas behind the shock front. For a strong shock, $P_1 \gg P_0$, in which case the pressure ahead of the shock may be ignored in the following discussion. The hydrodynamic equations may be solved analytically under assumptions (1 to 3), but the analysis is lengthy; so we consider a simplified model in which all the mass interior to the shock is assumed to lie within a shell of thickness Δr and radius $R(t)$. Now apply mass conservation to a region of radius R before the explosion, and after the explosion when the shell reaches radius R; then it follows that

$$M = \frac{4\pi R^3}{3}\rho_0 = 4\pi R^2 \rho_1 \Delta r, \qquad (25.3)$$

where M is the total mass originally within a sphere of radius R. Using the junction conditions (25.2), we find that

$$\frac{\Delta r}{R} = \frac{1}{3}\frac{\gamma-1}{\gamma+1}. \qquad (25.4)$$

For an ideal gas with $\gamma = 5/3$, $\Delta r/R \simeq 0.08$; so we may assume that the shell is thin [in fact, we could take $\Delta r \to 0$ and $\rho_1 \to \infty$, while still maintaining fixed values of ρ_0 and R in (25.3)].

Next, apply Newton's second law to the shell. The force acting on it is given by the time-rate of change in the shell's momentum Mu_1. This must equal the pressure difference across the shell times the area. Denoting the pressure on the inside face of the shell by $P_c = \alpha P_1$, where α is constant, setting $P_0 = 0$, and noting that $\Delta r/R \ll 1$, we find that

$$\frac{d}{dt}Mu_1 = 4\pi R^2 P_c = 4\pi \alpha R^2 P_1. \qquad (25.5)$$

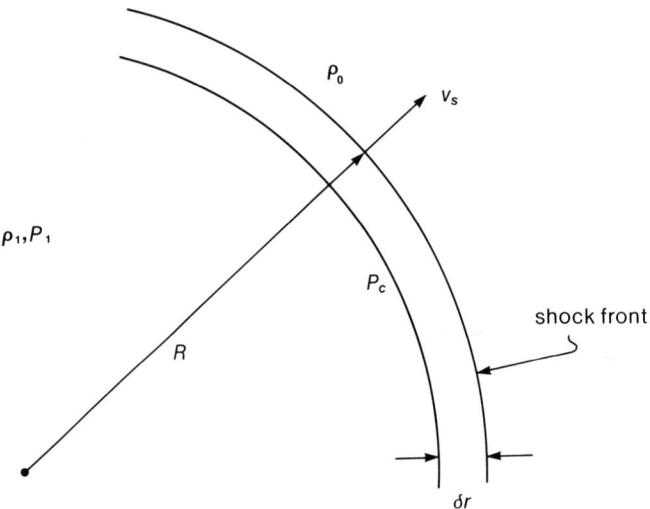

Figure 25.1. Expanding shock front. The gas pressure on the inside of the shell of thickness Δr is P_c.

Problem 25.1. Use the junction conditions, and the relation

$$v_s = dR/dt \qquad (25.6)$$

to show that the solution to (25.5) is given by

$$v_s = AR^{3(\alpha - 1)}, \qquad (25.7)$$

where A and α are constants.

The constants in (25.7) may be fixed by applying energy conservation to the system. The shell's kinetic energy is

$$E_k = \frac{1}{2} M u_1^2 \qquad (25.8)$$

and the internal energy within the shell is

$$U_c = \frac{4\pi R^3}{3} \frac{P_c}{\gamma - 1} = \frac{4\pi \alpha}{3} \frac{R^3 P_1}{\gamma - 1}. \qquad (25.9)$$

The junction conditions may then be used to show that the total energy of the nebula is

$$E = U_s + E_k$$
$$= \frac{4\pi}{3} \rho_0 A^2 \left[\frac{2\alpha}{\gamma^2 - 1} + \frac{2}{(\gamma + 1)^2} \right] R^{3(2\alpha - 1)}. \qquad (25.10)$$

The total energy E is a constant, and must equal the energy of the explosion E_T; so the exponent of $R(t)$ must vanish. It readily follows that $\alpha = \frac{1}{2}$ and

$$A = \left(\frac{E_T}{\rho_0} \right)^{1/2} \left[\frac{3(\gamma - 1)(\gamma + 1)^2}{4\pi(3\gamma - 1)} \right]^{1/2}$$
$$\equiv \frac{2}{5} \left(\frac{E_T}{\rho_0} \right)^{1/2} \xi_0^{5/2}. \qquad (25.11)$$

The last step defines the constant ξ_0. Finally, integration of (25.6) and (25.7) immediately yields

$$R(t) = \xi_0 (E_T/\rho_0)^{1/5} t^{2/5}, \qquad (25.12)$$

where the initial radius $R(0)$ (of order of the stellar radius) may be ignored. This is justified because the analysis above does not apply during the early stages when $R(t) \approx R(0)$. For $\gamma = 5/3$, the constant $\xi_0 =$ 1.062. It follows from (25.2) and (25.7) that

$$v_s \sim R^{-3/2}, \qquad P_1 \sim R^{-3}. \qquad (25.13)$$

The solution (25.12) depends on only two parameters; the total energy released by the explosion, E_T, and the initial density ρ_0 of the interstellar medium. To the extent that ρ_0 is known in the vicinity of a supernova remnant, the energy release E_T may be obtained from the nebula's radius, and its expansion velocity v_s. Furthermore, as shown by Problem 25.2, the age of the remnant may also be calculated. For example, the Veil nebula in Cygnus is observed to have bright filaments that form a nearly spherical shell of radius $R \approx 20$ pc, and are expanding at about 115 km/sec.

Problem 25.2. Show that the radius and expansion rate of a supernova remnant are related by

$$R = \frac{5}{2} v_s t. \qquad (25.14)$$

Suggest specific observations that would yield absolute values of R and v_s.

Problem 25.3. Show that the radius of a remnant, and the temperature within it are given by

$$R(t) = 6.1 \times 10^{14} \left(\frac{E_T}{E_0 n_0} \right)^{1/5} t^{2/5} \text{ cm}, \qquad (25.15)$$

$$T(t) = 6.5 \times 10^{20} \left(\frac{E_T}{E_0 n_0} \right)^{2/5} t^{-6/5} \text{ K}, \qquad (25.16)$$

where $E_0 = 10^{50}$ ergs, $n_0 = \rho_0/m_H$, and the time is in seconds.

According to (25.14) the nebula is about 70,000 years old. Assuming an interstellar mass density $\rho_0 \simeq 10^{-24}$ g/cm³, (25.12) may be differentiated, giving

$$v_s(t) = \frac{2}{5} \xi_0 (E_T/\rho_0)^{1/5} t^{-3/5}. \qquad (25.17)$$

Solving for E_T and using the observed expansion rate and derived age yields, for the energy release by the supernova, $E_T \simeq 10^{50}$ ergs, in agreement with theoretical models. Finally, using (25.16), we find the temperature within the nebula to be about 10^6 K, corresponding to thermal energies in the x-ray region. It is no surprise, then, that the Veil nebula is a source of soft x rays. During the expansion, the shock front has swept up more than 500 $M_\odot$ of interstellar gas and dust.

Problem 25.4. The preceding analysis also applies to expanding nebulae due to novae. Assume a typical expansion rate of 20 km/sec and a nebula radius of 1 pc, and find the age, thermal energy input by the nova, and the gas temperature within the nebula. In what region of the spectrum would the corresponding black-body spectrum fall?

Although this simple model does not give the exact radial dependence of quantities inside the expanding shell, it does reproduce the correct order of magnitude for such things as E_T, R, and the remnant's age. Figure 25.2 shows the radial behavior (schematically) of the exact solutions. Note the rapid drop in density and pressure behind the front.

The exact solution $R(t)$ is in fact (25.12) with a slightly different numerical constant ξ_0.

The expansion described in the preceding is essentially adiabatic, since very little radiation escapes from the region inside the shell. Late in the expansion, however, the radiation mean free path becomes large, and the system begins to cool rapidly. When this happens, the total energy within the nebula is no longer conserved. The momentum is, however; so the subsequent evolution is governed by the requirement that $Mu_1 \approx R^3 v_S$ = constant. This is immediately integrated to give

$$R(t) \sim t^{1/4}, \qquad (25.18)$$

which is more gradual than (25.12). Once this stage of the expansion is reached, the gas inside the front (which becomes a compression wave rather than a shock wave) cools rapidly.

25.2. FILAMENTARY STRUCTURE

A distinctive feature in supernova remnants is the intricate network of filaments bordering the expanding edge of the nebula, typical examples of which are seen in photographs of the Veil and Crab nebulae. Bright emission lines of OII, OIII, NII, and SII originate from these regions, as do fainter lines of H. The spectra are similar to the spectra of planetary nebulae. In fact, as in planetary nebulae, most (up to 90 percent) of the optical radiation arises from these emission lines. Furthermore, because of the wide range in ionization potential of these ions (see Table 6.1), it appears unlikely that black-body radiation is the source of the excitation.

The filaments probably arise from regions of greater than average density in the interstellar medium, which are further compressed by the expanding shock wave. Since they are denser than the surrounding gas, they cool more rapidly. In fact, temperatures of several times 10^4 K are typical within the filaments, nearly two orders of magnitude lower than the temperatures within the diffuse component of the nebula.

The structure of the filaments varies irregularly within periods of several years. Even in the oldest remnants, portions of filaments fade, reappear, and combine with others in an irregular fashion. They are also observed to be strong radio sources, with spectra

$$F_\nu \sim \nu^{-\alpha}, \qquad (25.19)$$

where F_ν is the observed radio flux, usually given in watts cm^{-2} Hz^{-1}. The spectral index α is a constant,

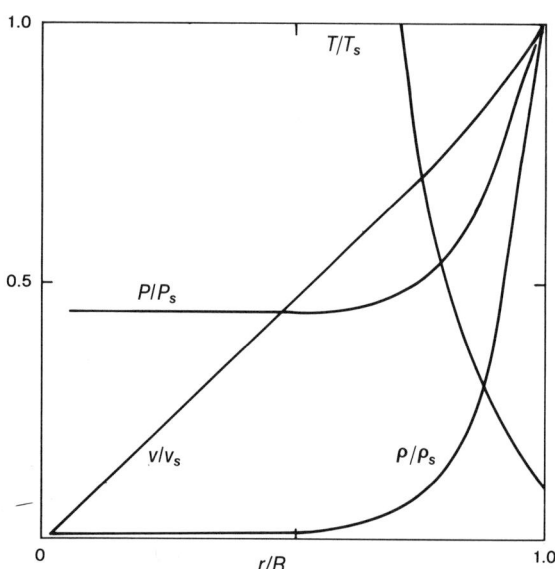

Figure 25.2. Temperature, pressure, density and gas velocity behind a spherically expanding shock front set up by instantaneous energy release at $r = 0$. Subscript s denotes value at the front.

and lies generally between 0.5 and 1 for supernova remnants; typically, $\alpha \simeq 0.8$. A spectrum of this type can not be produced over an extended frequency range by black-body radiation, but it can result from synchrotron radiation. As we will see later, this radiation requires magnetic fields and ultrarelativistic charged-particle motions (electrons). In fact, observations are consistent with the existence of magnetic fields threading through the filaments and having magnitudes as high as 10^{-3} gauss.

25.3. NONTHERMAL RADIO COMPONENT

The radio sources discussed in Section 20.7 were thermal in nature. Clearly, any hot plasma will emit some radiation at radio wavelengths. Thermal emission, however, makes a negligible contribution to the radio spectrum of supernova remnants. The brightness temperatures at meter wavelengths of typical remnants are $T_b \sim 10^8$ K or more, but spectral-line emission is consistent with electron temperatures T_e of order 10^4 K (which also fits the observed thermal x-ray spectrum). Furthermore, the thermal radio intensity due to black-body radiation at T_e is much less than the observed intensities. Therefore some mechanism other than thermal emission must be present. A typical radio spectrum is shown in Figure 25.3, which also shows the spectrum from the Crab nebula. Comparison with a typical thermal spectrum shows that these are decidedly nonthermal sources.

The best known source of nonthermal radiation likely to apply under conditions typical in diffuse plasmas, such as supernova remnants, is synchrotron radiation, which requires a supply of charged particles moving in magnetic fields. The ionized gas in magnetic filaments provides such an environment.

Calculating the synchrotron spectrum and energy loss is complicated by the fact that the electrons are highly relativistic, moving with speeds $v \approx c$. However, the following simplified analysis reveals the crucial features that show up in most cosmic radio sources, including radio emission from galaxies and quasars. We consider the motion of an electron with velocity v, moving in a magnetic field **B**. The arguments will be nonrelativistic, but we will note the relativistic corrections obtained from more nearly complete treatments. The motion of the electron is shown in Figure 25.4. The magnetic force on it is given by

$$\mathbf{F} = \frac{e\mathbf{v} \times \mathbf{B}}{c}. \qquad (25.20)$$

F is perpendicular to **B**; so the component of the electron's velocity along the field remains constant, but its motion in the plane perpendicular to **B** is circular. Equating the magnetic acceleration, F/m_e, to the centripetal acceleration gives

$$\frac{v^2}{r} = \frac{eB_\perp v}{m_e c}, \qquad (25.21)$$

where $B_\perp$ is the component of **B** normal to **v**. The frequency of the motion is $\nu = v/2\pi r$, which may be substituted into (25.21), yielding $\nu = eB_\perp/2\pi m_e c$. This would be the frequency of radiation from an electron moving nonrelativistically, and it would be radiated essentially isotropically. When $v \approx c$, three primary changes must be made. First, the radiation is concentrated in the forward direction, lying almost entirely within a narrow cone of angular half-width $\Delta\theta \sim (1 - v^2/c^2)^{1/2} = (m_e c^2/E_e)$, where the electron's energy is $E_e = (p_e^2 + m_e^2 c^2)^{1/2} c$ (Figure 25.4, a). It is therefore highly directional radiation. Second, the radiation is linearly polarized at right angles to **B**. Observations of supernova remnants indicate that polarizations as high as 10 percent are typical in the radio emission from the filaments. Finally, the spiraling electron will radiate at the fundamental frequency

$$\nu_m = \frac{eB_\perp}{2\pi m_e c}\left(\frac{E_e}{m_e c^2}\right)^2 \qquad (25.22)$$

and at harmonics of ν_m. The spectral distribution [probability that an electron of energy E_e will emit a photon of frequency ν, $p(\nu/\nu_m)$], is shown in Figure 25.5.

The power radiated by a nonrelativistic electron is given by (6.43); with the acceleration F/m_e from (25.20), this becomes

$$-\frac{dE_e}{dt} = \frac{2}{3}\frac{e^4 B_\perp^2}{m_e^2 c^3}\left(\frac{E_e}{m_e c^2}\right)^2$$

$$= 1.58 \times 10^{-15} B_\perp^2 \left(\frac{E_e}{m_e c^2}\right)^2, \qquad (25.23)$$

where the last factor is a relativistic correction similar to the last one in (25.22). Actually (25.23) is correct only for $E_e \gg m_e c^2$, and for photon energies $h\nu \ll E_e$. The observed cosmic-ray energy spectrum is accurately described by the power law

$$N(E_e)\,dE_e = KE_e^{-\gamma}\,dE_e, \qquad (25.24)$$

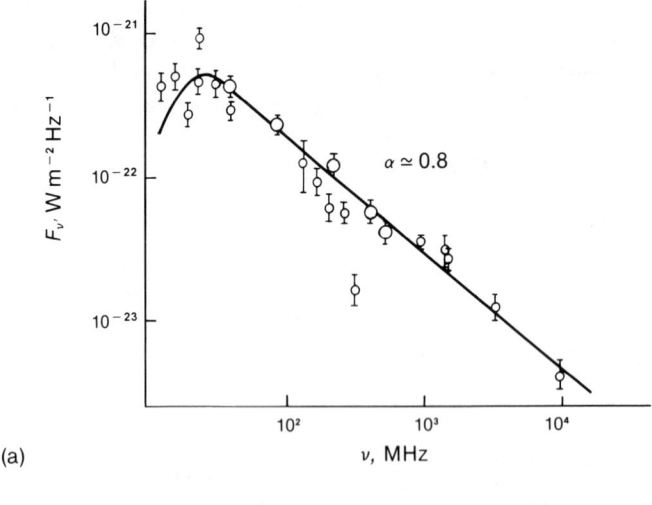

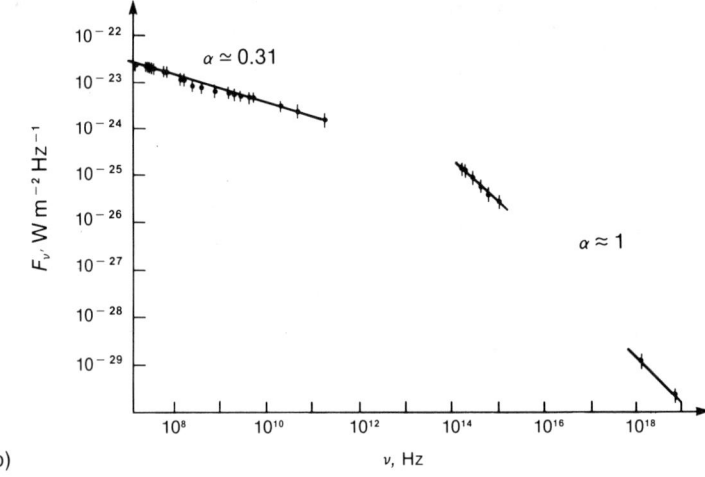

Figure 25.3. Radio spectra of two supernova remnants: (a) the Cassiopeia A nebula; and (b) the synchrotron spectrum of the Crab nebula. The spectral index and the power law α satisfy $\alpha = (\gamma - 1)/2$.

with K and γ constants. $N(E_e)$ is the electron number density per unit energy. High-energy electrons in supernova remnants are one source of cosmic rays; so it is reasonable to assume that their distribution is also of the form (25.24). The radio intensity from an extended source of thickness l along the line of sight is given by the integral of (25.23) and (25.24) over the spectrum $p(\nu/\nu_m)$ shown in Figure 25.5. Finally it must be integrated over the source region. We will approximate the spectrum by assuming that all the energy is radiated at frequency $\nu = 0.5\,\nu_m$ where the spectrum is a maximum. Thus, if the electrons are uniformly distributed in volume along the line of sight, and the gas is optically thin at radio frequencies, we have

$$I_\nu = -\frac{1}{4\pi} \int_0^l dr \int_0^\infty \frac{dE_e}{dt}$$
$$\times \delta(\nu - \nu_m) K E_e^{-\gamma} dE_e$$
$$\simeq -\frac{Kl}{4\pi} \int_0^\infty \delta(\nu - \nu_m) E_e^{-\gamma} \left(\frac{2}{3}\right)^2$$
$$\times \frac{e^4 B^2}{m_e^2 c^3} \left(\frac{E_e}{m_e c^2}\right)^2 dE_e. \qquad (25.25)$$

456 / DIFFUSE SUPERNOVA REMNANTS

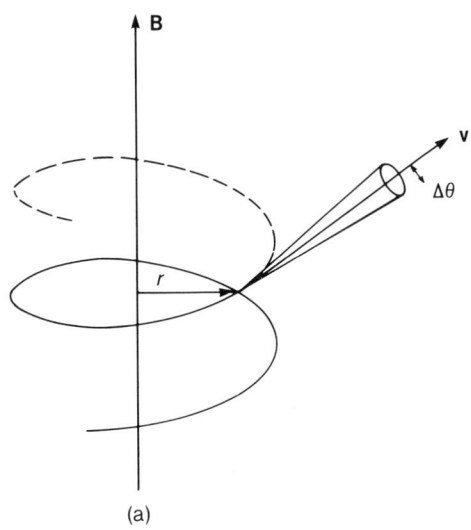
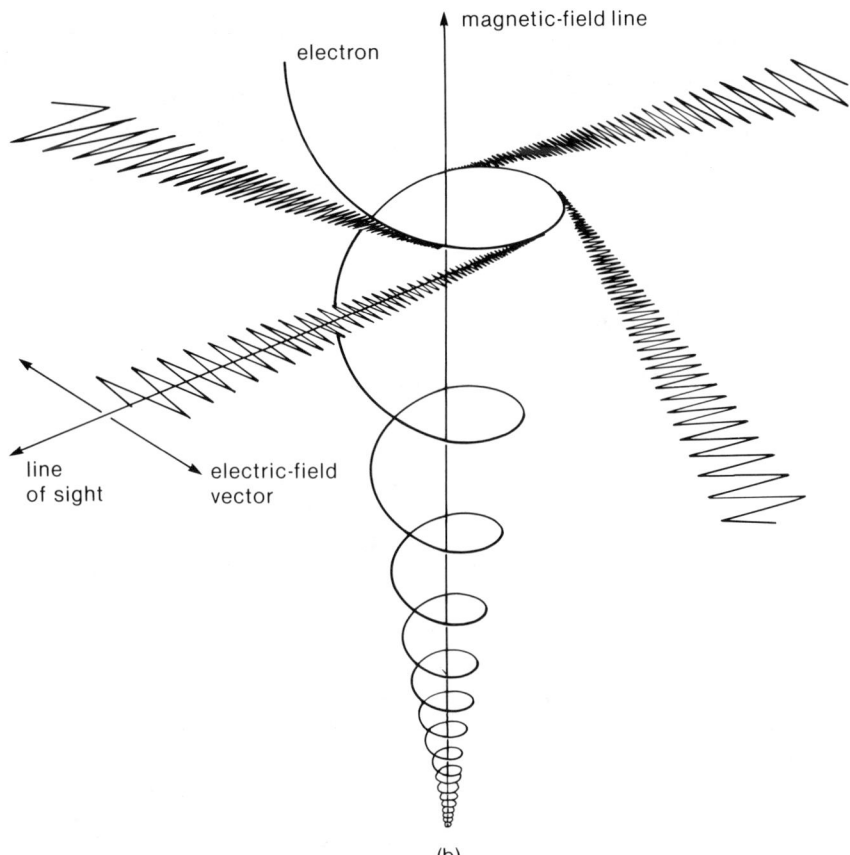

Figure 25.4. Synchrotron radiation from the electron accelerated by magnetic field: (a) angular cone containing most of the radiation; (b) polarization plane of electric field.

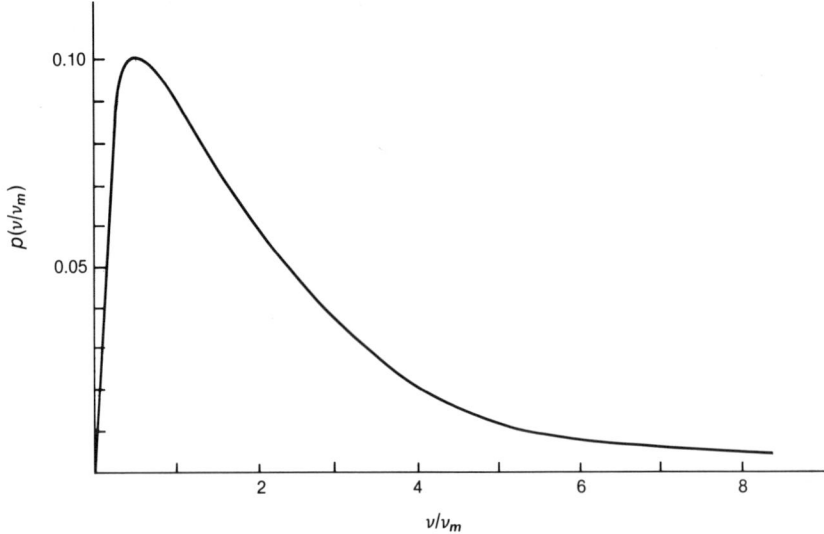

Figure 25.5. Power spectrum from synchrotron radiation.

To arrive at the last form we assume that **B** is uniform throughout the source, and take as its average magnitude normal to the electron velocities (randomly distributed) $B_\perp = (2/3)^{1/2} B$.

Problem 25.5. Assume that the velocities **v** of an ensemble of electrons are randomly oriented in space, and show that $B_\perp^2 = (2/3) B^2$, where $B_\perp^2$ is the average over the ensemble.

Noting that E_e is related to ν_m and $B_\perp$ by (25.22), the radio intensity is

$$I_\nu \sim Kl\, B_\perp^{(\gamma+1)/2}\, \nu^{-(\gamma-1)/2}. \qquad (25.26)$$

Defining the spectral index by $\alpha = (\gamma - 1)/2$, we see that the observed flux is

$$F_\nu \simeq \left(\frac{l}{d}\right)^2 \int_{\Delta\Omega} I_\nu\, d\Omega, \qquad (25.27)$$

where the source distance is d, and the integral is over the solid angle $\Delta\Omega$ subtended by the source at the observer. It is evident from (25.19) that $F_\nu \sim \nu^{-\alpha}$. For $\alpha = 0.8$, we have $\gamma \simeq 2.6$.

Problem 25.6. Carry out the derivation of the radio intensity (25.26).

If the correct spectral distribution is used, it is found that

$$I_\nu = \frac{e^3}{m_e c^2}\left(\frac{3e}{4\pi m_e^3 c^5}\right)^{(\gamma-1)/2}$$
$$\times Kl\, u(\gamma) B^{(\gamma+1)/2}\, \nu^{-(\gamma-1)/2}, \qquad (25.28)$$

where $u(\gamma)$ is a numerical factor of order unity.

The preceding analysis suggests that the source of radio emission in supernova remnants is synchrotron radiation from ultrarelativistic electrons spiraling around magnetic lines of force in the filaments. This mechanism explains in a natural way the power-law dependence (25.19) and the strong linear polarization of the radio waves. Protons may also contribute to the radio spectrum; however, because of their large masses relative to m_e, the corresponding intensity will be down by a factor of about 10^{-11}.

At low frequency the plasma begins to absorb radiation strongly, and the spectrum deviates from the linear form (25.19). This process, synchrotron self-absorption, is explained as follows. Over a small frequency band, the nonthermal electron distribution may be approximated by a Maxwellian distribution whose temperature is given by

$$3kT_e \simeq E_e \qquad (25.29)$$

where E_e is given by (25.22) at frequency $\nu = \nu_m$. The maximum intensity at temperature T is that of a black body, which for long wavelengths is

$$I_\nu \simeq \frac{2kT_b\nu^2}{c^2} = \left(\frac{8\pi m_e^3 c}{eB_\perp}\right)^{1/2} \nu^{5/2}. \qquad (25.30)$$

The last step uses (25.29) and (25.22). The final result does not correspond to a black body, since the temperature T_b is frequency-dependent. Evidence of the low-frequency cutoff ν_1 in F_ν is seen in the spectrum of Cassiopeia A (Figure 25.3). Measurements at low frequency of radio sources that show synchrotron self-absorption may be used to calculate the magnetic field strength from (25.30). Similar effects may also be observed in the radio spectrum of some extragalactic sources.

Problem 25.7. For Cassiopeia A, $l \simeq 2$ pc and its distance $d \simeq 3$ kpc. From Figure 25.3, $F_\nu \approx 10^{-22}$ watts m^{-2}Hz^{-1} at $\nu = 10^7$ Hz, where the source is thick. Estimate the magnetic field in the filaments from these data.

We may estimate the lifetime of the electrons that give rise to the radio spectrum from (25.22). Assuming a typical radio wavelength $\lambda \sim 3$ m $= c/\nu_m$, and $B = 2 \times 10^{-5}$ gauss, we find $E_e = 10^9$ eV. The rate at which an electron radiating in a magnetic field loses energy can be estimated from the energy-loss rate (25.23):

$$\tau \simeq -\frac{E_e}{dE_e/dt} \sim \frac{3m_e^4 c^7}{2E_e B_\perp^2 e^4}. \qquad (25.31)$$

This shows that 10^9 eV electrons in a field as strong as 10^{-4} gauss will radiate for about 10^7 years. Furthermore, a typical high-energy electron moving at nearly the speed of light will have traveled about 3×10^3 kpc in its spiraling motion through the filaments. This demonstrates how efficiently cosmic magnetic fields confine electrons in nebulae.

It also explains the high-frequency behavior of some sources, such as the Crab nebula (Figure 25.3).

It follows from (25.31) that the synchrotron lifetime of an electron of energy E_e is proportional to E_e^{-1} for a given source. Therefore, if the initial high-energy electron population in the remnant is not replenished, it must gradually change with time in such a way that the higher-energy states become depopulated first. In fact, the fractional change in E_e is $\Delta E_e/E_e \sim \Delta t/\tau$. Thus when $\Delta t \sim \tau$, states with E_e greater than E_2 will no longer contribute to the radiation to any significant degree. This sets a high-frequency cutoff ν_2 for the spectrum, which decreases with time.

Problem 25.8. Show that the high-frequency cutoff ν_2 is given approximately by

$$\nu_2 \approx \frac{m_e^5 c^9}{\pi e^7 B^3 \Delta t^2}, \qquad (25.32)$$

where Δt is the time elapsed between observation and electron ejection into the remnant.

The magnetic field and the distribution of relativistic electrons are the primary factors determining the structure of synchrotron sources. If the magnitude of the magnetic field is known, then the electron distribution can be calculated from the observed flux, the source dimension l, and the distance d. Unfortunately, independent measurements of B are seldom available for supernova remnants. One way around this difficulty is to relate the magnetic energy W_B in the remnant to the total energy in relativistic charged-particle motion W_e. The magnetic energy $W_B \approx B^2 V/8\pi$, where V is the volume of the remnant ($V = 4\pi l^3/3$). The energy in relativistic electrons is given by

$$W_e = V \int_{E_1}^{E_2} N(E_e) E_e \, dE_e$$
$$= V \int_{E_1}^{E_2} K E_e^{1-\gamma} \, dE_e, \qquad (25.33)$$

where the integration limits E_2 and E_1 define the interval over which the distribution (25.24) applies. For example, these may be taken as the high- and low-frequency cutoffs discussed earlier. The magnetic field is coupled to the plasma through the distribution of all relativistic charged particles (primarily electrons and protons). Thus, defining the total energy in relativistic charged particles by W_c, we might expect the energy

$$W = W_c + W_B \qquad (25.34)$$

to be a minimum. Although this is reasonable, it is not physically necessary. We do expect $W_c \lesssim W_B$, however, or the magnetic field would not be able to confine the charged particles for extended periods of time. Since the synchrotron spectrum is dominated by electrons [discussion following (25.28)], we must relate W_c to W_e, since only the latter is known observationally (Problem 25.9).

Problem 25.9. Show that, if $E_1 \gg E_2$ and $\gamma > 1$, the energy in relativistic electrons

$$W_e \sim \ell^2 \, I_\nu \, B^{-3/2}, \qquad (25.35)$$

Where ℓ is the source dimension and I_ν is the radio intensity. Use this to show that W is a minimum when $W_e \approx W_c$, where $W_e = kW_c$.

Unfortunately, there is no clear way of deriving this; so conventionally we take

$$W_e = kW_c. \qquad (25.36)$$

Fortunately, the results are relatively insensitive to k. Therefore, we assume

$$W_B = W_e/k. \qquad (25.37)$$

Problem 25.10. Use the results (25.35) to (25.37) to show that the magnetic field in the filaments of a supernova remnant is

$$B \sim \left(\frac{F_\nu}{k\, d\, \theta^3}\right)^{2/7}, \qquad (25.38)$$

where $\theta = \ell/d$ is the remnant's angular size.

As is shown in Problem 25.10, B may now be calculated from the observed flux F_ν, the object's distance, and its angular size. Furthermore, the exact value of B

Table 25.1
Charged particle energy in supernova remnants; k is the ratio of energy in relativistic electrons to the total energy in charged particles.

	$k = 1$	$k = 0.01$
Cassiopeia A:		
B (gauss)	2.5×10^{-4}	6.8×10^{-4}
W_c (ergs)	2×10^{48}	1.4×10^{49}
Veil nebula:		
B (gauss)	2×10^{-5}	5×10^{-5}
W_c (ergs)	2.5×10^{48}	1.7×10^{49}

depends only weakly on k, varying by only a factor of three for $10^{-3} \leq k \leq 10^{-1}$. We then invert (25.28) and obtain the constant k, giving the relativistic electron distribution. Proceeding in this way, from observations of the remnant Cassiopeia A and the Veil nebula we can derive the results in Table 25.1.

The evolution of nonthermal radio emission is partly due to the reduction in the number of high-energy electrons. However, the electrons that contribute the most are those whose lifetimes are of order 10^5 to 10^7 years, and significant changes in the total radio luminosity occur annually. For example, the observed rate $(dF_\nu/dt)F^{-1}$ for Cassiopeia A is about 0.012 yrs^{-1}, giving a time-scale orders of magnitude less than (25.31). In order to explain this rate of change, we must consider the magnetohydrodynamic behavior of the nebula.

The filamentary regions surrounding the nebula contains magnetic fields, charged particles, and gas. The charged particles (plasma) are effectively tied to magnetic lines of force, and collisions between electrons and neutral particles tend to keep the latter moving with the plasma. Furthermore, the plasma has a high conductivity; so the magnetic flux (23.92) will be conserved as the matter evolves dynamically. Therefore, as the nebula expands, the field will decrease:

$$B \sim l^{-2}. \qquad (25.39)$$

Because the electrons are strongly coupled to the field, their energy E_e will also decrease because of expansion, as shown in Problem 25.11.

Problem 25.11. Suppose the electron energy is $U = N_e \bar{E}_e$, where N_e is the number of electrons and $\bar{E}_e$

(a)

(b)

Figure 25.6. Crab nebula: (a) H_α (red light), showing volume emission from hot gas inside remnant shell; (b) photograph taken in blue light shows no filaments.

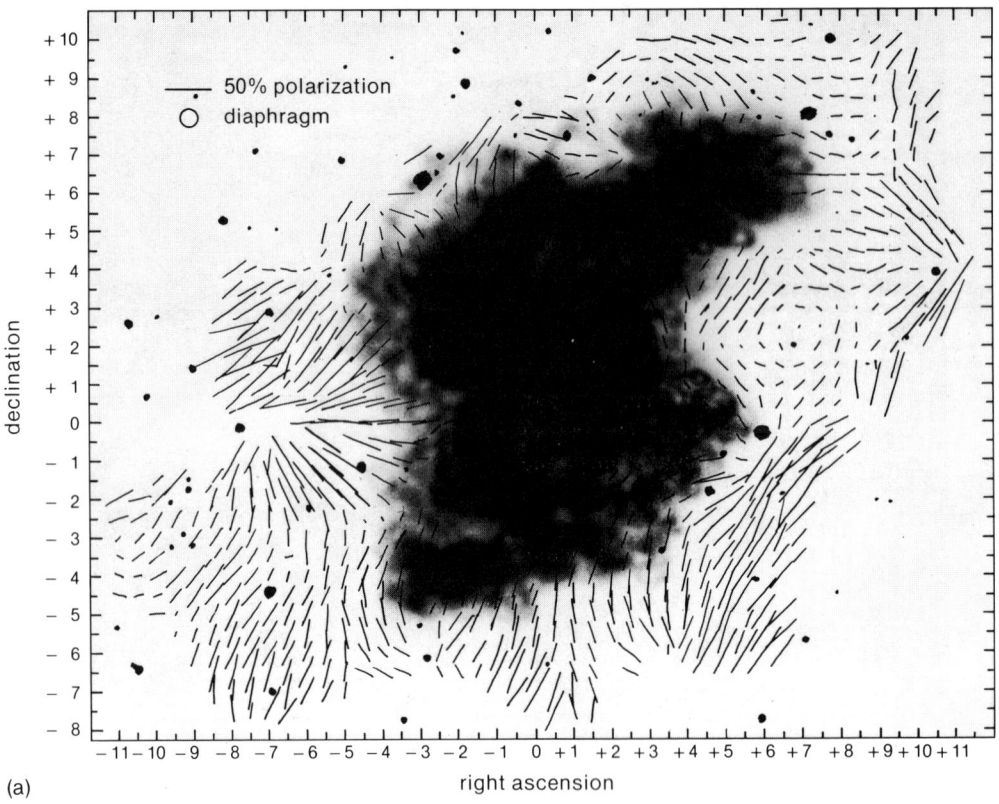

(a)

Figure 25.7. Optical and radio polarization of the Crab nebula: (a) optical polarization superimposed on image of the nebula taken in blue light; (b) radio polarization at 11 cm, in which the bar lengths are proportional to the polarized intensity.

an average energy. If the change in U is due to adiabatic expansion against the magnetic pressure $P_B \approx B^2/8\pi$, show from $dU = -P\,dV$ that

$$E_e \sim \ell^{-1}. \qquad (25.40)$$

Finally, we integrate (25.24) over energy, and note that $n_e \approx N_e/l^3$ to show that

$$K \sim l^{-2-\gamma}. \qquad (25.41)$$

Substituting (25.39) to (25.41) into the expression for the radio intensity (25.26) we find that I_ν decreases with increasing radius of the supernova remnant:

$$I_\nu \sim l^{-4(\alpha+1)}. \qquad (25.42)$$

Similarly, the observed flux

$$F_\nu \sim (l/d)^2 I_\nu \sim l^{-2(2\alpha+1)}. \qquad (25.43)$$

Clearly, the luminosity will also decrease with expansion, as does F_ν. Therefore we may expect $F_\nu \sim t^{-2(2\alpha+1)}$ during the early stages in young remnants, when their radius is proportional to their age. For Cassiopeia A, we find

$$\frac{1}{F_\nu}\left|\frac{dF_\nu}{dt}\right| = \frac{2(2\alpha+1)}{T} = 0.017 \text{ yr}^{-1},$$

assuming $\alpha = 0.8$ and an age $T = 300$ yrs. This exceeds the observed rate by about 25 percent. The error results because the nebula has begun to decelerate, so that its radius is not exactly linear in time, but increases at a slower rate.

462 / DIFFUSE SUPERNOVA REMNANTS

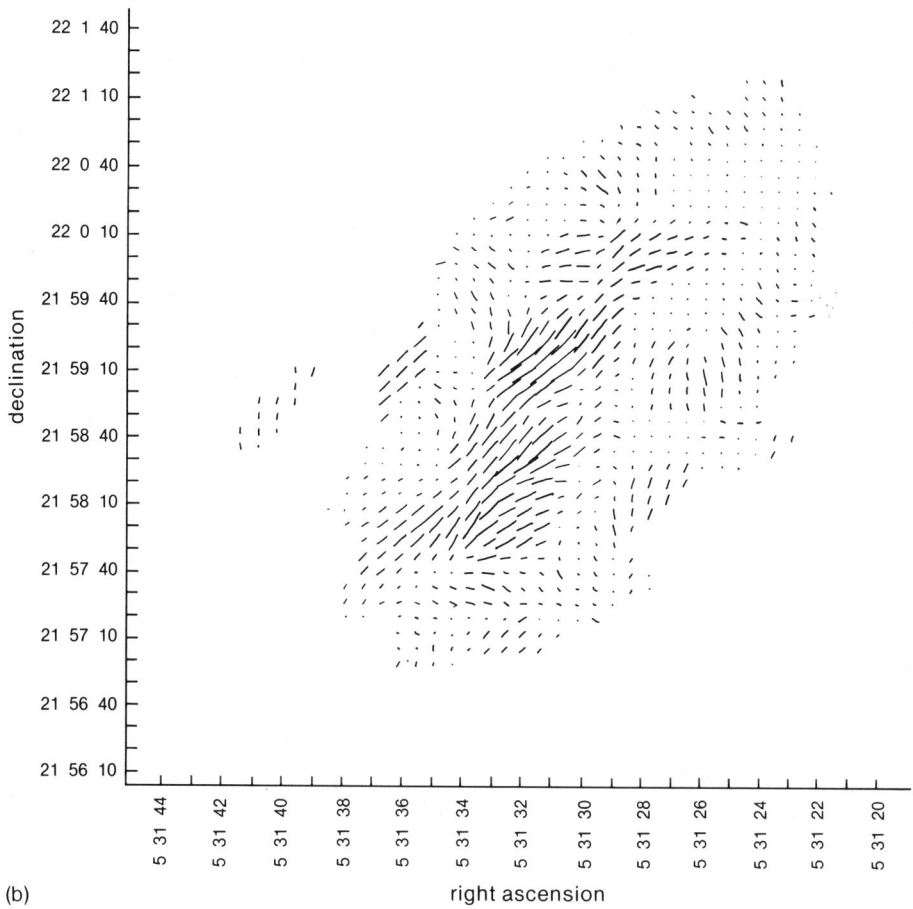

(b)

25.4. THE CRAB NEBULA

In many respects the Crab nebula appears to be unique among supernova remnants. It coincides with the site of the supernova observed in 1056, which was observed for nearly a year after. Historical data suggest that its light curve was Type I, and that its absolute magnitude at maximum was $M_v \simeq -18$. The nebula is observed to emit synchrotron radiation whose spectrum ranges from radio frequencies to very hard γ rays. It also contains near its center the youngest known pulsar.

The Crab nebula is not spherical, but appears ellipsoidal. As seen projected onto the plane of the sky, its semimajor and semiminor axes are about 4′ and 2′, respectively, and its average apparent rate of expansion is $0''.23$ yr^{-1}. The radiation emitted by the nebula consists of two primary components. When photographed in H_α, the filamentary structure characteristic of many supernova remnants is observed (Figure 25.6, a). Photographs taken in blue light, using filters designed to exclude line-emission features, show a smooth and generally structureless component, whose radiation originates from the entire volume of the nebula (Figure 25.6, b). In most remnants, as in planetary nebulae, line emission accounts for 80 to 90 percent of the brightness. However, in the Crab nebula most (90 percent) arises from within the volume of the nebula, with only about 10 percent due to line emission in the filaments. Furthermore, irregular fluctuations in brightness during periods of several years are observed in the filaments, a characteristic not typical of other remnants or gaseous nebulae.

The line spectra of the Crab nebula are typical of other nebulae, showing strong emission from O, N, and S, and fainter H_α emission. The Doppler shift observed in these lines indicates that the filaments are expanding at about 1,200 km/sec. Relative line strengths imply temperatures in the filaments of 20,000 K or more. Finally, the observed abundances of most elements, with the possible exception of He, appear to be normal. There is some evidence, however, that the relative abundance of (He/H) in the nebula may be three times its value in the interstellar medium.

The electromagnetic radiation from the nebula extends from meter-wavelength radio emission down to hard gamma rays at frequencies in excess of 10^{22} Hz. The spectrum (Figure 25.7) corresponds to synchrotron radiation with $\gamma = 1.6$ for $\nu \leq 10^{12}$ to 10^{14} Hz, and $\gamma \simeq 1$ at higher frequencies. The gaps in the spectrum in the ultraviolet ($\nu \approx 10^{16}$ Hz) and the infrared ($\nu \approx 10^{13}$ Hz) result from atmospheric absorption. The abrupt change in slope of the power spectrum near $\nu \simeq 10^{13}$ Hz is intrinsic to the nebula, and will be discussed later. The total synchrotron energy-loss rate from the nebula is estimated to be

$$\left(\frac{dE}{dt}\right)_{\text{synchrotron}} \simeq 1.2 \times 10^{38} \text{ergs}, \quad (25.44)$$

assuming that its distance is about 2 kpc (Problem 25.12).

The polarization of light from the nebula indicates that the filaments are permeated by fairly uniform magnetic fields that run along the filaments (Figure 25.7); analyses similar to those discussed in Section 25.3 indicate field strengths of 10^{-4} to 10^{-3} gauss.

The expansion of the Crab nebula is more complicated than that of the older supernova remnants discussed in Section 25.1, which are essentially expanding freely. In fact, the motion of the Crab nebula is due in part to the presence of the pulsar at its center. Consider the currently observed angular size $\theta \simeq 180'$ and expansion rate $\dot{\theta} \approx 0''.23$ yr^{-1} of the nebula. If $\dot{\theta}$ remained constant, the remnant's age could not exceed $\theta/\dot{\theta} \simeq 780$ yrs; but the supernova that formed the nebula occurred more than 900 years ago. Therefore the remnant's expansion must be accelerating. An analysis of the apparent shape and motion of the nebula, taking into consideration its nonspherical geometry, indicates that it is about 1.7 kpc from the Sun, and has an average diameter of about 2 pc and an average acceleration of 1.4×10^{-3} cm/sec^2.

Problem 25.12. Suppose the filamentary shell of the Crab nebula is spherical, and is expanding with constant apparent angular acceleration $d^2\theta/dt^2 = \alpha$. Its present apparent rate of expansion $d\theta/dt = \omega_0 = 0''.23$ yr^{-1}, and line emission from the filaments indicates a maximum Doppler shift of 1,200 km/sec. Find the nebula's distance, diameter, and linear expansion rate.

Accelerated expansion implies that energy is being fed into the nebula. This is further required by the power spectrum (Figure 25.3). The lifetime of synchrotron electrons is given by (25.31), which for the Crab nebula is roughly

$$\tau \simeq \frac{8.4 \times 10^3}{E_{e,9}} \text{ yr}, \quad (25.45)$$

where $B = 10^{-3}$ gauss and $E_{e,9}$ is in units of 10^9 eV. Electrons giving rise to the optical emission with $\nu \approx 10^{15}$ Hz have typical energies given by (25.22) of about 10^{11} eV and lifetimes $\tau < 10^2$ yrs. Furthermore, the lifetime for x-ray electrons ($E_e \approx 10^{12}$ to 10^{14} eV) is several years. Since the optical brightness of the nebula has remained essentially the same for at least the last 100 years, the supply of high-energy electrons ($E_e > 10^{11}$ eV) must continually be replenished. The abrupt change in γ for ν near 10^{12} to 10^{13} Hz would then indicate the difference in energy distribution of electrons originally injected by the supernova explosion, and those being continually supplied by the rotating pulsar today. Additional support for continuing energy injection comes from observations of the x-ray component, which is less than one degree in size, indicating that the electrons capable of emitting at these energies decay before they can reach the nebula's periphery. High-resolution photometry of the central regions around the pulsar shows a series of highly variable wisps, the inner of which moves outward and at right angles to the magnetic field with velocities exceeding 40,000 km/sec $\approx 0.1\,c$. The wisps are observed to oscillate with periods of several months, and probably represent plasma waves radiating away from the central pulsar.

We can estimate the energy input needed to maintain the observed acceleration of the filamentary structure by finding the time-rate of change of its kinetic energy, $E = 1/2 \int \rho v^2 \, dV$, which is

$$\frac{dE}{dt} = \frac{dE_a}{dt} + \frac{dE_p}{dt}$$

$$= \int \rho v a \, dV + \frac{1}{2} \int \rho v^2 \nabla \cdot \mathbf{v} \, dV. \quad (25.46)$$

The first term is the energy increase needed to accelerate a fixed mass of the shell; the second term represents the mass of interstellar gas accumulated by the leading edge of the expanding nebula. For the Crab nebula, the acceleration term $\dot{E}_a \simeq 1.6 \times 10^{38}$ ergs/sec and the second term $\dot{E}_p \simeq 1.7 \times 10^{38}$ ergs/sec. If the energy loss to synchrotron radiation (25.44) is included, the total power in the nebula is about 4.5×10^{38} ergs/sec. This yields a net energy release since formation in excess of 10^{49} ergs.

Problem 25.13. Show that dE_a/dt is given by

$$\frac{dE_a}{dt} = Mva + 2\pi m_H n_0 R^2 v^3 \quad (25.47)$$

for a spherically symmetric nebula, constant acceleration, and expansion velocity $v = dR/dt$. Assume ρ constant in the interstellar gas.

The preceding analysis emphasizes the net kinetic-energy change of the nebula, but says nothing about the physical processes that ultimately convert rotational energy of the pulsar into nebular acceleration. The force on the filamentary mass is due mostly to the relativistic electrons and the magnetic fields. The latter exert a pressure $B^2/8\pi$ across the shell that tends to move the plasma outward in a way that distributes the field as uniformly as possible. To this must be added the pressure P_c resulting from the relativistic electrons, which spiral along the magnetic field lines. Assuming spherical symmetry for the nebula, the net force on the filaments is

$$Ma \simeq 4\pi R^2 \left(P_e + \frac{B^2}{8\pi} \right). \quad (25.48)$$

At the periphery of the nebula $P_c \simeq B^2/8\pi$; the synchrotron emission of the filaments gives $B \approx 3 \times 10^{-4}$ gauss. Therefore (25.48) yields a total mass

$$M_{\text{filaments}} \simeq (BR)^2/a \approx 0.3 M_\odot, \quad (25.49)$$

which includes all matter swept up by the expanding nebula (about $4\pi m_H n_0/3 \simeq 0.1 \, M_\odot$) as well as the mass ejected by the supernova. Although the exact value in (25.49) is uncertain by a factor of two or so, it shows that the mass ejected by the supernova was no more than a few tenths of a solar mass, as expected for a Type I event.

Part 5

**GALAXIES AND
THE UNIVERSE**

Chapter 26

THE EXPANDING UNIVERSE

26.1. Redshift and Expansion

The spectrum of the Andromeda galaxy (M31), the largest galaxy near our Galaxy, exhibits a blueshift $\Delta\lambda/\lambda$ that when corrected for solar motion, corresponds to a velocity of approach of order 80 km/sec. Since M31 is a member of the Local Group of galaxies to which our Galaxy belongs, this motion is largely due to its dynamical interaction with other members of the group. As galaxies of increasingly greater distance from us are observed, we find some whose spectra are redshifted, and fewer that are blueshifted. For distances of several Mpc, all spectra appear redshifted. The most significant observation of modern cosmology is the result, established first by Hubble in the 1930s, that the redshifts of the galaxies are proportional to their distance from us, $z \propto r$. Interpreting z as a radial velocity, and assuming z to be small, we have

$$v = H_0 r, \qquad (26.1)$$

where H_0 is a constant of dimensions sec^{-1}. It is customarily measured in km sec^{-1} Mpc^{-1}. The best current observations place H_0 in the range

$$50 \le H_0 < 100 \text{ km sec}^{-1} \text{ Mpc}^{-1}. \qquad (26.2)$$

H_0 is known as Hubble's constant. For $H_0 = 50$ km sec^{-1} Mpc^{-1}

$$t_H = H_0^{-1} = 2.1 \times 10^{10} \text{ yrs}.$$

Strictly speaking the restriction to small z is important, since it is not clear what the Hubble relation might mean if z were large. For small z, $v/c \ll 1$, and the distance associated with (26.1) is taken to be

$$r \simeq \frac{cz}{H_0} = 6 \times 10^3 \, z \left(\frac{50}{H_0}\right) \text{ Mpc}. \qquad (26.3)$$

We will see later that this relation requires modification when z is large. The problem does not arise in practice, since the only objects for which we have independent estimates of distance, the galaxies, all have relatively small z. The main problem that arises is in the proper definition of distance, which may have several different values depending on precisely how it is defined. Roughly speaking, the problem becomes serious when the light-travel-time from an object becomes comparable to the time-scale for the expan-

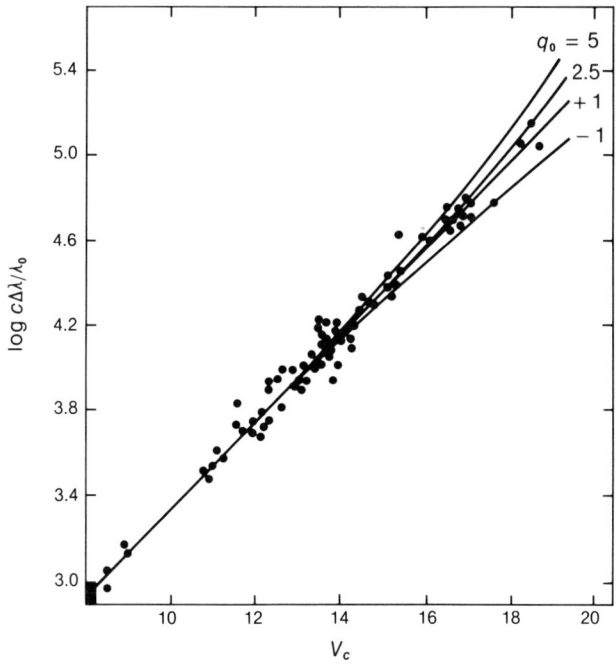

Figure 26.1. Redshift versus apparent visual magnitude for the brightest galaxies in 84 clusters. The natural wavelength is λ_0, and $c = 3 \times 10^5$ km/sec. The lines through the data correspond to different deceleration parameters q_0.

sion of the universe, or of some relevant subsystem of it (for instance, for the quasistellar objects).

Nevertheless, it is customary to derive a "distance" from a redshift by using the Hubble relation for galaxies, and for other objects that show redshifts but whose distance cannot be calculated by other means. Especially for the quasistellar objects, this procedure may produce erroneous results if the redshifts are partly gravitational or are caused by something other than the general expansion of the universe. Figure 26.1 shows the redshift as apparent visual magnitude of 84 galaxies, each of which is the brightest member of a cluster of galaxies. The curves through the observations are labeled by assumed values of the deceleration parameter (see below). The small box in the lower left-hand corner represents the region originally investigated by Hubble.

The redshift exhibited by the spectra of distant galaxies as described by (26.1) is widely accepted as evidence that matter in the universe is receding from us. A fundamental tenet of modern cosmology is that we do not occupy a special region of the universe, but that all locations are in fact equivalent. The implication of this tenet and the hypothesis that the galaxies are receding is that the universe as a whole is in a state of expansion.

A complete description of the dynamics of the universe requires the full machinery of general relativity. However, several fundamental features that will be useful when galaxies are discussed may be obtained within the framework of Newtonian cosmology. The expansion of the universe implied by Hubble's law can be ignored for nearby objects.

Problem 26.1. Two bound galaxies ($M \simeq 2 \times 10^{11}\ M_\odot$) move in mutual circular orbits. How large would their relative orbit be if their relative velocity equaled the recessional velocity of one with respect to the other?

However, expansion affects the appearance of distant objects and the evolution of distributions of matter spread over large volumes. The reasons for this include the finite speed of light: we see objects that are a distance r away as they were r/c years in the past, and their radiation will be redshifted. The basis for these effects will be considered in the following sections.

26.2. Newtonian Cosmology

Modern theories of cosmology are based either on Einstein's general theory of relativity, or on theories that can, along with Einstein's theory, be called relativistic. In special cases (for isotropic and homogeneous universes in particular) a set of equations describing evolving universes can be derived from Newtonian theory; these equations, if properly modified and interpreted, agree with more general theories. To this end, we assume that Newtonian mechanisms and Newton's theory of gravitation apply throughout the universe. The modifications necessary to recover relativistically correct results will be noted at the conclusion of this section.

Although many theories about the structure of the universe as a whole start from less restrictive assumptions, the commonest form of cosmology assumes that the spatial universe is at any instant in its history homogeneous and isotropic. That is, each piece of it looks the same (apart from local irregularities that are assumed to be small) as any other piece, and there are no preferred directions in space. The situation may change with time, so that the whole universe may expand or contract, but it is assumed to maintain spatial homogeneity and isotropy at all times. To describe such a situation, we find it most convenient to adopt comoving coordinates, i.e., coordinates whose values for a particular object, say, a galaxy, do not change as the universe expands or contracts. To obtain real physical distances, we must multiply each of these coordinates by some scale factor indicating how the universe expands or contracts with time. The comoving coordinates are similar to Lagrange coordinates of fluid mechanics.

The principle of homogeneity and isotropy, which is sometimes known as the cosmological principle, implies that the universe must appear the same (apart from local irregularities) to any observer in it. An observer is normally thought of as residing on a galaxy that moves with the general expansion of the universe. In the presence of clusters of galaxies (which amount to local irregularities), this is no longer quite satisfactory, but then we can place our observer at the center of mass of the cluster. In general, we have to assume that we can define a set of fundamental observers who move with the general expansion of the universe. Consider three observers, O,A,B. The velocity of A relative to O will be some function of position, say, $\mathbf{v}(\mathbf{a})$. Likewise, the velocity of B relative to O is a function $\mathbf{v}(\mathbf{b})$, and, by the principle of homogeneity and isotropy, this must be the same function of position as for A. Furthermore, the velocity of A relative to B must be $\mathbf{v}(\mathbf{a}-\mathbf{b})$. But, by the usual rules for combining velocities, the velocity of A relative to B must be $\mathbf{v}(\mathbf{a}) - \mathbf{v}(\mathbf{b})$. So the function $\mathbf{v}$ must satisfy the relation

$$\mathbf{v}(\mathbf{a}-\mathbf{b}) = \mathbf{v}(\mathbf{a}) - \mathbf{v}(\mathbf{b}). \tag{26.4}$$

The only function $\mathbf{v}$ that can satisfy this relation is one in which each component of $\mathbf{v}$ is a linear function of position:

$$v_i = \sum_j A_{ij} r_j, \tag{26.5}$$

A_{ij} being a tensor. So far we have assumed only homogeneity. If we also impose spatial isotropy, then A_{ij} must be diagonal with its three diagonal components equal, or

$$\mathbf{v} = H(t)\mathbf{r}, \tag{26.6}$$

where $H(t)$ is a scalar, the *Hubble parameter,* which is in general a function of time.

Notice that we have now shown that only a Hubble law (a linear velocity-distance relation) can satisfy the requirements of homogeneity and isotropy of space. However, we are now describing the universe from the Newtonian point of view of a distant, external observer who can record the instantaneous properties of the universe. This instantaneous state of the universe is sometimes called a *world map.* Now, if an observer in the universe tries to measure distances and velocities in order to verify this relation, he comes up against the problem that the light signals he uses to measure both velocity and distance have taken a finite time to reach him. Thus he sees a distant galaxy not as it is, but as it was at some time in the past, when its distance may have been different and when the Hubble parameter may have been different. So the calculation of what the universe will look like to an observer in it, the *world picture,* is quite complicated, and depends in general on the detailed model we have of the past evolution of the universe.

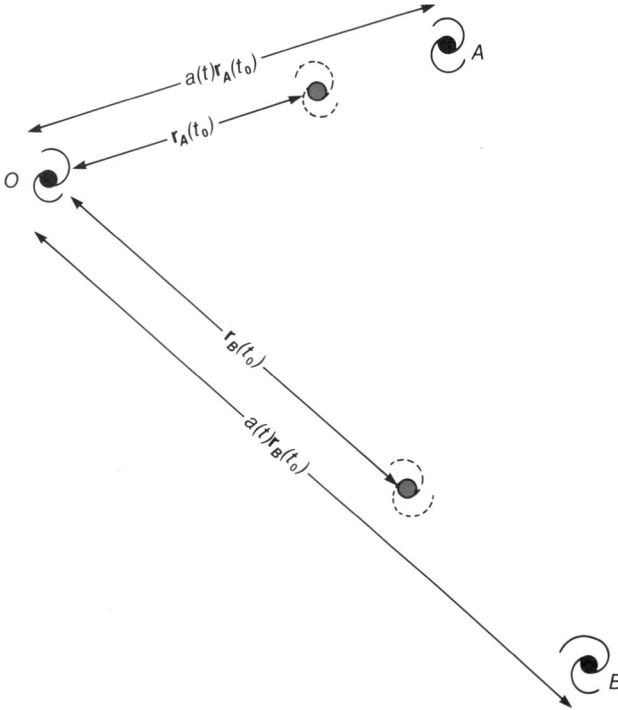

Figure 26.2. Coordinates of two galaxies in an expanding universe relative to observer O. Galaxy at $r_a(t_0)$ at $t = 0$ will be at $a(t)\, r_A(t_0)$ at time t.

Equation (26.6) may be written in the form

$$\frac{d\mathbf{r}}{dt} = H(t)\mathbf{r}. \tag{26.7}$$

If we define the dimensionless scale factor

$$a(t) \equiv \frac{r(t)}{r(t_0)}, \tag{26.8}$$

where $r(t_0)$ is the comoving coordinate relative to the observer O at time t_0 (Figure 26.2), we can rewrite (26.7) as

$$\frac{1}{a}\frac{da}{dt} = H(t). \tag{26.9}$$

It follows from (26.7) that once $a(t)$ is known, $r(t)$ may be obtained as a function of $r(t_0)$.

On the distance scales to which the principle of homogeneity and isotropy apply, the distribution of galaxies or clusters of galaxies can be approximated by a fluid of density ρ and pressure P. Therefore, if mass is conserved, and if $P \ll \rho c^2$, the density of matter in the universe must decrease as $r(t)^{-3}$, or

$$\rho(t)a(t)^3 = \rho(t_0) \equiv \rho_0, \tag{26.10}$$

where ρ_0 is the density at time t_0. This is a form of the continuity equation of fluid mechanics. For highly symmetric models of the universe, such as those discussed in this section, the assumption of mass conservation is valid. However, in more generalized relativistic models, mass or energy conservation need not hold. In such cases (26.10) need not hold.

The laws of gravitation are believed to act on all objects in the universe, and over all distance scales. It is therefore natural to assume that the recession of distant galaxies is due to the dynamical motion of the universe as a whole. This suggests that distant galaxies, whose velocities are given by (26.6) are being accelerated by the mass distribution of the universe, and that we can find an equation that describes their motion.

Consider a unit mass of the universe; its acceleration, which is produced by the forces acting upon it, is given by

$$\mathbf{F} = \frac{d\mathbf{v}}{dt}. \quad (26.11)$$

If we ignore pressure gradients (a valid assumption in the present epoch), the only force that is relevant is the gravitational force, which we express in terms of Poisson's equation,

$$\nabla \cdot \mathbf{F} = -4\pi G \rho. \quad (26.12)$$

It follows from (26.7) that

$$\frac{d\mathbf{v}}{dt} = \frac{d}{dt}[H(t)\mathbf{r}] = \mathbf{r}\left(\frac{dH}{dt} + H^2\right). \quad (26.13)$$

The divergence operator applied to this expression gives

$$\left[\frac{dH}{dt} + H^2\right]\nabla \cdot \mathbf{r} = -4\pi G \rho. \quad (26.14)$$

Now $\nabla \cdot \mathbf{r} = 3$, so this equation gives

$$\frac{dH}{dt} + H^2 = \frac{-4\pi G \rho}{3} \quad (26.15)$$

or

$$\frac{1}{H^2}\frac{dH}{dt} + 1 = -\sigma(t), \quad (26.16)$$

where we have defined the dimensionless density parameter

$$\sigma(t) = \frac{4\pi G \rho(t)}{3 H(t)^2}. \quad (26.17)$$

An alternative form of this equation is obtained by substituting (26.9) for $H(t)$:

$$\frac{1}{a}\frac{d^2 a}{dt^2} + \frac{4\pi G \rho}{3} = 0. \quad (26.18)$$

This equation implies that a static universe is impossible unless it is empty, since $d^2 a/dt^2 \neq 0$ if $\rho \neq 0$.

Problem 26.2. A static, nonempty universe can be obtained if the law of gravitation (26.12) is modified by the addition of a term λc^2 to the source term. Show that this yields a static universe with $\rho \neq 0$. The constant λ is the notorious cosmological constant.

Equation (26.18) can be integrated once if one multiplies through by the factor $2a \cdot da/dt$, and if one uses the equation of mass conservation (26.10) to obtain

$$2\frac{da}{dt}\frac{d^2 a}{dt^2} + \frac{8\pi G}{3}\rho_0 \frac{1}{a^2}\frac{da}{dt} = 0. \quad (26.19)$$

Each term is now a perfect differential, and a further integration gives

$$\left(\frac{da}{dt}\right)^2 - \frac{8\pi G}{3}\rho_0 \frac{1}{a} = \text{constant}. \quad (26.20)$$

Introducing the quantity k, defined so that the constant of integration is $-kc^2$, we find

$$\frac{1}{a^2}\left(\frac{da}{dt}\right)^2 - \frac{8\pi G}{3}\rho = -\frac{kc^2}{a^2}, \quad (26.21)$$

where the equation of continuity (26.10) has been used to rewrite the second term on the left-hand side. Formally (26.21) is identical to one of the Friedmann equations obtained from Einstein's equations for a homogeneous, isotropic universe. Thus Newtonian mechanics gives the same equation as does relativity theory. The fundamental difference is in the interpretation of the constant k and the density ρ. In Newtonian theory, ρ is the mass density, but in relativistic cosmology it is the total energy density divided by c^2 (see discussion in Section 16.3). Finally, in relativistic cosmology k is a measure of the curvature of space, whereas in the Newtonian model it appears as a constant of integration. The meaning of this constant of integration can be seen by multiplying equation (26.21) by $\frac{1}{2}r^2$, and using the definition of the Hubble constant,

$$\frac{v^2}{2} - \frac{GM(r)}{r} = -\frac{kc^2 r^2}{a^2}, \quad (26.22)$$

since the mass within radius r is

$$M(r) = 4\pi r^3 \rho / 3.$$

The left-hand side is just the total energy, kinetic plus potential, per unit mass of the matter at r. Therefore k is a measure of the total energy in Newtonian cosmology.

Finally we note that (26.21) does not completely specify the dynamics, since we do not yet know how $\rho(t)$ evolves with time. A second equation is therefore needed, and can be obtained by the following arguments. Denote the pressure of the matter in the universe by P, and the total energy (excluding gravitational energy) within a sphere of radius r by

$$E = \frac{4}{3} \pi \rho c^2 r^3.$$

Then, as the universe expands, work will be done by the pressure against surrounding fluid elements, and

$$\frac{dE}{dt} = -P \frac{dV}{dt}$$

or

$$c^2 \pi \frac{4}{3} \left(\frac{d\rho}{dt} r^3 + 3\rho r^2 \frac{dr}{dt} \right) + 4\pi P r^2 \frac{dr}{dt} = 0.$$

This may be rewritten, using (26.8) in the form

$$\frac{d\rho}{dt} + 3(\rho + P/c^2) \frac{\dot{a}}{a} = 0. \quad (26.23)$$

Equations (26.21), (26.23), and an equation of state $P(\rho)$ represent a complete mathematical description of a homogeneous and isotropic universe that is relativistically correct as long as k is interpreted as the spatial curvature, and ρc^2 as the total material energy density (including radiation, neutrinos, and the energy associated with pairs of elementary particles).

Problem 26.3. In a matter-dominated universe, the pressure is negligible compared to the local rest-mass density. Show that (26.23) reduces to the continuity equation (26.10) in this case. What is the corresponding result in the radiation-dominated epoch, where $P = (1/3) \rho c^2$?

26.3. GENERAL PROPERTIES OF COSMOLOGICAL MODELS

We now consider several consequences of the results developed in Section 26.2 for an isotropic, homogeneous, matter-dominated universe. Since the equations (26.21) and (26.23) are relativistically correct, we will, in keeping with relativistic theory, refer to k as the curvature constant, and understand by ρc^2 the total matter-energy density. We denote the current epoch by t_0, and all quantities refering to the current epoch ("now") by subscript zero. Some relativistic models include a cosmological constant (see Problem 26.2). We will assume throughout the discussions to follow that the constant is zero.

The rate at which the Hubble constant changes with time is of interest, since in most models it is expected to change as a result of gravitational slowing down of the expansion. From its definition, we have

$$\frac{dH}{dt} = \frac{d}{dt}\left(\frac{1}{a}\frac{da}{dt}\right) = H^2 \left(\frac{1}{H^2} \frac{1}{a} \frac{d^2 a}{dt^2} - 1 \right). \quad (26.24)$$

The quantity

$$q(t) = -\frac{1}{H^2} \frac{1}{a} \frac{d^2 a}{dt^2} \quad (26.25)$$

is called the deceleration parameter, so that

$$\frac{dH}{dt} = -H^2(q + 1). \quad (26.26)$$

Thus if $q = 0$,

$$a^{-1} \frac{d^2 a}{dt^2} = 0,$$

da/dt is constant, and the expansion is said to be uniform. Finally, note that for models of the Universe with a zero cosmological constant and zero pressure, the density parameter $\sigma(t)$ and the deceleration parameter $q(t)$ are identically equal to each other. Note that $q = 0$ does not mean that $H(t)$ is constant. It is clear from the definition of q that the left side of equation (26.16) is equal to $-q(t)$; so the current value of $H(t)$, which is the Hubble constant H_0, is observable, and in principle so is q_0. Unfortunately, the practical difficulties in observing q_0 are very great. Current estimates place q_0 in the range 0.5 to 1.0. The current average mass density of the universe can be related to q_0 and H_0 by (26.18):

$$\rho_0 = -\frac{3\ddot{a}}{4\pi Ga} = \frac{3H_0^2 q_0}{4\pi G}$$

$$= 10^{-29} q_0 \left(\frac{H_0}{50}\right)^2 \text{ g cm}^{-3}, \quad (26.27)$$

where H_0 is in km sec^{-1} Mpc^{-1}. A density of 10^{-29} g cm^{-3} is higher than seems to be observed for the local mean density of the universe, which is perhaps in the range 10^{-31} to 10^{-30} g cm^{-3}. However, this local mean density is extremely difficult to define in practice. In fact, the larger the volume of space sampled, the lower the mean density found, with no sign of leveling off to any constant value at some sample size. Also, the observed value of q_0 depends on the departure of the Hubble velocity-distance relation from linearity at large distances, which is difficult to measure observationally.

If the numbers are taken at face value, however, there appears to be a problem of "missing mass," in that the density of the universe needed to account for the observed value of q_0 is larger than the density of observed matter in the universe. The problem is similar to (and perhaps of the same order of magnitude as) the problem of the disagreement between the masses of clusters of galaxies calculated from the velocities of the galaxies in them and the mass of the visible material. If we divide (26.21) by $H^2(t)$, we find, for the matter-dominated epoch,

$$1 - 2q = \frac{kc^2}{H^2 a^2}, \quad (26.28)$$

which implies that only if $q_0 = \sigma_0 = \frac{1}{2}$ can space be flat (i.e., $k = 0$). It follows that the density required to close the universe, that is, to make it just finite, is

$$\rho_0 = \frac{3H_0^2}{8\pi G} = 4 \times 10^{-30} \left(\frac{H_0}{50}\right)^2 \text{ g cm}^{-3}.$$

In the matter-dominated epoch one can express k in terms of the constants H_0 and q_0,

$$-kc^2 = H_0^2 (1 - 2q_0); \quad (26.29)$$

so the Friedmann equation can be written entirely in terms of H_0 and q_0, as

$$\left(\frac{da}{dt}\right)^2 = \frac{H_0^2}{a} \left[2q_0 + a(1 - 2q_0)\right]. \quad (26.30)$$

Notice here that q_0 determines the actual shape of the function $a(t)$, but the Hubble parameter determines only the time-scale. For small changes in time, (26.31) describes the variation in $a(t)$ given the present values of H_0 and q_0. It can be shown, in fact, that the solution to the Friedmann equations is determined completely by H_0 and q_0.

Problem 26.4. For small values of redshift (Section 26.4) the scale factor $a(t)$ may be expanded about t_0. Show that

$$a(t) = a(t_0)$$
$$\times \left[1 + H_0(t - t_0) - \tfrac{1}{2} H_0^2 q_0 (t - t_0)^2 + \cdots \right]. \quad (26.31)$$

The current general behavior of Friedmann models can be extracted from (26.30). Since da/dt is now positive, a must be increasing. If the deceleration parameter $q_0 = \frac{1}{2}$, then the right-hand side of (26.30) is positive, and approaches zero as $t \to \infty$. For $q_0 < \frac{1}{2}$, it approaches the asymptotic value $H_0^2(1 - 2q_0)$. Finally, if $q_0 > \frac{1}{2}$, then the right-hand side of (26.30) decreases, becoming zero at some maximum expansion. Thereafter, da/dt changes sign, and the universe begins to collapse. The critical density dividing open from closed models is given by

$$\rho_c = \frac{3H_0^2}{8\pi G} = 4 \times 10^{-30} \left(\frac{H_0}{50}\right)^2 \text{ g cm}^{-3}.$$

It is possible to solve the Friedmann equations for $a(t)$, and thus for $r(t)$, for given values of k and the equation of state $P(\rho)$. The simplest model is one for which $P = 0$ and $k = 0$. This is the matter-dominated Euclidean (flat) universe (Einstein–de Sitter model). The two Friedmann equations for this case are

$$\frac{1}{a^2}\left(\frac{da}{dt}\right)^2 = \frac{8\pi G\rho}{3} \quad (26.32)$$

and (26.10), which follows immediately from (26.23).

Introducing the present matter density ρ_0, and using the mass-conservation law (26.10), we can rewrite (26.32) as

$$\frac{dt}{da} = \left(\frac{3}{8\pi G \rho_0}\right)^{1/2} a^{1/2}. \qquad (26.33)$$

Integrating gives

$$t = \frac{2}{3}\left(\frac{3}{8\pi G \rho_0}\right)^{1/2} a^{3/2} + \text{constant}. \qquad (26.34)$$

Denoting the time when $a = 0$ by $t = 0$, then this becomes

$$a(t) = \alpha t^{2/3} \qquad (26.35)$$

where

$$\alpha = (6\pi G \rho_0)^{1/3}.$$

The Hubble constant in the Einstein–de Sitter model is

$$H(t) = \frac{1}{a}\frac{da}{dt} = \frac{2}{3t}. \qquad (26.36)$$

But t is just the age of the universe measured from the instant when $a = 0$; thus the current age is given by $(2/3H_0)$. The deceleration parameter, q, is readily calculated as

$$q = -\frac{1}{H^2}\frac{1}{a}\frac{d^2a}{dt^2} = \frac{1}{2}. \qquad (26.37)$$

Recall that q must be ½ for k to equal zero. Finally, we note that the Einstein–de Sitter model is an example of a Big Bang model, which has an origin at a finite time when $a = 0$. In this particular model, the universe continues to expand indefinitely.

The matter-dominated Friedmann models for arbitrary k can also be obtained analytically. First we rewrite (26.30) in the form

$$\left(\frac{\dot{a}}{a}\right)^2 = H_0^2 \left[1 - 2q_0 + 2q_0\left(\frac{a_0}{a}\right)\right]. \qquad (26.38)$$

This can be solved for $q_0 > ½$ and $q_0 < ½$. The former case corresponds to a closed universe, the latter to an open universe. For example, the solution for a closed universe is given by the parametric form

$$\frac{a(t)}{a_0} = \frac{q_0}{2q_0 - 1}(1 - \cos\theta),$$
$$H_0 t = \frac{q_0(\theta - \sin\theta)}{(2q_0 - 1)^{3/2}}. \qquad (26.39)$$

This may be recognized as the equation for a cycloid (Figure 26.3). The maximum radius $a_{\max}$ occurs for $\theta = \pi$, and is given by

$$a_{\max} = \frac{2q_0 a_0}{2q_0 - 1}. \qquad (26.40)$$

The universe expands, reaches a maximum extent $a_{\max}$, and then recollapses in a time

$$t_c = \frac{2\pi q_0}{H_0(2q_0 - 1)^{3/2}}. \qquad (26.41)$$

The maximum radius depends on the average mass density ρ_0 through (26.18). For $q_0 = ½$, the universe is just bound and closed. The average (present epoch) mass density is then $\rho_0 = 5 \times 10^{-30} (H_0/50)^2$ g cm^{-3}. If $q_0 = 1$, then ρ_0 is increased by a factor of two, and the radius of the universe $a(t)$ is half its maximum value. The present age of the universe follows from (26.39), with $a(t) = a_0$, and is

$$t_0 = \frac{\pi/2 - 1}{H_0} = 1.18 \times 10^{10}\left(\frac{50}{H_0}\right) \text{ yrs.} \qquad (26.42)$$

Maximum expansion occurs when the universe is about $\pi H_0^{-1} = 6.5 \times 10^{10}$ years old, and the final collapse to $a = 0$ would occur about 13×10^{10} years after the Big Bang ($t = 0$).

Problem 26.5. Suppose that the universe is open, and $0 < q_0 < ½$. Define the development angle ψ by

$$\frac{a(t)}{a_0} = \frac{q_0}{1 - 2q_0}(\cosh\psi - 1), \qquad (26.43)$$

and show that the solution to the Friedmann equation (26.38) is

$$H_0 t = \frac{q_0(\sinh\psi - \psi)}{(1 - 2q_0)^{3/2}}. \qquad (26.44)$$

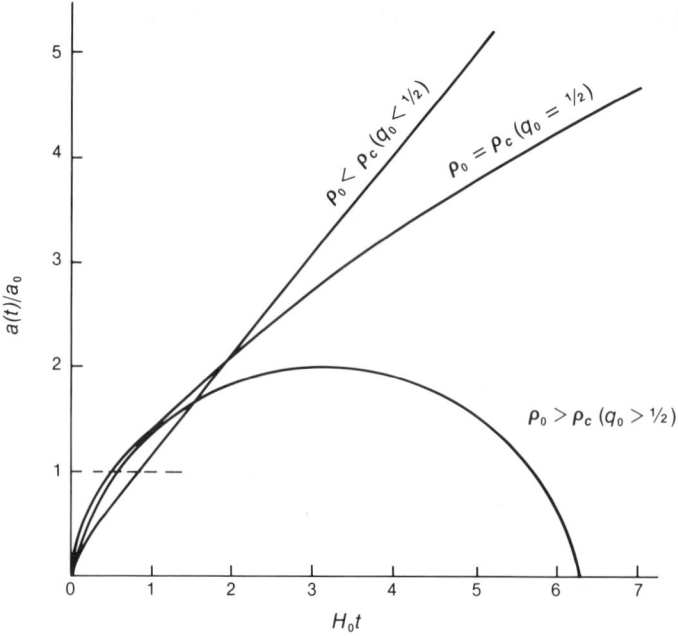

Figure 26.3. Plot of $a(t)/a_0$ versus $H_0 t$ for three Friedmann universes. Subscript zero denotes present values.

How old is this universe if the total mass density ρ_0 is due to the mass in galaxies ρ_G, and $H_0 = 50$ km sec^{-1} Mpc^{-1}? Take $\rho_G = 1.4 \times 10^{-31}$ g cm^{-3}. Explain physically why this value is larger than (26.42) for a closed universe.

The deceleration parameter may also be obtained for the Friedmann universes. They are

$$q(t) = \begin{cases} (1 + \cos\theta)^{-1} \text{ if closed,} \\ (1 - \cosh\psi)^{-1} \text{ if open,} \end{cases} \quad (26.45)$$

and clearly vary with time. In an open universe, $q(t)$ varies from ½ at $\psi = 0$ ($a = 0$) to zero as $\psi \to \infty$ ($a \to \infty$), and the deceleration gradually decreases as the expansion proceeds. For a closed universe, $q(t)$ is initially ½, increases as the expansion proceeds, and reaches its maximum value $a(t) = a_{\max}$. Thereafter $q(t)$ decreases, returning to its original value at t_c.

The present ages of the model universes we have described are consistent with the ages of stars, galaxies, and radioactive elements, in the sense that t_0 is greater than or comparable to them. The models should not be applied to the early or final epochs, because these are not matter dominated (recall that an equation of state $P = 0$ was assumed). Nevertheless, the matter-dominated Friedmann models are useful in describing present-day aspects of the universe. Figure 26.3 shows $a(t)/a_0$ plotted against $H_0 t$ for three cases. The curve labeled $\rho_0 > \rho_c$ is a closed universe for which $q_0 = 1$; the $\rho_0 = \rho_c$ curve is the Einstein–de Sitter model; and the $\rho_0 < \rho_c$ curve is an open universe with q_0 corresponding to a present mass density $\rho_G = \rho_0 = 1.4 \times 10^{-31}$ g cm^{-3}. The present epoch t_0 is given for each model by the intersection of the curve with the dashed line ($a = a_0$).

26.4. Cosmological Redshifts

The Friedmann equations developed in the preceding sections relate the function $a(t)$ to the time t, and, using (26.8), give the development of $r(t)/r_0$. In order to find out what properties of the universe are seen by an observer, we need to find out how light signals travel between two points. A photon travels with speed c as seen by any observer in the universe. Furthermore, the universe through which it travels is expanding at different rates at different times. Calculating the trajectory of a photon is therefore nontrivial. A general result may be obtained from the Friedmann equations (26.21) and (26.23) for a radiation-dominated universe, in which photons travel at the same rate as the expansion. Taking the derivative of $a(t)$, we find

$$\frac{da}{dt} = \frac{1}{r_0}\frac{dr}{dt}; \quad (26.46)$$

next we rewrite (26.21) in the form

$$\frac{da}{dt} = \left(\frac{8\pi G}{3}\rho - k\right)^{1/2} = \left(\frac{8\pi G\rho_0}{3a^2} - k\right)^{1/2}$$
$$= \frac{1}{a}\left(\frac{8\pi G\rho_0}{3} - \frac{kr^2}{r_0}\right)^{1/2} \quad (26.47)$$

using the solution to (26.23), $\rho a^4 = \rho_0$, and (26.8). Equating (26.46) and (26.47), and rewriting, gives

$$a\frac{dr}{dt} = \left(\frac{8\pi G}{3}\rho_0 r_0^2 - kr^2\right)^{1/2}.$$

Finally, rescaling the coordinate r, as

$$r \to (8\pi G \rho_0 r_0^2/3)^{1/2} r/c,$$

we have

$$\frac{c\,dt}{a(t)} = \frac{dr}{\sqrt{1-kr^2}}. \quad (26.48)$$

This result is relativistically correct if k is taken to be the sign of the curvature constant, or to be zero if the curvature is zero.

The path along which a photon travels is described by (26.48). Suppose two signals (wave crests, photons) are emitted from an object A at distance r, at times t_e and $t_e + dt_e$ (subscript e stands for emission), and suppose they are received at times t_0 and $t_0 + dt_0$, respectively. By integrating along the path of the ray for the first signal, we obtain

$$-c\int_{t_0}^{t_e}\frac{dt}{a(t)} = \int_0^r \frac{dr}{\sqrt{1-kr^2}}. \quad (26.49)$$

Now the important point to notice is that the position of galaxy A as measured by its coordinate r does not change during the expansion of the universe. Rather, r serves to identify which galaxy we are talking about. Physical separations (proper lengths in relativistic terms) are measured by $a(t)r$ rather than by r itself. Repeating the analysis for the second signal emitted yields

$$-c\int_{t_0+dt_0}^{t_e+dt_e}\frac{dt}{a(t)} = \int_0^r \frac{dr}{\sqrt{1-kr^2}}. \quad (26.50)$$

Note that the right-hand sides of (26.49) and (26.50) are identical, since they do not depend on t. Therefore we have

$$\int_{t_0}^{t_e}\frac{dt}{a(t)} = \int_{t_0+dt_0}^{t_e+dt_e}\frac{dt}{a(t)}$$
$$= \int_{t_0}^{t_e}\frac{dt}{a(t)} + \frac{dt_e}{a(t_e)} - \frac{dt_0}{a(t_0)}, \quad (26.51)$$

or

$$\frac{dt_0}{dt_e} = \frac{a(t_0)}{a(t_e)}, \quad (26.52)$$

or finally, in terms of the wavelength of light,

$$\frac{\lambda_{\text{obs}}}{\lambda_{\text{em}}} = \frac{c\,dt_0}{c\,dt_e}. \quad (26.53)$$

The redshift is conventionally expressed as

$$1 + z = \frac{a(t_0)}{a(t_e)}, \quad (26.54)$$

where z is defined as the cosmological redshift. It is positive provided that $a(t_0) > a(t_e)$, that is, provided the universe is expanding. Equations (26.48) and (26.54) are basic to the whole discussion of redshifts in cosmology. Notice that if we can look back to the time close to the beginning of the expansion (the Big Bang), when $a(t_e)$ was practically zero, the redshift becomes very large.

As an example, consider the redshift as a function of the time of emission of the light for the Einstein–de Sitter universe. Then $a(t)$ is given by equation (26.35), and

$$1 + z = (t_0/t_e)^{2/3}. \quad (26.55)$$

Problem 26.6. Assume that the universe is 10^{10} years old. How long ago was the light from a quasi-stellar object with redshift $z = 2$ emitted? (Assume a cosmological redshift for the quasar, and the Einstein–de Sitter model).

We should also note that for the Einstein–de Sitter model, (26.49) can be integrated directly to give the relation between time of emission t_e, time of reception t_0, and the coordinate r:

$$\frac{c}{\alpha}\int_{t_e}^{t_0} t^{-2/3}\, dt = \int_0^r dr,$$

$$\frac{3c}{\alpha}(t_0^{1/3} - t_e^{1/3}) = r. \qquad (26.56)$$

Similar relations may be worked out for the other Friedmann models.

Problem 26.7. Galaxies and clusters of galaxies may have developed out of density perturbations present at the time of hydrogen recombination ($z = 10^3$). Find the age of an Einstein–de Sitter universe ($k = 0$, $P \ll \rho c^2$) when this occurs. What is the matter density and the value of $H(t)$ at recombination in this model? Assume that $t_0 = 2 \times 10^{10}$ yrs.

Problem 26.8. Show that the Hubble constant at epoch z_i is given by

$$H_i^2 = H_0^2 \times [2q_0(1 + z_i)^3 + (1 - 2q_0)(1 + z_i)^2]. \qquad (26.57)$$

Then show that the mass density at that epoch is

$$\rho_i = \rho_{c,i} \frac{2q_0(1 + z_i)}{2q_0(1 + z_i) + (1 - 2q_0)}. \qquad (26.58)$$

From this show that if $\rho_0 > \rho_c$, then $\rho_i > \rho_{c,i}$ as well.

26.5. Cosmological Distances

Since the velocity-distance relation (26.1) is a basic, observable feature of the universe that has cosmological implications, we obviously need to calculate the apparent distance of a galaxy as seen by an observer in the universe. Unfortunately, although the concept of distance is relatively trivial in ordinary experience, particularly in flat space, where the velocity of light is, to all practical purposes, infinite, distance is much harder to handle in cosmology. For instance, do we take the distance of a galaxy to be what it is now, as seen by an observer at a large distance from both our Galaxy and the galaxy we are interested in, or do we take the distance at the time the light was emitted, again as seen by some external observer, or do we take some average of these? Do we base a distance on the apparent brightness of the object, or perhaps on the apparent size (angular diameter)? The distance that would be seen by a distant observer is sometimes called the *metric distance*. Of course, it is a function of time.

The answer is that (a) any of these definitions of distance is adequate, but (b) they do not all give the same result for the distance of a galaxy, and therefore (c) the concept of a unique distance that has any absolute physical meaning must be abandoned. This might perhaps have been anticipated from a basic result of relativity theory, that space and time cannot be separated, but must be considered together in a space-time continuum. What we need to do in practice is give an operational definition of distance based on how we propose to measure it. The most common of such distances is the luminosity distance, D_L, based on the apparent luminosity of the object, and defined by

$$D_L^2 = \frac{L}{4\pi l}, \qquad (26.59)$$

where l is the apparent luminosity and L the (assumed known) absolute luminosity. Obviously, D_L is equal to the ordinary distance in the classical limit. Another possible practical distance measurement is the angular-diameter distance, defined by

$$D_\phi = \frac{d}{\phi} \qquad (26.60)$$

where ϕ is the angular diameter and d is the linear diameter. Again, this is equivalent to the ordinary distance in the classical limit.

The problem is further complicated by the possible non-Euclidean geometry of space, but for now we will ignore this complication, and use the Einstein–de Sitter model as an illustration. A photon is emitted from galaxy A at time t_e and travels through space at velocity c until it encounters our Galaxy O (Figure

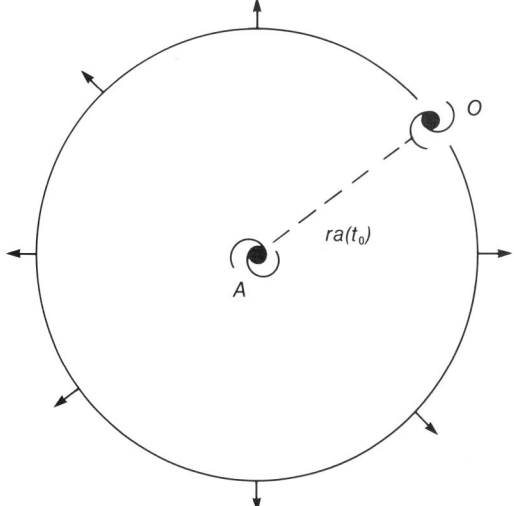

Figure 26.4. Expanding electromagnetic wave emitted from A at t_0 when Galaxy O was at r.

26.4). Since $k = 0$ for the Einstein–de Sitter universe, the physical distance between two points whose coordinate difference is r is just $a(t)r$. Thus the two galaxies A and O were a distance $a(t_e)r$ apart when the wave was emitted. During the time $t_0 - t_e$ that the photon takes to travel from A to O, the galaxies have moved further apart because of the expansion of the universe. Since light travels in straight lines, a wave front emitted at t_e will form the surface of a sphere whose radius is given by the distance (as seen by an external observer) of A from O, at the time of reception of the light, t_0. This quantity is $a(t_0)r$ [we include $a(t_0)$ explicitly, even though units for r may be chosen such that $a(t_0)$ is unity]. The surface area of the sphere is then $4\pi a(t_0)^2 r^2$, and the luminosity distance is just equal to the metric distance at the time of reception of the light; thus

$$D = a(t_0)r. \qquad (26.61)$$

There are also two other corrections that should be applied to the luminosity distance: (1) the apparent luminosity is usually measured in terms of energy; so there is a correction of $(1 + z)$ to take account of the fact that the redshift reduces the energy of each photon by this amount; (2) the rate at which photons are emitted is apparently changed because of the change in time-scale of all events as represented by (26.52). If these two corrections are incorporated into the definition of the luminosity distance, it becomes

$$\begin{aligned} D_L' &= (1 + z)a(t_0)r \\ &= \frac{a(t_0)}{a(t_e)} a(t_0)r \\ &= \frac{a(t_0)^2}{a(t_e)} r. \end{aligned} \qquad (26.62)$$

Further corrections are needed in practice because of the intrinsic variation of brightness as a function of wavelength in the spectrum of the galaxy, and the fact that observations are not made monochromatically, but involve integration over some finite wavelength interval. We will not discuss these corrections in detail here; they are usually lumped together and referred to as the K correction.

The angular-diameter distance involves a somewhat different geometry. At the time of emission of two photons from opposite ends of the galaxy (Figure 26.5), the direction of emission must be such that the photons arrive together at O at the same time, t_0. During the travel time, the galaxy A will have moved further away because of the expansion. However, light

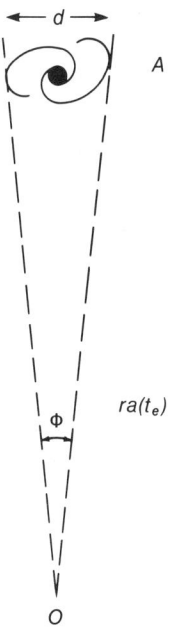

Figure 26.5. Geometry determining angular diameter of a distant galaxy.

26.5. COSMOLOGICAL DISTANCES / 479

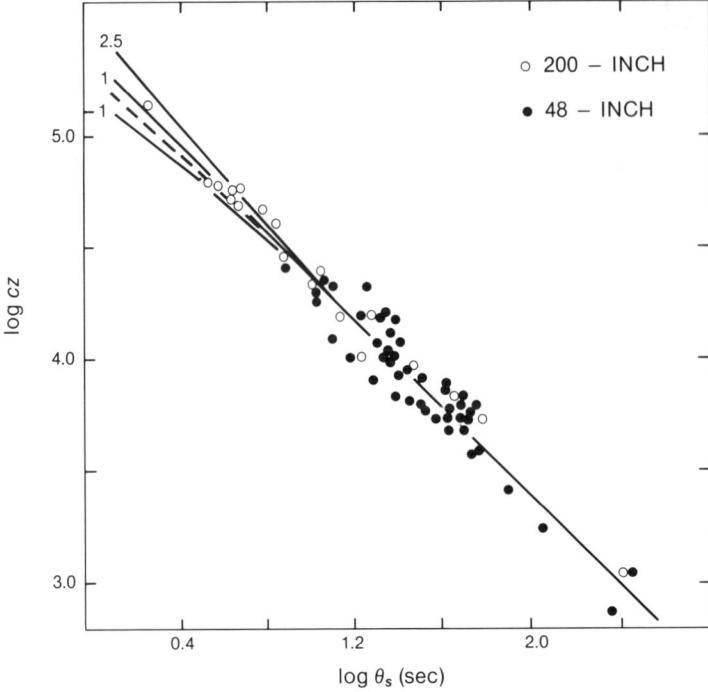

Figure 26.6. Redshift versus isophotal angular diameter (in seconds of arc) for bright galaxies. Numbers next to curves through data are labeled by q_0, and $c = 3 \times 10^5$ km/sec.

travels in straight lines; so the geometry of the situation is governed by the metric distance of the galaxy at the time of emission of the light, $a(t_e)r$. Therefore the angular-diameter distance D is equal to the metric distance at time of emission,

$$D_\phi = a(t_e)r. \qquad (26.63)$$

By eliminating r and t_e between (26.55), (26.56), (26.61), (26.62), or (26.63), we can express these various distances in terms of the redshift z. This gives the results

$$D_L = 3ct_0[1 - (1+z)^{-1/2}], \qquad (26.64)$$

$$D_L' = 3ct_0[(1+z) - (1+z)^{1/2}], \qquad (26.65)$$

$$D_\phi = 3ct_0[(1+z)^{-1} - (1+z)^{-3/2}]. \qquad (26.66)$$

Problem 26.9. Verify the expressions (26.64) to (26.66).

Notice that t_0 is related to the Hubble constant through (26.36) so that the constant $3ct_0$ is also equal to $2c/H_0$ for the Einstein–de Sitter universe. For small values of z, each of these expressions reduces to the approximation

$$cz = H_0 D, \qquad (26.67)$$

as they should. For the most part, the corrected luminosity distance D_L' is used in preference to D_L. The angular diameter distance goes through a maximum at $z = 1.25$, corresponding to a minimum in the angular diameter. It is a property of several cosmological models, with the same general behavior of $a(t)$ as in the Einstein–de Sitter model, to have a minimum in the angular diameter. A plot of observed redshift against isophotal angular diameters of bright galaxies in clusters is shown in Figure 26.6; the curves represent theory using Friedmann models, with q_0 as shown next to each curve. Notice that the scatter in the data of the most distant objects (smallest isophotal diameters) is too large to uniquely select out the value of q_0.

480 / THE EXPANDING UNIVERSE

Problem 26.10. Show that the Hubble relation $cz = H_0 D$ follows from the three distances above.

Problem 26.11. Estimate the minimum angular diameter expected for a galaxy like our own, assuming an Einstein–de Sitter universe.

Problem 26.12. A photon is emitted at time t and is observed at time t_0. Show that the travel time and redshift are related by

$$t_0 - t = H_0^{-1}[z - z^2(1 + q_0/2) + \cdots] \quad (26.68)$$

for an arbitrary Friedmann universe (see Problem 26.4). Use (26.59) and (26.62) to express the apparent and absolute luminosity of the source in terms of z and q_0:

$$\ell = \frac{H_0^2 L}{4\pi z^2 c^2}[1 + z(q_0 - 1) + \cdots]. \quad (26.69)$$

Finally, show that the apparent magnitude is given by

$$m = M + 25 + 5\log\frac{cz}{H_0}$$
$$+ 1.086\, z(1 - q_0), \quad (26.70)$$

where c/H_0 is in Mpc.

Several corrections must be applied to the observed redshift-luminosity relation, including one for the Sun's rotational motion about the Galactic center, and for the fact that photons emitted at frequency ν_e are observed at frequency ν_0. These are relatively straightforward. Less certain corrections are associated with intergalactic absorption, and the evolution of galaxies themselves. The amount and nature of intergalactic matter are still uncertain; assuming that the absorption coefficient is known, the optical depth τ_ν can be constructed for a specific model of the universe. From it, the effective luminosity may be obtained. The effects of evolution on the apparent luminosity of distant galaxies are even more uncertain. If all galaxies formed at about the same epoch, then the more distant a galaxy is, the younger it should appear. Denote the luminosity of a distant galaxy at time t by $L(t)$. Then the apparent luminosity at the time of emission can be expressed as

$$L(t_e) = L(t_0) + \left(\frac{dL}{dt}\right)_0 (t_e - t_0)$$
$$= L_0\left[1 - \frac{\dot{L}_0}{L_0}\frac{z}{H_0}\right], \quad (26.71)$$

where (26.68) has been used to eliminate $t_e - t_0$, and the effects of galactic evolution as they appear today are contained in the term $\dot{L}_0/L_0$. Observational data indicate that $\dot{L}_0/L_0$ may be as large as 0.4×10^{-9} yr^{-1}, which could bear significantly on the value of q_0. For example, the observed deceleration parameter q_{obs} is probably in the range 0.5 to 1.5, which means that when evolutionary effects are taken into account, q_0 could lie between -0.3 and 0.7.

Problem 26.13. Modify (26.69) to include the effects of galactic evolution as expressed by (26.71). Show that the observed deceleration parameter q_{obs} is given by

$$q_{obs} = q_0 - (\dot{L}_0/L_0)H_0. \quad (26.72)$$

If all corrections are lumped into a single term, the present value of the Hubble constant can be expressed in the form

$$5\log H_0 = 31.06 + M_B, \quad (26.73)$$

with H_0 in km sec^{-1} Mpc^{-1}. M_B represents the absolute magnitude of the brightest galaxies.

26.6. THE PRIMEVAL FIREBALL

The universe must have passed through a stage of extremely high temperature and density if the Big Bang hypothesis is correct. The physics under these extreme conditions surrounding the initial instants near $t = 0$ is not completely understood. However, we can gain some understanding of the early universe if (as seems reasonable) it may be assumed that the energy density near $t = 0$ decreases at least as rapidly as a^{-3}. Then (26.21) is

$$\dot{a}^2 = \frac{8\pi G}{3}\rho a^2 - kc^2 = \frac{8\pi G}{3a}\rho a^3 - kc^2$$

$$= \frac{8\pi G}{3a}(\rho_0 a_0^3) - kc^2 \simeq \frac{8\pi G}{3a}\rho_0 a_0^3.$$

To good approximation, the expansion of the early universe can be described by

$$(\dot{a}/a) = (8\pi G\rho/3)^{1/2}, \qquad (26.74)$$

so that the expansion rate at any instant should be of order

$$\tau_{\exp}^{-1} \sim (G\rho)^{1/2}. \qquad (26.75)$$

The most convenient parameter characterizing the early universe is its temperature $T(t)$. We will find that the universe was initially quite hot, in the sense that its entropy per baryon was large. For temperatures in excess of 10^{12} K, there was enough thermal energy to maintain large numbers of subnuclear particles and excited states. Many of these states would be unstable at lower density and temperature. Consider the expanding fireball (the very early universe) as it cooled below 10^{12} K. Extrapolating back from the present epoch, baryon number density $n \approx 10^{28}$ cm^{-3} when $T = 10^{12}$ K. This corresponds to a mass density of order $m_H n \approx 10^4$ g cm^{-3}; the corresponding energy density of radiation $\rho_R c^2 \approx aT^4$ gives $\rho_R \approx 10^{13}$ g cm^{-3}. Therefore the energy density appearing in the expansion time-scale for this epoch is essentially ρ_R, and

$$\tau_{\exp} \approx 10^{-4} \text{ sec}. \qquad (26.76)$$

The time-scales associated with the decay of exotic particles are very short compared with (26.76), and it is assumed that all baryonic matter is in the form of neutrons and protons by the time the universe has cooled to 10^{12} K. Since the average thermal energy per baryon is 100 MeV or so, no nuclei exist at this epoch. Furthermore, the internucleon separation is $r_0 \approx n^{-1/3} \approx 10^{-9}$ cm, so the neutrons and protons form ideal gases.

Electron-Positron Pairs and Radiation

If the universe is electrically neutral, then the number density of protons, n_p, requires an equal number density of free electrons, $n_e \approx n_p$. However, at $T \approx 2m_e c^2/k$, thermal e^+e^- pair production is copious:

$$\gamma + \gamma \rightarrow e^+ + e^-.$$

If we denote the average energy of a photon at temperature T by E_γ, and the reduced Compton wavelength $\lambdabar = \hbar c/E_\gamma$, then the number density of thermally produced e^-e^+ pairs is of order

$$n_{\text{pairs}} \approx \frac{1}{\lambdabar^3} \approx \left(\frac{kT}{\hbar c}\right)^3, \qquad (26.77)$$

where we have set $E_\gamma \approx kT$ to within factors of order unity. From this relation, and the assumption that $n_e \approx n_p \approx n$, the total nucleon number density, it follows that

$$\frac{n_{\text{pair}}}{n_e} \approx \frac{n_{\text{pair}}}{n} \approx \left(\frac{kT}{\hbar c}\right)^3 \frac{1}{n} \approx 10^8. \qquad (26.78)$$

Since the ratio T^3/n may be considered nearly constant, (26.78) should hold as long as the universe was hot enough to maintain e^-e^+ pairs.

The principal coupling between electron pairs and photons is Compton scattering,

$$\gamma + e^\pm \rightarrow \gamma + e^\pm,$$

in which energy exchange between radiation and matter occurs. The time-scale for $e^\pm$ and γ to equilibrate thermally is, to order of magnitude,

$$\tau_{\gamma e} \sim (n_{\text{pairs}} \sigma_T c)^{-1} \sim \frac{(\hbar c/kT)^3}{\sigma_T c} \approx 10^{-21} \text{ sec}, \qquad (26.79)$$

where σ_T is the Thomson cross section. Since $\tau_{\gamma e} \ll \tau_{\exp}$ during this epoch, matter and radiation will be coupled, having the same temperature.

Neutrinos

The principal constituents of the universe for $T \lesssim 10^{12}$ K so far include free electrons, nucleons, photons, and thermally produced e^+e^- pairs. To this list must be added the neutrinos and antineutrinos. These will be coupled to the electron and positron by thermal pair-production processes

$$e^- + e^+ \leftrightarrow \nu_e + \bar{\nu}_e \qquad (26.80)$$

and the analogue of Compton scattering; for example,

$$e^- + \nu_e \rightarrow e^- + \nu_e. \qquad (26.81)$$

The rate of (26.80) is strongly dependent on the temperature (see Chapter 13). The scattering (26.81) represents the primary mechanism coupling neutrinos and matter. At $T \approx 10^{11}$ K, the cross section for $e^- \nu_e$ scattering is of order $\sigma_{\nu e} \approx 10^{-42}$ cm^{-2}, and the time-scale for $e^- \nu_e$ scattering is of order

$$\tau_{e\nu} \sim (n_{\text{pair}} \sigma_{\nu e} c)^{-1} \approx 10^{-3} \text{ sec}. \quad (26.82)$$

For $T \lesssim 10^{11}$ K, $\tau_{e\nu} > \tau_{\text{exp}}$, and the neutrinos decouple thermally from the rest of the matter in the universe. Prior to decoupling, the neutrinos will have had a thermal spectrum characterized by the temperature $T(t)$. There is no clear theoretical evidence that neutrinos should be more plentiful than antineutrinos in the earliest epochs. It is usually assumed that slightly more ν_e than $\bar{\nu}_e$ existed prior to decoupling, in which case the neutrino chemical potential $\mu_\nu \ll kT$. The ν_e spectrum in subsequent epochs would be given by

$$n_{\nu_e}(E_\nu, t) = \frac{4\pi E_\nu^2}{c^3 h^2} \frac{1}{e^{E_\nu/kT(t)} + 1}. \quad (26.83)$$

When $T(t)$ reaches about 10^{10} K, the $e^- e^+$ pairs annihilate, and the energy released goes into the radiation field and (via Compton scattering on e^-) into the matter. Since the neutrinos have already decoupled, their temperature will be unaffected by pair annihilation; so the radiation-matter component of the universe will be hotter than the neutrino component. Examination of the entropy change during this process shows that

$$T_\nu(t) \simeq (4/11)^{1/3} T_\gamma(t) = 0.714 \, T_\gamma(t). \quad (26.84)$$

As we will see, this relation should still hold now if $T(t)$ is understood to be the radiation temperature.

Neutron-Proton Ratio

The ratio of free neutrons to free protons will also change during the expansion as a result of the weak interaction

$$n + \nu_e \rightarrow p + e^-. \quad (26.85)$$

Supposing that the free nucleons obey Boltzmann statistics, and taking the chemical potentials $\mu_\nu \approx \mu_e \approx 0$ for the neutrino and electron, we find that the relative number of n to p in thermal equilibrium is given by

$$\frac{N_n}{N_p} = \exp[-(m_n - m_p)c^2/kT]. \quad (26.86)$$

The cross section $\sigma_{p \rightarrow n}$ for the reaction (26.85) depends on E_ν^2, and is of order 4×10^{-43} cm^2 at 10^{10} K, and $n \approx 7 \times 10^{30}$ cm^{-3}. The reaction time-scale is of order

$$\tau_{n \rightarrow p} \approx (\sigma_{n \rightarrow p} n_\nu c)^{-1} \approx 10 \text{ sec}, \quad (26.87)$$

and is decreasing as $T(t)^{-5}$. Since radiation is still the dominant source of energy, the expansion rate

$$\tau_{\text{exp}} = (\dot{a}/a) \approx T(t)^{-2} \approx 5 \text{ sec} \quad (26.88)$$

when $T \approx 10^{10}$ K. It is at about this epoch that the n/p ratio is fixed. For $T > 10^{10}$ K, the decay rate $\tau_{n \rightarrow p}$ is so rapid that it is effectively instantaneous relative to τ_{exp}, and N_n/N_p is given by the equilibrium relation (26.86). For T much less than 10^{10} K, $\tau_{n \rightarrow p}$ is too slow for (26.85) to be significant. Detailed calculations indicate that the ratio is fixed between 10^{10} K ($N_n/N_p \approx 0.1$) and 5×10^8 K ($N_n/N_p \approx 0.2$).

Primeval Helium

No nuclei (apart from the proton) can survive photodissociation at the temperatures predicted for the very early universe. But as the expansion proceeds, lowering $T(t)$, thermonuclear reactions might eventually be expected to result in a distribution of more massive nuclei. If the early universe were relatively more dense than predicted by current conditions (so that interparticle separations at 10^{12} K, for example, were many times smaller than $r_0 \approx 10^{-9}$ cm), many-body effects might be expected to lead to nuclear statistical equilibrium (Section 12.1). However, at the densities predicted when $T \lesssim 10^{12}$ K, only two-body reactions are frequent enough to be important. The absence of stable nuclei with $A = 5$ and $A = 8$ effectively impedes the buildup of more massive nuclei. This does not present a serious problem, since nuclei heavier than $A = 4$ may be produced by stellar nucleosynthesis (Chapter 12). The set of reactions

$$p + n \rightarrow H^2 + \gamma, \quad (26.89)$$

$$H^2 + H^2 \rightarrow He^3 + n \rightarrow H^3 + p, \quad (26.90)$$

$$H^3 + H^2 \rightarrow He^4 + n, \quad (26.91)$$

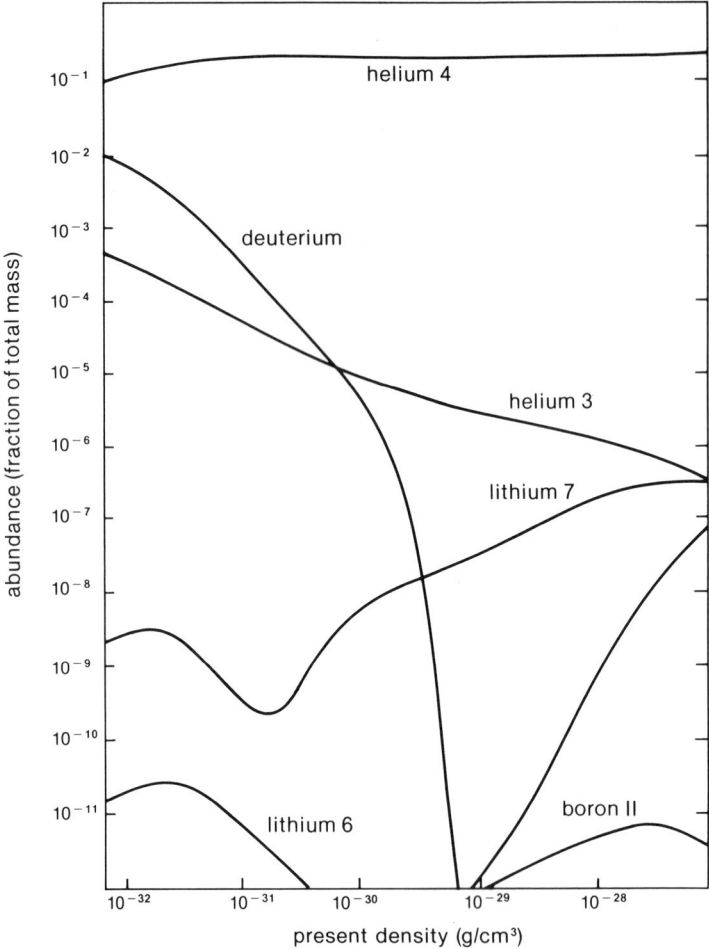

Figure 26.7. Abundance of light nuclei in the universe as a function of present mass density of the universe (relative to hydrogen).

can only occur when the temperature is in the range 10^{10} K to several times 10^8 K. These reactions are believed to be the primary steps that produce the He^4 and H^2 abundances observed throughout the universe. The first (26.89) is the slowest, and determines the overall rate of the network. The H^2 rapidly burns through to He^4. The equilibrium abundances of free neutrons X_n, free protons X_p, and deuterons X_D satisfy the Saha equation,

$$\frac{X_n X_p}{X_D} = \frac{4}{3} \frac{(2\pi kT)^{3/2}}{nh^3} \left(\frac{m_n m_p}{m_D}\right)^{3/2} e^{-Q/kT}, \quad (26.92)$$

where $Q = (m_n + m_p - m_D)c^2$, and the statistical weights of the neutron and proton are two. The deuteron's spin $s_D = 1$; so its statistical weight is 3. The total baryon number density n decreases with expansion, and the equilibrium concentration shifts from free nucleons to H^2. Model calculations that include reactions (26.89) to (26.91) predict that the primeval He^4 abundance Y lies in the range $0.23 \lesssim Y \lesssim 0.28$. Figure 26.7 shows predicted nuclear abundances of light elements as a function of the present mass density of the universe. The He^4 abundance is essentially constant, but the deuterium abundance decreases rapidly with increasing $\rho(t_0)$. Recent measurements of the ratio H^2/H^1 range from about 10^{-4} to 10^{-5}, corresponding to a deuterium abundance $2 \times 10^{-4} \gtrsim X_D \gtrsim 2 \times 10^{-5}$; from Figure 26.7 the present mass density would then be in the range $3 \times 10^{-31} \lesssim \rho(t_0) \lesssim 10^{-30}$ g cm^{-3}, which is consistent with the estimated mass density due to galaxies, $\rho_G \simeq 3 \times 10^{-31}$ g cm^{-3}, and predicts that $0.03 \lesssim q_0 \lesssim 0.06$. If all the H^2 in the universe is of cosmological origin, then the universe should be open.

Stellar hydrogen burning produces He^4, but most of this is locked up in dead stars. Observations of heavy element mass fractions show a strong correlation with stellar age, which is consistent with their production by stellar nucleosynthesis. The He^4 abundance, however, appears to be nearly independent of stellar age. For example, X_{He} is in the range 20 to 30 percent in the young Orion nebula (age $\approx 10^7$ to 10^8 years), in the Sun (age $\approx 4 \times 10^9$ years), and in a planetary nebula in the globular cluster M15 (age comparable to that of the Galaxy itself). These observations are most easily understood if the He^4 abundance is a remnant of the primeval fireball.

Deuterium production and its release into interstellar matter by stellar processes is believed to be improbable, since, once formed in stars, H^2 tends to burn completely to He. In fact, it is generally assumed that the deuterium in the universe is primeval in origin. To the extent that this is a valid hypothesis, cosmic H^2 is a useful indicator of the early conditions in the universe.

The expansion quenches nuclear burning at the point where about a quarter of the mass is He^4; the composition of the universe remains essentially unchanged as it cools toward the epoch of hydrogen recombination at $T(t_R) \approx 4{,}000$ K. During this period, radiation and matter remain coupled because of Comptonization, and equilibrium at a common temperature is maintained. Probably the most important phenomenon associated with this period of the expansion is the development of density perturbations. It has long been thought that some of these may eventually develop into protogalaxies (see Chapter 31).

Recombination

Once $T(t)$ reached about 4,000 K, the electrons and protons recombined to form neutral hydrogen; the reduction in electron number density resulted in a reduction in coupling between matter and photons. At recombination, the radiation would have had a blackbody spectrum at temperature $T(t_R) = T_R$, and its subsequent evolution would be largely unaffected by the matter content of the universe (this is strictly true only if absorption and emission by cosmic gas is ignored). In effect, it should retain its black-body nature for all subsequent epochs.

The present-day appearance of this relic of the early universe can be found by several simple arguments. Denote the number density of photons whose frequency is ν_R at recombination by

$$n_\gamma(t_R) = \frac{1}{e^{h\nu_R/kT_R} - 1}, \tag{26.93}$$

where $T_R = 4{,}000$ K. The same photon observed now would have had its frequency redshifted to

$$\nu = \nu_R \frac{a(t_R)}{a(t_0)}. \tag{26.94}$$

If the spectrum is to remain Planckian during the subsequent expansion,

$$n_\gamma(t_0, \nu) \sim \frac{1}{e^{h\nu/kT_\gamma(t_0)} - 1}. \tag{26.95}$$

This holds if the radiation temperature satisfies

$$T_\gamma(t_0) = T_\gamma(t_R) \frac{a(t_R)}{a(t_0)}. \tag{26.96}$$

In other words, T_γ decreases as $a(t)^{-1}$. The matter temperature T_m, which equals T_R at recombination, will also decrease, but at a different rate. After decoupling, the matter pressure is given by the equation of state for an ideal, nonrelativistic gas, $P \sim \rho T_m$; if the expansion is adiabatic, then the pressure must also satisfy $P \sim \rho^{5/3}$. Combining these two relations for P gives $\rho \sim T_m^{3/2}$. During this epoch the continuity equation is given by (26.10), so that $\rho \sim a^{-3} \sim T^{3/2}$, or

$$T_m(t_0) = T_m(t_R) \frac{a(t_R)^2}{a(t_0)^2}. \tag{26.97}$$

Problem 26.14. Find the rate of decrease of the neutrino temperature $T_\nu(t)$ following decoupling, when the spectrum is (26.83).

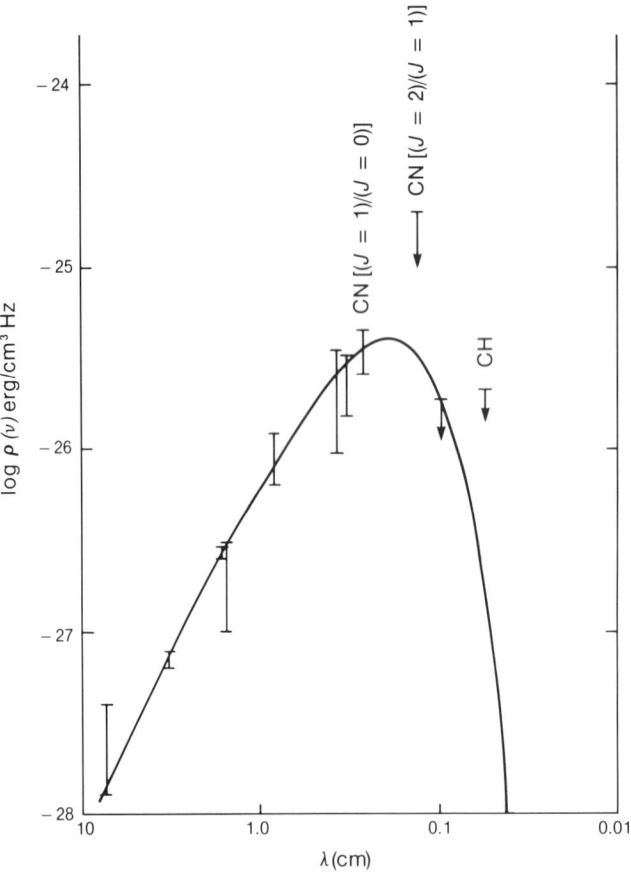

Figure 26.8. Cosmic background radiation.

The measurement of an isotropic black-body spectrum of apparent cosmic origin, and having a temperature $T \approx 2.7$ K is strong evidence for the Big Bang origin of the universe. The results of measurements from the radio to the infrared are shown in Figure 26.8, along with the energy density of radiation at 2.7 K. Figure 26.9 shows more recent results, which include balloon-based observations from the upper atmosphere, and which establish the Planckian nature of the spectrum. The black-body temperature indicates that the scale factor $a(t)$ has increased about 4,000 K/2.7 K $\approx$ 1,500 times since recombination. If the present average mass density is equal to 2×10^{-30} g cm^{-3}, then at recombination $\rho_R \approx 6 \times 10^{-21}$ g cm^{-3}, corresponding to $n_H \approx 3 \times 10^3$ cm^{-3}. In matter at these conditions, the entropy of the radiation per baryon (a parameter conserved during the subsequent expansion) is

$$\sigma(t) = \frac{4aT(t)^3}{3kn(t)} = \frac{4aT_R^3}{3kn_R} \approx 10^8 - 10^9. \quad (26.98)$$

Problem 26.15. A spherical volume contains N photons whose spectrum satisfies

$$n_\gamma(\nu) = \frac{8\pi\nu^2}{c^3} \frac{1}{e^{h\nu/kT} - 1} \text{ cm}^{-3}\text{ Hz}^{-1}.$$

Show that the photon spectrum in a subsequent epoch has the same form.

The Big Bang model predicts the existence of

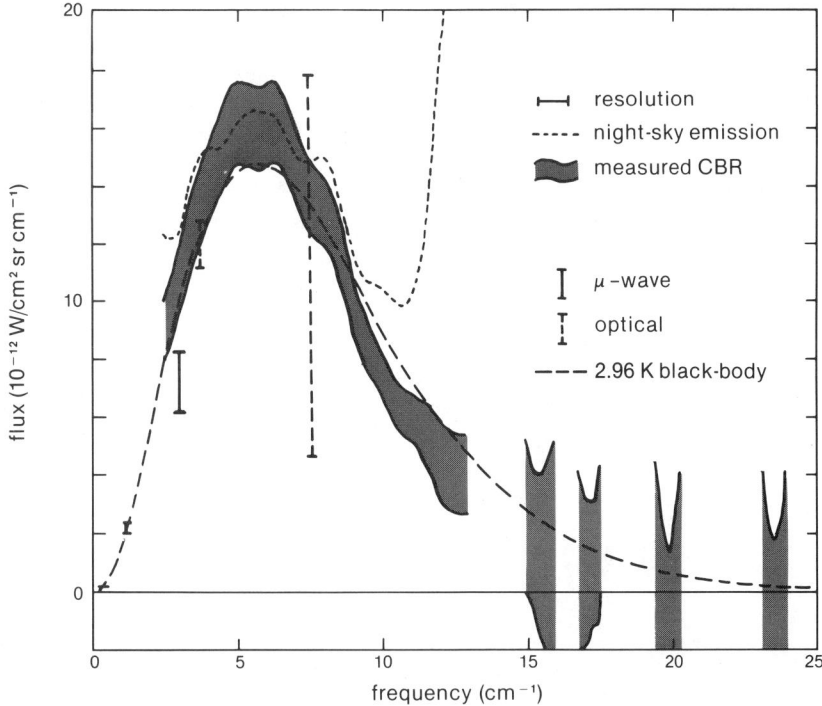

Figure 26.9. More recent measurements of cosmic background radiation.

several relics of the primeval fireball. These include the 2.7 K radiation background, the He^4 and H^2 abundances, a 1.8 K neutrino background, and the recessional motion of galaxies that increases with distance from the observer.

Problem 26.16. Explain why the universe prior to epoch $z \approx 10^3$ (recombination) is not accessible to observation by strictly electromagnetic methods. Discuss how earlier epochs could (in principle) be observed directly.

The isotropy and homogeneity inherent in the Friedmann models imply that the cosmic background radiation should appear isotropic. Attempts to detect anisotropy of cosmological origin have so far failed, indicating that the background is isotropic to about 10^{-3} K. The observations have revealed an unexpected effect, a smooth $\cos\theta$ variation that is interpreted as due to the relative motion of the local group of galaxies, with a velocity of about 550 km/sec.

26.7. THE MASS DENSITY OF THE UNIVERSE

The single most important parameter in the Friedmann model is the ratio

$$\Omega = \rho_0/\rho_c, \tag{26.99}$$

or, since ρ_c is fixed by Hubble's constant, the present energy density of the universe, $\rho_0 c^2$. According to the discussions in Section 26.3, the universe is open if $\Omega < 1$, closed if $\Omega > 1$. The current estimate of $H_0 = 55$ km sec^{-1} Mpc^{-1} gives $\rho_c = 5 \times 10^{-30}$ g cm^{-3}. The mass in galaxies is estimated to be $\rho_{gal} \approx 3 \times 10^{-31}$ g cm^{-3}, which is too small to close the universe. The universe contains other forms of energy besides baryons (which make up most of the mass in galaxies). The energy associated with the background radiation corresponds to a mass density $\rho_{3K} \approx 10^{-33}$ g cm^{-3}, and the mass density associated with neutrinos, (including those produced during stellar evolution, particularly during core collapse) is estimated to be $\rho_\nu \lesssim 10^{-33}$ g cm^{-3}. Evidence for intergalactic gas has been found in clusters of galaxies (Section 28.4); x-ray observations indicate that the gas contributes less than 10 percent of the mass in galaxies, or $\rho_{IG} \approx 10^{-34}$ g cm^{-3}. Finally, the masses of systems of galaxies (from pairs to rich clusters) predicted by the virial theorem requires that the galaxies contain hidden mass, from which $\rho_{\text{virial}} \lesssim 10^{-30}$ g cm^{-3}. The best estimates suggest that $\Omega \lesssim 1/50$; even if hidden mass is included, $\Omega < 1$, indicating that the universe is open. It has been argued that there may be additional mass in neutrinos that has escaped detection, and that long-wavelength gravitational radiation may fill the universe. There may also be dark matter of all kinds. In principle there could be enough mass associated with these components to close the universe, but as yet no data exist that convincingly bound their contribution to the total mass density.

Chapter 27

GALAXIES

We do not understand galaxies as well as we do their constituents, the stars and interstellar matter. Although the broad outlines of stellar evolution seem clear enough, we do not have a similarly clear picture of the sequence of stages making up galactic evolution. The Sun lies within one type of galaxy (a spiral), and we are able to observe part of its structure, and the behavior of some of its individual constituents, in detail. Unfortunately, only a portion of the Galaxy is visible to us; so we must turn to external galaxies to complete our knowledge of large-scale properties. Observations of our Galaxy and of external spiral galaxies supply complementary data about spirals in general. The constituents of galaxies include stars, gas and dust, starlight, cosmic rays, and magnetic fields. The relative amounts of each depend on the galactic type and on the location within the galaxy.

27.1. Galactic Morphology

Galaxies may be placed into one of four broad morphological classes, or Hubble types, largely on the basis of the galaxies' geometry as revealed by their luminosity distribution. The basic classes are: elliptical (E); lenticular (SO); spirals (S); and irregulars (I). About 1 percent of the galaxies observed cannot be placed in one of these categories, and are designated as peculiar (P). Each of the four basic Hubble types may be further subdivided, as we will discuss.

Elliptical galaxies range in shape from spherical to highly flattened systems. Since ellipticals do not contain a fundamental plane (whereas disk galaxies do), their absolute orientations are not known. What appears on photographic plates is a projected image. It is generally assumed that two of the three orthogonal axes of elliptical galaxies are equal. The possibility that some may actually be triaxial ellipsoids can not be ruled out (in this case the three principal orthogonal axes are unequal). We shall ignore this possibility, and assume that all ellipticals are in fact oblate spheroids. Denote the semimajor and semiminor axes of the projected ellipsoid by a and b, respectively. Then the nearest integer n obtained from $10(1 - b/a)$ describes a sequence, starting with (E0) and moving toward increasing flatness as n increases. In practice this classification scheme must be modified, since the ellipticity of galaxies is not strictly constant in radius. Conventionally, the assigned value corresponds to the maximum isophotal value, which is generally largest for isophotes near the galactic center. Ellipticals show no signs of disk structure, and exhibit luminosity

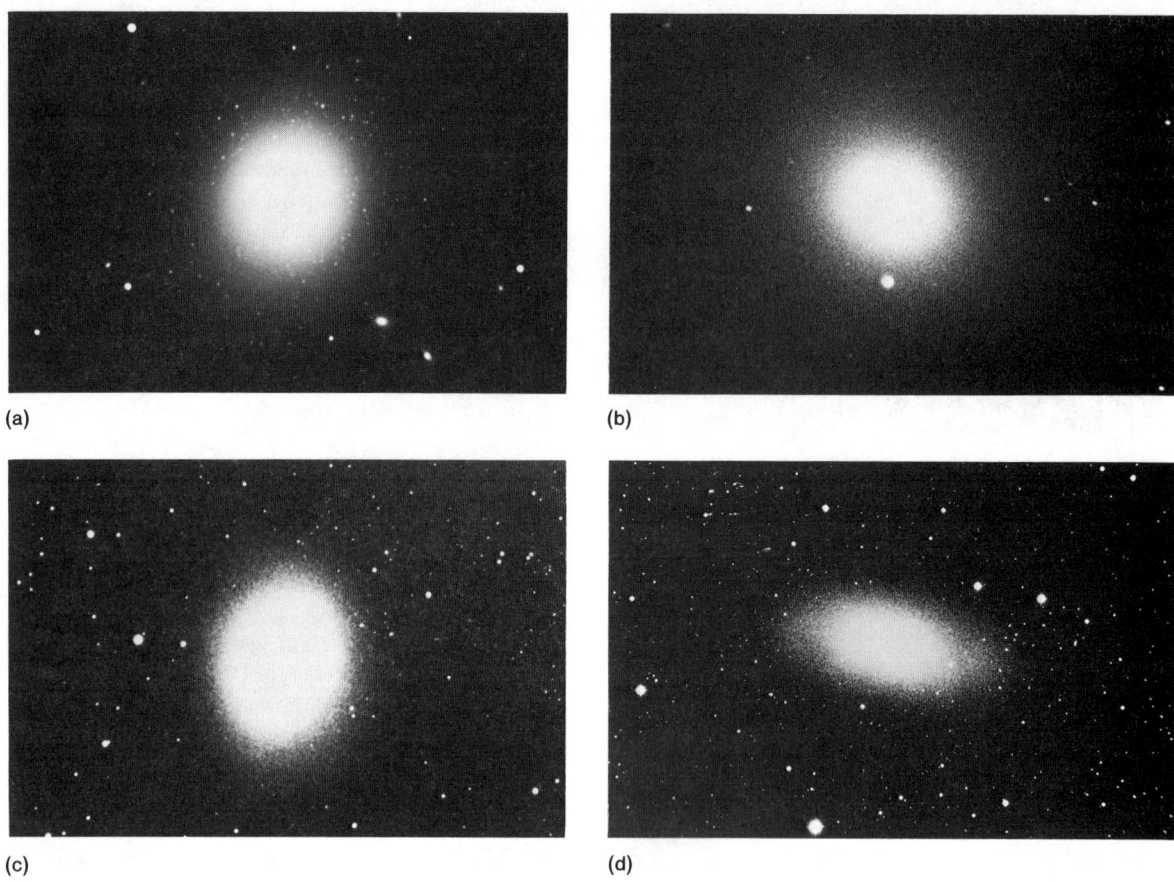

Figure 27.1. Elliptical galaxies: (a) NGC 4486, or M 87, type E0; (b) NGC 4472, type E1; (c) NGC 221, type E2; (d) NGC 205, type E6.

distributions (energy emitted per unit projected surface area per second) that vary smoothly from a compact nuclear region outward. The most highly flattened ellipticals have $b/a \simeq 0.3$ and are denoted E7. In all there are eight subclasses En. Galaxies of type E0 are called *spherical* galaxies, but remember that what appears in a photograph to be a spherical galaxy might be any En seen "face on." This likelihood can be estimated statistically by assuming that all ellipticals are randomly oriented. Normal ellipticals show no apparent structure. Figure 27.1 shows: NGC 4486 (M 87), an E0 (actually E0p, since it contains a jet) giant elliptical, the brightest in the Virgo cluster; NGC 4472, an E1; NGC 221, an E2 also in Virgo; and NGC 205, an E6 galaxy.

For $b/a \lesssim 0.3$, photographic plates show the existence of a fundamental plane normal to the major axes;

it resembles a thin disk. There is usually no evidence of spiral structure, but the luminosity distributions are similar to those observed in spiral galaxies. Systems having a fundamental plane but no observed spiral structure are called *lenticulars*. When seen nearly edge-on, they resemble a thin convex lens containing a nucleus. The lens is surrounded by an extended envelope. Ordinary lenticulars are denoted by S0; a typical example, NGC 1201, is shown in Figure 27.2(a). Plates of some lenticulars show a bar-shaped structure, which consists of stars and gas embedded in the fundamental plane. These are *barred lenticulars,* and are denoted by SB0. NGC 2859, shown in Figure 27.2(b), is an example. The transition from E to S0 types appears to be smooth.

Spiral galaxies show a thin disk containing dust and gas. The spiral structure, or arms, may originate near

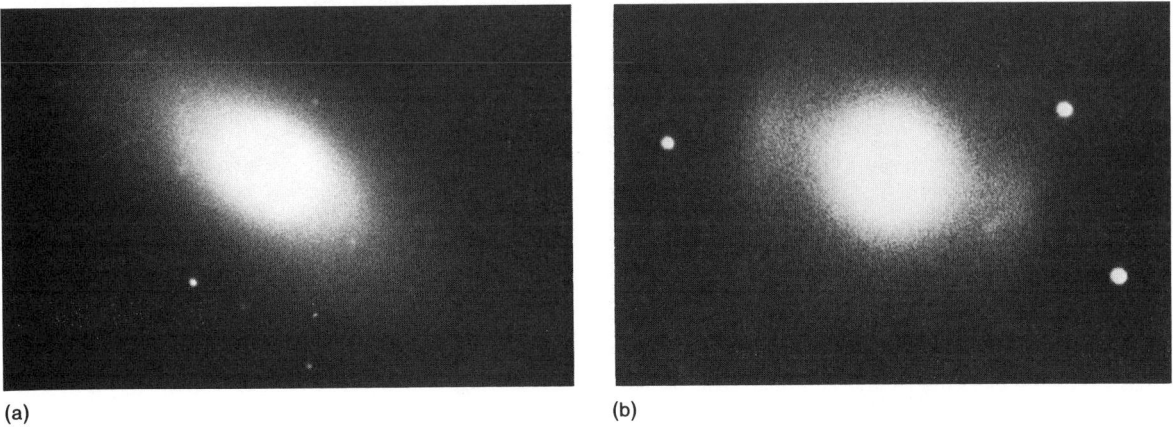

Figure 27.2. Lenticular galaxies: (a) NGC 1201, type S0; NGC 2859, type SB0.

Figure 27.3. Spiral galaxies: (a) NGC 2811, type Sa; (b) NGC 3031, type Sb; (c) NGC 628, type Sc.

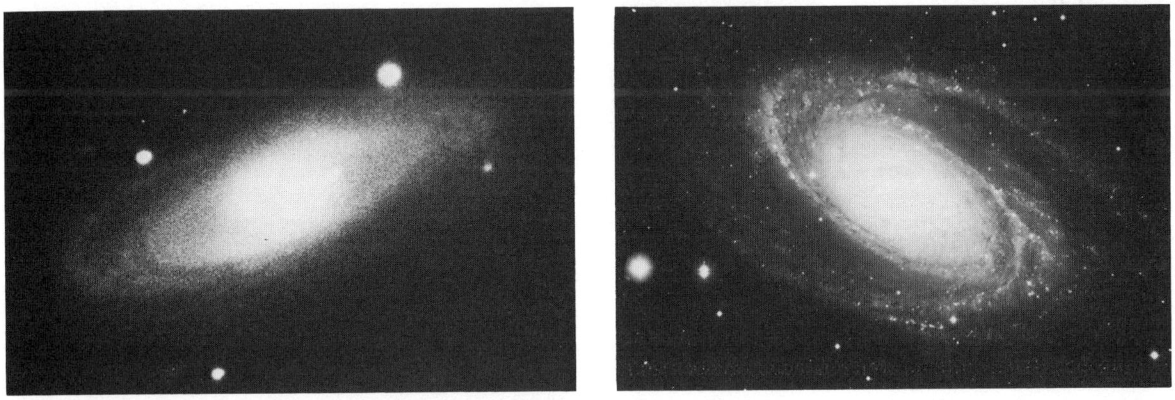

27.1. GALACTIC MORPHOLOGY / **491**

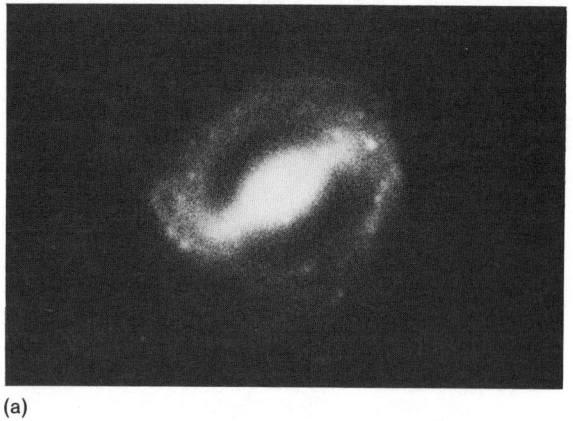

Figure 27.4. Barred spirals: (a) NGC 175, type SBa; (b) NGC 1300, type SBb; (c) NGC 2525, type SBc.

the nucleus (ordinary spirals, denoted S) or they may begin from the ends of a bar-shaped concentration of stars, dust, and gas (barred spirals, denoted SB). Each of the two families S and SB is further subdivided into types a, b, and c. Type a shows incipient spiral structure. The transition from b to c is characterized by a decreasing prominence of the nucleus, and less tightly wound spiral arms. The spiral structure in SB systems begins at the bar ends. In these systems, as in SBO galaxies, the bar lies in the plane of the disk and is concentric with the disk. Figure 27.3 shows representative ordinary spirals: NGC 2811, a multiarmed Sa; NGC 3031 (Andromeda), Sb; and NGC 628, an Sc. Figure 27.4 shows representative barred spirals: NGC 175, SBa; NGC 1300, SBb; and NGC 2525, SBc.

The relationship between the E, SO, and S types is (except for the subdivision of lenticulars into ordinary and barred) illustrated by Hubble's "tuning fork" diagram, Figure 27.5. Detailed photometry reveals much more structure than is suggested by this classification scheme, particularly for lenticulars and spirals. In some ordinary spirals, the arms appear to trail directly from the nucleus, as in NGC 598, in Figure 27.3(c); in others they originate tangentially from a bright ring surrounding the nucleus, as in NGC 2841. Similar phenomena occur in barred spirals. In NGC 1300, Figure 27.4(b), the arms trail from the bar ends, but in systems such as NGC 2523 (Figure 27.6), the bar terminates on a ring from which the arms appear to originate. Many galaxies originally believed not to contain recognizable structure, and therefore classed as irregular, have been found to contain weak spiral features. The large Magellanic cloud (LMC) contains weak spiral structure and a bar. The small Magellanic cloud (SMC) is a prototype for many irregular systems, known as Magellanic irregulars, which may be denoted by IM or IBM. The transition from barred spirals to Magellanic irregulars is represented by LMC, which is type SBM, and that from S to IM is

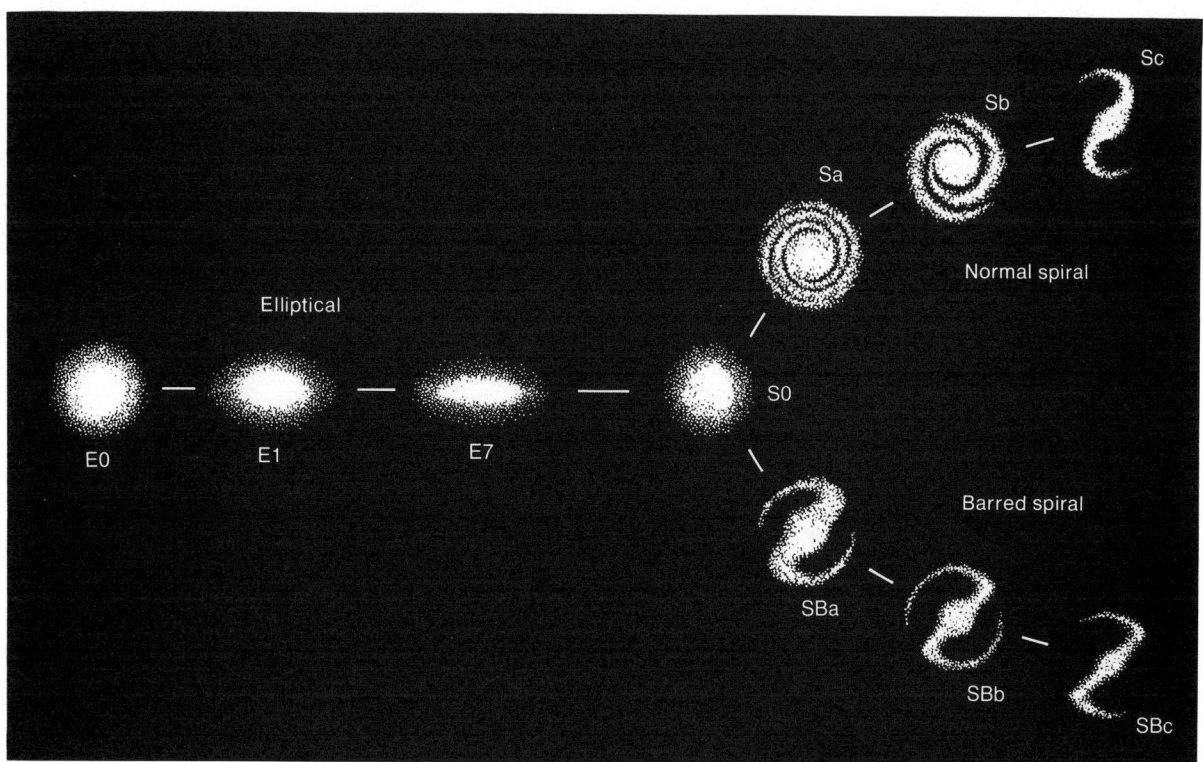

Figure 27.5. Schematic Hubble classification scheme for galaxies.

Figure 27.6. The galaxy NGC 2523, type SB(r).

27.1. GALACTIC MORPHOLOGY / **493**

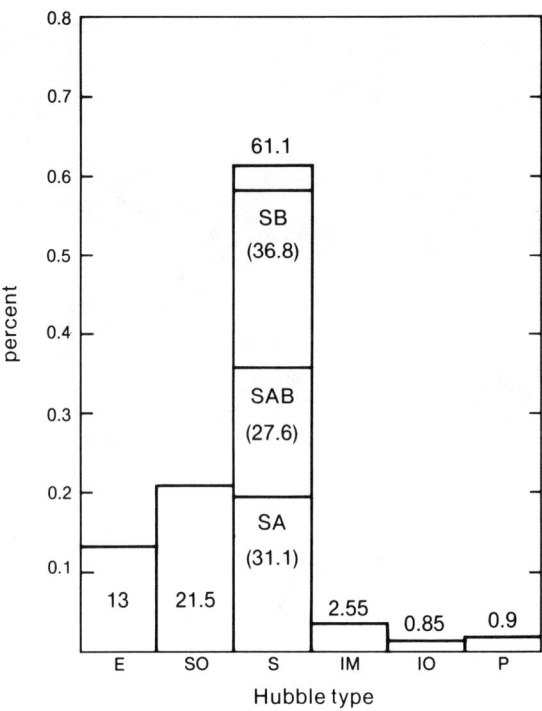

Figure 27.7. Frequency distribution of galaxies among Hubble types. The bin for spirals is divided into subclasses A, AB, and B; numbers in parentheses are the percentage of class S.

denoted SM. The remaining irregulars may be classed as IO. The distribution of observed galaxies between the basic classes E, SO, S, IM, IO, and P is shown in Figure 27.7

Finally, the transition from one Hubble type to another (E → SO; SO → S; S → SB, etc.) is, in all cases, regular and smooth. Extensions of the original Hubble classification scheme have been developed to take these recently discovered features into consideration.

Classification schemes are based largely on the apparent geometry of galaxies, and reflect underlying properties that appear to range continuously from one type to another. Observations of galaxies are limited primarily to photometric studies of light distribution, emission from neutral hydrogen 21.1-cm radiation), calculations of mass by means of the virial theorem or detailed studies of rotation curves in spirals, and integrated spectra. Each of these supplies valuable information about galactic structure, dynamics, and evolution.

27.2. SURFACE BRIGHTNESS OF GALAXIES

The total light output of a galaxy is important if its structure is to be understood. It is also important because large ellipticals represent one of the standard candles for distance measurements in cosmology. The total luminosity, or equivalently the magnitude, of a galaxy is difficult to define precisely and unambiguously for several reasons. First, galaxies do not have well-defined edges, and their extended regions are faint and have low luminosity gradients. Second, the light from a galaxy is redshifted by the expansion of the universe. This second effect is easy to allow for in principle, since the amount of redshift depends on the galaxy's distance. Unfortunately, distances to remote galaxies are themselves uncertain, and errors in distances directly affect galactic magnitudes corrected for redshift. The first effect—the absence of a well-defined edge—is conventionally compensated for by basing the magnitude on the amount of light emitted within a prescribed boundary of constant brightness (isophote). This may be taken as the limiting brightness of the night sky, about 26.5 magnitudes per square second of arc measured in the photographic band. Magnitudes measured in this way range from about -22 for giant ellipticals down to about -10 for dwarf ellipticals (dE). The range in absolute magnitude varies with type. Ellipticals range from $M_v \simeq -22$, the brightest normal objects in the universe, down to -9 for the dwarf elliptical Leo II in the local group. Spirals and lenticulars range from -22 down to about -16, and irregulars vary from about -18 down to -12. A range in absolute magnitude of -9 to -22 corresponds to a range in luminosity of $10^6 - 10^{11} L_\odot$.

Problem 27.1. Estimate the number of stars in a dE galaxy of absolute magnitude $M_v \simeq -9$.

In additon to total luminosity, a photoelectric scan of a galaxy yields information about its projected luminosity distribution, $I(r)$, defined as the apparent light intensity per unit area emitted by the galaxy. The luminosity distribution supplies information about the internal structure and dynamics of galaxies. Observations indicate that $I(r)$ is different for E, S0, and S types.

In E-type galaxies, $I(r)$ varies smoothly from the center outward, showing no major breaks or discontinuities. In fact, the observation of normal ellipticals can be quite well reproduced except near the nucleus by the expression

$$I(r)/I_0 = \begin{cases} (r/a + 1)^{-2}, \text{ for } r/a < 21.4, \\ 22.4\,(r/a + 1)^{-3}, \text{ for } r/a \geq 21.4. \end{cases} \quad (27.1)$$

Here a and I_0 are parameters that differ from one galaxy to another, and r is the distance measured from the center along the projected major axis. It is therefore the surface brightness of the two-dimensional image. For ellipticals less bright than $M_V \approx -17$, the observed distributions fall off more rapidly than indicated by (27.1). An alternative distribution that agrees with (27.1) to within the precision of current observations is

$$\log [I(r)/I(a)] = -3.3\,[(r/a)^{1/4} - 1]. \quad (27.2)$$

In this expression $r = a$ is the distance from the center within which half the light originates. The expression (27.2) fits all normal bright ellipticals ($M_V < -17$) well if the single parameter $I(a)$ is adjusted.

According to (27.2), the two-dimensional image of ellipticals reflects a single component system with a smoothly varying distribution of light. Most of the mass in ellipticals is believed to lie in stars, and stars supply most of the light. Therefore a knowledge of the three-dimensional (volume) luminosity is an important observational parameter. Because the orientation of observed ellipticals is not known (except statistically), the construction of volume distributions from projected surface distributions is difficult. For E0 galaxies with $I(r)$ given by the first term of (27.1), a volume luminosity at distance ρ from the center can be found, as

$$L(\rho) = \frac{L_0}{(\rho/\alpha + 1)^3}, \quad (27.3)$$

with α a scale factor. If we assume that the mass within ρ, $M(\rho)$, is proportional to $L(r)$, (27.3) suggests that E0 galaxies show a high degree of central concentration.

Photometry of lenticular systems reveals a composite structure, as shown schematically in Figure 27.8. The inner region is similar to the spheroidal distribution (27.2) for ellipticals. For $r/a > 0.1$, the approximation

$$\log [I(r)/I_0] = -\alpha r \quad (27.4)$$

fits the data. Lenticulars therefore appear to consist of a spheroidal nucleus, with the luminosity distribution typical of elliptical galaxies, embedded in an exponential disk. Three-dimensional models analogous to (27.3) for S0 galaxies are more difficult to construct, but suggest that $L(\rho)$ drops like $1/\rho^2$ outside the nucleus.

The luminosity distribution in S galaxies is complicated by the presence of spiral arms, patches of gas and dust, and orientation effects. Figure 27.9 shows photographic surface brightness plotted against apparent angular diameter in seconds of arc for a central cross section of NGC 628, an Sc galaxy about 9.2 Mpc distant, and seen face-on. The vertical scale gives the brightness in units of one star of apparent magnitude 26.5 per square second of arc. The figure shows the presence of spiral arms (three on either side of the nucleus) whose light distribution is highly irregular,

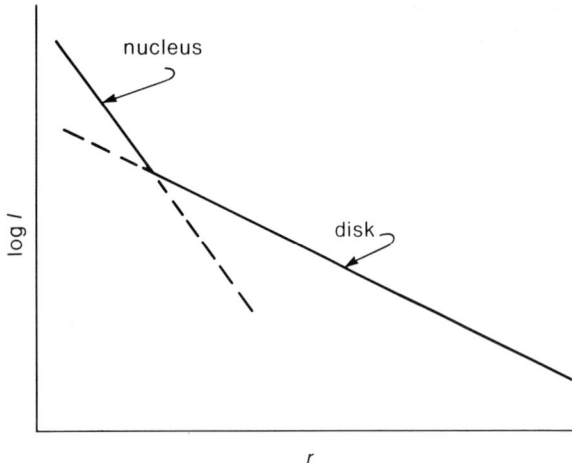

Figure 27.8. Projected luminosity distribution versus radius of components in lenticular galaxy (schematic).

and a smooth disk distribution (dashed curve). Differences in brightness within one spiral arm (a, b, or c) are large. Nevertheless, three basic components of $I(r)$ are distinguishable: (1) a disk component (dashed component in Figure 27.9); (2) a nuclear component; and (3) the spiral-arm component, obtained as the difference in brightness between the total light and (1). The nuclear and disk components are similar to the spheroidal and disk components in S0 types. In fact, the distributions when fit to ellipticals reproduce the observed distributions in the nuclei of spirals, and if the spiral component is removed, the underlying systems appear identical with the two component S0 galaxies described in the preceding.

The average distribution in photographic magnitude (pg) per square second for a sample of face-on Sc galaxies can be approximated by

$$m_{pg} = 21.3 + 5.5\, r/a \qquad (27.5)$$

for the disk, and for the nucleus by

$$m_{pg} = 20.15 + 50\, r/a. \qquad (27.6)$$

These expressions are plotted in Figure 27.10. The solid curve is the total observed distribution. The difference between it and the disk plus nuclear distribution represents light from the spiral arms.

In these Sc galaxies the disk contributes about 57 percent of the total light, the spiral arms about 41 percent, and the nucleus only about 2 percent.

Problem 27.2. The average luminosity distribution in late spirals can be obtained from (27.5) and (27.6), in the form

$$m_{pg} = \alpha_0 + \alpha_1(r/a). \qquad (27.7)$$

Assume that this holds for $(\alpha_0, \alpha_1) = (21.3, 5.5)$ for the disk, and $(20.15, 50)$ for the nucleus. Defining the total luminosity of each component as the light emitted within $r = a$, find the ratio L_N/L_D of total light emitted by the nuclear and disk components. It is sufficient to show first that

$$L \sim \frac{10^{-0.4\alpha_0}}{\alpha_1^2} \qquad (27.8)$$

for the total light from each component.

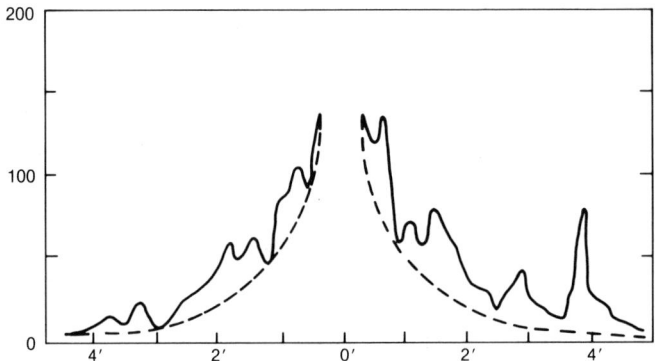

Figure 27.9. Photographic surface brightness versus apparent angular diameter (seconds of arc) for the Sc galaxy NGC 628.

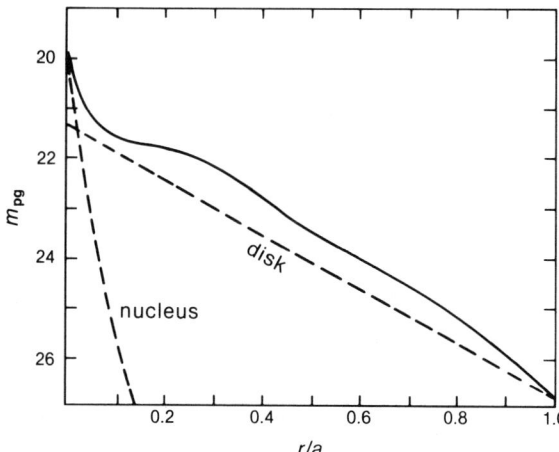

Figure 27.10. Photographic magnitude per square second of arc versus r/a for face-on spirals of type Sc (average of eight galaxies).

Photographs of SB galaxies suggest that the bar is a dominant feature. Photometry, however, usually indicates that it is a relatively small feature immersed in a nearly axisymmetric disk. For example, the average ratio of light emitted by the bar, L_B, to total light L_T for three SB galaxies is $L_B/L_T \simeq 0.12$, and the ratio of bar radius a_B to disk radius a_D averages $a_B/a_D \simeq 0.18$ in the same systems (the observed ratios in NGC 1313 are about average). The small value for a_B/a_D means that the asymmetry associated with SB galaxies is confined to the inner disk. In fact, most of the light comes from what appears to be an axisymmetric disk with exponential luminosity distributions like (27.4), as it does for ordinary spirals. To within current observational uncertainties, both S and SB galaxies have the same $I(r)$ curves outside the bar.

The bar itself appears to follow the $(r/a)^{1/4}$ distribution (27.2) typical of elliptical systems, particularly in barred spirals earlier than SBb. The bar is usually coincident with the disk center in early types. For SBc galaxies there is direct evidence that the bar center is displaced from the disk center by a small amount. The bar in LMC, an SBb type, is displaced about 0.5 kpc, which corresponds to 15 percent of the disk radius. The cause of this additional asymmetry is not clear.

Despite the uncertainties involved in the photometry of individual galaxies, E, SO, and S types can be clearly distinguished in terms of luminosity distributions. Ellipticals appear to be single-component (spheroidal) systems. This means that the luminosity density of the outer regions is not independent of that of the central regions. SO and S galaxies, on the other hand, exhibit at least two components, a central spheroidal distribution similar to that of ellipticals, and a separate disk component which is approximately exponential in r/a. Although there are some exceptions, the intensity scale I_0 in SO and S galaxies is nearly constant (21.65 magnitudes per square second of arc in the blue) despite differences in absolute magnitude of up to a factor of five. The length scale $1/\alpha$ of the exponential disk (27.4) ranges from 1 to 5 kpc in SO-Sb types, showing no apparent trend with type, but tends to decrease to $\lesssim$ 2kpc in Sc-Sm spirals.

Because ellipticals appear to have single-component luminosity distributions like (27.4), the amount of light emitted by their outer portions is not independent of the luminosity of their centers, and a change in central brightness results in an overall change in total brightness. Spirals, however, contain nuclear and disk components that may differ independently; so the ratio of light emitted by the nucleus to that from the disk may change in such a way that the total brightness remains constant. In fact, spirals of about the same absolute magnitude show a dramatic reduction in L_N/L_D as we go from SO through Sa to Sc. The difference between spheroidal systems and spheroidal plus exponential disk systems contains important clues about the dynamic and evolutionary properties of galaxies, as we will see later.

If the absolute magnitude M_V of a galaxy is known, then the empirical relation

$$M_V = -6.0 \log A + 7.14 \qquad (27.9)$$

gives a reasonable measure of its projected major axis A, in pc. The observed range $-9.1 \gtrsim M_V \gtrsim -22$ yields major axes between 7 kpc and 50 kpc. The smallest dE galaxies appear to be no larger than globular clusters, and have essentially the same absolute magnitude.

Problem 27.3. What is the estimated luminosity of the Galaxy? The absolute magnitude of M 31 is $M_V = -21.1$. What is its luminosity and radius?

Color Distributions in Galaxies

Color distributions in galaxies may be obtained by exposing specific color-sensitive plates to regions cen-

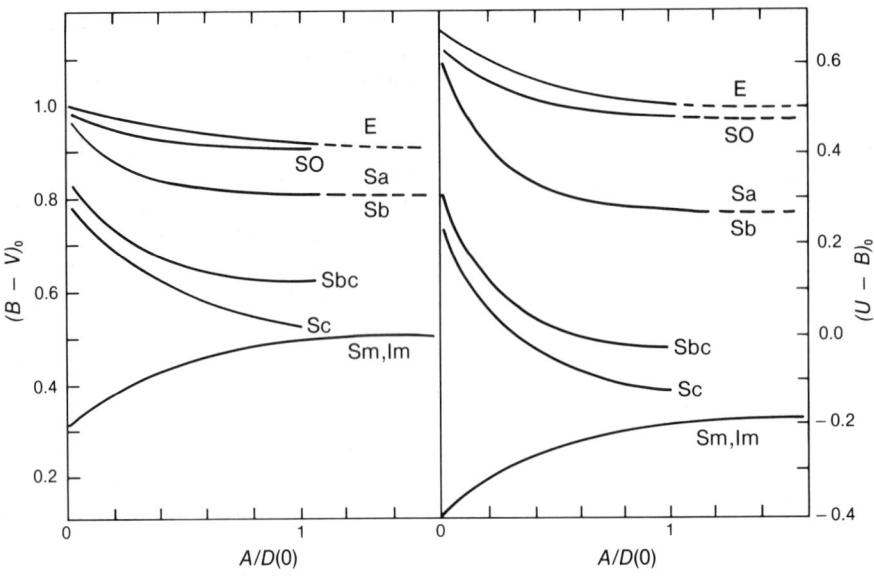

Figure 27.11. Mean normal color-aperture relations for galaxies. Here A is the aperture diameter, and $D(0)$ is the face-on diameter.

tered on a galactic nucleus and increasing the aperture diameter A. As A increases toward the face-on diameter $D(0)$ of the image, light from regions further from the nucleus is included in the exposure. Figure 27.11 shows normalized $B - V$ and $U - B$ color distributions, averaged over various galactic types. Increasing $A/D(0)$ means that more of the disk or nonnuclear light is being included. In all except SM and IM irregulars, the colors shift toward the blue as more light from nonnuclear material is included. The trend in E galaxies is small (only a few percent increase over the image), but becomes significant for the spirals.

Ellipticals show nearly constant color distributions, suggesting that the same stellar types dominate the luminosity throughout these systems. Giant ellipticals ($M_V \lesssim -16$ mag) have $B - V \simeq 0.9$; dwarf ellipticals ($M_V \gtrsim -14$) have colors resembling globular clusters ($B - V \simeq 0.6$). There is a rapid and rather smooth decrease in integrated $B - V$ for ellipticals with M_V in the range -14 to -16, as shown in Figure 27.12. This dependence of integrated color on M_V suggests that the stellar composition is dependent on absolute magnitude, and may mean that the processes of star formation are different in large and in small ellipticals. Or it may indicate that the galactic formation and evolutionary processes are different.

In both S0 and S galaxies, the colors of the spheroidal and disk components are different, and differ with type as shown in Figure 27.11. The nuclear regions are redder than the disk. The steady decrease in $B - V$ from Sa to Sc reflects the increase in relative numbers of young stars in later-type spirals. There is also a trend toward bluer light in the outer disk, an effect greatest in Sc types. In these galaxies the outer arms appear to contain younger stellar populations, and therefore more gas, than do the inner arms. In fact, the average $B - V$ color of the spiral arms in Sc galaxies (total color minus disk color) ranges from about $+0.4$ for $r/a \simeq 0.1$ to -0.1 beyond $r/a \simeq 0.6$. The stellar types dominating the light in spiral arms in Sc galaxies therefore range from early F to late B types. The average color of the disk, $B - V \simeq 0.5$, is characteristic of late F to early G-type stars. The average color $B - V \simeq 0.8$ of the spheroidal component is more nearly representative of early K types. The picture that emerges for Sc galaxies indicates a rich substructure of stellar types: the oldest and reddest stars having formed in the nucleus; a flat population of solar-like stars, called the disk population, more or less uniformly distributed; and a blue population of young to very young, recently formed stars in the spiral arms. Although these details are more difficult to resolve in early-type spirals (Sa-Sb), Figure 27.11 indicates a smooth transition. We may thus adopt the qualitative features of Sc galaxies as applicable to ordinary spirals in general.

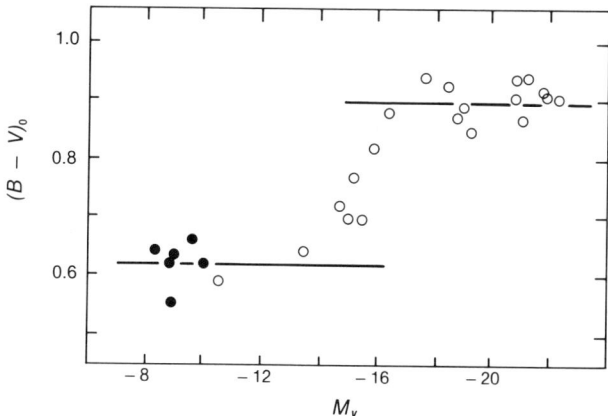

Figure 27.12. Intrinsic color index of elliptical galaxies (open circles) and globular clusters (filled circles) plotted against their absolute magnitude.

The nuclei (spheroidal components) of spiral and lenticular galaxies become increasingly blue in later types. The average shift in color is small but systematic (about 4 percent) in Sa-Sb, but reaches 60 percent for late Sc, indicating that the stellar populations dominating the nuclear light are correlated with spiral type.

The amount of obscuring matter present in galaxies may be estimated from studies of color versus apparent axial ratios b/a. Any gas or dust present in a galaxy will scatter and absorb starlight. Scattering on average does not change either the color or luminosity, but absorption will redden the apparent color. Reddening due to scattering can in principle be separated from cosmological redshift effects when the distance is known. The amount of reddening depends on how much obscuring matter is traversed by the light. In an E7 galaxy seen edge-on (smallest b/a value), the maximum amount of gas and dust would be traversed. As the line of sight becomes more nearly face-on ($b/a = 1$), a lesser amount will be encountered. Therefore, if a sample of E7 galaxies having random orientation is observed, and if they contain enough obscuring matter to modify the light, the observed color will redden according to the value of b/a, which ranges from 1.0 to its minimum value of 0.3. The analysis is complicated in practice by the probability that some galaxies that appear to be E6 may actually be E6 seen edge-on, or E7 titled somewhat to the line of sight. Within the limits of observations, there is no correlation between color and apparent axial ratio, indicating that ellipticals and S0 normally do not contain significant amounts of gas or dust.

Similar arguments may be applied to the color of nuclei in spirals for various inclination angles. The observations again indicate an absence of dust or gas in early types, but indicate its presence in later types.

Problem 27.4. Under what conditions could gas and dust exist within spiral nuclei and still yield colors independent of inclination for a given spiral type? Suggest a way to check this possibility observationally.

27.3. GALACTIC MASSES

The mass of a galaxy is a basic parameter for dynamical models, and represents an important constraint on the distribution of stellar types making up the total luminosity. The distribution of mass as a function of galaxy type is also important cosmologically. Galactic masses may be estimated from rotation curves (in single spirals) or stellar velocity dispersions (in ellipticals); from Kepler's third law in binary galaxies; and from the virial theorem applied to rich clusters of galaxies. The first two methods yield reasonably consistent mass estimates, but the third tends to predict values that appear to be systematically large. Each of these methods is considered below.

Velocity Dispersions (E Galaxies)

Stars in galaxies move in orbits determined by a mass distribution that appears to be smooth. Although stellar encounters (star-star scattering, as opposed to an individual star scattering off the background mass distribution) are possible, we will see in Section 28.2 that such events are rare in present-day systems. We therefore assume that the stars are in equilibrium motion and that the virial theorem is applicable. Then

$$2T + \Omega = 0. \quad (27.10)$$

The gravitational potential energy Ω is

$$\Omega = -\frac{1}{2}\sum_{i\neq j}\frac{M_i M_j G}{|\mathbf{r}_i - \mathbf{r}_j|} \equiv -\alpha\frac{M^2 G}{R}, \quad (27.11)$$

where the sum is over stars of mass M_i. The right-hand side is written in terms of the galactic mass M, a characteristic dimension $2R$ of the galaxy, and a constant α, which depends on the type. The total kinetic energy is the sum of rotational and random kinetic contributions:

$$T = \frac{1}{2}\sum_i M_i v_i^2 + T_{\rm rot} \equiv \frac{1}{2}M\langle v^2\rangle + T_{\rm rot}, \quad (27.12)$$

where $\langle v^2\rangle$ is the average of v_i^2 with respect to the galactic center. The term $T_{\rm rot}$ allows for rotational motion of the galaxy as a whole. Although this contribution is large for spiral disks, it is expected to be small in ellipticals. Therefore we consider the $\langle v^2\rangle$ term first. A simple model for $\langle v^2\rangle$ is suggested by observations of stellar motions in the spheroidal component in our Galaxy (recall that the spheroidal component in spirals resembles that in ellipticals). These stars generally have highly eccentric galactic orbits, which, to first approximation, resemble anharmonic oscillations passing through the Galactic center. Furthermore, their velocity distribution appears to be Gaussian, with dispersion σ^2.

Problem 27.5. Assume that stars in elliptical galaxies are normally distributed in each velocity component:

$$\phi_i(v_i) = \frac{1}{(2\pi)^{1/2}\sigma_i}e^{-v_i^2/2\sigma_i^2}. \quad (27.13)$$

Show that

$$\langle v^2\rangle = \sigma^2, \quad (27.14)$$

where $v^2 = v_1^2 + v_2^2 + v_3^2$ and $\sigma^2 = \sigma_1^2 + \sigma_2^2 + \sigma_3^2$.

Only the line-of-sight component of a star's velocity is observable in nearby galaxies, but if the velocity dispersions σ_i are all the same, then $\sigma^2 = 3\sigma_r^2$ where σ_r^2 is the dispersion in this component. Combining (27.11) and (27.12) in the virial theorem, and defining the rotational kinetic energy by

$$T_{\rm rot} = \frac{1}{2}\beta M\langle v^2\rangle, \quad (27.15)$$

we find for the total mass of the galaxy

$$M = \frac{3\sigma_r^2 R(1+\beta)}{\alpha G} = 7\times 10^9 R\sigma_r^2\frac{(1+\beta)}{\alpha} M_\odot, \quad (27.16)$$

where R is in kpc and σ_r is in 100 km/sec in the final expression. The constants α and β depend on the geometry (e.g., galactic type).

Problem 27.6. Derive (27.16). In a giant E0 galaxy with $R \simeq 50$ kpc, $\sigma_r \simeq 400$ km/sec. What is the galaxy's mass? Assume $\alpha = 1$ and $\beta = 0$. Are these reasonable estimates for α and β? How would the result change if more nearly realistic values were available?

Rotational motion has been observed in the central regions of some normal ellipticals, of order 50 km/sec per 0.1 kpc, but no direct evidence exists for rotation in the outer portions. Since the average flattening in E types is $1 - b/a \simeq 0.6$, these galaxies should not be rapidly rotating. For early types,

$$\epsilon \equiv 1 - b/a \ll 1,$$

a recent theoretical estimate of $T_{\rm rot}$ yields

$$\beta = 8\epsilon/(5 - 8\epsilon),$$

which, for the E2 galaxy M 32 ($\epsilon = 0.2$), gives $\beta = 0.47$.

An estimate of the parameter α appearing in the

potential energy is possible if we assume that the distribution of mass (e.g., stars) follows that of light,

$$L(r) = KM(r), \qquad (27.17)$$

where $L(r)$ is the three-dimensional luminosity interior to r as constructed from the two-dimensional observed luminosity distribution, such as (27.2). It can then be shown that for E0 galaxies

$$\Omega = -0.33 \, M^2 G / R_{1/2}, \qquad (27.18)$$

where half the light originates within a sphere of radius $R_{1/2}$. In En galaxies ($n \neq 0$), three modifications to (27.18) must be made. First, Ω is increased by a factor $(a/b)^{2/3}$. This is a volume-reduction effect, since increasing the ellipticity (for fixed mass) brings most mass elements closer together. Second, $R_{1/2}$ is replaced by $R'_{1/2}$, which is measured along the minor axis of the apparent ellipse. Third, since the absolute orientation of the galaxy is not known, a statistical factor multiplies the coefficient 0.33. This factor reflects the probability that the apparent axial ratio a/b seen in projection averages 16 percent less than the true ratio, assuming random distribution in space. For En we then have

$$\Omega = -0.35 \left(\frac{a}{b}\right)^{2/3} \frac{M^2 G}{R'_{1/2}}. \qquad (27.19)$$

Masses of elliptical galaxies obtained from velocity dispersions range from $3.6 \times 10^{10} \, M_\odot$ for M 32 (type E2) to $3.5 \times 10^{12} \, M_\odot$ for an E0p galaxy, NGC 4486.

Rotation Curves (Spiral Galaxies)

Population I stars near the Sun are observed to move in nearly circular orbits about the Galactic center. Typical rotational velocities are of order 200–300 km/sec, and their random motions (noncircular component) are usually of order 0 to 10 km/sec. Consequently, their kinetic energy is mostly rotational, and mass estimates for spirals must come from other models than (27.16). A very rough estimate may be found by using Kepler's third law. Suppose a star or HII region moves in a circular orbit of radius a about the Galactic center with speed v. Then, if its centripetal acceleration is balanced by the gravitational pull of matter within its orbit,

$$M = \frac{v^2 a}{G} = 2 \times 10^{10} M_\odot \left(\frac{v}{250}\right)^2 \left(\frac{a}{10}\right), \qquad (27.20)$$

with v in km/sec and a in kpc. The applicability of this method assumes that the observed star or HII region is far enough from the nucleus that its orbit encloses most of the galaxy's mass.

Although (27.20) is useful for order-of-magnitude purposes, it is limited by the fact that not all the enclosed mass is spherically distributed about the nucleus, and that an arbitrary amount of spherically distributed matter may lie outside the orbit a without influencing the star's motion. The first difficulty may be removed by observing the rotational velocity of disk matter (stars and HII clouds) as a function of radius from the nucleus to the outermost observable regions of the galaxy (the rotation curve). Assuming that the inclination angle of the disk is known, the velocity of matter in the disk may be found. A distribution of mass is then assumed, from which the expected velocity curve can be calculated. When a distribution is found that yields the observed velocity curve, it may be integrated to give the total mass interior to the last point of the rotation curve. Examples of this procedure are considered in detail in Section 29.3. This method circumvents the first difficulty with the simple estimate (27.20), but is nonetheless insensitive to the amount of mass exterior to the last observed point of the velocity curve. Both methods therefore yield lower limits to the mass of the galaxy.

The velocity-curve method may also be used for SB galaxies. However, it is useful only for those that have nearly axisymmetric luminosity (and thus, presumably, mass) distributions.

A modification using 21.1-cm velocity profiles from HI regions, which may extend well beyond the optical disk, may give more nearly complete mass estimates. This assumes, however, that HI regions throughout the disk move with the stellar and HII components. Recent observations of forbidden line emissions ([NII] and [SII]) from interstellar gas near hot stars in galactic disks extend measurements of velocity curves to distances of order 100 kpc from galactic nuclei. For example, the rotation curve for M 31 has been observed optically out to radial distances of about 20 kpc from the nucleus, and radio (21.1 cm) observations extend it to about 50 kpc. Both sets of data indicate that the rotational velocity is essentially contant ($v \sim$ 200 km/sec) for $r \gtrsim 16$ kpc. This implies that the mass distribution in this region is linear, $M(r) \sim r$. Optical studies, on the other hand, indicate that the surface brightness here is very low. Masses of spiral galaxies obtained from rotation curves out to about 10 kpc range from $1.3 \times 10^9 \, M_\odot$ for the Sc galaxy NGC 6503 to $3.4 \times 10^{11}\ M_\odot$ for the Sb galaxy M 31 (NGC 224) and give mass-to-light ratios M/L of order $10\ M_\odot/L_\odot$. If the rotation curve observed in M 31 remains flat to 100 kpc, then that galaxy's mass is of order $10^{12}\ M_\odot$, and $M/L \sim 100\ M_\odot/L_\odot$, a figure that is consistent with mass estimates from the virial theorem applied to systems containing spiral galaxies similar to M 31 and with mass estimates implied by the motion of galaxies in the local group.

Extensive surveys of forbidden line emission from disk matter in Sa and Sc galaxies indicate that $v(r)$ is essentially flat (or gradually rises) out to the optical limit. Several trends can be established from these observations: (1) $v(r)$ and v_{max} (and thus galactic mass) increase with increasing galactic luminosity L_G, with v_{max} in the range 100 to 300 km/sec; (2) the slope $dv(r)/dr$ near galactic nuclei increases with L_G; (3) for a given L_G, $v(r)$ is greater for Sa types than for Sc types at every point r; and (4) $dm(r)/dr \rightarrow$ constant at large distances. According to these results, the mass range for spiral galaxies is 6×10^9 to $2 \times 10^{12}\ M_\odot$.

Estimates of the mass of our Galaxy have recently been extended, based on orbital motion of CO clouds ($r \sim 16$ kpc), globular clusters ($r \sim 30$ to 60 kpc), and the assumed orbits of SMC ($r \sim 50$ kpc), LMC ($r \sim 70$ kpc), and satellite galaxies ($r \gtrsim 80$ kpc). All are consistent with $dm(r)/dr \rightarrow$ constant, with $v(r) \approx 220$ to 250 km/sec, and imply a mass of at least $10^{12}\ M_\odot$.

The masses obtained above are clearly lower limits. These observations raise a serious problem; because the mass-to-luminosity ratio decreases with increasing radius in the disk, the additional mass (nearly ten times that of previous estimates) can not be due to stars. Furthermore, the mass can not lie within the galactic disk, since it would result in an instability leading to bar formation. It has been suggested that spiral galaxies contain massive halos of extremely low luminosity matter, which, incidentally, helps to stabilize gaseous disks; but no such halos have been observed as yet. Theory requires that they consist of matter with a large mass-to-light ratio. Possible candidates include dead stars (perhaps white dwarfs), substellar masses formed during the initial stages of galaxy formation, and even large masses of neutrinos and black holes.

Masses of Double Galaxies

The methods used to measure masses of stars in binary systems have been adapted to give mass estimates for bound pairs of galaxies. In principle, the method is the same for both types of systems; in practice, only the galaxies' radial velocities v_r and projected angular separation a_0 can be measured. Generally the linear separation between galaxies believed to be physical pairs is less than 0.2 Mpc. Assuming a typical relative velocity to be 200 km/sec, and an average separation to be 0.15 Mpc, we can show the orbital period is comparable to 5×10^9 yrs, the estimated age of the Sun.

The geometry of a bound pair of galaxies is shown in Figure 27.13. Their true separation is a, and they move in circular orbits in the y, z-plane. The observer's position is in the direction $(\pi/2 - \phi, \pi/2 - \psi)$. The projected separation seen by the observer is $a_0 = a\cos\phi$. The difference between the line-of-sight velocities of the two galaxies is $\Delta v = v \cos\phi \cos\psi$, where v is their true relative velocity. Denoting the masses of the two galaxies by M_1 and M_2, and equating centripetal to gravitational forces, we find that

$$\frac{M_1 M_2}{M_1 + M_2} \frac{v^2}{a} = \frac{M_1 M_2}{a^2} G.$$

Eliminating a and v, we find the total mass of the two galaxies to be

$$M_1 + M_2 = \frac{(\Delta v)^2}{G} a_0 \sec^3 \phi \sec^2 \psi. \quad (27.21)$$

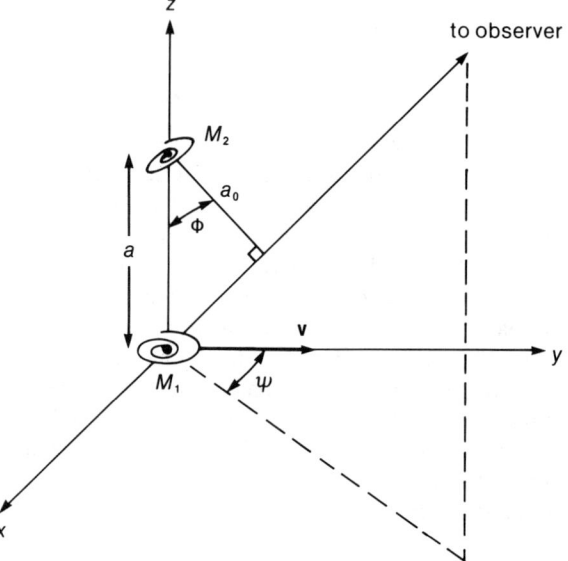

Figure 27.13. Geometry of a bound pair of galaxies.

Values for all quantities on the right-hand side of (27.21) can be found observationally, except for the angles ϕ and ψ that fix the pair's orientation. This limits the information that can be obtained for individual pairs.

The usefulness of (27.21) lies in what it tells us about the average masses of pairs. Assume that physical pairs are randomly oriented relative to an observer. Then the average

$$\int \sec^3 \phi \sec^2 \psi \frac{d\Omega}{4\pi} = 0.29, \quad (27.22)$$

where the integral is over the unit sphere. Averaging both sides of (27.21) over angles yields

$$M_1 + M_2 = 0.29 \frac{(\Delta v)^2 a_0}{G}. \quad (27.23)$$

Although this expression may be in error for a specific galaxy pair, it will yield average masses if applied to a large enough sample.

Problem 27.7. Discuss how one might go about finding average masses for specific galactic types using (27.23).

We saw in Section 26.1 that a galaxy at distance d is receding relative to our Galaxy with a speed $v = Hd$, where H is Hubble's constant. Then the angular separation of two galaxies $s = a_0 H/v$, and (27.23) may be rewritten as

$$M_1 + M_2 = (\Delta v)^2 \frac{vs}{H} \sec^3 \phi \sec^2 \psi$$

$$= 261 \, vs(\Delta v)^2 \, M_\odot. \quad (27.24)$$

The last form contains s in minutes of arc, and v and Δv in km/sec, and has been averaged over angles.

The advantage of (27.23) over (27.20) is that it reflects the entire mass of the system. It may also be applied to galaxies at such distances that rotation curves or velocity dispersions cannot be measured.

Average masses obtained from pairs are $3 \times 10^{10} \, M_\odot$ for Sc, SBc, and irregulars; $6 \times 10^{10} \, M_\odot$ for Sa-Sb and SBa-SBb; and $6 \times 10^{11} \, M_\odot$ for E and S0 types.

Galaxies in Clusters

When observed throughout a large enough region, most galaxies are found to belong to clusters containing between 10^2 and 10^3 individual members. The density contrast between a cluster and the surrounding region is large, suggesting that they are gravitationally bound systems. If they are bound and in equilibrium, the virial theorem may be used to find a value for the cluster mass M_{cl}. The arguments are similar to those used in relating masses of E galaxies to stellar velocity dispersions. Assuming that the cluster kinetic energy corresponds only to random motion, we may use (27.10) with

$$T = \frac{1}{2} \sum_i M_i \langle v_i^2 \rangle = \frac{3}{2} \sum_i M_i \langle v_{ri}^2 \rangle, \quad (27.25)$$

where M_i is the mass of a galaxy whose velocity with respect to the cluster center is v_i, with line-of-sight component v_{ri}. The cluster's gravitational potential energy is

$$\Omega = -\frac{1}{2} \sum_{i \neq j} \frac{M_i M_j G}{r_{ij}}, \quad (27.26)$$

where r_{ij} is the true distance between the galaxies labeled i and j. The observed distance $(r_{ij})_0 = r_{ij} \sin \theta$ is r_{ij} projected against the plane of the sky (Figure 27.14). We now replace the term r_{ij}^{-1} by its projected value averaged over angle, by noting that

$$\left\langle \frac{1}{r_0} \right\rangle = \left\langle \frac{1}{r \sin \theta} \right\rangle = \frac{1}{r} \int \frac{\sin \theta \, d\theta \, d\phi}{4\pi \sin \theta}$$

$$= \frac{2\pi^2}{4\pi r} = \frac{\pi}{2r}, \quad (27.27)$$

where the subscripts have been omitted for convenience. The left-hand side of (27.27) is the projected intergalactic distance averaged over angles. As in (27.23), this procedure may introduce considerable uncertainty in Ω_{ij} for a given pair, but when summed over all galaxies in the cluster, the average error becomes small. Substituting (27.25) to (27.27) into the virial theorem, we arrive at an expression that contains only observed quantities, and the galactic masses

$$3 \sum_i M_i \langle v_{ri}^2 \rangle = \frac{2}{\pi} \sum_{i \neq j} \frac{M_i M_j G}{(r_{ij})_0}. \quad (27.28)$$

applicable, and masses would be smaller. Some clusters appear to contain numbers with anomalously high velocities. However, some clusters, such as the Coma cluster, have a high degree of central condensation. If these systems are in fact disrupting, then they must have been formed more recently than 2×10^{10} years ago, as is currently believed. Furthermore, an explosive agent of gigantic proportions probably would be required to set them in motion.

Problem 27.8. What energy is needed to disperse a spherical cluster of galaxies of average mass $2 \times 10^{11}\ M_\odot$ containing 10^3 galaxies, if the initial cluster radius is 1 Mpc? How long will it take for the cluster to disperse?

27.4. Stellar Content of Galaxies

Color distributions, as described in Section 27.2, give some indication of the stellar types that make up various parts of galaxies. Based on color alone, we may conclude that E and Sa-Sb galaxies have a composite energy spectrum that resembles spectral type G5. Sc galaxies, on the other hand, can range from A5 to G5. Furthermore, the light from E galaxies and from the nuclei of Sa-Sb galaxies does not require early-type stars, but hot stars are needed for the centers of Sc and Ir galaxies, and in the disk of Sb-Sc and some Sa galaxies.

A more detailed description of stellar content is obtained from stellar-population synthesis studies, which attempt to relate the observed colors, absorption line, and emission line features to the underlying stellar composition. In practice, only stars brighter than $M_V \simeq -5$ can be studied in detail. For the nearest galaxies (LMC and SMC), this corresponds to main-sequence stars brighter than about O9 and giants brighter than B0. Accurate colors can be observed down to about $M_V \simeq 0$, which rules out main-sequence stars later than B8 and giants of type F0 to K3. Thus even in the nearby members of the local group, most of the stars that make up much of the mass and light of a galaxy are not observed.

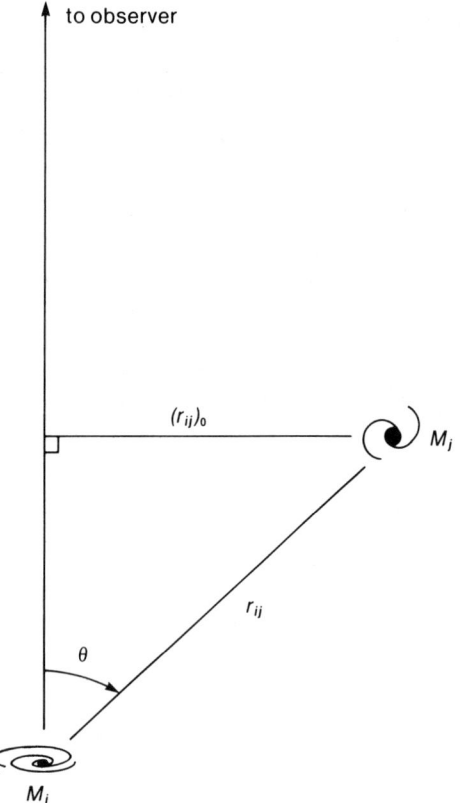

Figure 27.14. Observed galactic separation.

This may be used to find a value for the total cluster mass $\Sigma_i M_i$, from which estimates of the individual galactic masses (average values) M_i are obtained.

Mass averages for galaxies using (27.28) are found to be systematically large compared with estimates based on pairs, or on velocity dispersions and rotation curves in single galaxies. For example, the average mass of E galaxies in the Virgo E cluster is $3 \times 10^{12}\ M_\odot$; and the average mass per galaxy in the Hercules cluster, which contains many spiral types in addition to ellipticals, is $10^{12}\ M_\odot$. These values are about ten times larger than the averages just discussed. The reason for this discrepancy is not clear, but two alternatives are evident. First, clusters may contain additional mass that is not in the form of galaxies, such as intergalactic gas, or collapsed stars. If this alternative is to be accepted, theory must explain why less than 10 percent of the matter in the universe ends up in galaxies, and why the missing mass is unseen. A second alternative is to suppose that clusters are in fact unbound systems, in which case (27.28) is no longer

Problem 27.9. What is the faintest stellar type that might be resolved in SMC or LMC (distance = 50 kpc) and in M 31 (distance 680 kpc), assuming that the limiting apparent magnitude is $m_L = 24$ mag?

Stellar-population synthesis attempts to answer two fundamental questions about galaxies. First, what fraction of the stars come from each unresolved stellar class? Second, could a significant amount of mass in galactic nuclei be in nonstellar forms, such as H_2, white dwarfs, neutron stars, or black holes? The observational data for a specific galaxy or galactic nucleus to be modeled includes integrated color, and the major absorption and emission line features in the wavelength range 3300–11000 Å. The line features usually chosen are those most sensitive to stellar effective temperature, luminosity, and composition. These include lines due to Ca, Mg, Na, and H, and the molecular bands of CN, CH, MgH, CaH, and TiO. Spectra and integrated light are also taken for stars in the solar neighborhood, and in globular clusters in our Galaxy, to be used as basic building blocks for the population synthesis. The spectra and total light output of these "standard" stars are blended until the color, line features, and magnitude of the system under study are reproduced. In this way the absolute number of stars of various spectral types in the system may be estimated.

The spectrum of the bar in LMC, shown in Figure 27.15, illustrates how this is done. The assumed stellar population must reproduce the overall flux as a func-

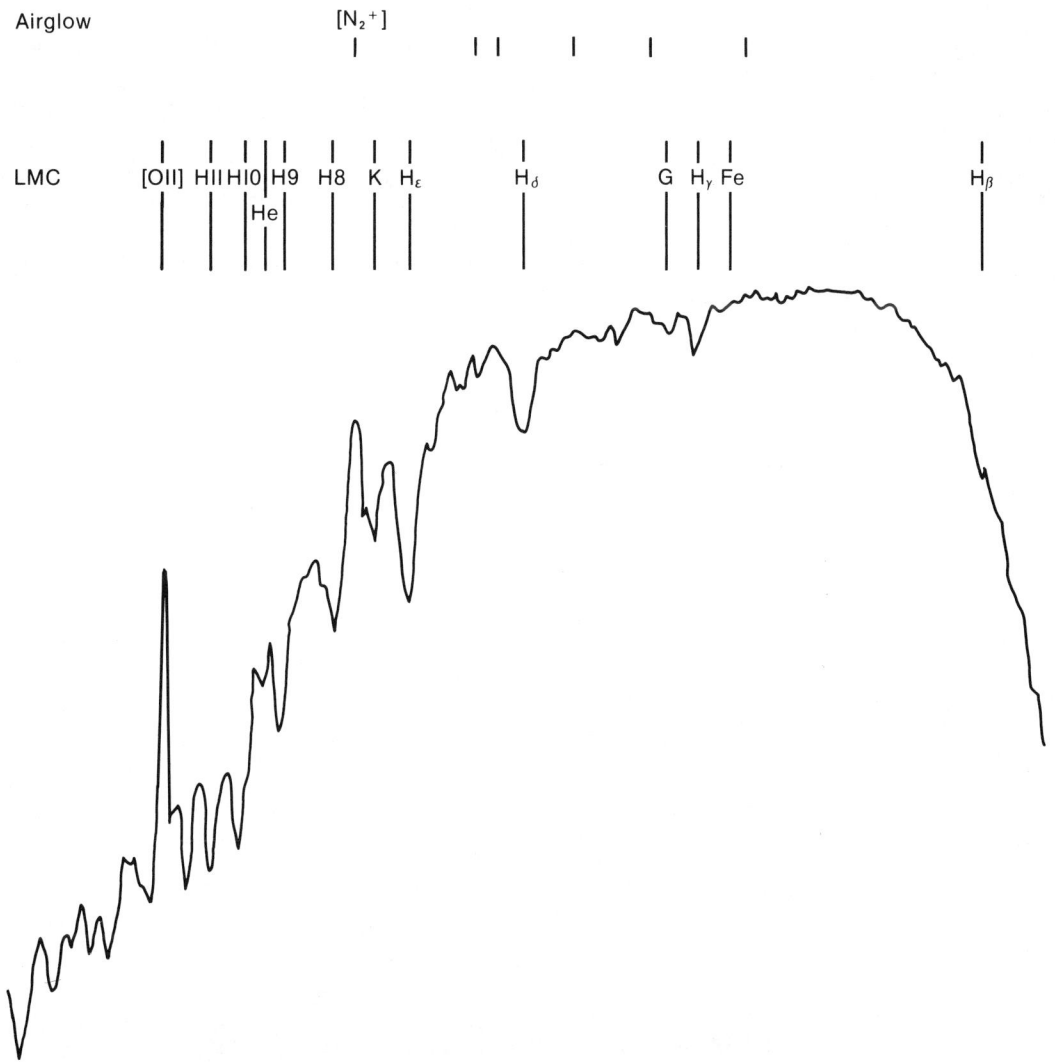

Figure 27.15. Spectrum of bar in the Large Magellanic Cloud (wavelength increasing to right).

tion of frequency, and it must account for the prominent line features as well. As is typical of many galaxies, the bar spectrum is highly composite. The HeI absorption line at λ3820 indicates the presence of stars earlier than B5; the H_γ and H_δ lines are typical of spectral types B through F; and the G band, and the FeI line at λ4384, arise from types G and K. The spectrum appears to be reasonably well reproduced by a stellar distribution similar to that in the solar neighborhood.

The spectra of nuclei in Sa-Sb galaxies, such as M 31, show significant differences. Their spectra are dominated by molecular bands, indicating large numbers of cool K and M stars. Figure 27.16 shows the percent of luminosity (left scale) and mass (right scale) for the nucleus of M 31 obtained by stellar-population synthesis. Two features are immediately evident: the spectral types that contribute most to the nucleus' light are those that contribute least to its mass. In fact, early-type dwarfs (main-sequence stars), subgiants, and giants of type G0-M0 dominate the luminosity but contribute less than 2 percent of the mass. Most of the mass is in late-type dwarfs (M0V-M8V), which contribute only about 5 percent to the total light. This is typical of galactic nuclei and of E galaxies. Another characteristic of population-synthesis models is a need for giants and subgiants to have strong-lined (possibly super-metal-rich) spectra.

Since the mass of each stellar type used in the population synthesis is known, the total (stellar) mass of the nucleus is predicted. For M 31, the nuclear mass is found to be $6.8 \times 10^8 \, M_\odot$, or about 2.1×10^{-3} times the total mass of the galaxy. If M 31 is typical of Sb galaxies as a whole, we would conclude that very little of the total mass lies in the nucleus of spiral galaxies.

Models of nuclei in Sc galaxies show a shift toward earlier spectral types, with F0-K5 supplying most of the light. Unlike the nuclei of Sa-Sb galaxies, which contain virtually no early-type stars (more than 50 O9V stars would give a spectrum too rich in blue light in M 31, for example), about 0.5 percent of the light in the nuclear region of the Sc galaxy NGC 618 appears to be from OBV stars (about 10^4 stars, or about 0.02 percent by mass). In typical Sc nuclei the mass is more evenly distributed over late stellar types (intermediate FV); in NGC 628 roughly 30 percent of the mass supplies 80 percent of the light. Models of Sc nuclei produce masses that average about 1 percent of the

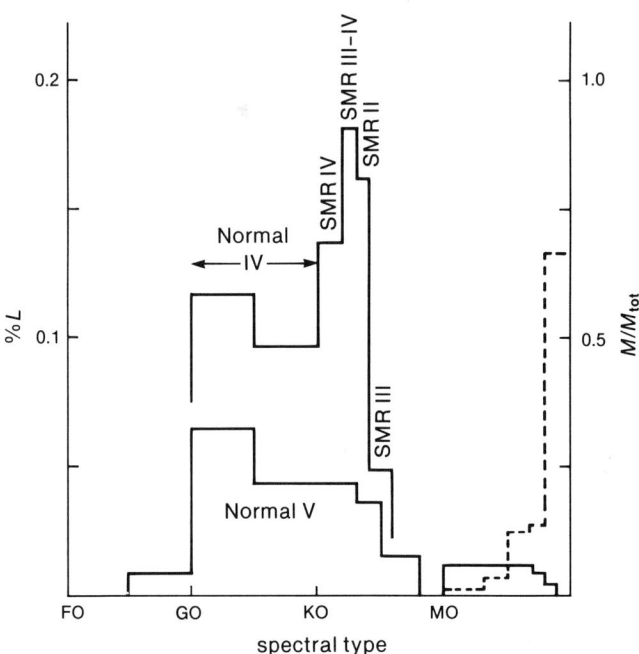

Figure 27.16. Stellar mix for the nucleus of M 31, showing percent of contribution to luminosity (left) and to mass (right).

total galactic mass, though individual values may vary by a factor of 10. For NGC 628, $M_{\text{nucl}} \simeq 2 \times 10^{-3} M_{\text{tot}}$, and for NGC 1084, $M_{\text{nucl}} \simeq 7 \times 10^{-2} M_{\text{tot}}$.

The relative number of bright stars to massive ones is reflected in part by the mass-to-luminosity (light) ratio M/L, where each is in solar units. The ratio of mass to visible light for several stellar types is given in Table 27.1. As expected, it increases with later type along the main sequence, and is relatively small for giants. The ratio M/L for galaxies ranges from about one for spirals and irregulars to 80 or 100 for giant ellipticals. The stellar mix for the nucleus of M 31 shown in Figure 27.17 corresponds to $M/L \simeq 40$, but for the nucleus of NGC 628 (type Sc) the stellar mix corresponds to $M/L \simeq 0.7$. We will see that M/L varies with morphological type.

Table 27.1
Stellar mass-luminosity ratio.

Spectral type	M/L_V	$M/M_\odot$
B0 V	0.005	18
A0 V	0.07	3.2
F0 V	0.22	1.7
G0 V	0.74	1.1
K0 V	2.1	0.78
M0 V	22	0.47
M5 V	131	0.21
G0 III	0.08	2.5
K0 III	0.07	4.0
M0 III	0.05	6.3

Problem 27.10. Find the mass to blue light ratio M/L_B for the stars in Table 27.1. Explain the systematic differences between the results and the ratio M/L_V as a function of spectral type.

Observed M/L ratios impose useful constraints on stellar types in galaxies. Assuming that the entire mass of a galaxy is in stars, we may express the mass-to-light ratio as

$$M/L = \sum_i M_i N_i \bigg/ \sum_i N_i L_i, \qquad (27.29)$$

where M_i is the mass of a star of spectral type i, L_i is its luminosity, and N_i their number. The following example shows immediately that a significant fraction of the total light can come from a small mass fraction.

Assume that only stellar types M8 V and K0 III exist in the nucleus of a galaxy with observed $M/L =$

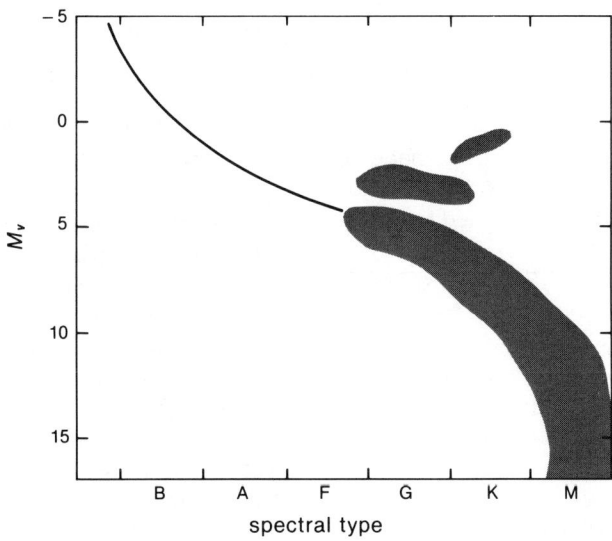

Figure 27.17. Schematic HR diagram based on stellar synthesis of the nucleus of M 31.

40 (the model resembles qualitatively the nucleus of M 31 shown in Figure 27.17), and with total mass $M_{tot} = 10^{10} \, M_\odot$. Then the total luminosity $L_{tot} = (L/M) M_{tot} = 2.5 \times 10^8 \, L_\odot$. The sum in (27.29) is now over stellar types M8 V and K0 III only, and can be solved for the number of M8 V stars:

$$N_M = \frac{M_K L_{tot} - M_{tot} L_K}{M_K L_M - M_M L_K}, \qquad (27.30)$$

where subscripts K and M denote the two spectral classes. The definition of total mass $M = \Sigma_i M_i N_i$ has been used to eliminate N_K:

$$N_K = \frac{M_{tot} - M_M N_M}{M_K}. \qquad (27.31)$$

Using the data in Table 27.1, and noting that $M_N = 0.1 \, M_\odot$, $L_M = 3.4 \times 10^{-5} \, L_\odot$, we find

$$N_M = 9.98 \times 10^{10},$$
$$N_K = 4.33 \times 10^6.$$

This implies that less than 0.2 percent of the mass is in giants, although they supply 98.8 percent of the light (compare Figure 27.16).

Problem 27.11. Derive (27.29) and verify the results above. What is the maximum number of B0 V stars that could be added to the model in place of M8 V stars without changing either N_K or the total luminosity?

Problem 27.12. Consider an E galaxy made up only of M0 V stars. The observed $M/L \simeq 22$. How many B0 V stars would need to be added to reduce M/L to 2? Assume $M = 10^{10} \, M_\odot$.

As the examples above suggest, the ratio M/L varies with galactic type. Ellipticals have mass-to-light ratios in the range 10–80 with an average of about 70. Most spirals and irregulars lie in the range 0.1 to 20, with an average near $M/L = 1.0$.

Stellar-population synthesis is most useful for finding the content of galactic nuclei, where there is usually relatively little gas or dust. Late-type spirals (Sc and SBc) and irregular galaxies do show evidence of gas in their nuclei. It then becomes necessary to allow explicitly for the mass of HI and HII clouds, and for the luminosity from hot HII regions when modeling these nuclei. As a further complication, the presence of gas suggests that star formation may be occurring as well.

The HR diagram for the nucleus of M 31 is shown in Figure 27.17. The main sequence has burned down to late F stars, indicating that star formation occurred up to the last 3×10^9 years, but has since ceased. The nucleus of M 31 is therefore a relatively old system, resembling an old open cluster.

The nuclear region of NGC 628 suggests a stellar population whose HR diagram is shown in Figure 27.18. The number of G-K giants is relatively larger than in the stellar model for M 31, representing about 3 percent of the total mass. The emission spectra at low wavelength requires a small admixture of early-type main-sequence stars, starting with O5-O9 V. Although most of the main sequence is populated, there is a rapid decrease in the number of stars earlier than F6. This suggests that a major stage of star formation occurred earlier than 3×10^7 years ago, as in the nucleus of M 31. The presence of O5-O9 V stars, however, indicates that star formation has continued up through at least the last 10^5 years, but that its rate has decreased. In NGC 628, early supergiants contribute only about 2 percent of the total light, but in NGC 2903, another Sc galaxy, early-type supergiants (B0-A2 I) contribute more than 20 percent of the total light and more than half the ultraviolet. Furthermore, in some Sc nuclei (such as NGC 2903), star formation is probably still occurring.

27.5. GENERAL CHARACTERISTICS OF GALAXIES

Several general characteristics of our galaxy, and of other galactic types, are summarized in Tables 27.2 and 27.3. These represent what are believed to be among the more important properties of normal galaxies that theory must attempt to explain. The primary constituents of the Galaxy are stars, gas and dust, magnetic fields, starlight, and cosmic rays (Table 27.2). For most applications the last two components may be neglected. The role of magnetic fields is also often neglected, although this neglect may not always be justified. Values in the solar neighborhood are flagged; they should apply to similar regions elsewhere in the disk of our Galaxy.

Less specific values are available for other galactic

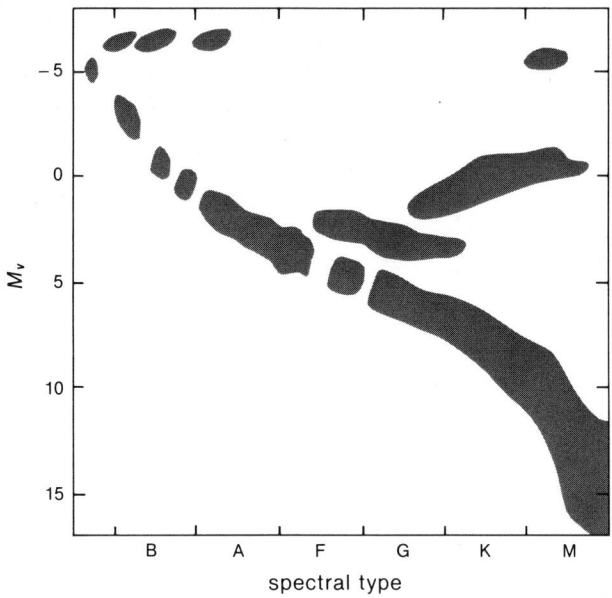

Figure 27.18. Schematic HR diagram of the nucleus of NGC 628.

types. Table 27.3 contains representative values for the luminosity, total mass, ratio of gas mass to total mass, and mass-to-luminosity ratio for major galactic types. The relative distribution of galactic types was illustrated in Figure 27.7. The data summarized in the tables can be used to simplify the theoretical approach to galactic structure and dynamics. For example, the analysis of spiral galaxies consists broadly of two aspects: those properties that involve the behavior of one or a collection of stars moving in the background of the galaxy as a whole (stellar dynamics); and the behavior of both gaseous and stellar collective modes (spiral density waves). In principle these apparently different approaches simply represent two alternative ways of looking at long-range gravitational phenomena involving a large number of bodies. For many situations of interest in normal galaxies, the ratio $2MG/Rc^2$ (which measures the importance of general relativistic corrections to Newtonian gravitation) is small. Using data from Tables 27.2 and 27.3 we find that $2MG/Rc^2 \simeq 5 \times 10^{-7}$. Consequently, Newtonian mechanics and gravitation are usually adequate. However, in some (or perhaps all) galactic nuclei, and in some globular clusters, general relativity could play an important role, even in so-called normal galaxies. For example, x-ray and gamma-ray bursts from the galactic center and from several globular clusters could indicate the presence of massive black holes there. Black holes, or the actual collapse to a black hole, may also play a significant role in the dynamics of peculiar galaxies.

Table 27.2
Properties of the Galaxy (Type Sb ?).

Disk radius	12 kpc[a]
Disk thickness (½ density point)[b]	
gaseous component	0.14 kpc
stellar component	0.31 kpc
Radius of extended spherical system	15 kpc
Total mass	$1.4 \times 10^{11} M_\odot$
Sun's distance from Galactic center	10 kpc
Ratio of gas to stellar mass density[a]	0.3
Typical dispersion velocity[a]	
stellar motion	30 km/sec
molecular motion	1 km/sec
turbulent gas motion	8 km/sec
Energy density (ergs/cm³)	
galactic rotation	1.3×10^{-9}
magnetic field (B ~ 10^{-5} gauss)	4×10^{-12}
turbulent gas motion	0.5 to 35×10^{-12}
cosmic rays	10^{-12}
starlight	7×10^{-13}

[a] >80 kpc based on recent data (Section 27.3).
[b] Values in the solar neighborhood.

Table 27.3
Mass, luminosity, and gas content of galaxies.

Galactic type	$\dfrac{L}{10^{10} L_\odot}$	$\dfrac{M}{L}$	$\dfrac{M}{10^{10} M_\odot}$	$\dfrac{M_{gas}}{M}$
E	0.2	50–80	10–10^2	$\leq 10^{-6}$
S0		50		
Sa		30		
Sb	0.2	20	4	0.05
Sc	0.1	10	1	0.1
I	0.04	5	0.2	0.2

A cross section of the Galaxy showing lines of constant density (in units of mass density near the Sun), as well as the distribution of observed globular clusters, is shown in Figure 27.19. The Sun's position at about 10 kpc from the center is also shown. The globular clusters and the older Population II stars in the halo follow Keplerian orbits, some of high eccentricity, about the Galactic center. The disk stars follow nearly circular orbits within the plane, and show relatively little motion normal to the plane. The motion in the disk supplies one of the more important pieces of observational data about spiral galaxies, the rotation curve, which is known not only for our Galaxy, but for a number of nearby galaxies as well.

The formation of a spiral galaxy can be envisioned in the following way. An initial protogalactic cloud becomes unstable and begins to contract. Given sufficient angular momentum, the cloud contracts rapidly,

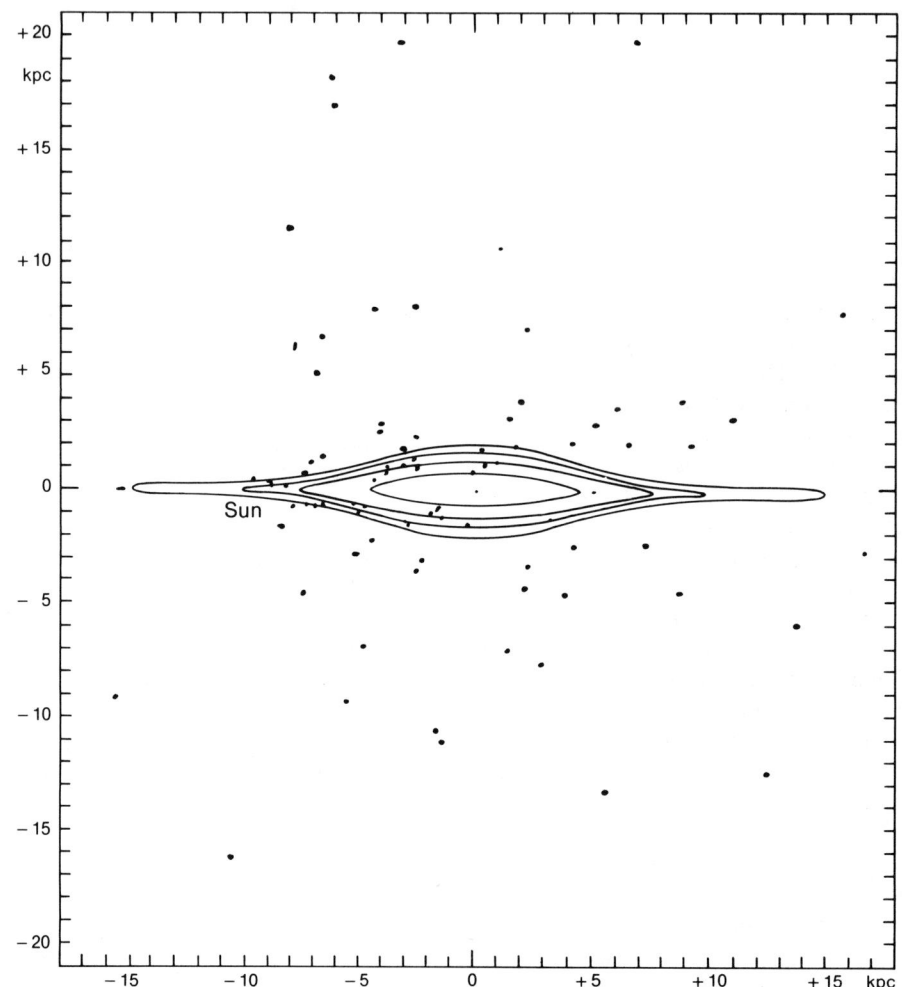

Figure 27.19. Projected location of globular clusters and isodensity contours for our Galaxy. The Sun is located in the disk about 10 kpc from the center.

perpendicular to its angular momentum axis, until the centripetal acceleration at each point balances the gravitational acceleration. Collapse continues along the rotation axis until turbulent velocities build up enough pressure to halt collapse. At this stage the initial spherical, slowly rotating cloud has become a rapidly rotating, thin disk. Some stars may have formed out of the protogalactic gas prior to, or during, the initial slow collapse stage. These will remain behind as the gas collapses, forming a more or less spherical distribution of old stars, presumably in aggregates that are now identified as globular clusters. Turbulence in the disk gas aids the next stage of gas cloud fragmentation to form new stars. Some turbulence may end up as peculiar velocities of the disk stars; some may remain in the interstellar gas. Typical values are given in Table 27.2 for the solar neighborhood. This scenario appears reasonable for the formation of spirals, but it raises several basic questions. Clusters of galaxies are observed to contain ellipticals as well as spirals, whose ages (based on the existence of very old stars) are probably comparable. How then do ellipticals form? Are they the product of protogalactic clouds whose initial specific angular momentum is less than some critical value? Finally, the development of an initial instability capable of producing a galaxy in the observed mass range during the rapid expansion phase of the early universe presents a fundamental problem that has yet to be overcome.

Chapter 28

DYNAMICS OF STELLAR SYSTEMS

The universe contains a diverse collection of stellar systems, which differ in size and content. The smallest are typified by stellar associations that contain 10^2 stars or more, and the largest by giant elliptical galaxies. Table 28.1 summarizes data on the most common stellar systems. The motion of each star in a system is described by a force equation (equation of motion) that includes the gravitational potential due to all other stars in the system (the effect of the interstellar gas is generally small, and can often be neglected). A complete description of the stellar motion in an isolated galaxy can in principle be extracted from a set of coupled differential equations, one for each star in the system. Unfortunately, there are far too many stars in each of these systems to permit their individual motions to be followed in detail. One way around this difficulty is to follow groups of stars, and to consider average properties of the system as functions of position and time. This is the goal of stellar dynamics, which attempts to describe the average spatial distributions of matter (stars and gas) in stellar systems, and its relation to the average (or smoothed) gravitational potential and stellar velocity distribution.

28.1. Stellar Dynamics

Unlike hydrodynamics, stellar dynamics is a statistical theory, in which the particles (stars) have a negligible likelihood of undergoing direct collisions. In fact, in the relatively dense galactic center, the average stellar number density is about 10^4 stars per pc^3, and their average velocities are of order 200 km/sec. Taking $\pi R_\odot^2$ as a typical stellar cross section, we find the time between collisions to be $\tau \sim (nv\sigma)^{-1} = 3 \times 10^{14}$ years. Outside galactic nuclei, and in other stellar systems, the direct collision time is much larger. Dynamic steady states, where average values contain no explicit time-dependence, and nonsteady states, which do depend on time, must be considered. Only for nonsteady states do evolutionary effects arise directly.

Before developing a statistical approach to stellar dynamics, we will review the gravitational equation as it applies to spherical and axial systems. When relativistic effects are small and the mass distribution is slowly changing, the gravitational force on a star can be derived from the potential $\Phi(r)$, which is time-independent, and which is described for the distribution of all forms of matter by Poisson's equation

$$\nabla^2 \Phi = 4\pi G \rho. \qquad (28.1)$$

Table 28.1
Characteristics of stellar systems.

Stellar group	Number of stars	Dimension (optical radius)	Content
Stellar association	10^2 to 10^3	3 to 10 pc	Young stars, gas and dust
Globular clusters	10^4 to 10^6	10 to 50 pc	Old stars (Pop. II)
Dwarf galaxies	10^6 to 10^8		
Galactic nuclei	10^4		
Irregular galaxies (Irr II)	10^8 to 10^{10}	2 to 10 kpc	All stellar types; abundant gas and dust. Spectra dominated by types A–F.
Spiral galaxies	10^9 to 10^{12}	7 to 50 kpc	All stellar types; gas and dust. Types A–K dominate spectra.
Elliptical galaxies	10^9 to 10^{13}	10 to 200 kpc	Old stars (Pop. II). Essentially no dust; little gas.

The mass density ρ includes stars, gas, and dust. Although the last two components usually represent a small fraction of the total mass, they must be retained for an analysis of spiral structure. The gravitational force acting on a star at r is given by

$$\mathbf{F} = -\nabla \Phi. \tag{28.2}$$

In cylindrical coordinates applicable to our Galaxy,

$$\nabla^2 \Phi = \frac{1}{r} \frac{\partial}{\partial r}\left(r \frac{\partial \Phi}{\partial r}\right) + \frac{1}{r^2} \frac{\partial^2 \Phi}{\partial \theta^2} + \frac{\partial^2 \Phi}{\partial z^2}, \tag{28.3}$$

and

$$F_r = -\frac{\partial \Phi}{\partial r}, \quad F_\theta = -\frac{1}{r} \frac{\partial \Phi}{\partial \theta},$$
$$F_z = -\frac{\partial \Phi}{\partial z}. \tag{28.4}$$

It should be remembered that the Newtonian theory of gravitation is linear, and that superimposing solutions to (28.1) and (28.2) creates another solution of (28.1). Finally, we note for later use (in Chapter 30) the form of Poisson's equation for an infinitesimally thin disk. Formally, the mass density can be expressed as

$$\rho(\mathbf{r}) = \sigma(r, \theta)\delta(z), \tag{28.5}$$

$$\int_{-\infty}^{\infty} \delta(z - a)g(z)\, dz = g(a), \tag{28.6}$$

where $\sigma(r, \theta)$ is the surface density (units of mass/area), and (28.6) defines the Dirac delta function $\delta(z)$; the function $g(r, \theta, z)$ is arbitrary.

Problem 28.1. What is the gravitational force inside a cloud of dust and gas of uniform density surrounding a star of mass M?

A statistical description of stellar dynamics begins with the concept of a stellar distribution function $f(\mathbf{r}, \mathbf{v}, t)$, which gives the distribution of stars in coordinate space and velocity space. By definition $f(\mathbf{r}, \mathbf{v}, t)\, d\mathbf{r}\, d\mathbf{v}$ is the number of stars in a spatial volume element $d\mathbf{r}$, whose velocities lie between $\mathbf{v}$ and $\mathbf{v} + d\mathbf{v}$, at time t. A stellar distribution function could be defined for each stellar type (either population or spectral type, or both); however, we will consider only systems containing identical stars, whose mass we denote by m. The distribution function $f(\mathbf{r}, \mathbf{v}, t)$ is a scalar function of the six independent coordinates r_i and v_i. The six-dimensional space whose coordinates are (r_i, v_i) is referred to as phase space for the star. The dynamical state of a star is completely specified mechanically if its position and velocity at time t are given. A point in phase space therefore specifies the state of a star at a given instant t, and its dynamic motion traces out a trajectory in phase space. Some regions of phase space may not be accessible to a star. For example, if the star's energy E and angular momentum $\mathbf{J}$ are conserved, then $\mathbf{v}$ and $\mathbf{r}$ are constrained by

$$E/m = \tfrac{1}{2}v^2 + \Phi(r),$$

$$\mathbf{J}/m = \mathbf{r} \times \mathbf{v}.$$

Finally, f is assumed to be nonnegative, and to approach zero as any of the coordinates r_i or v_i approaches $\pm \infty$. Physically, this means that the system is of limited extent.

An immediate consequence of the definition of f is that the number of stars per unit volume at $\mathbf{r}$ is

$$n(\mathbf{r},t) = \int d\mathbf{v}\, f(\mathbf{r},\mathbf{v},t)$$

$$= \int_{-\infty}^{\infty}\int_{-\infty}^{\infty}\int_{-\infty}^{\infty} f(\mathbf{r},\mathbf{v},t)\, dv_x\, dv_y\, dv_z, \quad (28.7)$$

where the integration over all velocity space has been shown explicitly. The total stellar mass density is

$$\rho_* = mn(\mathbf{r}, t), \quad (28.8)$$

and the total number of stars in the system at time t is

$$N_* = \int d\mathbf{r} \int d\mathbf{v}\, f(\mathbf{r}, \mathbf{v}, t) = \int d\mathbf{r}\, n(\mathbf{r}, t). \quad (28.9)$$

In the expression for N_*, the integral is over all space. Finally, the total mass density appearing in Poisson's equation includes the density of gas and dust ρ_g, and is

$$\rho = \rho_* + \rho_g. \quad (28.10)$$

Note that only ρ_* is determined by the distribution function above, although it is the total density ρ that enters into Poisson's equation. As we will see, Φ appears in the equation that determines $f(\mathbf{r}, \mathbf{v}, t)$; so all these quantities are connected.

Consider the time rate of change of the distribution function, remembering that it depends on six independent variables and the time:

$$\frac{df}{dt} = \frac{\partial f}{\partial x^i}\frac{dx^i}{dt} + \frac{\partial f}{\partial v^i}\frac{dv^i}{dt} + \frac{\partial f}{\partial t}. \quad (28.11)$$

Summation over repeated indices i is assumed. The analogy between this expression and the Lagrangian derivative (21.1) in hydrodynamics should be noted. For a system of masses to which (28.11) applies, the velocity and acceleration are defined by

$$v^i = dx^i/dt, \quad a^i = dv^i/dt. \quad (28.12)$$

Substituting these expressions into (28.11) yields the Boltzmann equation

$$\frac{df}{dt} = v^i \frac{\partial f}{\partial x^i} + a^i \frac{\partial f}{\partial v^i} + \frac{\partial f}{\partial t}$$

$$= v^i \frac{\partial f}{\partial x^i} - \frac{1}{m}\frac{\partial \Phi}{\partial x^i}\frac{\partial f}{\partial v^i} + \frac{\partial f}{\partial t}. \quad (28.13)$$

In the last expression, we have used (28.2) to replace the acceleration by F^i/m. The left-hand side of the Boltzmann equation df/dt must be specified to complete the description of the system.

If the change in velocity distribution is due to binary interactions between the constituents of the system (in this case, gravitational interactions between stars), then one may try to express the time rate of change in terms of scattering processes. This amounts to setting

$$\frac{df}{dt} = \left(\frac{df}{dt}\right)_{\text{collision}}. \quad (28.14)$$

After the right-hand side of (28.14), the collision term, is specified, then it and (28.13) may be equated. Two extreme cases of astrophysical interest are relatively easy to work with.

In the first we assume that there are no interactions or collisions between the stars (so the collision term vanishes: $df/dt = 0$), and that the changes in f depend on changes in velocity and acceleration, the latter often being the most important. A simple interpretation of the result $df/dt = 0$ arises from considering the behavior of an element of volume (six-dimensional) in phase space, $dV = d\mathbf{r}\, d\mathbf{v}$. The noninteracting or collision-free assumption is equivalent to the statement that the unit volume of phase space is preserved throughout the system's evolution. Its shape will change continuously, eventually becoming unrecognizable in extreme cases. Nevertheless, the volume of the element remains constant. The collision-free hypothesis is quite good for many applications to galactic dynamics, since the average separation and relative velocity of most stars makes collisions or scattering by individual stars quite rare. This does not mean that the stars are assumed to behave as free particles, since the right-hand side of (28.13) contains the acceleration, which is related directly to the gravitational potential via (28.14). In other words, although individual stellar collisions, or near encounters (scattering), are not allowed, the individual stars nonetheless are constrained to move in response to the galactic potential Φ.

Problem 28.2. Consider the collisionless Boltzmann equation $df/dt = 0$ for particles in a constant, uniform gravitational field in one dimension. Use separation of variables to solve for the dependence of $f(x, v, t)$ on x and v. Discuss the physical implications of the solution.

The opposite extreme occurs when there are frequent encounters, some of which may involve near-collisions. Frequent encounters cause stars originally in a unit volume of phase space to be scattered out of that volume, and others not originally in it to be scattered into it. The ultimate effect of such action is to randomize all velocities in dV. It can be shown, though the argument is complicated, that the resulting distribution function $f(\mathbf{r}, \mathbf{v}, t)$ is of the form

$$\overline{\int f(\mathbf{r}, \mathbf{v}, t)\, d\mathbf{r}} = f(\mathbf{v})$$

$$= \frac{\exp\left[-\left(\dfrac{v_x^2}{2\langle v_x^2 \rangle} + \dfrac{v_y^2}{2\langle v_y^2 \rangle} + \dfrac{v_z^2}{2\langle v_z^2 \rangle}\right)\right]}{(2\pi)^{3/2}(\langle v_x^2 \rangle \langle v_y^2 \rangle \langle v_z^2 \rangle)^{1/2}}, \quad (28.15)$$

where the bar denotes a time average. This is a Gaussian or normal distribution. We note that a Gaussian distribution in velocity components implies a Maxwellian distribution in speed if $\langle v_x^2 \rangle = \langle v_y^2 \rangle = \langle v_z^2 \rangle$. The assumption of frequent encounters is applicable as a first approximation to globular clusters (their inner regions).

A normalized distribution function gives the probability that certain combinations of velocities and coordinates can occur for a given system. For example, (28.15), is normalized, in the sense that

$$\int f(\mathbf{v})\, dv_x\, dv_y\, dv_z = 1. \quad (28.16)$$

Now suppose that $\langle v_x^2 \rangle = (30 \text{ km/sec})^2$. We may ask for the probability that a star has a velocity in the $\hat{x}$ direction in excess of 100 km/sec. The answer is obtained by calculating

$$P(v_x < 100 \text{ km/sec}) = \int_{-\infty}^{\infty} dv_y \int_{-\infty}^{\infty} dv_z \int_{-100}^{100} f(v)\, dv_x$$

$$= 2\int_0^{100} e^{-v^2/2\langle v_x^2 \rangle} \frac{dv_x}{\sqrt{2\pi \langle v_x^2 \rangle}}$$

$$= 2 \times 0.4996 = .9992, \quad (28.17)$$

which is the probability that a star has a velocity component $|v_x| < 100$ km/sec. The second to the last step can be obtained from standard integral tables for Gaussian distributions. The probability that a star has $|v_x| > 100$ km/sec is therefore $1 - 0.9992 = 8 \times 10^{-4}$, which is less than 0.1 percent.

A distribution function represents a steady state if it contains no explicit time dependence, that is, if

$$\frac{\partial f}{\partial t} = 0. \quad (28.18)$$

This represents a form of symmetry (time invariance). The existence of additional symmetries may be imposed in like manner. For example, an axially symmetric state (coordinates r, θ, z) with rotation axis z satisfies

$$\frac{\partial f}{\partial \theta} = 0. \quad (28.19)$$

Problem 28.3. It is frequently assumed that the Galaxy: (1) is in a steady state; (2) has a smooth mass distribution; and (3) is axially symmetric in the disk. Discuss observational evidence bearing on each of these assumptions.

The virial theorem may be applied to bound, self-gravitating stellar systems. Suppose that a spherical star distribution consists of N stars, each of the same mass m. The total kinetic energy is

$$T = \frac{1}{2}\sum_i m v_i^2 = \frac{m}{2}\sum_i v_i^2 = \frac{M}{2}\langle v^2 \rangle, \quad (28.20)$$

where $M = Nm$, and the sum includes all stars in the system. The gravitational potential energy is

$$\Omega = \frac{1}{2} \sum_{i \neq j} \frac{m^2 G}{r_{ij}} \equiv \frac{m^2 G}{2} N(N-1) \left\langle \frac{1}{r} \right\rangle$$
$$= \frac{M^2 G}{2} \left\langle \frac{1}{r} \right\rangle. \qquad (28.21)$$

The total number of terms in the sum is $N(N-1)$, and we will assume that $N \gg 1$. We denote by $\langle 1/r \rangle$ the harmonic mean radius of the system. Using the virial theorem for an equilibrium system, $2T + \Omega = 0$, (28.20), and (28.21), we find that

$$\langle v^2 \rangle = \frac{MG}{2} \left\langle \frac{1}{r} \right\rangle. \qquad (28.22)$$

The escape velocity of a star from the system is

$$v_{e,i}^2 = 2 \sum_{i \neq j} \frac{mG}{r_{ij}} = 2mG \sum_{i \neq j} \frac{1}{r_{ij}},$$

and its average over the entire system

$$\langle v_e^2 \rangle = \frac{1}{N} \sum_i v_{e,i}^2 = \frac{2mG}{N} \sum_{i \neq j} \frac{1}{r_{ij}}$$
$$= 2mG(N-1)\left\langle \frac{1}{r} \right\rangle \simeq 2MG \left\langle \frac{1}{r} \right\rangle. \qquad (28.23)$$

The harmonic mean radius may be eliminated between $\langle v^2 \rangle$ and $\langle v_e^2 \rangle$ to obtain

$$\langle v_e^2 \rangle = 4 \langle v^2 \rangle; \qquad (28.24)$$

despite the simplified derivation above, this relationship is valid in general, if $\langle \ldots \rangle$ is understood to be a mass average. If the stellar velocity distribution $f(\mathbf{v})$ is Maxwellian, then the fraction of stars in a spherical system that have $v > \langle v_e^2 \rangle^{1/2}$, and should therefore escape, is given by

$$\xi_e = \int_{2\langle v^2 \rangle^{1/2}}^{\infty} f(v) \, dv. \qquad (28.25)$$

The lower limit is twice the root-mean-square velocity in the Maxwellian distribution [see (28.15)]. The integral can be evaluated from tables, with the result that $\xi_e = 7.4 \times 10^{-3}$. Although this fraction is small, it can have a significant effect on the system's evolution (see Section 28.3).

Problem 28.4. The function $f(v)$ in (28.25) gives the distribution of stars whose velocity has magnitude v; for a Maxwellian distribution show that

$$f(v) = 4\pi v^2 (\alpha/\pi)^{3/2} e^{-\alpha v^2}, \qquad (28.26)$$

where $\langle v^2 \rangle^{1/2} = (3/2\alpha)^{1/2}$ is the root-mean-square velocity defined by

$$\langle v^2 \rangle = \int_0^\infty f(v) v^2 \, dv.$$

Then show that $\xi_e = 7.4 \times 10^{-3}$. The integral

$$\int_{6^{1/2}}^\infty x^2 e^{-x^2} \, dx = 3.27 \times 10^{-3}$$

cannot be solved in closed form.

28.2. RELAXATION TIMES AND STELLAR ENCOUNTERS

Stellar dynamics is basically a N-body problem in the presence of Newtonian gravitation. As we have seen, some headway may be made as long as only two-body interactions occur. Examples are known, however, in which three-body effects can dramatically alter the nature of a system. One example is the sling-shot effect, in which three stars, each having normal velocities, interact mutually in such a way that most of the relative motion of the systems is given up to one of them. The remaining two continue to move, though with very little velocity, but the third may be ejected from the system with large velocity.

A star's motion in a stellar system where two-body forces dominate will be affected primarily in two ways. First, there is the overall motion due to the smooth potential of the system as a whole; second, there are local encounters between the star and one or a few nearby stars. Usually the first will establish an overall galactic orbit; the latter (two-body encounters) will produce deviations from the galactic orbit (direct collisions are usually rare).

Problem 28.5. Estimate the likelihood of two stars colliding if they have randomly distributed orbits in the Galactic disk. Assume typical interstellar separations to be about 10 pc. How could you improve this estimate to incorporate the possibility of gravitational scattering during near encounters?

The relative importance of stellar encounters may be ascertained by a slight modification of the discussion of the Coulomb scattering of charged particles in Section 18.2 Stellar encounters lead to deviations of stellar trajectories from their incident line of motion (see Figure 18.3, in which m_A and m_B now denote stellar masses). Defining the deviation angle due to gravitational scattering by θ, we find that

$$\tan \frac{\theta}{2} = \frac{(m_A + m_B)G}{v_\infty^2 s}, \quad (28.27)$$

where s is the impact parameter and v_∞ is the relative velocity of the two stars at large separation. Using parameters typical of stellar motion in the disk of our Galaxy ($s = 1$ pc, $v_\infty = 20$ km/sec, and $m_A = m_B = M_\odot$), we find that $\tan(\theta/2) \simeq \theta/2 = 5''$. Consequently, (28.27) may be approximated by

$$\theta \simeq \frac{2(m_A + m_B)G}{s v_\infty^2}. \quad (28.28)$$

The sum of θ^2 for each stellar encounter over all encounters N measures their effect on an individual star. Where the sum reaches unity, the encounters have changed the course of the star's motion by about 60°:

$$\sum_{i=1}^{N} \theta_i^2 = (1 \text{ rad})^2. \quad (28.29)$$

When (28.29) holds, the star no longer moves as if its motion were determined by the smooth galactic potential. The time required for this to occur defines a dynamic relaxation time τ_D, given by

$$\tau_D = \frac{v^3}{4\pi G^2 (m_A + m_B)^2 n}, \quad (28.30)$$

where n is the stellar number density. The time τ_D establishes the rate at which two-body gravitational encounters slow down a test star injected into a field of stars. This effect is referred to as dynamical friction.

Problem 28.6. Assume that the stellar number density n is constant, that the average deflection angle θ per encounter is given by (28.28), and that encounters primarily change the star's velocity without altering its speed. Under these assumptions, show that in time τ

$$\sum_i \theta_i^2 = \frac{4\pi n(m_A + m_B)^2}{v^3} G^2 \tau, \quad (28.31)$$

where v is the star's speed.

A more nearly exact result can be obtained in analogy to Coulomb scattering (Section 18.2). The deflection time for the charged particles contains the Coulomb logarithm (18.32), which results because distant encounters are screened in an electrically neutral plasma. Gravitationally interacting masses are not screened by intervening matter, because the gravitational charge $m\sqrt{G}$ is of one sign only. Without a cutoff for large-impact parameters, the change in stellar momentum analogous to (18.29) diverges. For gravitational scattering, the cutoff usually adopted lies between the interstellar separation $R \approx n^{-1/3}$ and the system's diameter. The deflection time is reduced relative to the preceding simple estimate, and is given by

$$\tau_{\text{rel}} \simeq \frac{1.09 \tau_D}{\ln [R \langle v^2 \rangle / (m_A + m_B) G]}. \quad (28.32)$$

This estimate is more appropriate for globular clusters than is (28.30). In (28.32), v is in km/sec, R is in pc, and the stellar masses are in solar units. For $\langle v^2 \rangle = (20 \text{ km/sec})^2$, stars of one solar mass, and $R = 10$ pc, $\tau_{\text{rel}} = 0.08 \tau_D$. Table 28.2 shows representative values of the relaxation time for various stellar systems.

The relaxation time represents an estimate of the time needed for a system to attain statistical equilibrium as a result of stellar encounters. Two additional time-scales are also useful. The first is the stellar mixing time, τ_m.

Consider two stars, in different orbits, but near one another. As they continue to follow their individual orbits, they tend to move away from one another and merge into the stellar background. The time required for this to occur is about half a period, or the time it takes one star to cross the stellar system,

$$\tau_m \simeq 2R / \langle v^2 \rangle^{1/2}, \quad (28.33)$$

where the stellar velocity is averaged over the system. In a time order τ_m, the stellar population will have thoroughly mixed.

Table 28.2
Stellar relaxation time and number density in several stellar systems.

Group	τ_{rel} in years	n in $M_\odot/\text{pc}^3$
Stellar associations (Pleiades)	2×10^6	1.2
Solar neighborhood	10^{12}	0.049
Globular clusters	3.6×10^8	75.0
Galactic center	10^6	10.
Elliptical galaxy (excluding nucleus)	10^{14}	10

The other useful time-scale is a dynamical evolution time-scale τ_{DE}, which may be defined as the time required for overall changes in the system that are driven by relaxation processes, or by external effects (tidal interactions due to neighboring stellar systems), or both. Dynamical evolution may involve star loss from the system (stellar evaporation), collapse, or both. Table 28.3 contains order-of-magnitude values of the three time-scales for galactic clusters, globular clusters, and elliptical galaxies. These estimates suggest that all but the youngest galactic clusters have thoroughly mixed, relaxed, and evolved, with the oldest having essentially dissolved into the galactic disk. Globular clusters, which are typically 10^9 yrs old, are expected to be mixed and relaxed, but have probably not evolved very far. Elliptical galaxies are expected to be mixed, but should have not yet relaxed (as a result of close encounters) or evolved. The general tendency of richer stellar systems to relax more slowly relative to their mixing rates than do less rich systems can be illustrated by rewriting τ_m and τ_{rel} in terms of the mean stellar density and the total number of stars N in the group. Ignoring constants,

$$\frac{\tau_{rel}}{\tau_m} \approx \frac{N}{\ln N/2}, \qquad (28.34)$$

Table 28.3
Stellar mixing time, relaxation time, and dynamic relaxation time for several stellar systems.

Stellar system	Time-scale (years)		
	τ_m	τ_{rel}	τ_{DE}
Galactic clusters	10^6	10^7	10^9
Globular clusters	10^6	10^8	10^{10}
Elliptical galaxy	10^8–10^9	10^{14}	10^{14}

which increases with stellar number. Physically this means that in rich, well-mixed systems (globular clusters, galactic nuclei, and elliptical galaxies), the relaxation processes act slowly on stellar motions. Other time-scales than the preceding probably apply to elliptical galaxies and to clusters of galaxies (see Section 28.4).

Problem 28.7. Derive the ratio τ_{rel}/τ_m and find a value for the constant of proportionality.

28.3. GLOBULAR CLUSTERS

Globular clusters are extremely old, spherical, or nearly spherical stellar systems, containing typically 10^4 to 10^6 stars. About 100 are known to be members of our Galaxy, making up a spherical halo centered on the Galactic nucleus, and extending to the edge of the disk. Individual clusters move about the Galactic center in highly eccentric orbits, with periods of order 10^8 yrs. Globular clusters have also been observed in nearby galaxies of all types. Their absolute magnitude ranges from -5 to -9 mag; their stellar content is typical of extreme Population II, and the most luminous are red giants. Globular clusters also contain RR-Lyrae variables (often called cluster variables), which can be used to establish their distance. At least one globular cluster is known to contain a planetary nebula (NGC 7078), and several are known to contain white dwarfs. The existence of these evolved objects, and the apparent absence of young stars, and of gas or dust, indicates that the clusters are quite old.

A cluster's age is normally obtained from the location of the main-sequence turn-off point, which typically occurs near absolute magnitude 3.5 (see discussion of color-magnitude diagrams in Chapter 3), giving a cluster age of order 10^{10} yrs. Cluster radii vary from 10 to 50 pc, and the average stellar number density over the entire cluster is typically 0.4 pc^{-3}. In the central core of this cluster, $\langle n \rangle \approx 10^2$ to 10^3 pc^{-3}.

The time-scales discussed in Section 28.2, and summarized in Table 28.3, contain the key to explaining the structure and evolution of globular clusters. For a typical cluster containing $N \approx 10^5$ stars, relaxation times are typically 10^8 yrs, and the ratio of relaxation to mixing times $\tau_{rel}/\tau_m \simeq 300$ or larger; so the systems are well mixed. Since cluster ages are of order 10^{10} yrs, stellar relaxation will have had sufficient time to influence the cluster's structure. We consider first how stellar encounters determine the relaxation processes and the clusters's structure, and set the stage for evolutionary changes. As a first step, we should ask what general conclusions follow from the application of statistical principles to a system containing $N \approx 10^5$ self-gravitating stellar masses. For simplicity, let it be assumed that the cluster is spherical, and that each cluster star has the same mass m (we will discuss in the following the generalization to a distribution of masses), and that the cluster's total energy is conserved. The initial distribution of stellar energies will be set by the conditions attending the formation of the cluster; whatever that may have been, the result of binary encounters is to redistribute the cluster's internal energy in a way that maximizes its entropy as the system evolves toward a steady state. The steady state corresponding to the maximum entropy is Maxwellian, and the probability that a star has the energy per unit mass E is

$$p(E) = Cg(E)e^{-\alpha E}, \qquad (28.35)$$

where C is a normalization constant, $g(E)$ is the statistical weight of the state with energy E, and α is a positive constant. Two difficulties with this equilibrium state are immediately evident. First, for stars bound to the cluster, $E < 0$, and $p(E)$ diverges as E becomes more negative (that is, as the star becomes more tightly bound to the cluster). Second, a star that is only slightly bound to the cluster ($E \lesssim 0$) can, as the result of an encounter with another star, be scattered into an orbit with $E > 0$. The statistical weight for such states is proportional to the spatial volume that can contain stars with this energy. But this is in fact the entire universe; so $g(E)$ must be considered infinite for $E \gtrsim 0$. Thus, the steady state toward which stellar encounters drives the cluster is characterized by $p(E)$, which diverges both at large negative energies and for all positive energies. The difficulties at the lower end of the energy spectrum can be side-stepped; the simple approach breaks down there because the finite size of the stellar radii have not been taken into account, and because relativistic effects will arise at high core densities. The formation of tightly bound binary systems in the cluster core could also have a significant influence on the steady state there. Until conditions are reached where these effects become important, stellar encounters will attempt to drive the system toward a state whose velocity distribution is Maxwellian, but the system will never quite reach this state.

According to the virial theorem, the average stellar escape velocity is related to the root-mean-square velocity by $\langle v_e^2 \rangle^{1/2} = 2 \langle v^2 \rangle^{1/2}$, which is (28.24). Therefore a stellar velocity distribution that approaches a Maxwellian will have a fraction ξ_e of its stars at greater than escape velocity (28.25). In roughly one relaxation time τ_{rel}, $N\xi_e$ stars, which lie in the high-energy tail of the velocity distribution, will escape from the cluster. Crudely speaking, the stellar loss rate might be expected to go as

$$\frac{dN}{dt} = -\frac{\xi_e N}{\tau_{rel}}, \qquad (28.36)$$

where ξ_e and τ_{rel} will depend on position in the cluster. The distribution of velocities for stars actually bound to a cluster has been found to be given approximately by

$$f(E) \approx e^{-3E/\langle v^2 \rangle} - 1, \qquad (28.37)$$

which is a truncated Maxwellian distribution. For

$$E > \langle v^2 \rangle = 2MG\langle 1/r \rangle,$$

the core is essentially isothermal. Therefore,

$$r_c \equiv \left\langle \frac{1}{r} \right\rangle^{-1} = \frac{2MG}{\langle v^2 \rangle} \qquad (28.38)$$

serves as a natural definition of the core radius. The energy gained by stars with $E \gtrsim 0$ is balanced by the energy loss as the cluster core contracts and becomes more tightly bound. According to the virial theorem, the total energy of the cluster

$$E_{tot} = -\frac{M}{2}\langle v^2 \rangle = \frac{M^2 G}{2r_c}. \qquad (28.39)$$

Therefore, if E_{tot} becomes more negative (more tightly bound) as a result of star loss, then the core must contract. The subsequent evolution requires more detailed analysis than the preceding simple statistical arguments, but it is clear that the internal structure of the cluster will change, resulting in new relaxation times and a continued tendency for the stellar distribution to evolve toward a Maxwellian state with further loss of stars. In effect, relaxation and star loss lead to a succession of near equilibrium states of progressively greater core concentrations and a lower total mass of bound stars.

Problem 28.8. Estimate the dynamical evolutionary time-scale for a globular cluster using the preceding results.

The surface-brightness distributions of globular clusters are spherical, or nearly so. However, because they are members of large galaxies, they are subject to an external gravitational force. Although the external field of the galaxy may not have a significant effect on the brightness distribution, it can greatly influence the cluster's structure and evolution. Because of the gravitational tidal force acting on a cluster, a radius r_T exists beyond which a star will be stripped away from the cluster. The tidal force will also tend to promote stars to larger radii. Both effects tend to increase the rate of evolution of the cluster.

The relation between r_T, the cluster mass M_c, and the galactic mass M_g can be found easily for a spherical cluster at ratius r moving in a circular orbit of radius R_0 about the galactic nucleus. In a frame of reference centered on M_g and moving with the cluster, the force on a member star is

$$F = m\left(R\omega_0^2 - \frac{v^2}{R}\right). \quad (28.40)$$

We ignore stellar motion within the clusters and assume that each star moves with angular velocity $\omega = v/R$ about M_g. The tidal force acting across the cluster is

$$\Delta F \simeq \left(\frac{dF}{dR}\right)_{R=R_0} \Delta R$$
$$= m\left(\omega_0^2 + \frac{v^2}{R^2} - \frac{2v}{R}\frac{dv}{dR}\right)_{R=R_0} r$$
$$= 2m\omega_0 r \left(\omega_0 - \left(\frac{dv}{dR}\right)_{R_0}\right),$$

where $v_0 = \omega_0 R_0$ is the cluster's velocity with respect to the galactic center, and $\Delta R \approx r$ is the cluster radius. The velocity of a star at R is given by $v^2 = MG/R$, so that $(dv/dR)_0 = -\omega_0/2$ (notice that we are ignoring the internal motion of stars in the cluster). Therefore

$$\Delta F \simeq \frac{3mv_0^2}{R_0^2}r = \frac{3mM_gGr}{R_0^3}. \quad (28.41)$$

This represents the difference between the gravitational force exerted by the galaxy on a star a distance r away from the cluster's center and that on the cluster's center of mass. The gravitational force between the star and the globular cluster (assuming that the star is on the edge of the cluster) is

$$F_{cl} = \frac{mM_cG}{r^2}. \quad (28.42)$$

Stars for which $\Delta F > F_{cl}$ will therefore be removed from the cluster. Combining (28.41) and (28.42), we obtain the tidal radius of the cluster,

$$r_T = R_0(M_c/3M_g)^{1/3}. \quad (28.43)$$

This serves as a natural definition of the outer radius of a globular cluster. A more precise analysis shows that the galactic field will define certain escape regions near the cluster (recall the discussion of gravitational equipotential surfaces in Section 17.2).

Let us summarize the conclusions that follow from the preceding simplified analysis. First, there appear to be three parameters that determine the structure of a globular cluster at any instant: the total number of stars, N; the core radius r_c, given by (28.38); and the tidal radius r_T, given by (28.43). The tidal radius depends on the position of the cluster in its orbit around the galaxy, and will vary from one cluster to another. The remaining two depend on the stellar distribution function and the number of stars N left in the cluster. Figure 28.1 shows the apparent surface density for clusters obtained from a simple model that incorporates features similar to those we have discussed. The apparent surface density is obtained by projecting the actual spatial density onto the plane of the sky, which can be most readily compared to observations. The numbers next to each curve give log (r_T/r_c) for the model. Evidently a larger radius gives a larger cluster. A rough correlation may be made between stars in certain velocity groups and the structure at various positions in a cluster. The low-velocity stars in the distribution function $f(E)$ should be found primarily in the high-density core of the cluster, whereas the high-velocity stars spend most of their time in the low-density halo. Relaxation forces the distribution function to be Maxwellian for low velocities, but can not at high velocities. Therefore, deviations from steady state show up essentially in the halo; clusters that have had time to relax have essentially isothermal cores. In fact, the cores of all relaxed stellar

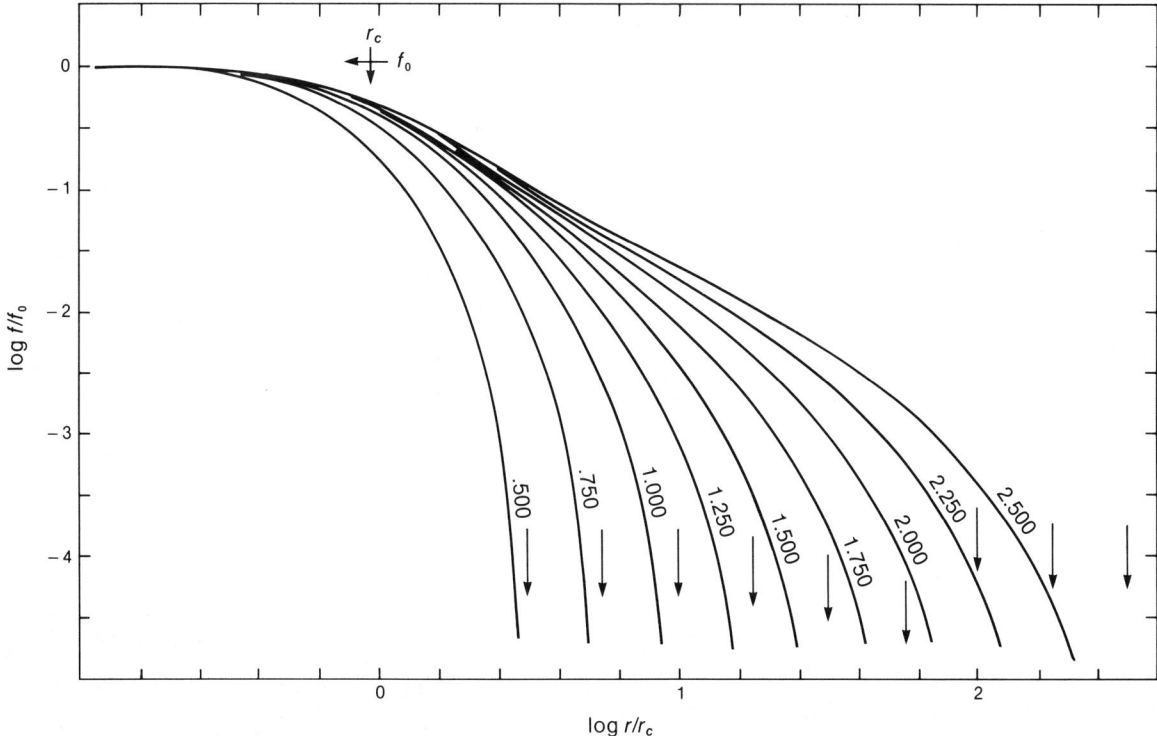

Figure 28.1. Projected surface-brightness distribution of model globular clusters. The central distribution is f_0, and r_c is the core radius. Each curve corresponds to a value of $\log(r_T/r_c)$; the arrow gives the model's limiting radius.

systems with large N should have the same structure. Figure 28.1 shows that, although the extent of the halo varies strongly with r_T, the shape of the core is essentially always the same.

Perhaps the most surprising result is the fact that these models fit the observational data. Figure 28.2 shows results for M13 and NGC 5053. The following parameters have been obtained for ten concentrated globular clusters using observational data and theoretical models similar to the ones shown in Figures 28.1 and 28.2. The cluster masses range from 1.6 to 11.2 × 10^5 $M_\odot$, and their average mass-to-luminosity ratio $(M/M_\odot)/(L_V/L_\odot) = 1.7$. The average central velocity dispersion $\langle v_r^2 \rangle^{1/2} = 12$ km/sec, and the resulting core radii $r_c \simeq 0.5$ pc. For these compact clusters, $\log r_T/r_c$ averages 1.8.

Despite the excellent agreement between theory and observation, a simple statistical model cannot be expected to describe the cluster's internal structure or its detailed evolution. More detailed methods are needed to attack these issues. One approach follows the motion of a large number of sample stars, each one of which represents some fraction of the total number of stars in the cluster. Encounters between these stars may be followed, and the cluster's evolution followed in detail. These studies confirm the general conclusions obtained by using simple models and fundamental statistical principles. In particular, a core develops that contains about half the cluster mass, and evolves toward an essentially Maxwellian velocity distribution that is isotropic. As the energy of a star approaches zero, $f(E)$ deviates from Maxwellian, and is given by (28.37), except for $E \approx 0$, where it falls off more rapidly. The stars promoted to energies $E \lesssim 0$ gradually move outward, establishing a cluster halo, where they move in nearly radial orbits. Figure 28.3 shows three computed profiles of stellar number density as a function of distance from the cluster center. According to these evolutionary calculations, the initial core contraction is accompanied by the formation of the cluster halo. The initial configuration (shown by the dash-dot curve) is a sphere of constant stellar number density. As the cluster evolves, the central density increases, and the halo expands. Significant stellar loss

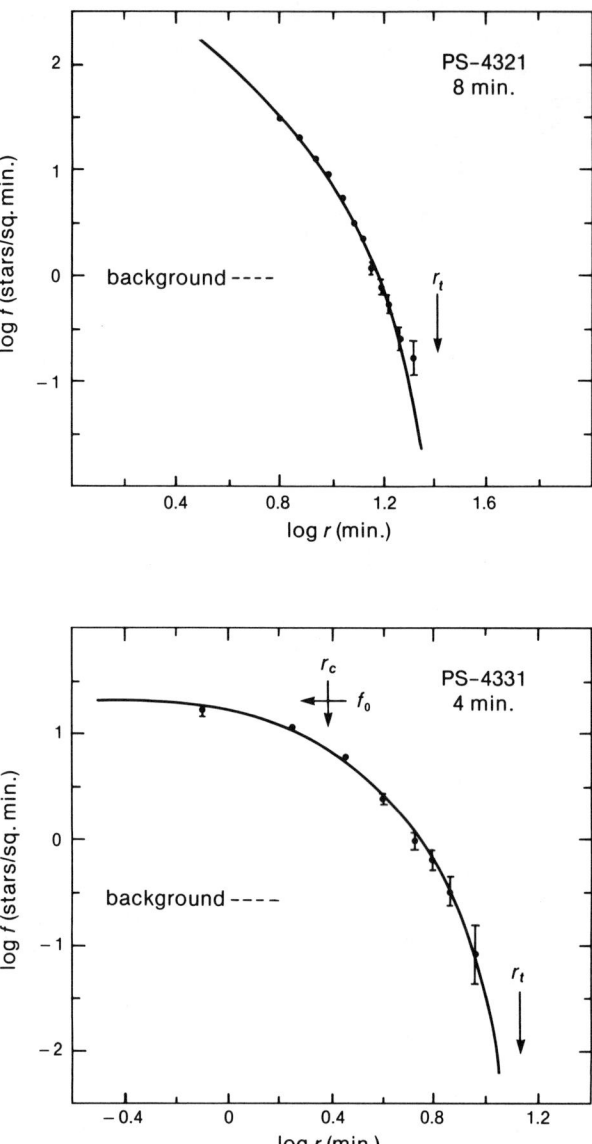

Figure 28.2. Star counts in the globular cluster M13, compared with theoretical curve for $\log(r_T/r_c) = 1.5$. The lower figure gives star counts in globular cluster NGC 5053, compared with a theoretical curve for $\log(r_T/r_c) = 0.75$. The dashed line is the background level.

appears to occur late in the evolution, after the halo has been well established, and results from the transfer of small amounts of energy to stars with E only slightly less than zero prior to encounter. Figure 28.4 shows selected radii as a function of time (the numbers next to each curve give the fractional cluster mass contained within r during the contraction). The initial core contraction is roughly uniform, and the energy loss as it becomes more tightly bound goes into forming the halo.

Near the end of the calculation, the rate of contraction in the inner core accelerates (see late time behav-

522 / DYNAMICS OF STELLAR SYSTEMS

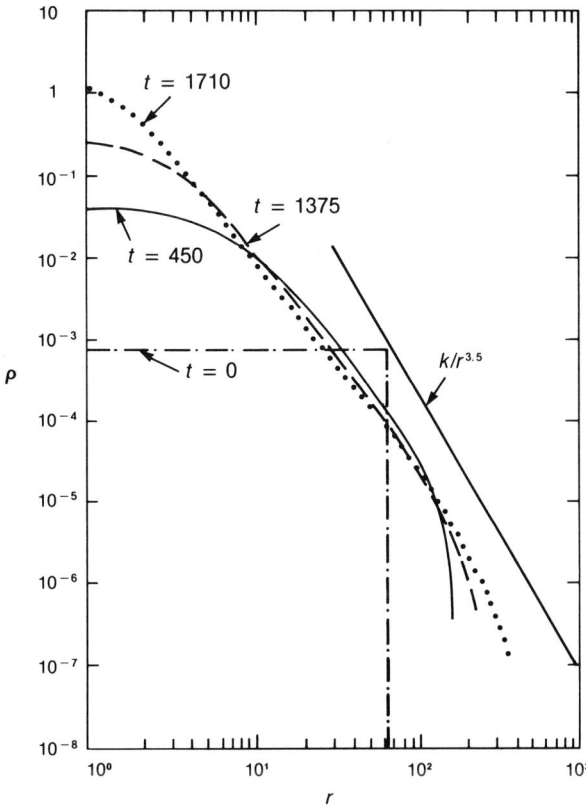

Figure 28.3. Stellar density distribution at different times in the evolution of a cluster model. The initial model (dash-dot) is a sphere of constant density in equilibrium, with all stars in circular orbits.

ior in Figure 28.4), and the qualitative character of the evolution changes. In fact, if the rate is extrapolated to $r = 0$, the inner-core collapse would be complete in 10 to 20 τ_{rel}, and could involve 2 percent or more of the total cluster mass. The increased energy loss accompanying this collapse goes into structural rearrangement at intermediate radii, and causes the cluster to expand (see Figure 28.4). The onset of core collapse is gradual, but appears to have set in when $\rho_0/\rho_{1/2} \approx 10^3$, where ρ_0 is the cluster's central density, and $\rho_{1/2}$ is the density at the radius where $m(r) = M_c/2$.

Violent phenomena are likely to occur if core collapse persists. As the stellar number density increases, the likelihood of stellar collisions increases; since the stellar relative velocities should be small compared to the escape velocity from the stellar surface, collisions should result in coalesence. The final outcome of collapse could include the formation of a massive black hole. The x rays from globular clusters could come from such an object. Whether or not the ultimate collapse of the core is averted by processes not included in current treatments of stellar dynamics is a matter of debate.

Real clusters contain a distribution of stellar masses rather than stars of a single mass. Statistically, all stars should approach energy equipartition in a time comparable to τ_{rel} for the lightest stars. In the process, heavier stars lose kinetic energy to the lighter ones, and must therefore sink toward the center of the cluster, and mass stratification develops. Figure 28.5 shows the results of numerical models that include three stellar components chosen to represent Population II. Mass stratification implies a gradual reddening in starlight from the cluster center to its halo; unfortunately, the change in $B - V$ expected from the center to halo is only a few hundredths of a magnitude.

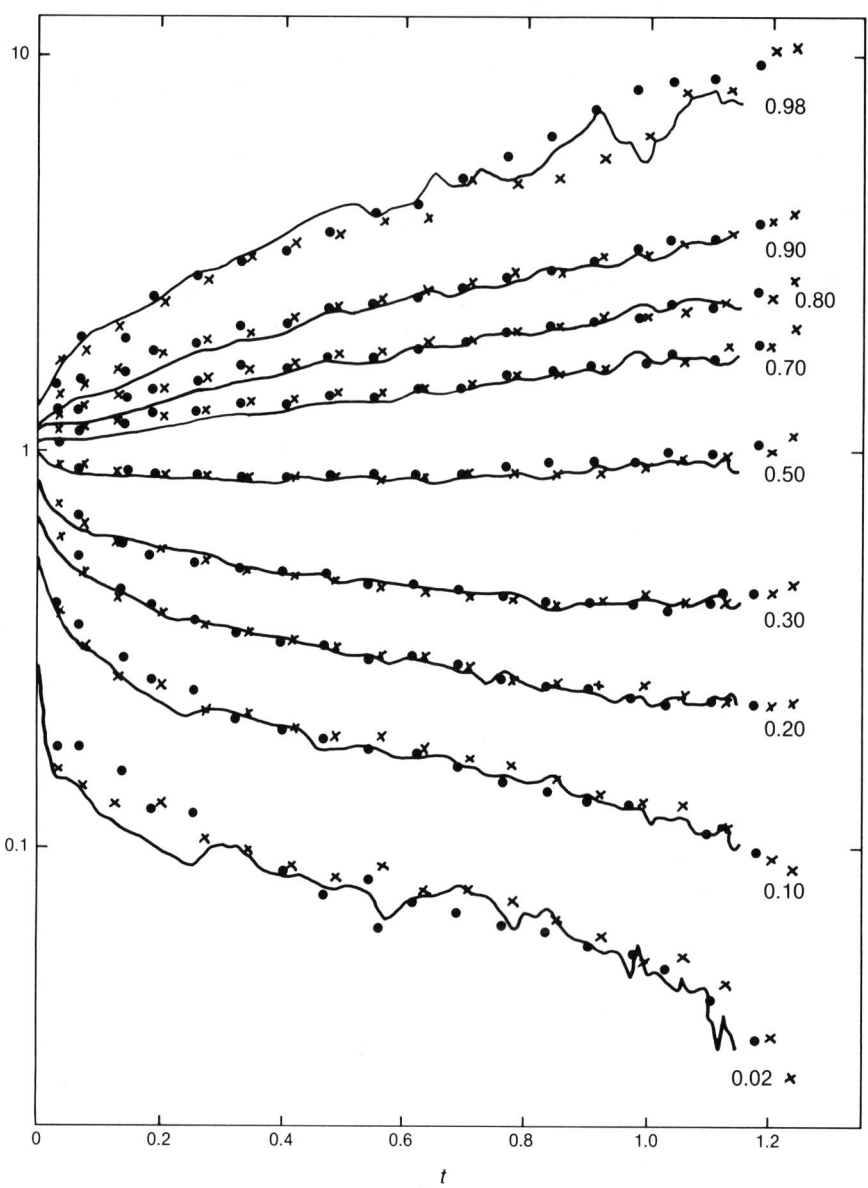

Figure 28.4. Radius containing fixed fractions of cluster mass versus time for model shown in Figure 28.3.

28.4. Clusters of Galaxies

The spatial distribution of galaxies over distances measured in megaparsecs (Mpc = 10^6pc) does not appear uniform; rather, galaxies on this scale are observed to form clusters of varying size and richness. Detailed studies of the richest clusters reveal that their structure closely resembles that of globular clusters of stars. In fact, many of the methods of stellar dynamics have been applied to the structure of clusters of galaxies.

Clusters of galaxies form two distinct groups. The regular clusters are spherically symmetric, centrally condensed, extremely rich, containing in excess of 10^3 member galaxies brighter than $M_V = -16$. Regular clusters consist almost entirely of elliptical and lenticu-

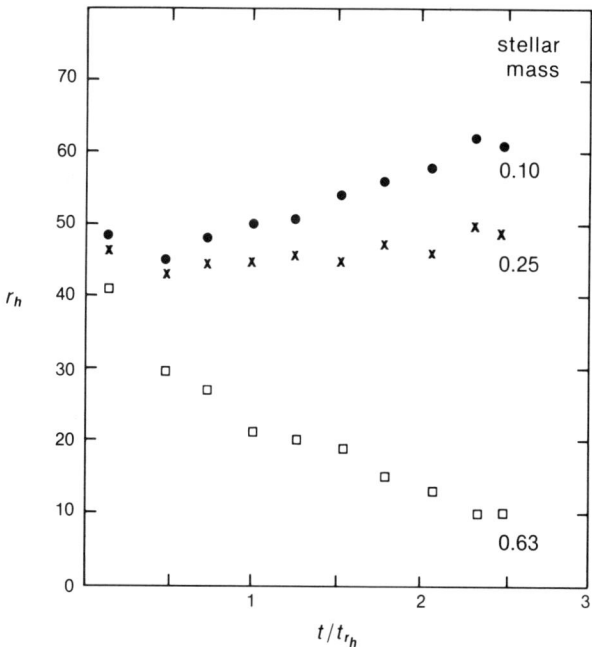

Figure 28.5. Mass stratification in cluster model containing three stellar-mass components. Points represent the radius that contains half the mass for each of the three stellar-mass groups versus time. Stellar mass is in units of $M_\odot$.

lar galaxies; their brightest members are giant ellipticals, for which $M_V \approx -23$ to -24. It is believed that the largest regulars contain more than 10^4 galaxies (many of which are smaller dwarf ellipticals). The paradigm for this class is the Coma cluster.

The second group consists of irregular clusters, which are characterized by an amorphous distribution of galaxies. Although they show no evidence of central condensation, they may contain several small condensations distributed throughout the cluster. Irregular clusters range in size from small objects like our local group (containing about twenty members), to rich clusters, such as the Virgo cluster, containing more than 10^3 members. Irregular clusters contain galaxies of all types, and appear to be more common than the regular clusters.

There are significant differences between the processes that determined the structure of regular clusters of galaxies and those for stellar globular clusters. Nevertheless, these two groups share some features in common. Figure 28.6 shows the projected surface density of galaxies in the Coma cluster vs. distance from the cluster center. The curve is the surface density of an isothermal gas sphere. Two points should be noted: first, the degree of central condensation; second, the apparent fact that the distribution is very nearly isothermal, except in the outer regions, where it falls rapidly with increasing distance. These data are similar in character to the data for stellar globular clusters. In particular, the velocity distribution of galaxies in Coma is expected to be Maxwellian, with $\langle v_\parallel^2 \rangle^{1/2} \approx 10^3$ km/sec ($v_\parallel$ is the line-of-sight velocity of the member galaxies, and the brackets denote a time average).

Supposing the Coma cluster to be gravitationally bound, we may use the virial theorem to estimate its mass. The result,

$$M_{\text{coma}} \simeq 3 \times 10^{15} \, M_\odot, \qquad (28.44)$$

and the observed luminosity,

$$L_{\text{coma}} \simeq 1.4 \times 10^{13} \, L_\odot, \qquad (28.45)$$

imply a mass-to-luminosity ratio of 200 $M_\odot/L_\odot$. The distance to the cluster is about 135 Mpc, and its radius is probably about 4 Mpc.

The Coma cluster appears to have evolved to a state

28.4. CLUSTERS OF GALAXIES / 525

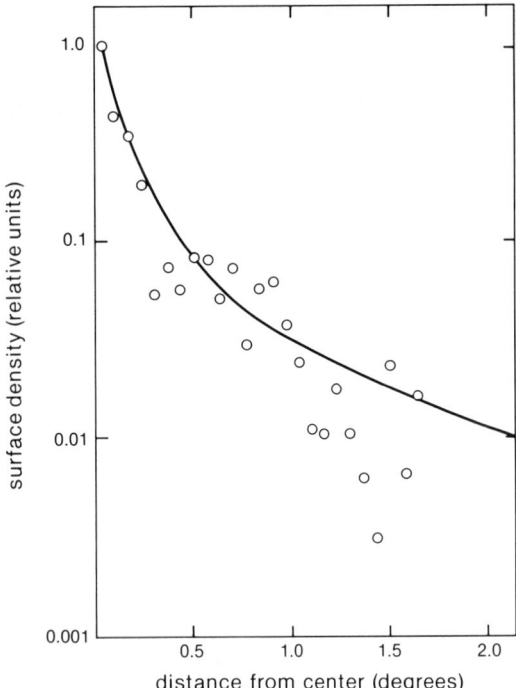

Figure 28.6. Observed surface density of galaxies in the Coma cluster versus distance from the cluster center. Curve is for an ideal isothermal gas sphere.

change of E_g, the energy of a galaxy whose mass is M_g, because of changes in the local gravitational field $\Phi(\mathbf{r}, t)$ is

$$\frac{dE_g}{dt} = -M_g \frac{\partial \Phi}{\partial t}, \tag{28.46}$$

where

$$E_g = M_g \left(\frac{1}{2} v^2 - \Phi(\mathbf{r}, t) \right), \tag{28.47}$$

and v is the galaxy's instantaneous velocity. As the background field $\Phi(\mathbf{r}, t)$ changes dynamically, the energy of each galaxy changes, and the system evolves toward an equilibrium state. The relaxation time associated with the energy-loss rate (28.46) may be expressed as

$$\tau_v \equiv \left\langle \frac{(d\epsilon/dt)^2}{\epsilon^2} \right\rangle^{-1/2}, \tag{28.48}$$

where $\epsilon = E_g/M_g$, and the brackets denote an average over the system. Rewriting (28.46) in terms of the energy per unit mass, $\dot{\epsilon} = \partial \Phi/\partial t$, and inserting into (28.48), gives

$$\tau_v = \left\langle \frac{(\partial \Phi/\partial t)^2}{\epsilon^2} \right\rangle^{-1/2}.$$

Suppose that the change in Φ is comparable to Φ itself, and that this change takes place in a time comparable to the free-fall time for the cluster $\tau_\mathrm{ff} \sim (G\bar{\rho})^{-1/2}$. Then, to order of magnitude, $\epsilon \approx \Phi$ and $\partial \Phi/\partial t \approx \Phi/\tau_\mathrm{ff}$, so that $\tau_v \approx \tau_\mathrm{ff}$. In fact, the arguments can be refined to yield

$$\tau_v \simeq \frac{3}{4} \left(\frac{R^3}{MG} \right)^{1/2}, \tag{28.49}$$

where R is the cluster's radius and M is its total mass. Adopting the mass (28.44) and the radius 4 Mpc, we find that $\tau_v \simeq 2 \times 10^9$ yrs, which is probably short enough for relaxation to have determined the velocity distribution of the galaxies in the cluster. Unlike relaxation due to close encounters, the rate of violent relaxation is independent of galaxy mass; therefore mass stratification need not occur.

The relaxation indicates the time-scale needed for the cluster to reach equilibrium, but says nothing about the nature of the equilibrium state. Violent

of equilibrium, just as globular clusters have, but the relaxation processes must have been quite different. Estimates of the relaxation time due to binary encounters between galaxies in regular clusters gives

$$\tau_\mathrm{rel} \sim 10^{16} - 10^{18} \text{ years},$$

which is clearly too long for those processes to have played a role in the evolution. The process most likely to have influenced their structure is violent relaxation. The Coma cluster is believed to be some tens of billions of years old. It has been argued that as protogalaxies formed, the mass in the cluster was already gravitationally bound, and that the gravitational field of the cluster as a whole underwent dynamical variations. As a result, the orbits of the newly formed galaxies within the cluster changed significantly as the cluster condensed. The motion of galaxies in the rapidly time-varying background field of the cluster led to energy exchange, and to the relatively rapid establishment of an equilibrium state.

A simple estimate of the time-scale for violent relaxation may be made as follows. The time-rate of relaxation is a much more difficult process to evaluate theoretically than relaxation by close encounters. However, there is reason to suspect that the state toward which clusters of galaxies are driven is also Maxwellian.

Problem 28.9. Estimate τ_v for the globular clusters and for elliptical galaxies (for the latter, take $M = 2.5 \times 10^{12}\ M_\odot$ and $R = 250$ kpc). What role would you expect violent relaxation to play in these systems?

Problem 28.10. Assume that a dynamically evolving cluster is near equilibrium, so that the virial theorem ($2T + \Omega = 0$) applies on average. Show that

$$\tau_v = \frac{3}{4}\left\langle \frac{(\partial \Phi/\partial t)^2}{\Phi^2} \right\rangle^{-1/2}.$$

Then apply the time-dependent virial theorem (22.6) to show that the fundamental mode of vibration about the equilibrium state has angular frequency

$$\omega = (MG/2\alpha R_0^3)^{1/2},$$

where $I = \alpha MR^2$, and R_0 is the equilibrium radius of the cluster. Use these results to show that

$$\tau_v = \frac{3}{4}\left(\frac{2\alpha R_0^3}{MG}\right)^{1/2}.$$

As noted in the preceding (and in Section 27.3), masses for clusters of galaxies may be obtained from the virial theorem. An alternative method employs galaxy counts and an estimate of the mass of each galactic type. This approach is clearly biased in favor of large, bright galaxies in a cluster, but these are presumably the most massive. Galaxy counts systematically produce cluster masses M_{count} that are smaller than the mass M_{virial} obtained from the virial theorem. For example, $M_{count} \simeq 6 \times 10^{14}\ M_\odot$ for the Coma cluster, which is about one-fifth M_{virial} (28.44). Despite the uncertainties inherent in the methods leading to M_{count}, it is generally concluded that $M_{virial} > M_{count}$ by a significant amount.

We recall that M_{virial} includes not only mass in galaxies, but all forms of matter, whereas M_{count} can at best include only galaxies. This observation has led to speculation that clusters contain hidden mass. One form that hidden mass could take is intergalactic gas. Indirect evidence for the existence of gas between galaxies in rich clusters comes from observations of radio galaxies. Many of these objects consist of a large central galaxy (often peculiar) flanked by two large lobes from which most of the radio emission arises. The radio lobes consist of high-energy charged particles ejected in opposite directions from the central galaxy. These particles move along magnetic field lines that also originate in the central galaxy, and radio emission occurs from synchrotron processes in the lobes. Because of its orbital motion in a cluster, a radio galaxy will travel through any intergalactic gas, producing a backward drag on the charged particles and the magnetic field to which they are coupled. Figure 28.7 shows radio-emission contours from the lobes of NGC 1265 in the Perseus cluster, superimposed on the optical image containing the central galaxy. The galaxy, located in the lower central part of the figure, is moving downward with a velocity (along the line of sight) estimated to be about 2×10^3 km/sec relative to the mean motion of the other galaxies in the cluster.

Many regular clusters of galaxies, including the Coma cluster, emit a diffuse x-ray spectrum with luminosities in the range of 10^{43} to 10^{45} erg/sec. The x-ray luminosity varies smoothly across the cluster, which is consistent with emission from an intergalactic gas (Figure 28.8). The most likely mechanism is thermal *Bremsstrahlung* from a hot dilute gas. The emissivity applicable to the gas in clusters of galaxies is given roughly by

$$4\pi j(\nu) = 6.79 \times 10^{-38} \frac{n_e^2 e^{-h\nu/kT}}{\sqrt{T}} \quad (28.50)$$

in ergs cm^{-3} sec^{-1} Hz^{-1}, and with T in K. The spectrum observed for diffuse x rays from clusters fits (28.50) if $T \approx 10^8$ K, and number densities near the cluster center are of order 10^{-3} cm^{-3}. The latter implies a mass density of order 2×10^{-27} g cm^{-3}, which is roughly 10^3 times the average mass density of the universe. A gas temperature of order 10^8 K corresponds to a root-mean-square thermal velocity $\langle v^2 \rangle^{1/2} \approx 10^3$ km sec^{-1}. This is comparable to the observed root-mean-square velocity of galaxies in the Coma cluster, as would be expected if both components were in equilibrium. The distribution of gas would resemble the mass distribution of galaxies, so that knowledge of $n_{gal}(r)$ can be used to estimate $N_{gas}(r)$, and from it the total mass in intergalactic gas.

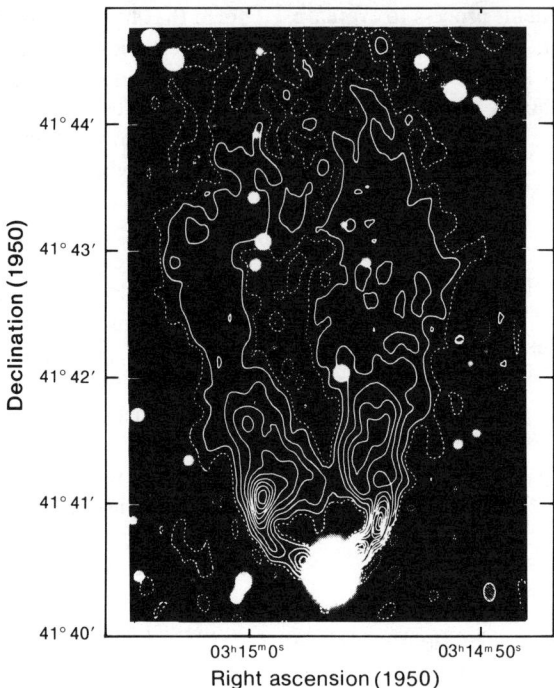

Figure 28.7. Radio-emission contours from galaxy NGC 1265 in the Perseus cluster. The 5,000 MHz map of the radio tail is superimposed on a blue print of the optical galaxy NGC 1265.

In the Coma cluster

$$M_{gas} \simeq 5 \times 10^{15} M_\odot$$

is comparable to the total mass in galaxies. However, the mass $M_{count} + M_{gas}$ is not sufficient to bind the cluster. The existence of intergalactic gas is nevertheless well established.

Problem 28.11. If intergalactic gas cools primarily by *Bremsstrahlung* losses, how long would it take for the gas in the Coma cluster to cool? Assume that the electrons are relativistic, with energy density $u_e = 3 n_e kT$.

The differences in x-ray surface brightness characterized in Figure 28.8 may yield a clue to understanding the evolution of galaxies and clusters. The more centrally condensed clusters (as determined from their x-ray emission) have high x-ray luminosities (10^{44}– 10^{45} ergs/sec), and intergalactic gas temperatures of order 10 keV or more (1 keV = 1.16×10^7 K). Clusters typified by A1367 (Figure 28.8a) have lower luminosities ($L_x \approx 10^{43}$ ergs/sec) and intergalactic gas temperatures of a few keV. In these systems the gas is localized near galaxies, having been recently emitted by or stripped away from them. This would explain the observed clumpiness in x-ray brightness and suggests that these systems have not evolved much since their formation. In the centrally condensed clusters, the gas is presumably associated with the smooth gravitational field of the entire cluster. The relatively small core radii (typically 0.25 Mpc) argue that these clusters have evolved dynamically, which is consistent with their larger observed x-ray luminosities.

Problem 28.12. Suppose that the gas in clusters such as A1367 (Figure 28.8a) is associated with individual galaxies, while the gas in centrally condensed clusters (A476) is associated with the cluster's gravitational potential field. Show that the former

528 / DYNAMICS OF STELLAR SYSTEMS

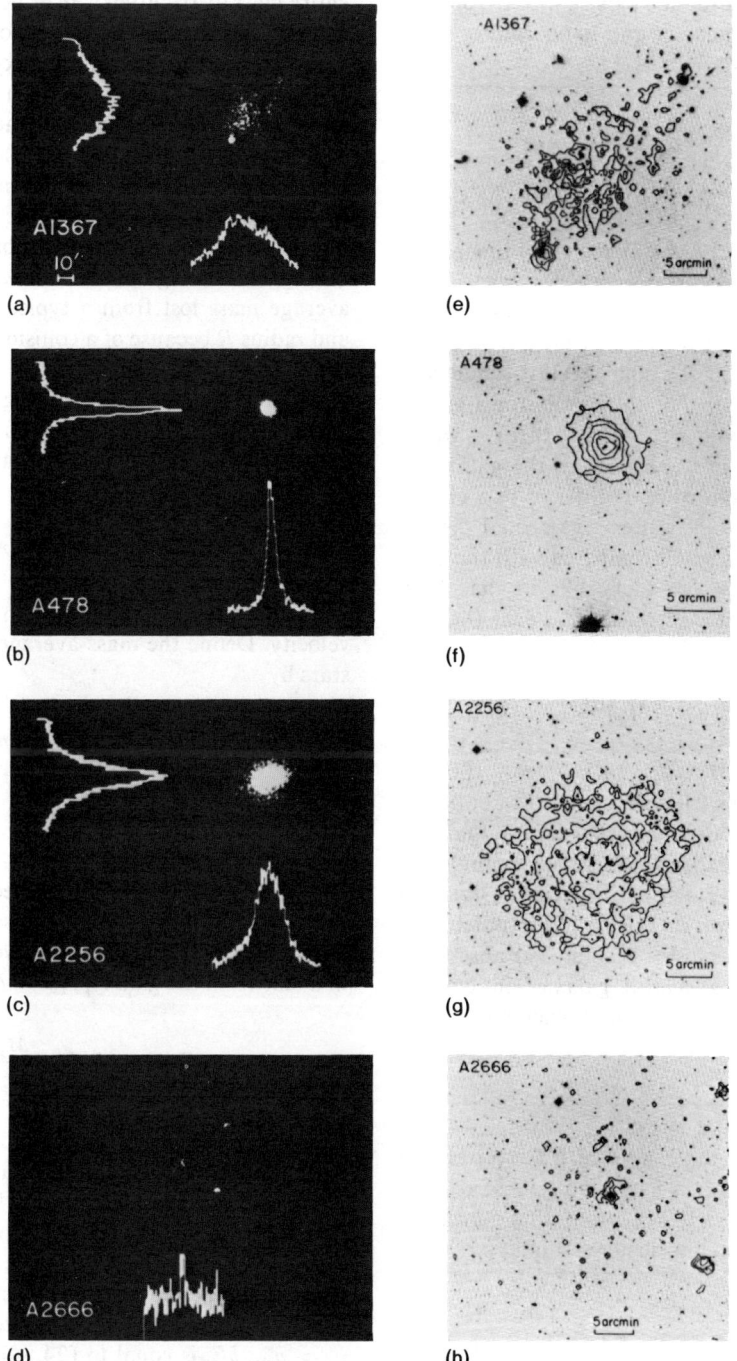

Figure 28.8. Projected x-ray emission across clusters of galaxies measured from the Einstein Observatory. Profiles show emission across two perpendicular axes: (a) broad emission clumped around individual member galaxies; (b) emission sharply peaked around central galaxy; (c) smooth, centrally enhanced, but less peaked than (b); (d) irregular (poor) cluster. Inset is cluster image in the visible spectrum. Lower part of figure gives isointensity of x-ray image superimposed on optical image, showing: (e) nonspherical, multiple source nature; (f) central source; (g) less peaked but condensed source; (h) irregular source.

28.4. CLUSTERS OF GALAXIES / **529**

should exhibit lower x-ray luminosities, and estimate the ratio of the intergalactic gas temperature in the two classes of clusters.

X-ray studies of clusters of galaxies at cosmological distances ($0.1 \lesssim z \lesssim 0.7$) indicate that they are similar to nearby clusters.

The missing mass required by the virial theorem to bind clusters could lie in black holes, distributed throughout the cluster or in their center. A limit to the mass of a black hole located in the center of a cluster can be set by considering the tidal effects it would have on neighboring galaxies. The tidal radius of a galaxy near a black hole is of order $r_T \sim R(M_B/M_g)^{1/3}$, where R and M_g are the radius and mass of the galaxy. If intergalactic separations were of order r_T in a cluster center, the galaxies would suffer severe tidal disruption within time-scales small compared with the cluster's age. No evidence for such behavior has been observed; so intergalactic separations r_0 near the center

$$r_0 > R(M_B/M_g)^{1/3}. \quad (28.51)$$

For the Coma cluster $M_g \simeq 2 \times 10^{12} M_\odot$, $R \simeq 100$ kpc, and $M_B \simeq 3 \times 10^{15} M_\odot$ (the mass needed to build the cluster), tidal disruption should occur for intergalactic separations of order 1 Mpc or less. Galaxy counts yield a higher central number density. In fact, distributions similar to the one shown in Figure 28.6 predict r_0 as small as 0.1 Mpc at the cluster's center. The absence of evident activity in the central galaxies in clusters argues against missing mass in giant black holes.

Problem 28.13. What is the most massive black hole that could lie at the center of a Coma cluster without tidally disrupting neighboring galaxies? How large would such a black hole be, and what qualitative effects might it have on the cluster's structure?

A large number of smaller black holes could be distributed throughout the cluster without producing tidal distortion. Because the distributions observed in galaxy counts and implied by relaxation models appear to agree, the distribution of mass in black holes should, apart from a scale factor, be similar to the spatial distributions of galaxies in a cluster. Further, if the entire mass distribution is in or near statistical equilibrium, the average space velocity of a black hole should be comparable to that of a galaxy ($v_B \simeq 10^3$ km/sec). Consequently, collisions between black holes and galaxies will occur throughout the cluster, and each collision will scatter stars out of the galaxies. A limit to the average mass of black holes in a cluster may be set by the requirement that the fractional mass loss by a typical galaxy because of collisions be less than unity throughout the age of the cluster. First consider the average mass lost from a typical galaxy of mass M_g and radius R because of a collision with a black hole of mass M_B. This may be most easily estimated by envisioning the black hole moving into and through the galaxy. The stars scattered out of the galaxy by collision with a black hole carry away a kinetic energy

$$\delta E = \sum_i m_* v_i^2 = m_* \sum_i v_i^2, \quad (28.52)$$

where the sum is over the stars having at least escape velocity. Define the mass-averaged velocity of ejected stars by

$$\langle v^2 \rangle_{\text{esc}} = \frac{\sum_i m_* v_i^2}{\sum_i m_*} = \frac{m_* \sum_i v_i^2}{\delta M_g}, \quad (28.53)$$

where δM_g, the total mass of ejected stars, is just the mass loss by the galaxy because of the collision. If $\langle v^2 \rangle_{\text{esc}}$ is approximated by the average escape velocity of a star from the galaxy, $v_{\text{esc}}^2 \approx M_g G/R$, then

$$\delta E \simeq v_{\text{esc}}^2 \delta M_g \approx \frac{M_g G \delta M_g}{R}. \quad (28.54)$$

The change in energy associated with stars scattered with velocities $v_* > v_{\text{esc}}$ in the galaxy by the incident black hole can be approximated by the product of the scattering cross section, $\sigma_s = \pi a^2$, and the stellar energy flux past the black hole during the collision interval, integrated over the stellar energy distribution. The total energy corresponding to stars that are given $v_* > v_{\text{esc}}$, δE, is equal to (24.54). To order of magnitude, this is

$$\delta E = \pi a^2 (\rho v^2) v \, \delta t, \quad (28.55)$$

where ρv^3 is the flux of stellar kinetic energy relative to the black hole, and δt is the collision time. Assuming that $M_b \gg m_*$, and that the star's velocity in the galaxy is small compared with the collision velocity v, the gravitational force of the black hole will deliver an impulse to each star with which it collides, knocking it from the galaxy. The cross section for this process should be, to order of magnitude,

$$\pi a^2 \approx \pi (M_b G/Rv^2)^2.$$

Denoting the stellar mass density by $\rho \approx M_g/R^3$, and the collision time by $\delta t \sim R/v$, the kinetic energy carried away by ejected stars (28.55) is approximately

$$\delta E \approx \left(\frac{M_b G}{v^2}\right)^2 \left(\frac{M_g}{R^3}\right) R. \quad (28.56)$$

Equating (28.54) and (28.55) and canceling common factors, we find the fractional change in galactic mass per collision to be of order

$$\left(\frac{\delta M_g}{M_g}\right)_{\text{coll}} \approx \frac{M_b^2 G}{M_g v^2 R}. \quad (28.57)$$

The collision rate between a single galaxy and the distribution of black holes in the cluster is

$$\frac{1}{\tau} \approx n_B v \sigma_g, \quad (28.58)$$

where n_B is the black-hole number density, and the galaxy's collision cross section is $\sigma_g \approx \pi R^2$. Finally, the average fractional mass-loss rate for a typical galaxy is the product $(\Delta m/m)_c \tau^{-1}$ averaged over the spatial distribution of galaxies in the cluster:

$$\frac{1}{m_g}\frac{dm_g}{dt} \approx \left\langle n_B v \sigma_g \left(\frac{\Delta m}{m}\right)_c \right\rangle. \quad (28.59)$$

If the mass in black holes is to bind the cluster, then

$$n_B \approx (M_{\text{cl}}/M_B)(3/4\pi R_{\text{cl}}^3),$$

where R_{cl} is the cluster radius. The maximum average mass of a black hole is obtained by requiring that

$$\frac{d \ln m_g}{dt} < t_0^{-1}, \quad (28.60)$$

where t_0 is the age of the cluster. Replacing the right-hand side of (28.59) by average quantities, we can use the preceding result to find the order-of-magnitude constraint on M_B:

$$M_B \simeq \frac{M_g}{M_{\text{cl}}} \frac{v t_0}{G} \frac{R_{\text{cl}}^3}{R}. \quad (28.61)$$

This overestimates the mass; the average over the galaxy distribution implicit in (28.59) is actually strongly peaked near the center (see Figure 28.6); so a radius smaller than R_{cl} should be used in (28.61). Empirical distribution functions indicate that

$$M_B \lesssim 3 \times 10^8 \, M_\odot \, (M_g/M_\odot)(L_\odot/L_g), \quad (28.62)$$

where L_g is an average galactic luminosity, and $t_0 \approx 2 \times 10^{10}$ years. Thus $M_B \lesssim 3 \times 10^{10} \, M_\odot$ could supply the necessary mass to bind clusters without collisionally depleting their galaxies of stars.

Although the postulated existence of black holes of intermediate galactic mass appears to solve the problem of hidden mass in clusters, theory has not yet made plausible the mechanism by which so many (about 10^5 for the Coma cluster) massive black holes could be formed along with galaxies in the early universe.

Problem 28.14. What effect would dynamical friction be expected to have on the motion of a star moving initially in circular orbit in a globular cluster and in an E0 galaxy?

Chapter 29

AXIALLY SYMMETRIC GALAXIES

The distribution of galaxies among morphological types represented in Figure 27.8 indicates that few (probably no more than a few percent) are spherically symmetric. In fact, most galaxies are believed to be axially symmetric. The most impressive features of spiral galaxies (both normal and barred) represent dramatic apparent deviations from axial symmetry; however, there is growing theoretical and observational evidence that these features represent relatively small perturbations of the gas and stellar-mass distribution. In fact, significant progress in explaining these features has been made by assuming that S and SB systems are very nearly axially symmetric.

The present structure of elliptical galaxies (and probably of the spherical component in spirals) and the processes that lead to their formation cannot be treated separately, because the two-body relaxation time in these systems is greater than the age of the universe; so these systems cannot have evolved much since their formation. The most widely held theories of galaxy formation assume that localized increases of density develop in the early universe, and that at least some of these survive until hydrogen recombination. If these perturbations are gravitationally self-bound, then they will eventually contract, even though the universe as a whole continues to expand. The morphological type that presumably results after the protogalaxy reaches a quasistatic state probably depends on the protogalaxy's gas content during the final collapse, as well as on its angular momentum and proximity to neighboring galaxies during formation. Elliptical galaxies are not observed with a/b greater than about 2 (a and b are the major and minor axes), whereas spirals are typified by a/b of 10 to 20. The range in shape En of ellipticals may be attributed to the nature of the initial perturbation, and to the presence of neighboring protogalaxies. The radial variations in brightness characteristics of all ellipticals are probably a result of the formation process, and may indicate that protogalaxies that become ellipticals contained very little gas. Finally, as noted in Chapter 27, there is strong evidence that the spheroidal components of spirals are similar in nature to elliptical systems. In this chapter we will ignore the structure of the galactic nucleus.

29.1. Elliptical Galaxies

The surface brightness $I(r)$ of elliptical galaxies appears to be independent of aspect ratio a/b; $I(r)$ varies smoothly from the center outward, and resembles the surface brightness of globular clusters. In fact, the three-parameter models, which have been found to fit globular-cluster brightness distributions so successfully (see, for example, Figure 28.2), also fit ellipticals. Figure 29.1 shows $I(r)$ for the ellipticals NGC 3379 and NGC 4472 fit by a single globular-cluster model. Despite the similarity in surface brightness between ellipticals of different aspect ratio, there is a marked variation in isophotal ellipticity as a function of radius. Another general characteristic, which should bear directly on the formation process, is the presence of radial composition gradients, which indicate that the abundance of heavy elements decreases from the center of the galaxy outward. These gradients show up primarily in the strength of the CN band and metallic absorption bands. Absorption bands are typical of late-type stars, and their existence has been interpreted as evidence that the stars in the outer regions of the ellipticals formed after those further toward the center were formed.

The early stages of galaxy formation will be discussed in Chapter 30. For the moment, we will assume that a localized isothermal density perturbation of sufficient magnitude to be gravitationally bound has formed. We will also adopt a big-bang model of the origin of the universe (see, for example, the Friedmann models in Chapter 26), and denote the initial instant $a = 0$ by $t = 0$. According to current theories, the initial protogalactic perturbations begin to grow at about the time of H recombination, which occurs at epoch $z = 10^3$. At this epoch in the early expansion of the universe, electromagnetic radiation and matter decouple. In some respects the condensation of protogalaxies resembles protostar formation, where gravitationally bound matter becomes separated from background matter. One major difference, however, is that the (assumed) uniform density background from which galaxies form is itself in a rapid state of expansion. The mass in the perturbation (or protogalaxy) must therefore be great enough to overcome the outward-directed kinetic motion of the matter in the perturbation. We will suppose this to be true, and consider the subsequent behavior of the protogalaxy and its implications for the structure of the elliptical systems. Finally, the distribution of matter between stars and interstellar gas will play an important role in the evolution. It is often assumed that star formation occurred rapidly enough following recombination that nearly all the gas condensed out into stars, or was swept out of the protogalaxy, by the time of maximum collapse. As a first approximation, we will assume that the matter lies entirely within stars and will ignore gas.

Early Evolution

Consider the epoch of recombination $t = t_i$ ($z_i = 10^3$), when the matter density of the universe was $\rho_{e,i}$, and imagine that the density increase is centered at $r_i = 0$, is spherically symmetric, and is of radius R_i (see Figure 29.2). If we locate our frame of reference at $r_i = 0$, the Friedmann models can be applied (with slight modification) to describe the subsequent expansion of the perturbation.

The nature of the expansion (open or closed) for the exterior is qualitatively unimportant for the following arguments; for simplicity we will assume it to be an open universe ($q_0 < 1/2$), for which

$$\rho_{e,i} < \frac{3H_i^2}{8\pi G} \equiv \rho_{c,i}. \qquad (29.1)$$

The Hubble parameter at recombination is $H(t_i) = H_i$; it may be expressed in terms of H_0, q_0, and z_i using (26.57). We will also assume that for $t > t_i$ the universe is matter-dominated. Focus attention on spherical mass shells of radius r_i at $t = t_i$. Since the perturbation is spherically symmetric, and the exterior is taken to be isotropic and homogeneous, the mass exterior to a given shell centered at $r_i = 0$ will have no effect on that shell's motion. Furthermore, the gravitational force acting on a given shell depends only on the total mass contained within that shell (a result that is also relativistically correct for the assumed symmetry about $r_i = 0$). But the density appearing in the Friedmann equation (26.21) is just the mass interior to r_i, $m(r_i)$, divided by the volume $4\pi r_i^3/3$. In other words, the scale factor for an expanding universe with the density perturbation shown in Figure 29.2(a) is given by the solution of

$$\frac{1}{a^2}\left(\frac{da}{dt}\right)^2 - \frac{8\pi G}{3}\bar{\rho}(r) = -\frac{kc^2}{a^2}, \qquad (29.2)$$

with the mean density

$$\bar{\rho}(r) \equiv \frac{1}{V}\int_0^r \rho(r')\,dV' = \frac{3}{r^3}\int_0^r \rho(r')r'^2\,dr'. \qquad (29.3)$$

The mean density is shown schematically in Figure 29.2(b). The continuity equation obtained from

29.1. ELLIPTICAL GALAXIES / 533

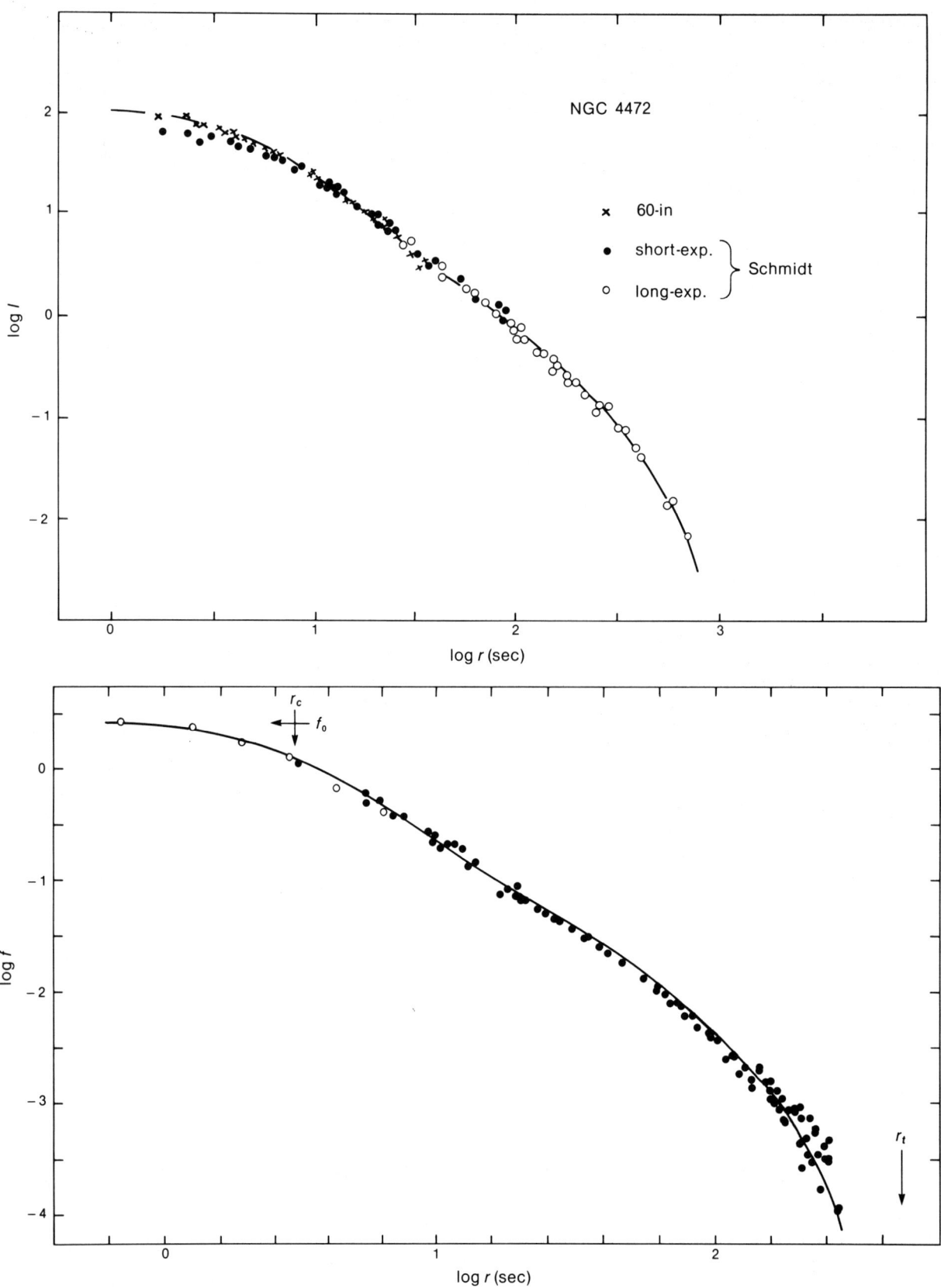

Figure 29.1. Surface brightness of elliptical galaxies versus distance from center for (a) NGC 4472 and (b) NGC 3379 (type E1). The solid curve is cluster model.

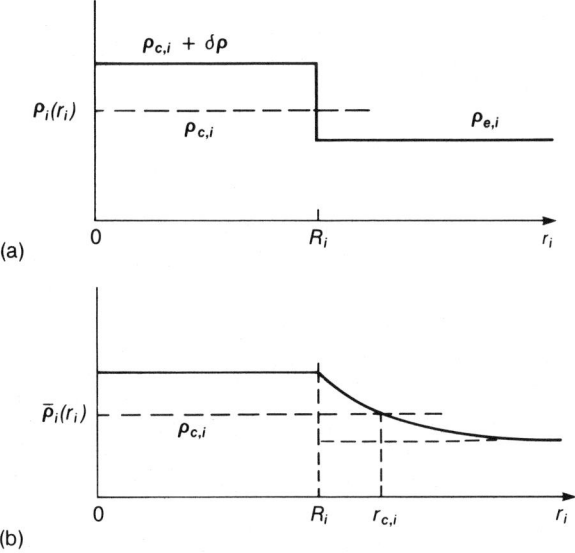

Figure 29.2. Protogalactic density increase: (a) initial density versus radius; (b) average density inside radius r_i obtained from (a).

(26.23) for the mean density is

$$\bar{\rho}(r)a^3 = \bar{\rho}(r_i) \equiv \bar{\rho}_i. \quad (29.4)$$

This implies that our length scale is set by $a(t_i) = a_i = 1$. Finally, the density inside the perturbation is assumed to exceed the critical density at recombination.

Proceeding as in Section 26.3, we may rewrite Friedmann's equation (29.2) in the form

$$\left(\frac{da}{dt}\right)^2 = H_i^2\left(\frac{2q_i}{a} + 1 - 2q_i\right). \quad (29.5)$$

All quantities bearing subscript i are evaluated at epoch $Z_i = 10^3$. They are defined in analogy with their present-epoch counterparts discussed in Section 26.3, except that ρ_0 is replaced by $\bar{\rho}_i$; in particular,

$$2q_i = \bar{\rho}_i/\rho_{c,i}. \quad (29.6)$$

Therefore the solutions obtained earlier apply here also. If q_0 is replaced by q_i for r_i such that $\bar{\rho}_i > \rho_{c,i}$, the solution to (29.5) is, for $t \geq t_i$,

$$a(t,r_i) = \frac{q_i}{2q_i - 1}(1 - \cos\theta), \quad (29.7)$$

$$H_i t = \frac{q_i}{(2q_i - 1)^{3/2}}(\theta - \sin\theta). \quad (29.8)$$

Since q_i varies with density, the scale factor may be different for different mass shells r_i. We have indicated this by exhibiting explicitly its dependence on the parameter r_i in (29.7).

Equations (29.7) and (29.8) should not be applied for times small compared with t_i when the perturbation is forming. The qualitative behavior of the solution is shown in Figure 29.3; the uppermost curve represents the expansion of the universe as a whole, and the lower curve represents the density perturbation, which separates from the background at t_i (the present, t_0, lies off scale to the right). At t_i the perturbation is expanding; its initial radius is R_i, and even though it is gravitationally bound, it increases in radius until reaching its maximum extent $a(t_c/2, r_i)R_i$ at time $t_c/2$. By this time its expansion has decreased to zero, and the perturbation begins to contract. If unhindered, the contraction would proceed to zero radius at t_c given by (29.7) with $\theta = 2\pi$:

$$t_c = \frac{2\pi q_i}{H_i(2q_i - 1)^{3/2}} \approx \frac{\pi \rho_{c,i}^{3/2}}{H_i(\bar{\rho}_i - \rho_{c,i})^{3/2}}$$

$$= \frac{\pi}{H_i}\left(\frac{\rho_{c,i}}{\delta\rho}\right)^{3/2}. \quad (29.9)$$

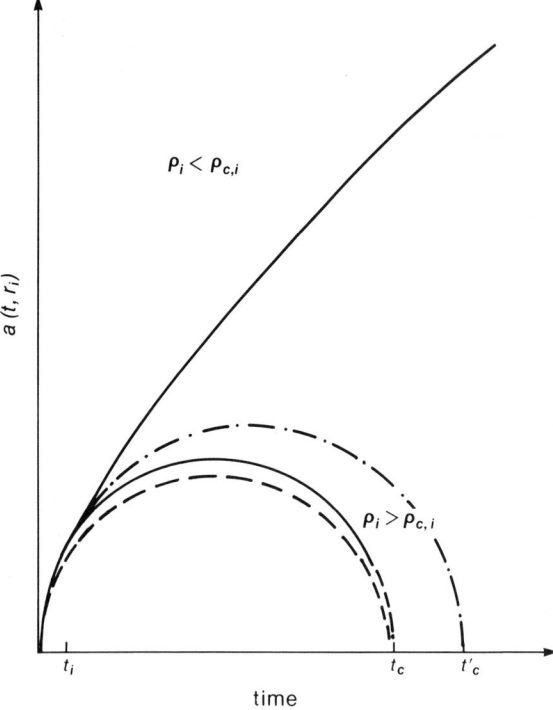

Figure 29.3. Radius versus time for protogalactic perturbation in expanding universe. Upper solid curve is Hubble background expansion; lower solid curve is outer radius R_i of protogalaxy proper. Dash-dot curve is matter between R_i and $r_{c,i}$ (Figure 29.2) that is bound gravitationally to the protogalaxy.

Here we have assumed that

$$|\bar{\rho}_i - \rho_{c,i}| \ll \rho_{c,i}.$$

The radius of a mass shell located at r_i at recombination ($t = t_i$) has changed to

$$r(t) = a(t,r_i) r_i$$

at time t because of expansion. The variation of a with r_i due to the variation in $\bar{\rho}_i(r_i)$ shown in Figure 29.2 has several important consequences. First, for

$$r(t) < R(t) = a(t,R_i) R_i,$$

the mean density $\bar{\rho}(r)$ is a constant, and it follows from (29.9) that t_c is independent of r_i. Thus all shells inside the initial perturbation (dashed curve in Figure 29.3) reach r_{max} at the same time, and collapse at the same time (this is exactly what occurs in a standard closed Friedmann model). This region can be taken as the protogalaxy proper, and its total mass is given by

$$M_i = \frac{4\pi}{3} R_i^3 (\rho_{c,i} + \delta\rho) \simeq \frac{4\pi}{3} R_i^3 \rho_{c,i}, \quad (29.10)$$

since we assume $\delta\rho \ll \rho_{c,i}$. Next, consider a mass shell in the range $R_i < r_i < r_{c,i}$ at epoch t_i. Although the density here is less than $\rho_{c,i}$, the mean density within r_i is greater than critical; so these shells are also bound to the protogalaxy.

Problem 29.1. Show that the collapse time for the protogalaxy, t_c, and the collapse time t'_c for the bound matter at r'_i, where $R_i < r'_i < r_{c,i}$, satisfy

$$\left(\frac{t'_c}{t_c}\right) = \frac{\delta\rho}{(\bar{\rho}_{e,i} - \rho_{c,i}) + (\rho_{c,i} - \bar{\rho}_{e,i} + \delta\rho) R_i^3 / r_c'^3}, (29.11)$$

where $\bar{\rho}_i(r_i)$ is defined as in (29.3), and $\rho_i(r_i)$ is as shown in Figure 29.2.

But the mean density in this region is less than the mean density in the protogalaxy; so the collapse time t_c given by (29.9) is greater for $r_i < R_i$. As a consequence, the mass contained between R_i and $r_{c,i}$ continues to expand after the protogalaxy has reached maximum extent, and then falls back onto it at a later time (see Figure 29.3). The ratio of t'_c to t_c is given by (29.11), assuming that $\rho_{c,i} - \bar{\rho}_{e,i}$ and $\delta\rho$ are small compared with $\rho_{c,i}$. The mass associated with this infalling matter can be estimated from (29.11) in a straightforward manner, by noting that at time $t = t_c$, all mass that was originally within $R_i < r'_i < r_{c,i}$ will have collapsed onto the protogalaxy. But this mass is just

$$M = 4\pi \int_{R_i}^{r'_i} \rho_i(r)\,dr\,r^2 = \frac{4\pi \bar{\rho}_{e,i}}{3}(r_i'^3 - R_i^3)$$

$$= M_i \left(\frac{\bar{\rho}_{e,i}}{\rho_{c,i}}\right)\left(\frac{r_i'^3}{R_i^3} - 1\right),$$

where M_i is given by (29.10). Eliminating $(r_i'^3/R_i^3)$ with the help of (29.12), and noting that $\bar{\rho}_{e,i} \approx \rho_{c,i}$, we find that the mass added to the protogalaxy due to infall by time $t = t'_c$ is

$$\delta M = M_i \frac{(t/t_c)^{2/3} - 1}{1 + \frac{\rho_{c,i} - \rho_{e,i}}{\delta \rho}\left(\frac{t}{t_c}\right)^{2/3}}$$

$$= M_i \frac{(t/t_c)^{2/3} - 1}{1 + (1 - 2q_0)(H_0 t/2q_0)^{2/3}}. \quad (29.12)$$

The final form results if (29.10), (26.57), and (26.58) are used to replace t_c and the densities at epoch z_i, which appear in the denominator, by q_0 and H_0 (which are presumably known). Finally, mass shells originally at $r_i > r_{c,i}$ are unbound (relative to the perturbation), and will continue to expand well after the perturbation has collapsed (uppermost curve in Figure 29.3).

Problem 29.2. In an Einstein–de Sitter universe $q_0 = \frac{1}{2}$ and $\rho = \rho_c$. Derive the expression $M(t)$ for a perturbation $\delta \rho \ll \rho_{c,i}$ in an Einstein–de Sitter universe where $q_0 = \frac{1}{2}$ ($\rho_{e,i} = \rho_{c,i}$).

Final Collapse

The collapse of the perturbation does not proceed to a singularity, as suggested by the model above. Instead, energy transfer between stars that move under the influence of the rapidly changing background gravitational field of the protogalaxy is believed to impel the system toward a state of equilibrium (recall the discussion of violent relaxation in Section 28.4). As equilibrium is approached, the initially inward motion is transformed into random stellar motion with a roughly Maxwellian velocity distribution, modified by a high-energy cutoff as in globular clusters. Numerical studies indicate that the spatial distribution is nearly isothermal throughout the central regions. Violent relaxation probably establishes quasiequilibrium within several crossing times (the time required for a typical star to cross the galaxy). Suppose that there is no energy dissipation during the collapse, so that the total energy of the system is conserved, and denote the radius of maximum expansion of the protogalaxy by $R(t_c/2) = R_0$. At this stage there is negligible kinetic energy, and the total energy of the protogalaxy is its binding energy,

$$E_0 = -\frac{3}{5}\frac{M_i^2 G}{R_0} = \Omega_0. \quad (29.13)$$

After reaching maximum expansion, the protogalaxy collapses to a state of virial equilibrium with radius R, and total energy

$$E = T + \Omega, \quad (29.14)$$

where T is the kinetic energy of stellar motion, and $\Omega = -3M^2G/5R$. The maximum radius R_0 and radius when virial equilibrium has been established, R, are related by the virial theorem, $2T + \Omega = 0$, and conservation of energy, $\Delta E = E - E_0$, during dissipationless collapse:

$$E = T + \Omega = -\frac{1}{2}\Omega + \Omega = \frac{1}{2}\Omega = \Omega_0$$

or, in terms of radii,

$$R = R_0/2. \quad (29.15)$$

The development of elliptical galaxies (the preceding analysis assumed spherical symmetry throughout) may be due to tidal interactions with neighboring protogalaxies, which presumably form at about the same time. If protogalactic perturbations develop asymmetrically, as seems plausible, then tidal interactions with neighboring masses will exert a torque on the protogalactic perturbation that will give it a net angular momentum. Depending on the relative orientations of the protogalactic perturbation and its neighbors, an angular velocity ranging from zero to $\omega_c \approx M_i G/R_i^3$ may be imparted to the perturbation. The range in ellipticity actually observed in ellipticals is consistent with this picture. Imagine that a protogalaxy acquires a net angular momentum J around the time of maximum expansion. Then collapse along the rotation axis will proceed approximately as though the system were spherical with $R = R_0/2$, but collapse in the equatorial plane will be opposed by the rotation. In the extreme case where rotation balances gravity ($\omega = \omega_c$), the final radius in the equatorial plane is just R_0, and the aspect ratio for the galaxy is $a/b = R_0/(R_0/2) = 2$. This corresponds to an elliptical of type E5. Actually, not all protogalactic perturbations would be expected to acquire the maximum angular momentum from tidal interactions; instead, a distribution of possible angular momenta would produce galaxies with aspect ratios varying between about one-half and one, corresponding to ellipticals of type E0–E5. We recall that very few ellipticals of type E6 or later are observed.

The surface-brightness distributions observed for ellipticals are presumed to reflect the stellar mass distribution. The collapse and relaxation of the spherical region initially within R_i in Figure 29.2 has been studied numerically; in all cases the radial density of the relaxed state $\rho(r) \sim r^{-4}$ in the outer region, rather than $\rho(r) \sim r^{-3}$, as implied by observed surface-brightness distributions. The bound matter within the region $R_i < r_i < r_{c,i}$ will also contribute to the mass distribution of the galaxy. When it is included, the envelope develops a distribution that falls off as $\rho(r) \sim r^{-3}$. Finally, the outer portions of the envelope will be stripped away by the tidal field of neighboring galaxies, producing a cutoff in the surface brightness as observed. Detailed models that include tidal interactions and cosmological infall appear to reproduce the envelope structure of elliptical galaxies. Because this structure results largely from processes that are the same for all protogalaxies (infall and tidal interactions), it should be independent of type, as is observed.

Problem 29.3. Consider three colinear protogalactic masses in an expanding Einstein–de Sitter universe. Show that the tidal force acting on stars in the middle protogalaxy falls off as t^{-2}.

Gas Dynamic Effects

Protogalaxies are likely to contain some gas, either of primordial origin or from stellar mass loss. The initial conditions of the gas are not well known, but it can be argued that the initial irregularities and subsequent dynamic evolution of the collapsing protogalaxy produce substantial turbulent motion. Collisions between interstellar gas clouds can dissipate energy very efficiently, and the evolutionary time-scales for a gaseous component can be quite rapid. The *Bremsstrahlung* cooling time for a thin, hot interstellar gas (see Problem 28.11) in which $n_e \approx 1$ and $T_e \approx 10^7$ to 10^8 K is roughly $\tau_{\text{Brem}} \approx 5 \times 10^7$ yrs. The free-fall time for the protogalaxy (if $R_0 \approx 2R = 200$ kpc and $M = 2.5 \times 10^{12} M_\odot$) is at least ten times longer. In any event, $\tau_{\text{Brem}} \ll \tau_{\text{ff}}$, and the contracting gas does not heat up significantly.

Cool-gas collapse tends to establish a highly condensed mass distribution. The rate of star formation is expected to increase with increasing gas density; so the concentration of stars will increase inwardly, as will the abundance of heavy elements. The resulting metallicity varies with distance from the galactic center, as indicated by observations. Energy release accompanying the formation of supernovae and HII regions will tend to heat the gas, which works to counter continued infall and may, if great enough, set up a galactic wind capable of sweeping gas out of the galaxy. This effect increases with decreasing galactic mass, and could explain why less massive elliptical galaxies have less prominent nuclei and smaller metallicity gradients than do more massive ones. It may also explain why smaller ellipticals have a lower overall metal abundance. Numerical models indicate that, for elliptical galaxies less massive than 10^9–10^{10} $M_\odot$, all gas will be expelled by this process. It has also been suggested that gas collapse, temporarily interrupted by successive stages of star formation and nucleosynthesis, may lead to the development of highly condensed metal-rich galactic nuclei.

A great deal of theoretical and observational work is needed before the formation and structure of elliptical systems can be understood in detail. Nevertheless, the influence of cosmic expansion, and gas dynamic effects are likely to play an important role in understanding these phenomena.

29.2. SPIRAL GALAXIES

The formation of spiral and lenticular galaxies is complicated by the presence of interstellar gas, which shows up as an optically dominant thin disk. Because elliptical galaxies show little rotation and no disk, it is natural to assume that the specific angular momentum of a protogalaxy plays an important role in determining whether the galaxy becomes a spiral or an elliptical. The dynamics of the formation process must also be considered, however. The formation process must be capable of producing the three distinct components observed in all spirals: (1) an extended halo, devoid of gas, whose stars have large velocity dispersion; (2) a centrally condensed, often bright, nuclear component containing stars, gas, and dust; and (3) a thin disk containing mostly young stars, gas, and dust. In this last component, the dispersion of stellar velocities is generally small. The presence of a metallicity gradient in the halo of our Galaxy, and the relatively compact nuclear bulge observed in many spirals, can be reproduced at least qualitatively by the methods developed for elliptical systems that include gas and an initial angular momentum distribution.

A plausible qualitative scenario for the formation of the three components of spiral galaxies begins with a rotating perturbation in an expanding universe. If the initial specific angular momentum is great enough, collapse will proceed most rapidly along the axis. Each epoch of star formation will occur in a spheroid of greater eccentricity, until the final stage, containing the youngest stars, settles out as a thin disk. Assuming that angular momentum can be transported efficiently away from the central regions, successive stages of infall could produce a highly compact nuclear region, as in ellipticals. Cosmological infall, which might well continue after the galaxy forms, could also add gas to the disk. As compelling as this scenario might be, efforts to test it computationally are frustrated at several levels. The initial distribution of angular momentum is not known, nor is it clear what role neighboring galaxies play. The formation of a rotating disk, and probably of a compact, nearly spherical nucleus, requires angular momentum transport from the gas during collapse. Finally, many of the gas-dynamic processes discussed for ellipticals at the end of Section 29.1 may also be important, as well as others unique to disk systems embedded in a stellar halo.

It should be noted that the evolution of spiral galaxies in clusters of galaxies may be strongly affected by the presence of hot intergalactic gas in at least two ways. First, if the gas is of high enough density and temperature (n_{gas} greater than several times 10^{-3} cm^{-3} and T_{gas} greater than several keV), interstellar gas in the galaxy will be heated, will expand, and will escape from the galaxy, becoming indistinguishable from the intergalactic medium. This could happen faster than the rate at which the gas can be replenished by stellar processes. Second, as is generally the case, the spiral galaxy is moving through the intergalactic medium with velocities in excess of approximately 10^3 km/sec, the relative motion of the interstellar and intergalactic gas will strip the galaxy.

Fortunately, we live within a spiral system, and can construct a reasonably clear picture of all three components observationally (local observations can be supplemented by studies of neighboring spiral systems, such as M31, that are near enough to permit resolution of many local details). In the remainder of this chapter we will develop the basic tools needed to describe our Galaxy and to compare theory and observations from our vantage point in the disk.

Kinematics

In galaxies having an axis of rotational symmetry, polar coordinates (r, θ, z), with origin at the galactic center, are particularly useful for theoretical developments. Observational data about our Galaxy, however, is measured relative to a point in the disk ~ 10 kpc from the center, which moves with angular velocity $\Omega(R_0) \simeq 25$ km sec^{-1} kpc^{-1} = 8.3 × 10^{-16} sec^{-1} about the center. Positional data relative to the Sun are often given in Galactic coordinates (l, b). The angle l is measured in the Galactic plane, (we choose $l = 0$ toward the center), and b is the angle above the plane (Figure 29.4). The Sun's angular momentum vector (and the Galaxy's angular momentum) point in the $b = 90°$ direction. Finally, it is convenient to separate the motion of a star in the plane of the Galaxy into two parts. The first is its circular velocity $\Omega(r)/r$. This is the linear velocity that the star would have if it moved in a circular orbit of radius r about the Galactic center. The second part is the star's peculiar velocity, which has Cartesian components (Π, Θ, Z), and is measured relative to the frame of reference moving with velocity $\Omega(r)/r$ at r (Figure 29.4).

The frame of reference to which a star's peculiar velocity is referred is called the local standard of rest (LSR). There are actually two natural definitions of this reference frame, and these, though not identical, agree quite closely. The first is the dynamical LSR, and is the one introduced in the preceding paragraph. It is determined by the net Galactic gravitational acceleration at point r. Thus, if the acceleration at r is denoted by $g(r)$ in magnitude and points radically inward, then the velocity of the dynamical LSR is given by

$$\mathbf{V}_{dyn} \equiv (0, \sqrt{rg(r)}, 0). \quad (29.16)$$

A second local standard of rest is the kinematic LSR. This reference frame moves with the average motion of the stars in the immediate neighborhood of the point. For a neighborhood whose radius is small relative to the distance r to the Galactic center, the components of the kinematic LSR velocity are given by

$$\mathbf{V}_{kin} \equiv (\bar{u}, \bar{v}, \bar{w}), \quad (29.17)$$

where $\bar{u} = N^{-1} \Sigma_i u_i$, $\bar{v} = N^{-1} \Sigma_i v_i$, $\bar{w} = N^{-1} \Sigma_i w_i$, ($u_i$, v_i, w_i) is the peculiar velocity of the ith star in the neighborhood, and N is the total number of stars in the neighborhood.

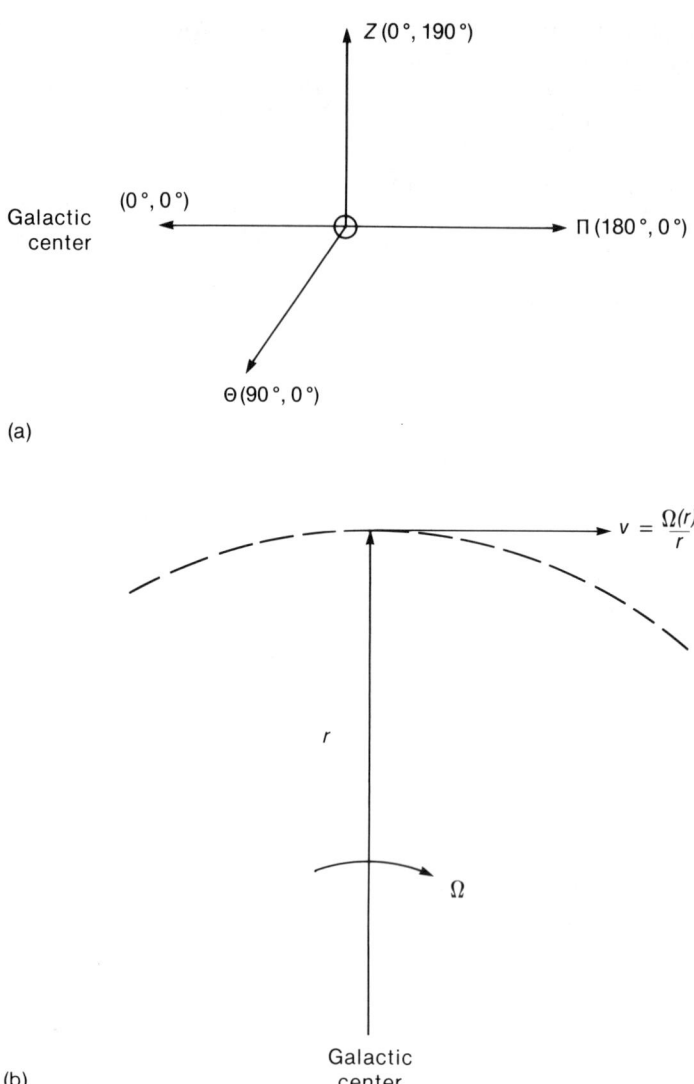

Figure 29.4. Galactic coordinates: (a) Galactic coordinates (l,b) and stellar peculiar velocity components (Π, Θ, Z) relative to solar neighborhood; (b) geometry of solar orbit about Galactic center.

Although (29.16) and (29.17) agree quite closely, there is a systematic tendency for the kinematic LSR to lag behind the dynamical LSR. The adopted value of the motion of the Sun's LSR, as based on observations of globular clusters and members of local groups of galaxies, is

$$\Theta = 250 \text{ km/sec}, \quad l = 90° \quad b = 0°. \quad (29.18)$$

The components of the Sun's peculiar velocity, and its resulting direction relative to this LSR, are:

$$\begin{aligned}
\Pi &= -10.4 \text{ km/sec}, \\
\Theta &= 14.8 \text{ km/sec}, \\
Z &= 7.3 \text{ km/sec}, \\
l &= 56°, \\
b &= 23°,
\end{aligned} \quad (29.19)$$

and the magnitude $(\Pi^2 + \Theta^2 + Z^2)^{1/2} = v_\odot = 19.5$ km/sec. This may be considered typical for stellar peculiar motion. According to (29.19), the Sun is moving out of the Galactic plane and toward smaller r. Notice that the peculiar velocity is a small fraction (about 8 percent) of the circular velocity (29.18).

Problem 29.4. Use the data above to estimate how close the Sun would move toward the Galactic center in one complete revolution, assuming that the radial component of its velocity remained unchanged with time. Give at least one reason why the Sun might not be expected to move this far in one period.

Stellar Dispersion Velocities

The stars in the solar neighborhood can be divided into two kinetic groups in terms of their velocities. The first group is characterized by young stars associated with the disk. Because of their peculiar velocities, stars initially localized in a given neighborhood tend to become dispersed as they orbit the Galactic center (differential rotation adds to this effect also). In general, all stars are expected to have some peculiar velocity, but of a magnitude usually small compared with the average circular velocity in the plane. Observational studies, particularly of stars in the solar neighborhood, indicate that to first approximation peculiar velocities are normally distributed, and that their mean component is zero. Let us define the number of stars per unit volume having peculiar velocity with component (u, v, w) along a set of orthonormal axes as

$$n(u, v, w) \, du \, dv \, dw = \frac{N\pi^{-3/2}}{\sigma_1 \sigma_2 \sigma_3}$$

$$\times \exp\left[-\left(\frac{u^2}{\sigma_1^2} + \frac{v^2}{\sigma_2^2} + \frac{w^2}{\sigma_3^2}\right)\right] du \, dv \, dw. \quad (29.20)$$

The total number of stars per unit volume in the neighborhood N is given by

$$N = \int du \, dv \, dw \, n(u, v, w). \quad (29.21)$$

The quantities $(\sigma_1, \sigma_2, \sigma_3)$, which measure the magnitude of the stellar dispersion, define the velocity ellipsoid

$$\frac{u^2}{\sigma_1^2} + \frac{v^2}{\sigma_2^2} + \frac{w^2}{\sigma_3^2} = 1. \quad (29.22)$$

One axis of the velocity ellipsoid is oriented at right angles to the Galactic plane; the other two lie in it. Of these two, the longest points nearly toward the Galactic center. General arguments show that if the Galactic disk were exactly axisymmetric, then the angle between the longest principal axis of the ellipsoid and the Galactic center (the vertex deviation) would be zero. The fact that the angle is not zero shows that the assumption of axial symmetry is not exact. We will see later that the theory of spiral density waves requires a small symmetry-breaking term in the gravitational potential that should modify the velocity dispersion. The velocities entering into (29.20) and (29.22) are measured relative to the principal axes of the ellipsoid. If they are measured relative to another set of axes (such as the Cartesian set in the r, θ, z direction), then the exponent of (29.20) will include cross terms uv, uw, and vw as well. The normal distribution (29.20) holds for stellar velocities less than about 65 km/sec.

Stars having peculiar velocities greater than about 65 km/sec form a separate kinematic group. These high-velocity stars are not members of the disk, but represent stars from the halo that are passing through the $z = 0$ plane, often with velocities of 200 km/sec or more. Almost always, when $|v| > 65$ km/sec, the v component is negative, supporting the interpretation that they are halo stars passing through the rotating disk. Finally, no observed velocities have positive Θ in excess of 65 km/sec.

Problem 29.5. Use the observed upper limit to Θ of 65 km/sec in the direction $(l, b) = (90°, 0°)$ to estimate a lower bound for the mass of the Galaxy. Carefully state assumptions made in obtaining your estimate, and note that v_{max} is measured relative to the local standard of rest. Why is the result a lower limit?

Problem 29.6. The average vertex deviation as observed for each spectral class in the solar neighborhood for main-sequence and red giant stars is shown in Figure 29.5. To what extent do the results support the claim that most young stars are members of the disk population, whereas older stars belong to both the disk and the halo regions?

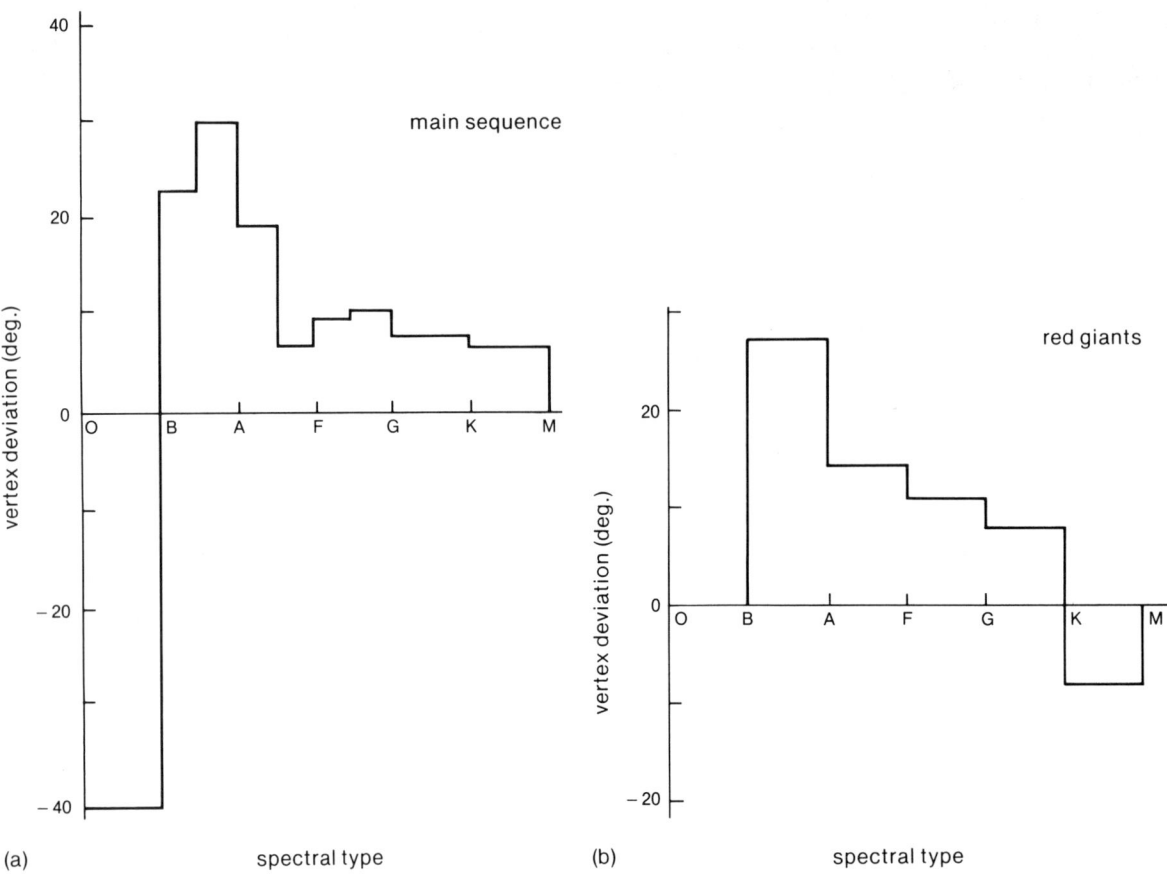

Figure 29.5. Vertex deviation (a) for main-sequence stars and (b) for red giants, versus spectral type.

29.3. ROTATION CURVES

A knowledge of the circular velocity of a test mass as a function of its position in the Galactic disk is important in constructing models of the Galaxy's mass distribution, for spiral density-wave theory, and as a means of interpreting observational data.

The qualitative motion in the disk may be obtained from Kepler's laws. If all the mass were concentrated in the nucleus, then, as the size of the test mass's orbit increased, its period and linear velocity would decrease $(r \sim P^{2/3} \sim v^{-2})$. However, the mass is not confined to the nucleus, but is spread throughout the disk and halo; as the radius of the orbit increases, so does the amount of mass within it. This increase in interior mass means that the velocity does not fall off as rapidly as for Keplerian orbits. In fact, if the mass density were uniformly distributed in a spherically symmetric manner, the velocity would increase linearly with radius $(v \sim \sqrt{2MG/R^3}\, r)$. A gradually decreasing mass density from center to rim is easily seen to imply a variation in velocity between these two extremes.

The actual variation in linear velocity of disk matter as a function of distance r from the center, $\Theta(r)$, can be determined from optical and radio observations. Direct optical measurements are limited to the immediate neighborhood of the Sun (a few kpc in radius); Cepheid variables may be used to extend it to cover the region $10 \lesssim r \lesssim 13$ kpc. The remaining regions are observed at radio wavelengths (primarily at 21.1 cm, HI; but also at wavelengths for CO and other interstellar molecules). Current estimates of $\Theta(r)$ in km/sec are shown by the curve in Figure 29.6. The radius is in kpc and the Sun's location is at $r_\odot = 10$ kpc. The region corresponding to $r \lesssim 3$ kpc is not completely understood, although there is strong evidence of rapidly

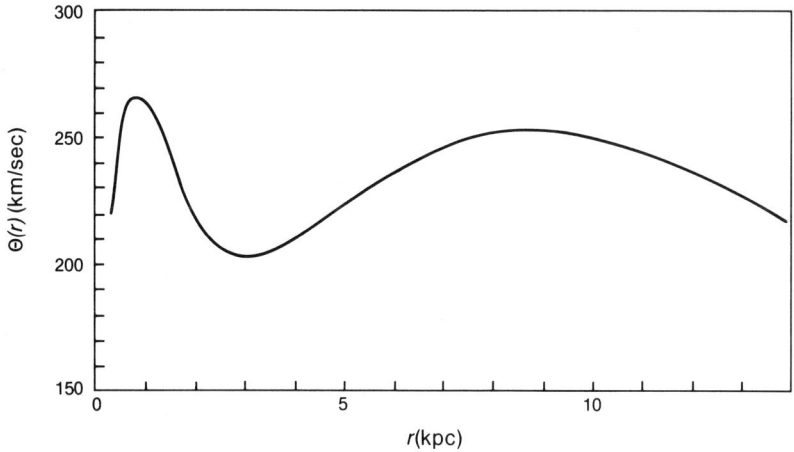

Figure 29.6. A recent model of rotation curve $\Theta(r)$ in our Galaxy.

expanding rings of interstellar hydrogen at about 3 kpc (the 3 kpc expanding arm), and very little (if any) evidence of neutral hydrogen within this region.

We can represent $\Theta(r)$ for $r > 3$ kpc and $r \lesssim 13$ kpc quite well by

$$\Theta(r) = 67.76 + 50.06r - 4.0448r^2 + 0.0861r^3, \quad (29.23)$$

where r is in kpc and Θ is in km/sec.* In particular, (29.23) reproduces (29.18), as well as the two constants of differential Galactic rotation (Oort's constants) A and B, defined by

$$A - B = \Theta_0/r_0, \quad (29.24)$$

$$A + B = -(d\Theta/dr)_{r=r_0}, \quad (29.25)$$

where $\Theta_0 \equiv \Theta(r_\odot)$ and $r_\odot$ is the position of the Sun. The constants A and B are directly related to observational data in the solar neighborhood. Current estimates are $A = 15$ km sec^{-1} kpc^{-1} and $B = -10$ km sec^{-1} kpc^{-1}.

It should be immediately evident from Figure 29.6 that the angular velocity

$$\Omega(r) = \frac{\Theta(r)}{r} \quad (29.26)$$

is not constant, but decreases with increasing radius. As a result the material in the disk is undergoing differential rotation, with outer portions rotating more slowly than the inner ones. One of the more important pieces of information that $\Theta(r)$ gives us about the Galactic plane is the gravitational acceleration acting on a test mass. In principle, the rotation curve determines the motion of gas and stars about the Galactic center, and as such should yield information about the forces and Galactic potential for $Z = 0$. Care must be exercised, however, since observationally $\Theta(r)$ is obtained from at least two distinct bodies of data, i.e., observations of nearby stars and of HI regions. When using $\Theta(r)$ to describe stellar motions throughout the plane, we tacitly assume that stars and gas move in the same manner. This may not always be true, however, since gas motion may be significantly affected by the Galactic magnetic field as well as by gravitation. Stellar motions are almost certainly determined by the background gravitational field, and should be insensitive to large-scale magnetic fields. It is therefore conceivable that the disk gas could obey a slightly different rotation law. In general, we will not worry further about such distinctions, but will use $\Theta(r)$ to describe both stellar and gaseous motions in the disk.

Problem 29.7. Suppose that the disk of the Galaxy is permeated by a weak, large-scale magnetic field ($B \lesssim 10^{-5}$ gauss) having a spiral structure as shown in Figure 29.7. The spiral pattern rotates rigidly with

---

*This model (and subsequent results based on it) does not include the possible increase in galactic mass ($m(r) \sim r$), implied by recent observations (see Section 27.3).

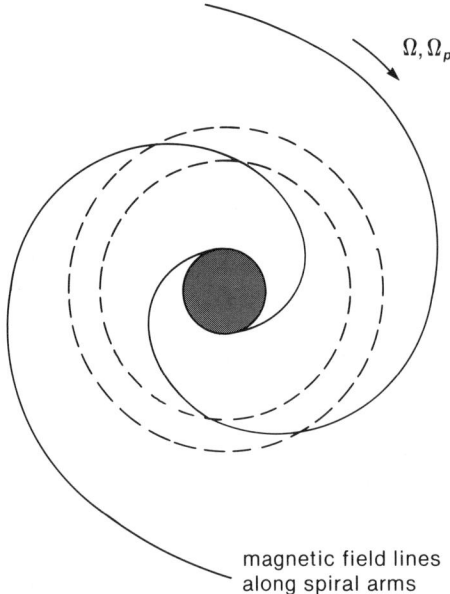

Figure 29.7. Spiral structure along Galactic magnetic field lines (schematic) in our Galaxy (see Problem 29.7).

angular velocity Ω_p as shown. The stars follow circular orbits (dashed) with angular velocity $\Omega(r)$ (differential rotation) in the same direction as the pattern, but with $\Omega(r) > \Omega_p$. Describe what you expect to happen to the rotating interstellar gas (specifically consider HI and HII).

Problem 29.8. Two stars move in circular orbits in the Galactic plane with orbital radii $r_2 > r_1$. Their orbital speeds are Θ_1 and Θ_2, and their periods P_1 and P_2. Show that the time in which star 1 will gain one-half lap on star 2 in their motion around the Galactic center is

$$T = \frac{1}{2} \frac{P_1 P_2}{P_2 - P_1}.$$

Assuming the values $r_1 = 4$ kpc, $\Theta_1 = 210$ km/sec, $P_1 = 0.98 \times 10^8$ yrs, $r_2 = 10$ kpc, $\Theta_2 = 250$ km/sec, and $P_2 = 2 \times 10^8$ yrs, how long would large-scale structure in the galactic disk survive under differential rotation?

Problem 29.9. Use the results of Problem 29.8 and the Galactic rotation curve to find the period of time needed for two stars initially 0.1 kpc apart at $r_1 = 10$ kpc and $r_2 = 10.1$ kpc to find themselves on opposite sides of the Galaxy?

29.4. FORCE LAWS

The techniques developed in Section 28.1 may be used to investigate the dynamics of axially symmetric galaxies, and in particular the dynamics of disks. Starting with the collisionless Boltzmann equation, a set of hydrodynamic equations can be derived that lead to force laws in, and perpendicular to, the Galactic plane. Mass distributions that produce reasonably simple force laws may be derived from these. Some of the results developed here will be used in subsequent discussions of spiral density waves. For simplicity, all stars are assumed to have the same mass, m.

Boltzmann's equation and Poisson's equation are coupled for stellar systems because each contains the acceleration of the masses. We may combine them by eliminating the acceleration in favor of the potential as in (28.13). Noting that $\mathbf{a} = \mathbf{F}/m$, we can rewrite (28.2) as

$$a^i = -\frac{1}{m}\frac{\partial \Phi}{\partial x^i}. \qquad (29.27)$$

Since each star has mass m, we will take this to be our unit mass ($m = 1$); when this is used in (28.13), we find

$$\frac{\partial f}{\partial t} + v^i \frac{\partial f}{\partial x^i} - \frac{\partial \Phi}{\partial x^i}\frac{\partial f}{\partial v^i} = 0. \qquad (29.28)$$

The coordinates used here are Cartesian (x, y, z), as are the velocity components, and summation over repeated indices is assumed.

Equation (29.28) contains the equations of hydrodynamics of a stellar system. In order to obtain them in standard form, we define the velocity average of an arbitrary quantity Q by

$$\langle Q \rangle = n^{-1} \int Q(\mathbf{r}, \mathbf{v}, t) f(\mathbf{r}, \mathbf{v}, t)\, d\mathbf{v}. \qquad (29.29)$$

The integral is over all velocity space $d\mathbf{v} = dv_x\, dv_y\, dv_z$, and n is the stellar number density (28.7). Note that n may depend on both $\mathbf{r}$ and t. The hydrodynamic equations are obtained by multiplying (29.28) by various powers of the velocity components and their products with one another, and integrating over velocity space. Specifically, the total time-derivative of $\langle \rho \rangle$ yields the continuity equation, and of $\langle v^i \rangle$ yields the three momentum equations (equations of motion).

For example, suppose that the average of df/dt is

taken. Using (29.28) and (29.29), we find

$$\int d\mathbf{v}\frac{df}{dt} = \int d\mathbf{v}\frac{\partial f}{\partial t}$$
$$+ \int d\mathbf{v}\, v^i \frac{\partial f}{\partial x^i} - \int d\mathbf{v}\, \frac{\partial \Phi}{\partial x^i}\frac{\partial f}{\partial v^i}. \quad (29.30)$$

This quantity vanishes in the absence of collisions. Observe that $\partial/\partial t$, $\partial \Phi/\partial x^i$ and $\partial/\partial x^i$ may be taken outside the integrals, because $\mathbf{x}$ and $\mathbf{v}$ are independent coordinates. When this is done, (29.30) becomes

$$\frac{\partial}{\partial t}\int f\, d\mathbf{v} + \frac{\partial}{\partial x^i}\int v^i f\, d\mathbf{v}$$
$$- \frac{\partial \Phi}{\partial x^i}\int \frac{\partial f}{\partial v^i}\, d\mathbf{v} = 0. \quad (29.31)$$

The first integral is n, and the second is the average $n\langle v^i \rangle$ of the ith component of the stellar velocity. Consider the x component: the last term may be written

$$\int \frac{\partial f}{\partial v_x}\, d\mathbf{v} = \int \frac{\partial f}{\partial v_x}\, dv_x\, dv_y\, dv_z$$
$$= \int dv_y \int dv_z \int \frac{\partial f}{\partial v_x}\, dv_x$$
$$= \int dv_y \int dv_z \int df. \quad (29.32)$$

The last integral is just $f(v_x, v_y, v_z, \mathbf{r}, t)$ evaluated for $v_x = \pm\infty$, which vanishes for the systems of concern here. Thus the last term is identically zero, and (29.31) becomes

$$\frac{\partial n}{\partial t} + \frac{\partial}{\partial x^i} n\langle v^i \rangle = 0. \quad (29.33)$$

This will be recognized as the usual form of a continuity equation, with ρ replaced by n. Equation (29.33) simply states that the change in local number density of stars results from their movement either into or out of the region under consideration. The continuity equation is unaffected by the background gravitational field Φ.

The equations of motion are obtained in a similar manner, starting with $v_i df/dt$ integrated over all velocities. Proceeding as before, we can show that

$$\int v_i \frac{df}{dt}\, d\mathbf{v} = \frac{\partial}{\partial t} n\langle v_i \rangle$$
$$+ n\frac{\partial \Phi}{\partial x^i} + \frac{\partial}{\partial x^j} n\langle v_i v_j \rangle = 0. \quad (29.34)$$

Problem 29.10. Supply the intermediate steps for the derivation of (29.34). Show in particular that

$$\hat{e}_k \int d\mathbf{v}\, v_i \frac{\partial}{\partial v^k} f = -\hat{e}_i n, \quad (29.35)$$

where $\hat{e}_k$ are unit vectors, and n is given by (28.9). This last step is easily shown by using integration by parts.

Note that (29.34) represents three equations ($i = 1, 2, 3$) corresponding to equations of motion along each orthogonal direction in coordinate space. The first term contains the acceleration, and the second term represents the gravitational force acting on the stars. The last term represents a pressure. In fact, if we define the pressure tensor

$$P_{ij} = n\langle u_i u_j \rangle \quad (29.36)$$

and express the stellar velocity v_i (defined relative to an inertial frame fixed, say, with respect to the Galactic center) as

$$v_i = \langle v_i \rangle + u_i, \quad (29.37)$$

with u_i representing the ith Cartesian component of the peculiar velocity, then straightforward algebra leads to the result

$$n\frac{\partial}{\partial t}\langle v_i \rangle + n\frac{\partial \Phi}{\partial x^i} + \frac{\partial P_{ij}}{\partial x^k}$$
$$+ n\langle v_k \rangle \frac{\partial}{\partial x^k}\langle v_i \rangle = 0. \quad (29.38)$$

In obtaining (29.38), we use the continuity equation, and the identity in Problem 29.11.

Problem 29.11. Using (29.37) show that

$$\langle u_i \rangle = 0. \quad (29.39)$$

Equation (29.38) is an alternative expression of equations of motion. Its similarity to the fluid equations of motion should be noted. The result (29.38) contains both a pressure gradient $\partial P_{ii}/\partial x_i = \nabla_i P$, where P is the usual scalar pressure, as well as stresses corresponding to $i \ne j$ terms. For an isotropic medium, the stresses vanish. In the Galaxy, however, they need not vanish.

Axially Symmetric Systems

The preceding discussions assumed Cartesian coordinates for both **r** and **v**. Normal Galactic systems, with the exception of irregulars, are symmetric at least relative to the axis of rotation (we ignore the usually small effects of spiral density waves). In these systems it is more convenient to use cylindrical coordinates (r, θ, z) than Cartesian coordinates (x, y, z). Although the components of **v** are still Cartesian, a new set (Π, Θ, Z) will be adopted that are rotated relative to (v_x, v_y, v_z), so that

$$\Pi = dr/dt, \quad \Theta = r\, d\theta/dt, \quad \text{and} \quad Z = dz/dt$$

at the point **r** (see Figure 29.8). The components (Π, Θ, Z) run from $-\infty$ to $+\infty$. In the new coordinates the collisionless Boltzmann equation is

$$\frac{\partial f}{\partial t} + \Pi \frac{\partial f}{\partial r} + \frac{\Theta}{r}\frac{\partial f}{\partial \theta} + Z\frac{\partial f}{\partial z}$$
$$+ \frac{\partial \Pi}{\partial t}\frac{\partial f}{\partial \Pi} + \frac{\partial \Theta}{\partial t}\frac{\partial f}{\partial \Theta} + \frac{\partial Z}{\partial t}\frac{\partial f}{\partial Z} = 0. \quad (29.40)$$

Newton's second law in cylindrical coordinates is

$$\ddot{r} = \dot{\Pi} = r\dot{\theta}^2 - \frac{\partial \Phi}{\partial r} = \frac{\Theta^2}{r} - \frac{\partial \Phi}{\partial r}, \quad (29.41)$$

$$r\ddot{\theta} + 2\dot{r}\dot{\theta} + \frac{1}{r}\frac{\partial \Phi}{\partial \theta} = \dot{\Theta} + \frac{\Pi \Theta}{r} + \frac{1}{r}\frac{\partial \Phi}{\partial \theta} = 0. \quad (29.42)$$

The last step uses

$$r\ddot{\theta} + \dot{r}\dot{\theta} = \dot{\Theta} \quad \text{and} \quad \dot{r}\dot{\theta} = \Pi\Theta/r.$$

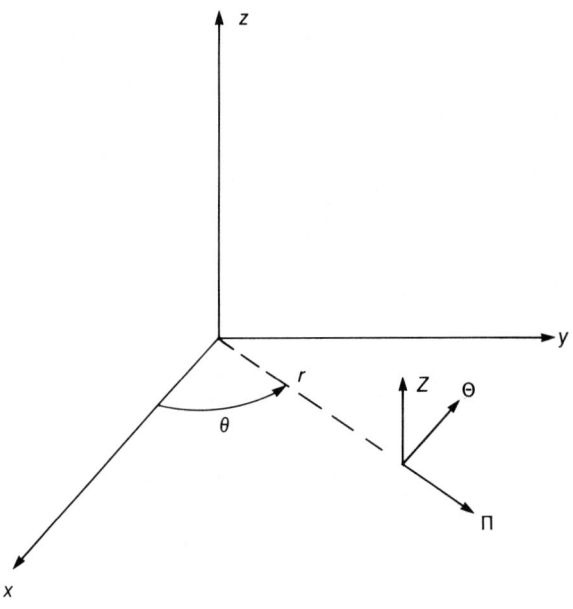

Figure 29.8. Transformation from Cartesian to polar coordinates in Galactic disk in (r, θ) plane.

Finally, the z component is

$$\ddot{z} = -\frac{\partial \Phi}{\partial z} = \dot{Z}. \quad (29.43)$$

The last three equations are solved for $\dot{\Pi}$, $\dot{Z}$, and $\dot{\Theta}$, which are then substituted into (29.40). The Boltzmann equation in cylindrical coordinates is

$$\frac{\partial f}{\partial t} + \Pi \frac{\partial f}{\partial r} + \frac{\Theta}{r}\frac{\partial f}{\partial \theta} + Z\frac{\partial f}{\partial z} + \left(\frac{\Theta^2}{r} - \frac{\partial \Phi}{\partial r}\right)\frac{\partial f}{\partial \Pi}$$
$$- \left(\frac{\Pi\Theta}{r} + \frac{1}{r}\frac{\partial \Phi}{\partial \theta}\right)\frac{\partial f}{\partial \Theta} - \frac{\partial \Phi}{\partial z}\frac{\partial f}{\partial Z} = 0. \quad (29.44)$$

The hydrodynamic equations in cylindrical coordinates are obtained from (29.44) by the same procedures used to obtain (29.33) and (29.34). The continuity equation is

$$\frac{\partial n}{\partial t} + \frac{\partial}{\partial r} n\langle \Pi \rangle + \frac{1}{r}\frac{\partial}{\partial \theta} n\langle \Theta \rangle$$
$$+ \frac{\partial}{\partial z} n\langle Z \rangle + \frac{n\langle \Pi \rangle}{r} = 0, \quad (29.45)$$

where the average $\langle \ldots \rangle$ is defined as in (29.29), with $f(\mathbf{r}, \mathbf{v}, t)$ expressed in the coordinates $(r, \theta, z, \Pi, \Theta, Z)$. The equations of motion are

$$\frac{\partial}{\partial t}n\langle\Pi\rangle + \frac{\partial}{\partial r}n\langle\Pi^2\rangle + \frac{1}{r}\frac{\partial}{\partial\theta}n\langle\Pi\Theta\rangle + \frac{\partial}{\partial z}n\langle\Pi Z\rangle$$
$$-\frac{n\langle\Theta^2\rangle}{r} + \frac{n\langle\Pi^2\rangle}{r} + n\frac{\partial\Phi}{\partial r} = 0, \quad (29.46)$$

$$\frac{\partial}{\partial t}n\langle\Theta\rangle + \frac{\partial}{\partial r}n\langle\Pi\Theta\rangle + \frac{1}{r}\frac{\partial}{\partial\theta}n\langle\Theta^2\rangle + \frac{\partial}{\partial z}n\langle Z\Theta\rangle$$
$$+ \frac{n}{r}\frac{\partial\Phi}{\partial\theta} + \frac{2n}{r}\langle\Pi\Theta\rangle = 0, \quad (29.47)$$

$$\frac{\partial}{\partial t}n\langle Z\rangle + \frac{\partial}{\partial r}n\langle\Pi Z\rangle + \frac{1}{r}\frac{\partial}{\partial\theta}n\langle\Theta Z\rangle + \frac{\partial}{\partial z}n\langle Z^2\rangle$$
$$+ n\frac{\partial\Phi}{\partial z} + \frac{n}{r}\langle\Pi Z\rangle = 0. \quad (29.48)$$

The preceding equations are completely general. However, they are of primary interest here for axially symmetric and stationary systems, for which all quantities are independent of time ($\partial/\partial t$ terms vanish) and of angle θ ($\partial/\partial\theta$ terms vanish). If, in addition, it is assumed that no streaming motion exists in either the radial or the z direction, then

$$\langle\Pi\rangle = \langle Z\rangle = 0 \text{ (no streaming motion).} \quad (29.49)$$

When (29.49) is satisfied, the continuity equation for axially symmetric, stationary systems (29.45) is trivially satisfied. Under these assumptions (29.46), (29.47), and (29.48) become

$$\frac{\partial}{\partial r}n\langle\Pi^2\rangle + \frac{\partial}{\partial z}\langle\Pi Z\rangle + \frac{n}{r}(\langle\Pi^2\rangle - \langle\Theta^2\rangle)$$
$$= -n\frac{\partial\Phi}{\partial r}, \quad (29.50)$$

$$\frac{\partial}{\partial r}n\langle\Pi\Theta\rangle + \frac{\partial}{\partial z}\langle\Theta Z\rangle + \frac{2n}{r}\langle\Pi\Theta\rangle = 0, \quad (29.51)$$

$$\frac{\partial}{\partial r}n\langle\Pi Z\rangle + \frac{\partial}{\partial z}n\langle Z^2\rangle + \frac{n}{r}\langle\Pi Z\rangle$$
$$= -n\frac{\partial\Phi}{\partial z}. \quad (29.52)$$

The last three equations form the basis of our discussion of the force laws in the Galaxy. They may be written in terms of observational velocities in the neighborhood of the Sun, which then yield predictions about stellar densities and the galactic potential.

Problem 29.12. If two velocity components v_i and v_j in a system are independent, then the distribution may be written as

$$f(v_i, v_j) = f(v_i)f(v_j). \quad (29.53)$$

Assuming this to be true, show that $\langle v_i v_j\rangle$ vanishes identically.

Problem 29.13. As a first approximation to the distribution of stellar velocities normal to the Galactic plane, take $f(\mathbf{v}, \mathbf{r}) = n(z)f(Z)$ for small z. It is intuitively reasonable to assume that the total number of stars with speeds between Z and $Z + dZ$ in the region between z and $z + dz$ is proportional to the time such stars spend there. Show that these assumptions lead to the identity

$$n(z)f(Z) = n(0)f(Z_0), \quad (29.54)$$

where Z_0 is the velocity along $\hat{e}_z$ at $z = 0$. Assume that the star's total energy $\frac{1}{2}Z^2 + \Phi$ is a constant during its motion.

Force Laws

Simple approximations to the Galactic force laws parallel and perpendicular to the plane can now be obtained. The force normal to the plane is most easily estimated, at least for small values of z/R, where R is the disk radius. For small motions the radial and perpendicular velocity components will be independent. The results of Problem 29.12 then show that $\langle\Pi Z\rangle = 0$. Substituting this into (29.52) we find

$$\frac{\partial}{\partial z}n\langle Z^2\rangle = -n\frac{\partial\Phi}{\partial z}. \quad (29.55)$$

This equation is difficult to solve as it stands. However, if $\langle Z^2\rangle$ is assumed to be a constant, $\langle Z^2\rangle = \langle Z_0^2\rangle_0$, where subscript zero denotes the value of Z_0 at $z = 0$, where the gravitational potential vanishes (all energy is kinetic), then (29.55) may be integrated immediately to give

$$n(z) = n_0 \exp\{-\Phi(z)/\langle Z_0^2\rangle\}. \quad (29.56)$$

Since the number density $n(z)$ and the perpendicular component of stellar velocities in the plane ($z = 0$) can be measured, at least in the solar neighborhood, (29.56) can be used to obtain the gravitational potential normal to the disk. This simple model cannot be expected to be in close agreement with observations out to very large z, but it does illustrate a very significant fact. Since Φ is the potential in Poisson's equation, it represents the gravitational field due to all forms of matter (stars, gas, dust) in the Galaxy. Thus, by using selected stars as tracers (gravitational test masses), the extent of the net Galactic potential may be estimated.

Problem 29.14. Use (29.54) to show that the stellar distribution function corresponding to (29.56) is Gaussian, that is,

$$f(Z) = \frac{1}{\sqrt{2\pi \langle Z_0^2 \rangle^{1/2}}} \exp(-Z^2/2\langle Z_0^2 \rangle). \quad (29.57)$$

By calculating $\langle Z^2 \rangle$ show that (29.57) is consistent with the assumption that $\langle Z^2 \rangle$ is constant.

Problem 29.15. A star is at a height z above the Galactic plane, which is assumed to have uniform density ρ_0. Show that the perpendicular component of the force is, for $z > 0$,

$$F_z = -4\pi G\rho_0 z. \quad (29.58)$$

The force law (29.58) illustrates several features for the dynamics perpendicular to the plane. Substituting (29.58) into the last of (29.20) to obtain Φ and using the result in (29.56), we find

$$n(z) = n_0 \exp(-\alpha^2 z^2), \quad (29.59)$$

where

$$\alpha^2 = 2\pi G\rho_0/\langle Z^2 \rangle.$$

The number density is thus Gaussian.

The distribution (29.57) must be improved in order to obtain reasonable agreement with observations. One successful approach is to replace (29.57) by a sum of Gaussian distributions, each having its own dispersion $\langle Z_i^2 \rangle$ representative of the known stellar population groups. Typical examples might include Population I,

Disk Population, and Halo Population stars. Following this scheme, but including a larger selection of subgroups, we obtain the force law shown in Figure 29.9 for the vicinity of the Sun. Indeed, for $z < 200$ pc, the force is linear to good approximation, as estimated in (29.58). For larger distances, the force levels off, becoming relatively constant above 1.5 kpc.

Problem 29.16. The force law (29.58) may be used to estimate the density in the galactic plane, ρ_0. Assume conservation of energy for a star moving along the z axis, and show that

$$\rho_0 = Z_0^2/4\pi G z_{max}^2, \quad (29.60)$$

where z_{max} is the maximum distance traveled by the star away from the plane. Data on observed Z_0 and z_{max}, which is obtained from the observed force law, are given in Table 29.1.

The force law in the solar neighborhood leads to an estimated total mass density

$$\rho \simeq 0.15 \, M_\odot/pc^3. \quad (29.61)$$

The observed mass density is about $0.08 \, M_\odot/pc^3$ ($0.05 \, M_\odot/pc^3$ due to stars, and $0.03 \, M_\odot/pc^3$ due to HI, HII, and He). Evidently about half the amount obtained dynamically has remained undetected. A resolution for this dilemma is not known. It might be thought that the difficulty lies in our improper use of

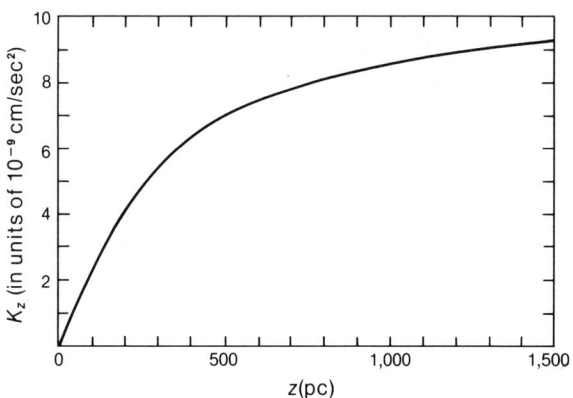

Figure 29.9. Gravitational force per unit mass in 10^{-9} cm sec^{-2} normal to the Galactic disk versus distance (pc) from the disk near the Sun.

Table 29.1
Distance data for force law.

Z_0 (km/sec)	z_{max} (pc)
9	100
37	500
60	1,000

Poisson's equation, which, in its full form is

$$\frac{\partial F_r}{\partial r} + \frac{F_r}{r} + \frac{\partial F_z}{\partial z} + \frac{1}{r}\frac{\partial F_\theta}{\partial \theta} = -4\pi G\rho. \quad (29.62)$$

In obtaining the simple estimate (29.58), we did not include the first two terms on the left-hand side (the last vanishes for axial symmetry). A simple analysis shows that these omitted terms are in fact small in the solar neighborhood, so that the approximations leading to (29.60) are not bad.

Problem 29.17. Suppose the radial force is that needed to keep stars in circular orbit in the plane: $F_r = -\Theta_c^2/r$, where Θ_c is given by (29.23). Show that the first two terms in (29.62) are unimportant for finding a value for ρ in the solar neighborhood.

The force law in the Galactic plane may be estimated from observed rotation curves $\Theta(r)$. This amounts simply to assuming Keplerian orbits, and taking

$$F_r = -\Theta_c^2/r, \quad (29.63)$$

with the rotations and circular velocities equated: $\Theta_c = \Theta(r)$. In other words, if the motion of a stellar group that is believed to move in nearly circular orbits around the Galactic nucleus is known, then (29.63) gives the resulting centripetal force acting on each member of the group (recall $m = 1$). For example, if (29.23) is used for $\Theta(r)$ in (29.63), the force law shown in Figure 29.10 is obtained.

29.5. GALACTIC MASS DISTRIBUTION

Finding values for F_r and F_z makes it possible to construct models of the Galactic mass distribution. In the simplest approach, a Galactic potential $\Phi(r, z)$ is constructed from a superposition of elementary geometric forms, such as spherical, spheroidal, and point masses (hollow shells have also been used to simulate density variations). Each form will contain one or more parameters, which are obtained as follows. The forces F_r and F_z are obtained from $\Phi(r, z)$. Equation (29.63) may then be used to derive the circular velocity Θ_c, which is then compared with an observed rotation curve, such as (29.23). Next the predicted F_z is compared with the observed force perpendicular to the Galactic plane in the solar neighborhood. The parameters in Φ are adjusted to maximize agreement. If the agreement is reasonably good, the model should give a lower bound for the total mass of the Galaxy.

The preceding method clearly depends on how carefully (and convincingly) the initial model for $\Phi(r, z)$ is chosen. In practice, the number of parameters (equivalently, the number and complexity of geometric shapes) assumed is limited by the amount of reliable data about various regions of the Galaxy. Although the method may amount to little more than curve fitting, it does supply results that are essential as input for other branches of Galactic physics, such as spiral-density-wave theory.

Problem 29.18. Why should the program outlined in the preceding for fitting theoretically obtained components F_r and F_z to observational data yield only a lower limit to the Galactic mass? Is this limitation observational or theoretical in nature?

To illustrate the procedure, suppose there are two primary mass components, a point mass m_c, and a spherical distribution with uniform density ρ_0, radius R_0, and total mass $m_s = 4\pi R_0^3 \rho_0/3$. The potential then depends on the three parameters m_c, m_s, and R_0. These may be fixed by assuming that the Sun moves in a circular orbit, with radius R_0 and speed $\Theta_0 = 250$ km/sec, and that $(d\Theta/dr)_{r=R_0} = -5$ km sec^{-1} kpc^{-1}. The latter value is obtained from observations in the solar neighborhood. Note that although additional mass may be spherically distributed outside the Sun's orbit, it has no effect on these parameters. The total force acting on a test mass at $r < R_0$ is

$$-F = \frac{m_c G}{r^2} + \frac{m_s G r}{R_0^3}. \quad (29.64)$$

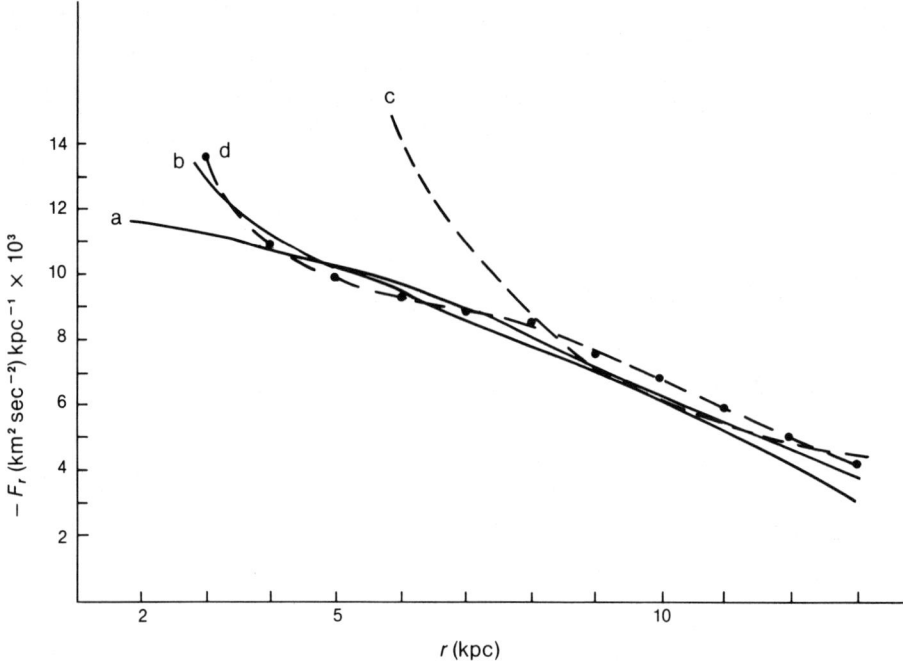

Figure 29.10. Radial force per unit mass in the Galactic disk (km^2 sec^{-2} kpc^{-1}). Curve is a fit to observed rotation curve; b is point mass plus inhomogeneous spheroids; c is point mass plus homogeneous spheroids; d (dash-dot) is recent model. In solar neighborhood $F_r \approx 6{,}250$ km^2 sec^{-2} kpc^{-1}.

The force on the Sun is given by (29.63), with $r = R_0$. At any point the circular velocity is given by (29.63). For the Sun, this gives

$$\Theta_0^2 = (m_s + m_c)G/R_0, \qquad (29.65)$$

which is readily solved for the total mass $m_c + m_s$. The result is $m_{tot} = 1.41 \times 10^{11}\ M_\odot$. From (29.63) and (29.64) it is easily shown that

$$\left(\frac{d\Theta}{dr}\right)_{R_0} = \frac{G}{\Theta R_0^2}\left(m_s - \frac{1}{2}m_c\right). \qquad (29.66)$$

Using the observed value for the left-hand side, we find that

$$m_s - m_c/2 = -2.81 \times 10^{10}\ M_\odot.$$

When this is combined with the estimated total mass, we find

$$m_c = 1.12 \times 10^{11}\ M_\odot,\ m_s = 2.86 \times 10^{10} M_\odot. \qquad (29.67)$$

In this simple model, nearly four times as much matter lies in the central point mass (nucleus) as in the dispersed portion. If the rotation curve is compared to the observed values, the agreement is reasonable only in the rather narrow range $9 < r < 11$ kpc. The problem lies in the fact that F_r is gradually decreasing throughout most of the Galaxy, as shown in Figure 29.10, and the combination of forces used in (29.63) can approximate this behavior within only a relatively narrow region. This limitation results largely from the use of a spherical mass distribution, whereas the Galaxy is actually rotationally flattened, so that the density actually decreases with increasing radius. Using a variable density improves the fit for $r > 10$ kpc, at least in the simple example above, while replacing a spherical distribution with a spheroidal one does not improve the fit for $r < 10$ kpc.

Acceptable models can be obtained if the mass distribution is taken to be spheroidal with variable density. The density is constant on spheroidal shells and decreases outward. Basically a density law is assumed, from which the rotation curve is obtained and fitted to the known curve. The rest of the procedure is as outlined earlier. A recent Galactic model constructed in this way is shown in Figures 29.11 and 29.12. These show $\Omega(r)$, the epicyclic frequency $\kappa(r)$ to be defined in Section 29.6, the projected mass density $\sigma(r)$, and the total mass contained within a spheroidal shell of semimajor axis r. About 25 percent of the mass lies outside the Sun's orbit ($R_0 = 10$ kpc). The spheroids are extremely flat, with $e = 0.9988$. Roughly half the total mass lies outside the spheroidal surface through the Sun, and the point mass at the origin is about $0.07 \times 10^{11} \, M_\odot$. The total mass is $1.8 \times 10^{11} \, M_\odot$. It is found that the density variation for $r > 10$ kpc is of the form r^{-4}.

Problem 29.19. Carry out the calculation leading to (29.66).

29.6. NONCIRCULAR ORBITS

In the preceding sections, stellar motions have been considered only in an average or statistical sense. Now consider the motion of a single star under the influence of the background Galactic potential. If the star moves in response to a spherically or axially symmetric potential, its simplest motion is circular, with speed $\Theta_c(r)$. The presence of symmetry-breaking perturbations, due either to large-scale departures from axial symmetry or to local effects caused by stellar encounters, leads to noncircular orbits of various degrees of complexity. In this section only those orbits that are obtained by small perturbations away from circular motion are considered. This analysis leads naturally to the concept of epicyclic motion.

The density of gas and stars away from the disk is quite small. In fact, for distances $|z|$ greater than about 0.14 kpc the gas density is less than half its $z = 0$ value and is dropping rapidly; the corresponding distance for stars is about 0.31 kpc. Most of the stars outside the disk presumably lie in globular clusters, or are part of the spherical halo. These stars or groups tend to move in nearly Keplerian orbits of varying eccentricity about the nucleus. The perturbations of the kind mentioned are not likely to affect such stars significantly. We will therefore concentrate on the behavior of stellar orbits confined, in the unperturbed state, to the Galactic disk.

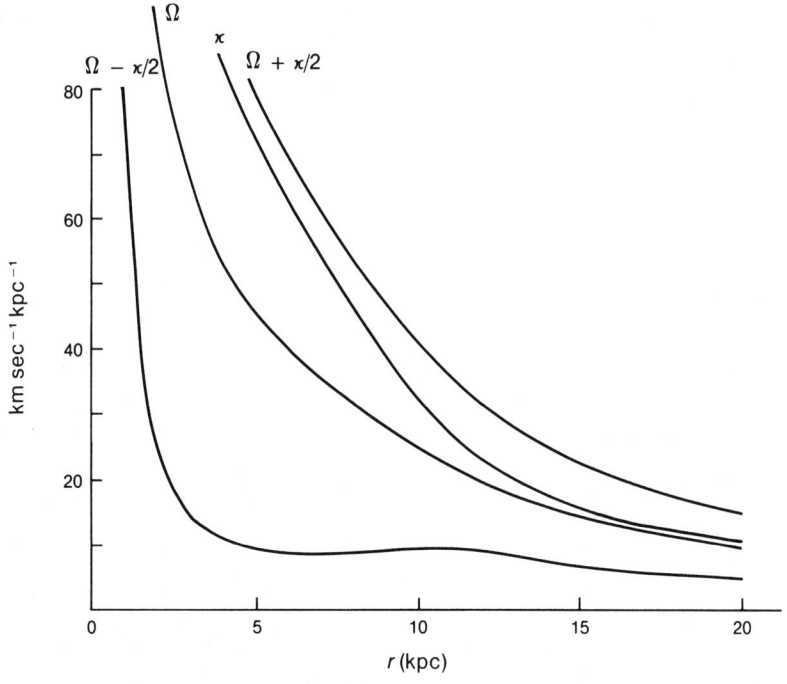

Figure 29.11. Rotation curve (Schmidt model) for our Galaxy, in km sec^{-1} kpc^{-1}, and epicyclic frequency.

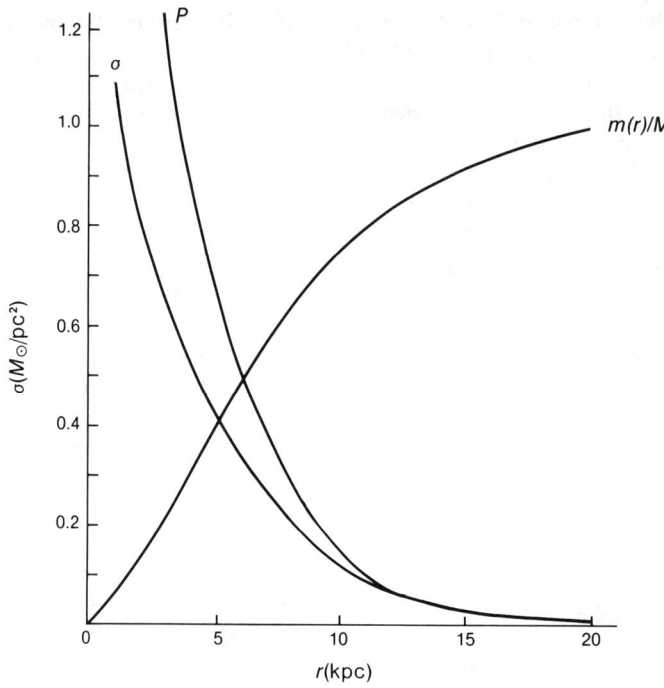

Figure 29.12. Projected surface density $\sigma(M_\odot/\text{pc}^2)$, mass density $(M_\odot/\text{pc}^3)$, and mass fraction obtained from Schmidt model of Figure 29.11. The mass fraction varies from 0 to 1, and $M = 1.47 \times 10^{11} M_\odot$.

A small perturbation will, in general, give the star a small added velocity (its peculiar velocity) in some arbitrary direction. The velocity may be decomposed into components (ξ, η) in the radial and angular direction. Consider the radial component first. The most convenient frame of reference to use here is one moving with the star. In this frame the net force per unit mass acting on it is

$$\ddot{r} = F_r + r\dot{\theta}^2, \qquad (29.68)$$

where F_r is given (29.63). The angular velocity $\dot{\theta} = \Theta/r$. The unperturbed motion results when $\ddot{r} = 0$, in which case F_r is just the centripetal acceleration and the radius is constant. The effect of a small perturbation can be incorporated by replacing the constant radius r_0 by

$$r = r_0 + \xi, \qquad \dot{r} = \dot{\xi}, \qquad \ddot{r} = \ddot{\xi}. \qquad (29.69)$$

Then (29.68) becomes

$$\ddot{\xi} = -\frac{\Theta_c^2}{r} + \frac{\Theta^2}{r}. \qquad (29.70)$$

Notice that the circular velocity Θ_c enters through F_r. Since ξ is small relative to r, we expand

$$r^{-1} \sim (1 - \xi/r_0)/r_0$$

and $\Theta_c(r)$ in the Taylor series about r_0:

$$\Theta_c(r) = \Theta_c(r_0 + \xi) \simeq \Theta_c(r_0) + \left(\frac{d\Theta_c}{dr}\right)_{r=r_0} \xi. \qquad (29.71)$$

Finally, assume that the initial radial velocity is Π_0. Conservation of angular momentum gives $\Theta(r) r =$

552 / AXIALLY SYMMETRIC GALAXIES

$\Theta_c(r_0)r_0$, which may be used in the approximate form

$$\frac{\Theta(r)}{r} = \frac{\Theta_c(r)_0 r_0}{r^2} \approx \frac{\Theta_c(r_0)}{r_0}\{1 - 2(\xi/r_0)\}, \quad (29.72)$$

along with (29.71) in (29.70) to obtain

$$\ddot{\xi} = \frac{(1 - \xi/r_0)}{r_0} \times \left\{\frac{r_0^2 \Theta_0^2}{r^2} - \left(\Theta_0 + \xi \frac{d\Theta_0}{dr}\right)^2\right\}. \quad (29.73)$$

All quantities of order ξ^2 or smaller have been dropped, and $\Theta_0 \equiv \Theta_c(r_0)$. When (29.73) is expanded to first order, we find

$$\ddot{\xi} = -\kappa^2 \xi, \quad (29.74)$$

where the epicyclic frequency κ is defined by

$$\kappa^2 = \frac{2\Theta_0^2}{r_0^2}\left[1 + \frac{r_0}{\Theta_0}\left(\frac{d\Theta_c}{dr}\right)_{r_0}\right]$$

$$= 4\Omega^2\left[1 + \frac{r_0}{2\Omega}\left(\frac{d\Omega}{dr}\right)_{r_0}\right]. \quad (29.75)$$

The last form is obtained by using $r\Omega = \Theta_c(r)$ evaluated at r_0. In the solar neighborhood, $\kappa = 31.6$ km sec^{-1} kpc^{-1}.

The differential equation (29.74) represents simple harmonic motion about r_0 with frequency $2\pi\kappa$. The solution, which satisfies the initial conditions at $t = 0$, $\xi(0) = 0$, and $\dot{\xi}(0) = \Pi_0$, is easily shown to be

$$\xi(t) = \frac{\Pi_0}{\kappa}\sin\kappa t, \quad (29.76)$$

where κ is evaluated at r_0. The star performs small radial oscillations about the unperturbed orbit defined by the local circular velocity $\Theta_c(r_0)$, with period

$$P_{\text{rad}} = 2\pi/\kappa. \quad (29.77)$$

It is readily verified from (29.75) and the rotation curve (29.23) that $P_{\text{rad}} = 1.9 \times 10^8$ yrs near the Sun. This corresponds to about 0.79 times the star's period of revolution, P_{rev}, around the Galactic center. Since P_{rad} and P_{rev} are not commensurate (P_{rev} is not an integer multiple of P_{rad}), the perturbed orbit is not closed, and the star wanders gradually within an annular ring about the galactic center.

Problem 29.20. A star orbits a point mass M in a circular orbit of radius R under the influence of a radial force $F_r = -MG/r^n$. Show that small radial perturbations result in closed orbits if $n = 2$ (Coulomb force). What happens to the orbit if $n > 3$?

Problem 29.21. Find the maximum width of the band centered on the Sun's orbit that would be occupied by stars in the solar neighborhood whose peculiar velocities had radial components of magnitude $|\Pi_0| \leq 30$ km/sec.

The star's perturbed velocity component in the angular direction η is most easily obtained by noting that for constant angular momentum

$$r^2 \frac{d\theta}{dt} = r\Theta = r_0\Theta_0. \quad (29.78)$$

Recall that $\Theta_0 = \Theta_c(r_0)$. The instantaneous radius, assuming a small change as in (29.69) may be used with (29.78) to show that

$$\dot{\theta} = \frac{r_0\Theta_0}{r^2} \approx \frac{\Theta_0}{r_0} - \frac{2\Theta_0}{r_0^2}\xi. \quad (29.79)$$

The change in the angular velocity is given by the last term in (29.79):

$$\Delta\dot{\theta} = -\frac{2\Theta_c}{r_0^2}\xi = -\frac{2\Theta_0\Pi_0}{r_0^2\kappa}\sin\kappa t, \quad (29.80)$$

where (29.76) has been used in the last step. To obtain the tangential velocity $\dot{\eta}$, (29.80) is multiplied by $r = r_0$ to give the lowest-order result

$$r\dot{\theta} \simeq r_0\dot{\theta} = \dot{\eta} = -\frac{2\Theta_0\Pi_0}{r_0\kappa}\sin\kappa t. \quad (29.81)$$

This is integrated to yield the tangential velocity

$$\eta(t) = \frac{2\Theta_0\Pi_0}{r_0\kappa^2}\cos\kappa t. \quad (29.82)$$

The ratio of amplitudes for the two motions (29.77) and (29.82) is given by

$$\frac{\text{ampl. } \eta}{\text{ampl. } \xi} = \frac{2\Theta_0}{r_0\kappa}. \quad (29.83)$$

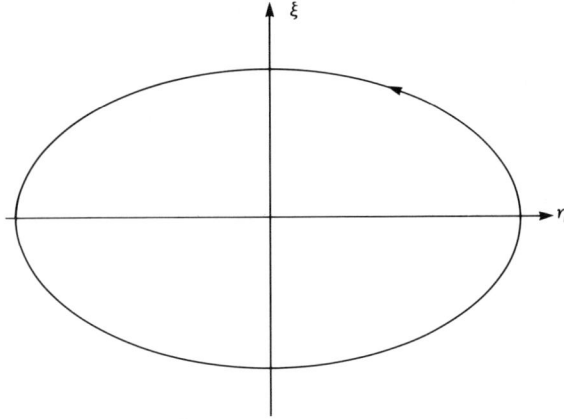

Figure 29.13. Epicyclic motion as seen by observer traveling at constant radius with circular velocity Θ_0 at $(\xi, \eta) = (0, 0)$.

In the solar neighborhood this ratio is 1.58. The extent of the tangential motion is therefore about half again as large as the radial extent. The motion as seen by an observer traveling at constant r_0 with the circular velocity Θ_0 would appear to trace out an orbit like that shown schematically in Figure 29.13. The center of the figure moves, relative to a fixed frame of reference, with linear speed Θ_0. As seen by an inertial observer, the star, which moves with retrograde motion, has velocity components $(\Pi, \Theta_0 + \dot{\eta})$, where $\dot{\xi} \equiv \Pi$. The relative orbit shown here is called the epicyclic orbit.

A simple physical argument shows why retrograde epicyclic motion arises. Consider two stars at r_0, one of which travels with the local circular velocity in a circular orbit. Imagine that the second star begins to undergo harmonic oscillations radially. As it moves outward it slows down tangentially because of the reduced gravitational force acting on it. It therefore falls behind the first star. As it passes within the circular orbit at r_0, it speeds up tangentially because the gravitational force increases in magnitude. The perturbed star now begins to overtake the one moving in a circular orbit. The alternating speedup and slowdown of the perturbed star produces the epicyclic orbit described by (29.82) and (29.83).

Problem 29.22. The force acting on a star at a distance z above or below the disk of our Galaxy is given by (29.58) for small z. Show that the period of oscillation is given by

$$P_z = (\pi/G\rho_0)^{1/2}. \qquad (29.84)$$

Compare the amplitude of harmonic motion normal to the disk with the amplitude radially as given by (29.76). Assume the $z = 0$ stellar velocity to be $Z_0 = 30$ km/sec and parameters representative of the solar neighborhood.

Chapter 30

SPIRAL STRUCTURE

Nearly two-thirds of the 1,500 bright galaxies known to date show definite spiral structure, and evidence obtained from 21.1-cm radiation within our Galaxy strongly suggests that it too possesses spiral structure. The spiral pattern often extends throughout most of a galaxy's visible disk, which typically has a length scale of 10 to 15 kpc or more. This smooth pattern may be broken or disrupted over regions of order 1 kpc in extent. These irregularities may be due in part to supernova outbursts, or possibly to local variations in the galactic magnetic field. When these local irregularities are ignored, a highly regular pattern, extending over the entire disk, often emerges.

In developing a simple model of spiral structure, we can ignore the detailed structure of the galactic halo and of the nucleus. Although matter in the halo may not be uniformly distributed, we may as a first approximation take it to be spherically symmetric about the nucleus. Then the motion of an element of matter in the disk at a distance r from the nucleus will be influenced only by the halo matter within a sphere of radius r, and according to Gauss' law (applied to the Newtonian gravitational potential) its response will be the same as if this halo mass were added to the nucleus. The greater r is, the more halo mass must be added to the nucleus, and this mass will contribute to the radial force law. The important point, however, is that the spherically symmetric component of the halo may be incorporated into disk models via the disk potential.

The disk consists primarily of three components; stars, dust, and gas, and the galactic magnetic field. The stellar component contains nearly all spectral types and classes, but the young, hot OB-type stars, although only a fraction of those present, are by far the most conspicuous observationally, partly because of their intrinsic brightness, but also because many are still found in or near dense HI clouds, from which they presumably formed. The intense ultraviolet radiation from these stars produces bright HII regions, which, because of their unusual size, are among the brightest visible objects in the disk. Consequently, the combination of bright young stars and gas is the most prominent optical feature, though it is a relatively small fraction of the disk matter.

Problem 30.1. Compare the luminosity of a typical B star on the main sequence with that of a typical HII region.

The amount of interstellar gas in spiral galaxies is probably between 10 and 30 percent (by mass) of the stellar component, and extends only about half as far away from the plane of the disk as does the stellar component. Interstellar gas and dust, much of it in the form of clouds, responds to the background gravitational field of the galaxy, but contributes little to it. Gas and dust motion may also be influenced by large-scale galactic magnetic fields, which couple to ions and electrons, and to magnetic grains. Magnetic fields in the disk, which show up in polarization measurements of starlight, probably do not exceed several times 10^{-6} gauss in the Galaxy. Observations indicate a locally irregular behavior, but suggest that the fields tend to run along the spiral arms. The major energy content of the disk of our Galaxy is given in Table 27.2.

Two major problems are associated with the existence of spiral structure in disk galaxies. First, a mechanism is needed to set up spiral arms once the disk has formed. This mechanism should be applicable to Sa, Sb, and Sc types, and possibly to Magellanic systems as well. Second, a mechanism must enable the pattern, once established, to persist for times comparable to the age of galaxies (up to 2×10^{10} years).

30.1. Difficulties: Streaming Motion and the Winding Dilemma

Because spiral patterns are so prominent, it is not surprising that the earliest models treated them as material arms, separated from one another by nearly empty space, which rotated about the nucleus like the spokes of a wheel. According to this model, the stars, gas, and dust in the disk were bound by their mutual gravitational field into arms. As the disk matter rotated about the nucleus, so did the arms. Observations of the rotation curves for representative spirals that show strong differential rotation pose serious problems for this interpretation. Because the outer regions move more slowly than those nearer the center, a material arm, once formed, should tend to wind up as the system rotates. This has already been shown for our Galaxy in Problem 29.8, where the time required for spiral structure to dissolve because of differential rotation was estimated to be about 2×10^8 years.

This important conclusion may be expressed in a form applicable to most spiral galaxies. Consider a section of spiral arm and the radius vector from the galactic center C to the point P on the arm (Figure 30.1), and a similar point Q nearby. The angle between

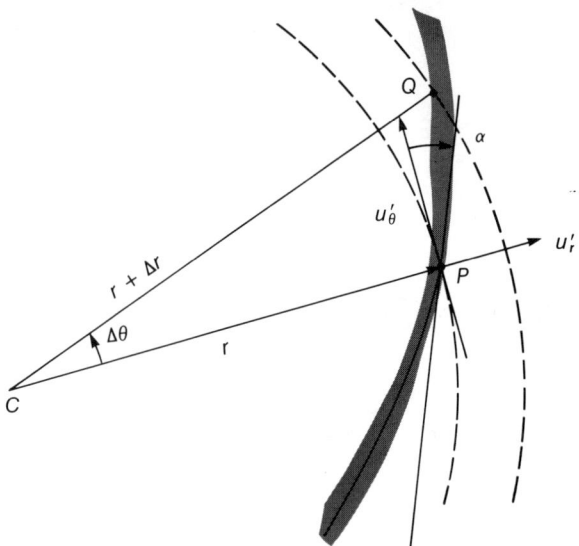

Figure 30.1. Motion of a star in a spiral arm (shaded), a distance r away from the galactic center C.

the tangent to the arm and the tangent to a circular orbit through the arm is defined to be α. Inspection of Figure 30.1 shows that as $\Delta\theta$ approaches zero, the radial separation of P and Q, Δr, also approaches zero, and the ratio $\Delta r / r \Delta\theta$ approaches $\tan\alpha$:

$$\tan\alpha = \frac{1}{r}\frac{dr}{d\theta}. \qquad (30.1)$$

Expanding $\Omega(r)$ in a Taylor series about r, we find the difference in angular velocity between a star at P and one at Q to be approximately $\Delta\Omega = (d\Omega/dr)\Delta r$. If the two stars were originally on the same radius vector ($\Delta\theta = 0$), then they would be on radius vectors separated by $\Delta\theta$ after a time T_{sp}, where

$$T_{sp}\Delta\Omega \simeq T_{sp}(\partial\Omega/\partial r)\Delta r = \Delta\theta. \qquad (30.2)$$

Using (30.1) to eliminate Δr, and noting that the period for a star at P to complete one cycle about the galactic nucleus is $T_{rot} = (2\pi/r\Omega)$, we find that

$$\frac{T_{sp}}{T_{rot}} = [2\pi \tan\alpha(\partial \ln\Omega/\partial \ln r)]^{-1}. \qquad (30.3)$$

Since $\Omega \sim 1/r$ throughout most of the disk in typical spiral galaxies, and taking $\alpha \simeq 6°$, as is typical of Sb types like our Galaxy, we find that (30.3) gives

$$T_{sp} \simeq 1.4\, T_{rot}. \tag{30.4}$$

For the Galaxy, $T_{rot} \simeq 2 \times 10^8$ years; so T_{sp} is of order 3×10^8 years. Although T_{sp} gives the time for two nearby stars initially at the same galactic longitude to become separated by $\Delta\theta$ because of differential rotation, it is comparable to the time needed for them to appear to belong to different spiral arms. Therefore T_{sp} is a reasonable measure of the permanence of a spiral feature due to a stellar-density contrast in a differentially rotating disk. Notice that (30.3) is applicable to any spiral galaxy, and that (30.4) is also approximately valid as long as $\Omega \sim 1/r$ throughout most of the disk, and as long as the spiral pattern is not too loose ($\alpha \ll \pi/2$).

Our Galaxy, and the nearby spiral M 31, are about 10^{10} years old, and typical stars in their disks have probably completed 50 revolutions about the galactic nucleus in that time. Clearly spiral patterns could not have persisted in these systems unless some mechanism maintained the structure, or re-formed it about once every 10^8 years.

Problem 30.2. One way around the issue of spiral structure is to argue that it represents one evolutionary stage of the galactic disk, and that it occurs only once per galaxy. Assume that our Galaxy and M 31 are typical spirals, and discuss the viability of this hypothesis. What fraction of observed spiral galaxies would be expected to show spiral structure?

The existence of young stars of about this age in associations within well-formed arms makes this interpretation difficult to accept.

Another model that arose out of early studies of stellar motion near the sun, suggests that stars stream along the arms, adjusting their motion so as to avoid the winding dilemma. In this model the assumption of nearby circular motion about the galactic nucleus by disk matter is abandoned, and a net outflow of mass in the radial direction must occur. Nevertheless, the spiral structure of the disk remains invariant under axial rotations by 2π and the spiral pattern moves rigidly with angular velocity Ω_s. Suppose the matter in a spiral arm flows outward along the spiral feature with velocity components u'_r and u'_θ (shown in Figure

30.1) as measured with respect to the spiral feature. Then if u_θ is the θ component of the matter's peculiar velocity (relative to an inertial frame),

$$u'_r = u'_\theta \tan \alpha = (u_\theta - r\Omega_s) \tan \alpha.$$

The implications of this mass flow may be seen by applying the continuity equation of hydrodynamics $\partial \rho / \partial t + \nabla \cdot (\rho \mathbf{v})$ to the disk. Before doing so, we rewrite the continuity equation in cylindrical coordinates (r, θ, z). Since the term $\partial \rho / \partial t$ remains the same, consider the divergence:

$$\nabla \cdot \rho \mathbf{v} = \mathbf{v} \cdot \nabla \rho + \rho \nabla \cdot \mathbf{v}. \tag{30.5}$$

The velocity

$$\mathbf{v} = \hat{e}_r u'_r + \hat{e}_\theta u'_\theta + \hat{e}_z u'_z,$$

where $u'_z = u_z$ is the peculiar velocity normal to the plane of the disk, and the unit vectors $\hat{e}_i$ are defined in the usual way. The first term in the divergence is easily shown to be

$$u'_r \frac{\partial \rho}{\partial r} + \frac{u'_\theta}{r} \frac{\partial \rho}{\partial \theta} + u'_z \frac{\partial \rho}{\partial z} = \mathbf{v} \cdot \nabla \rho. \tag{30.6}$$

The second term is more involved. Remembering that ∇ operates on both the components of $\mathbf{v}$ and the unit vectors (for example, $\partial \hat{e}_r / \partial \theta = \hat{e}_\theta$), we can easily show that

$$\rho \nabla \cdot \mathbf{v} = \rho \frac{\partial u'_r}{\partial r} + \frac{\rho}{r} \frac{\partial u'_\theta}{\partial \theta} + \rho \frac{\partial u'_z}{\partial z} + \frac{u'_r \rho}{r}. \tag{30.7}$$

The last term arises from the change in $\hat{e}_r$ with θ. Combining equations (30.6) and (30.7) gives the continuity equation in cylindrical coordinates:

$$\frac{\partial \rho}{\partial t} + \frac{1}{r} \frac{\partial}{\partial r} r u'_r \rho + \frac{1}{r} \frac{\partial}{\partial \theta} \rho u'_\theta + \frac{\partial}{\partial z} \rho u_z = 0. \tag{30.8}$$

The continuity equation in the form (30.8) states that the time-rate of change of the mass of a small volume element $r \Delta r \Delta \theta \Delta z$ is nonzero only if the mass flux through the outer boundary of the element is nonzero.

Problem 30.3. Integrate (30.8) over the range $0 \leq \theta \leq 2\pi, -\infty \leq z \leq \infty$, and $R \leq r \leq R + \Delta R$. If the

Table 30.1
The parameters characteristic of six spiral systems, as obtained from the density-wave theory. For the Galaxy, assuming $M = 13.1 \times 10^{10}\ M_\odot$, the theory predicts $\Omega_p = 13.5$ km sec^{-1} kpc^{-1}, and $\alpha = 6.9°$. R_c is the radius of corotation.

NGC	Classification		Distance (Mpc)	Inclination angle of disk (deg)	Radius of visible disk (kpc)	Radius of outermost HII region (kpc)	Total mass ($10^{10}\ M_\odot$)	$\dfrac{R(.5M)}{R_c}$	Ω_p (km sec^{-1} kpc^{-1})	α (deg)
	M Hubble	de Vaucouleurs								
224 (Andromeda)	31 Sb	SA(s) b	0.69	77.0	13.0	22.5	17.8	0.50	18.0	9.5
598	33 Sc	SA(s) cd	0.72	55.0	4.8	5.2	1.3	1.24	32.0	19.6
3031	81 Sb	SA(s) ab	3.2	55.0	8.9	16.2	12.5	0.57	26.0	11.8
5055	63 Sb	SA(rs) bc	7.3	58.6	6.5	7.3	5.8	0.31	32	4.0
5194 (Whirlpool)	51 Sc	SA(s) bcp	4.0	35.0	5.0	5.5	3.3	0.58	42.0	11.7
5457	101 Sc	SAB(rs) cd	6.9	22.0	15.5	15.3	14.5	0.70	13.0	15.7

galaxy's disk is a bound system whose mass is localized, and whose spiral structure is periodic with period π (double-armed spiral), show that

$$\frac{dM}{dt} = -\int \mathcal{F}(R+\Delta R)\, dS_2 + \int \mathcal{F}(R)\, dS_1,$$

where $\mathcal{F}(r) = u'_r \rho$ is the mass flux in the radial direction, and dS_1 and dS_2 are elements of surface area on a cylinder of radius R and $R + \Delta R$, respectively. Interpret this result.

Now apply the continuity equation to the disk of a spiral galaxy. Since we are primarily interested in total mass flow in the disk, integrate (30.8) over θ and z. The third term vanishes when integrated over angle. If we denote by a bar quantities averaged over angle and z,

$$\overline{Q} \equiv \int_0^{2\pi} d\theta \int_{-\infty}^{\infty} dz\, Q(\theta, z),$$

(30.8) becomes

$$\frac{d\overline{\rho}}{dt} + \frac{1}{r}\frac{d}{dr}\overline{(r\rho u'_r)} = 0; \qquad (30.9)$$

here we are assuming no mass loss or gain because of flow normal to the galactic plane. Since observations in our Galaxy indicate that u_θ and $\overline{\rho}$ are nearly constant in r over much of the disk, we may approximate the last term in (30.9) by

$$\frac{1}{r}\frac{d}{dr}\overline{r\rho(u_\theta - r\Omega_s)}\tan\alpha \approx (u_\theta - 2r\Omega_s)\overline{\rho}\tan\alpha$$

for galaxies in which α is also constant. The quantity $(u_\theta - 2r\Omega_s)\tan\alpha \equiv T_{st}$ has units of time, and we may rewrite (30.9) as

$$\frac{d\ln\overline{\rho}}{dt} = -T_{st}^{-1}, \qquad (30.10)$$

whose solution is $\overline{\rho}(t) = \overline{\rho}(0)\exp(-t/T_{st})$. We see that T_{st} represents the time for the density averaged over the disk to decrease by a factor $1/e$.

Problem 30.4. Find a rough value for T_{st} (order of magnitude), and show that if $\overline{\rho} \approx 0.08\, M_\odot/\text{pc}^3$, as is typical for matter near the Sun, then the disk must be losing $\sim 500\, M_\odot/\text{yr}$. Assume a disk volume of $(15)^2\, \text{kpc}^3$.

Since observations yield u_θ typically on the order of 230 km/sec (average for $b \le r \le 18$ kpc) in our Galaxy, (30.10) implies that the disk would lose up to 500 $M_\odot$/yr if its spiral structure were maintained by stars streaming along the arms. The disk density would therefore change considerably during about $1/50$ the age of the Galaxy. Consequently the streaming model requires addition of matter to the galactic disk, presumably at a comparable rate. However, there appears to be little observational evidence for this much mass loss from the galactic nucleus, or infall of intergalactic gas onto the disk. Although 21-cm observations of the region near the galactic nucleus suggest that mass is being injected into the disk, its rate appears to be 1 or 2 $M_\odot$/yr at most.

Problem 30.5. Estimate the mass flux onto the disk of the galaxy that the streaming model would require to keep the local density constant. Assume for simplicity that ρu_z is constant over the disk.

The streaming model avoids the winding dilemma, but in the process introduces what appears to be an unusually high rate of radial mass flow from the disk. It appears unlikely that the nucleus in our galaxy can supply the mass necessary for a constant local density, but it can be argued that mass may be added to the disk from the halo (See Problem 30.5). Nevertheless, the streaming model requires a mechanism capable of setting up radial mass flow, and the observations of a weak galactic magnetic field that may lie within the spiral arms offers a clue about how radial mass motion might arise in our Galaxy. To illustrate this, we suppose that the field lies in the plane of the disk and re-enters the galactic nucleus along the z axis. Charged particles that leave the nuclear region with random velocities will eventually become tied to the magnetic field of a spiral arm and will drift toward the outer edge of the disk. If the magnetic field along the spiral arm is B and the velocity of matter v, then the virial theorem implies that the kinetic and magnetic energy densities are approximately equal:

$$\frac{B^2}{8\pi} \approx \frac{1}{2}\rho v^2. \qquad (30.11)$$

To order of magnitude, $v \approx 230$ km/sec, and $\rho \simeq 2 \times 10^{-24}$ g/cm³, which gives $B \simeq 10^{-4}$ gauss. This exceeds observed Galactic fields by a factor of 20 to 100. More nearly realistic models of magnetic fields confined to spiral arms require slightly weaker field intensities, but still in excess of 10^{-5} gauss. The introduction of magnetic effects removes the problem of how mass motion occurs, but appears to raise other difficulties, not the least of which is a need for fields that exceed the observed limits.

30.2. Density Wave Theory: Physical Picture

The models discussed in Section 30.1 attempt to explain spiral structure by assuming that the stars, gas, and dust in the disk remain within a spiral arm. We now consider a model in which this assumption is relaxed. The spiral arms are considered to be areas of increased density in the galactic disk through which stars, gas, and dust move. The spiral arms in this model are density waves, which set up a rigidly rotating pattern on the differentially rotating disk. According to the model, once a density wave has been established, it sets up a local minimum in the gravitational field of the disk. Stars and other disk matter revolving about the nucleus will adjust their motion in response to the local field in such a way that they linger near the region of gravitational potential minimum.

The physical basis for the density-wave model is illustrated by analyzing the response of gas, dust, and stars to a pre-existing, rigid spiral gravitational field superimposed on a smooth, differentially rotating disk. Figure 30.2 shows the disk rotating in a clockwise direction. Light lines represent the circular orbits that would exist if there were no spiral gravitational field. The minimum of the perturbing spiral gravitational potential is shown by the heavy line, and rotates rigidly with angular velocity in the same direction as the stars, gas, and dust, which move with the local rotation velocity (to first approximation). The result is a rigidly rotating pattern on a differentially rotating disk. To see this, consider a small sample of stars and gas near b on an unperturbed orbit. Since the sample is roughly equidistant from the two nearest arms, the net gravita-

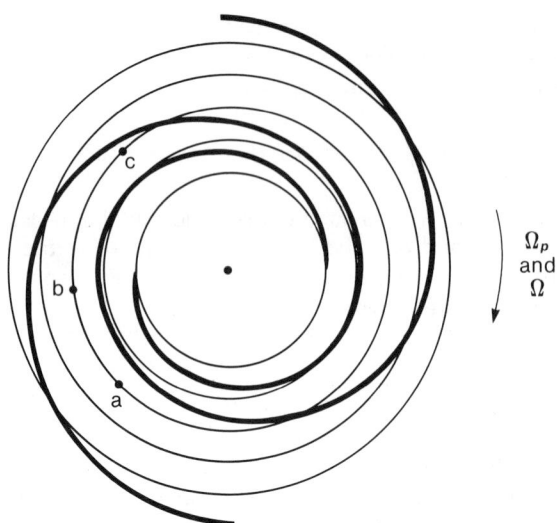

Figure 30.2. Stellar angular velocity Ω and spiral pattern angular velocity Ω_p in galactic disk.

tional force because of the increased density of the arms is zero, and the sample tends to move in a circular orbit. Continuing clockwise around the galactic center, the sample tends to move closer to the outer arm, until, in the vicinity of point c, the gravitational force of the outer branch dominates. This produces a net outward force, and the matter moves to slightly larger orbits, where, according to Kepler's law, its linear velocity is smaller. Similar arguments indicate that stars or gas leaving a spiral arm near a take longer to reach the point b as they revolve about the nucleus. As a result, stars, dust, and gas pile up along the spiral valley of the gravitational potential. This process tends to be self-reinforcing, the added matter tending to maintain the existing potential minimum as other matter moves out into the interarm region. The stars and gas are thus envisioned as moving in perturbed orbits, with the spiral gravitational potential representing the perturbation, much like the circular plus epicyclic orbit discussed earlier. In fact, as will be seen later, the frequency with which the disk matter meets a spiral gravitational potential minimum is determined by the local rotational angular velocity of the matter with respect to the disk, $\Omega - \Omega_p$, and the epicyclic frequency κ. The pattern corresponds to a slight increase in stellar and gas density in the vicinity of the potential minimum of a small spiral gravitational field superimposed on the smooth, rotationally symmetric disk, and is referred to as a spiral density wave.

The density-wave model attributes spiral structure to collective modes of a gravitating disk in which the disk matter moves through the pattern while contributing to it, and avoids the difficulties associated with the winding dilemma, with radial mass loss from the disk, and with the need for magnetic fields in excess of observed limits. In the following sections we will see that the model helps explain the correlation between the apparent decrease in galactic nuclear mass that accompanies the increase in disk content and openness of the spiral pattern observed in the transition from Sa to Sc galaxies. Finally, the density-wave model predicts that local increases in gas density may be great enough to trigger star formation along the inner edges of spiral arms. Despite these promising features, the density-wave theory is not free from difficulties. First, it is not clear how the initial perturbations are set up in the disk. It has been argued that local instabilities, nongravitational forces in the disk, or tidal effects due to nearby galaxies are responsible. Second, it must be demonstrated that spiral perturbations can be maintained for periods comparable to the age of the galaxy. Third, although an established spiral density-wave pattern appears to work for normal spiral galaxies, it is less clear that the approach will work for barred spirals. Since recent observations of galactic morphology indicate a smooth transition from normal to barred spirals, it is not clear why the spiral patterns in both groups should not be described by a single theory of density waves. Although these major issues are currently unsolved, the density-wave theory appears to be an attractive approach to spiral structure, and does offer insight into the astrophysics of disk matter.

30.3. Spiral Density-Wave Theory: Formulation

We will consider here the theory of density waves in axisymmetric disks, and illustrate it with a series of simple models. Although the approach is theoretical, we will use observational characteristics of our Galaxy and of nearby spiral galaxies to motivate mathematical approximations.

The basic model that we will discuss assumes that a small spiral perturbation already exists within an axisymmetric galactic disk; we will develop the response of the stars, gas, and dust to this perturbation. The model is based on a magnetohydrodynamic description of the gas and dust, a statistical treatment of stellar motion using the Boltzmann transport equation, and the solution of Poisson's equation for the total galactic gravitational field. The approach is complicated not only by its intrinsic nonlinearity, but also by the fact that the densities, gravitational potential, and stellar distribution function must be obtained self-consistently.

Most progress to date has been obtained from a linearized form of the model, which is nevertheless self-consistent. The linearized model, which assumes that the perturbations in gravitational potential and mass density are small, is motivated in part by observations of our Galaxy that indicate the density contrast between the arms and the interarm regions to be less than 10 percent. Magnetic fields are also neglected, since they play a minor role in determining the overall pattern of the disk. To see this, consider the various forces acting on a fluid element in the disk:

$$\rho \frac{d\mathbf{v}}{dt} = -\nabla P - \rho \nabla \Phi$$
$$- \frac{1}{8\pi} \nabla B^2 - \frac{1}{4\pi} (\mathbf{B} \cdot \nabla) \mathbf{B}. \quad (30.12)$$

This is the momentum equation of magnetohydrodynamics, and states that the net force density $\rho \, d\mathbf{v}/dt$ acting on a fluid element is due to pressure gradients and externally imposed gravitational and magnetic forces. Now consider the relative magnitudes of the terms on the right-hand side of (30.12). To order of magnitude, the pressure gradient $|\nabla P| \approx P/l$, where l is a characteristic length scale for pressure variations. Similarly, the magnetic-force density terms are of order B^2/l'. Observations of galactic magnetic fields suggest that they vary smoothly over distances of several kpc, which is typical of the length scale for variations in disk pressure and gravitational fields. Therefore, we may set $l' \approx l$ in discussions of large-scale disk structure. Pressure in galactic disks arises primarily from the gas component. Observations in our galaxy indicate typical gas temperatures $T \sim 10^2$ K and turbulent velocities of magnitude $a_0 \approx 60$ km/sec in spiral arms. Therefore, the turbulent pressure $\sim \rho a_0^2$ dominates, and the ratio of magnetic to pressure gradient force densities is roughly

$$\frac{|\nabla B^2|}{|\nabla P|} \approx \frac{B^2}{P} \simeq \frac{B^2}{\rho a_0^2}. \quad (30.13)$$

Typically $B \approx 2 \times 10^{-6}$ gauss, and $\rho \simeq 2 \times 10^{-24}$ g/cm^3, so that the ratio in (30.13) is about 0.06. A similar analysis shows that the magnetic terms are small relative to the force density $\rho\nabla\Phi$, but that $|\rho\nabla\Phi| \approx |\nabla P|$. Detailed studies indicate that the incorporation of magnetic fields lying along the spiral arms with $|\mathbf{B}| \lesssim 2 \times 10^{-6}$ gauss does not significantly alter a quasistationary density-wave pattern obtained by models in which $B = 0$.

Finally, since dust contributes a small mass fraction of the gas-dust system in the disk, and since in the absence of a magnetic field the two components react in the same way to gravitational perturbations, we will suppose that the disk consists of stars and gas only.

The following outline summarizes the approach used in building a model of spiral density waves, and serves as a guide for the next five sections: (1) the form of the gravitational field (including the small spiral perturbation) is assumed; (2a) the equations of stellar dynamics then require a redistribution of the disk stars; and (2b) the equations of hydrodynamics require a redistribution of the gas component; (3) the last two effects taken together give the net redistribution of matter in the disk; (4) this is equated with the mass distribution required to maintain the original spiral perturbation in (1), as determined by Poisson's equation. The last step determines the values of the free parameters in the theory, and leads to a self-consistent model of galactic spiral structure that can be compared with observations.

Before we proceed with the discussion of density waves, a review of the basic geometric properties of simple spirals will be useful later. Most of these follow from the general relation

$$n\theta = \Psi(r), \quad (30.14)$$

where $\Psi(r)$ is a monotonically increasing function of radius. The integer $n > 0$ though not essential here, is introduced for later convenience. A segment of a typical spiral curve is shown in Figure 30.3. We have already shown that the angle between a unit vector in the θ direction and the tangent to the spiral α is given by (30.1), or in terms of $\Psi(r)$,

$$\tan\alpha = n[r\, d\Psi/dr]^{-1} \equiv \frac{n}{kr}. \quad (30.15)$$

This defines the quantity k as the rate of change of Ψ with radius. The angle α is the pitch angle of the spiral. Early-type spiral galaxies (Sa and Sb) are characterized by relatively small pitch angles over most of the disk. For example, α is estimated to be between 4 and 6

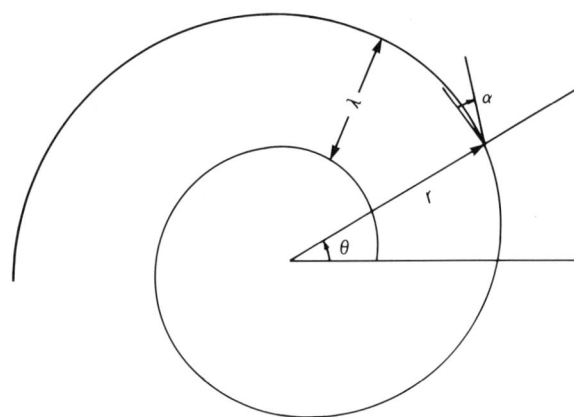

Figure 30.3. Logarithmic spiral curve, showing pitch angle α, and interarm separation λ.

degrees in the solar neighborhood. Spiral galaxies of these types are tightly wound. The distance λ separating two nearby segments of a spiral (see Figure 30.3) is given by

$$n\Delta\theta = \Psi(r + \lambda) - \Psi(r) = 2\pi n.$$

When λ/r is small (as is typical in the intermediate and outer portions of many spiral galaxies), the quantity

$$\Psi(r + \lambda) \simeq \Psi(r) + k\lambda,$$

and we find

$$\lambda = 2\pi n/|k|. \quad (30.16)$$

Problem 30.6. The logarithmic or equiangular spiral is obtained by setting

$$\Psi(r) = \Lambda \ln(r/r_0). \quad (30.17)$$

(a) Show that the pitch angle for this spiral is constant.

(b) Assume that the spiral structure of the Galaxy is equiangular, with $\alpha = 4.5°$, and that it has two arms. This is equivalent to setting $n = 2$, as shown in the text. Find the interarm spacing.

The logarithmic spiral described by (30.17) is a simple example illustrating the properties of spiral curves. It is also a useful approximate fit to theoretical models of spiral structure in our Galaxy. In fact, taking $r_0 = 5$ kpc and $\Lambda = 9.9$, we find that (30.17)

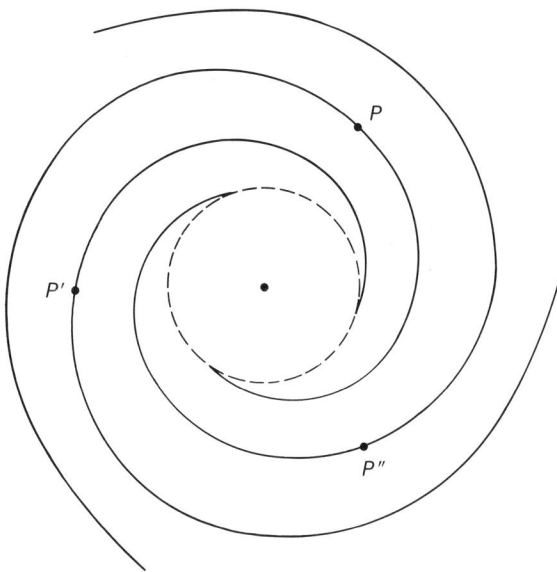

Figure 30.4. Three-armed spiral pattern. The density profiles at similar points along arms are equal.

reproduces the shape of the spiral arms in the Galaxy for $r \gtrsim 4$ kpc as obtained from density-wave theory. The simple form (30.17) is also useful in estimating the spiral properties of Sa and Sb galaxies.

Finally, consider a physical quantity expressed in the polar coordinates (r, θ, z) of the disk, such as the density $\rho(r, \theta, z)$. Suppose that the contours of constant density are spirals in the disk, and that there are n branches (Figure 30.4 shows the case in which $n = 3$). The density is therefore constant along lines satisfying (30.14), and the density remains unchanged under rotations by 2π. However, if there are n arms, it also remains unchanged under rotations by $2\pi/n$. Therefore the density must depend on r and θ in such a way that $n\theta - \Psi(r)$ is a constant:

$$\rho(r, \theta, z) = \rho(n\theta - \Psi(r), z).$$

Figure 30.4 illustrates that rotations by multiples of $2\pi/3$ carry the point P into the similar points P' and P'' lying on the other branches of the spiral pattern.

Gaseous Disks: Linear Theory

We now return to the density-wave model outlined in the first part of this section, in which the disk consists of a stellar and a gaseous component. Although there is more disk mass in stars than in gas ($\rho_{\text{gas}} \approx 0.3\,\rho_{\text{stars}}$, on average), we will see that most of the gas participates in density-wave motion, whereas only those stars having low velocity dispersions participate. In effect, the stars are the predominant source of spiral perturbations in the gravitational field of the disk, but the gaseous component responds most strongly to it. Therefore we may obtain a reasonably simple model of spiral density waves by using only the gaseous component, but with an increased mass density reflecting the presence of the stellar component. We now carry out the steps that were outlined above for the gaseous component only; that is, we ignore step (2a). We will find that the resulting model reproduces most of the properties (number of arms; interarm spacing; and relationships between disk gas content and degree of openness of spiral arms) typical of Sa and Sb galaxies. Two additional approximations will also be made. First, only thin disk models are treated. These are characterized by a disk thickness $z_{\max}/R_D \approx 0.3$ kpc/ 15 kpc = 0.02. Second, the gravitational field, density, and gas-velocity perturbations are assumed to be small.

The unperturbed state is obtained by projecting the mass density of an axisymmetric differentially rotating model onto the $z = 0$ plane. The velocity of an element of gas at r is given by $\Omega(r)r$. In the perturbed state, such an element has a peculiar velocity whose components (relative to an inertial observer at rest with respect to the center of the disk) are given by

$$\text{gas velocity} = (u, v + r\Omega), \qquad (30.18)$$

where u and v are small perturbing velocities, as yet unknown. The surface density (g/cm^2) is

$$\sigma = \sigma_0(r) + \sigma'(r, \theta, t); \qquad (30.19)$$

its integral over the disk surface gives the disk mass. The unperturbed density σ_0 depends only on r. In practice it is taken from existing models of axisymmetric galaxies without spiral structure, such as one discussed in Section 29.5, and is therefore initially known. Given the volume density $\rho(r, \theta, z, t)$, σ_0 is obtained by integrating over all values of z. The quantity σ' represents the density perturbation and will ultimately be identified as the spiral pattern. The motion of gas in response to an imposed spiral perturbation in the gravitational field is described by the gas hydrodynamic equations, written in fixed cylindrical coordinates (r, θ). These have been discussed in Chapter 21.

The mass continuity equation has already been discussed, and is given by (30.8) without the last term, if we replace ρ by σ, and set $u'_r = u$ and $u'_\theta = v + r\Omega$:

$$\frac{\partial \sigma}{\partial t} + \frac{1}{r}\frac{\partial}{\partial r}(ru\sigma) + \frac{1}{r}\frac{\partial}{\partial \theta}\sigma(v + r\Omega) = 0. \quad (30.20)$$

This relates the time-rate of change of the mass $\sigma r \, \Delta r \, \Delta \theta$ contained in a "volume" element $r \, \Delta r \, \Delta \theta$ to the mass flux in the radial (second term) and the angular (third term) directions. The second hydrodynamic equation is the momentum equation in the radial direction, and is derived in Section 21.4. It is

$$\sigma \frac{\partial u}{\partial t} + \sigma u \frac{\partial u}{\partial r} + \sigma \frac{(v + r\Omega)}{r}\frac{\partial u}{\partial \theta} - \sigma \frac{(v + r\Omega)^2}{r}$$

$$= \frac{\partial}{\partial t}(\sigma u) + \frac{1}{r}\frac{\partial}{\partial r}(r\sigma u^2)$$

$$+ \frac{1}{r}\frac{\partial}{\partial \theta}[\sigma u(v + r\Omega)] - \sigma \frac{(v + r\Omega)^2}{r}$$

$$= -\frac{\partial P}{\partial r} - \sigma \frac{\partial \Phi}{\partial r}. \quad (30.21)$$

The second line follows when the continuity equation is used to replace $\sigma \partial u/\partial t$ with $\partial \sigma u/\partial t$. The r component of the fluid element's momentum density is σu, and $\mathcal{F}_{rr} = \sigma u^2$ and $\mathcal{F}_{r\theta} = \sigma u(v + r\Omega)$, where $\mathcal{F}_{ij}$ is the rate at which the jth component of the momentum crosses a unit area in the ith direction. The term $\sigma(v + r\Omega)^2/r$ is the r component of the centripetal force density acting on the element. It is obvious that (30.21) is just Newton's second law in the r direction. Finally, the θ component of Newton's second law is

$$\sigma \frac{\partial v}{\partial t} + \sigma u \frac{\partial}{\partial r}(v + r\Omega) + \sigma \frac{(v + r\Omega)}{r}\frac{\partial v}{\partial \theta} + \sigma u \frac{(v + r\Omega)}{r}$$

$$= \frac{\partial}{\partial t}\sigma(v + r\Omega) + \frac{1}{r}\frac{\partial}{\partial r}(r\mathcal{F}_{\theta r})$$

$$+ \frac{1}{r}\frac{\partial}{\partial \theta}\mathcal{F}_{\theta\theta} + \sigma u \frac{(v + r\Omega)}{r}$$

$$= -\frac{1}{r}\frac{\partial P}{\partial \theta} - \frac{\sigma}{r}\frac{\partial \Phi}{\partial \theta}, \quad (30.22)$$

where $\mathcal{F}_{\theta r} = \mathcal{F}_{r\theta}$, and $\mathcal{F}_{\theta\theta} = \sigma(v + r\Omega)^2$ is the rate of transfer of the θ component of the fluid momentum across a surface normal to the θ direction, and $\sigma(v + r\Omega)$ is the θ component of the momentum density. The two terms on the right-hand side of (30.21) and (30.22) are the components of the pressure gradient and the gravitational force in the disk.

If the gas is in turbulent motion, with the average magnitude of the turbulent velocity denoted by a_0, then the turbulent fluid contributes a pressure to the fluid. If the turbulence is random, then the pressure it exerts across a surface is given by the product of the average momentum density σa_0 and the turbulent speed a_0. We have already seen that this type of turbulent pressure exceeds the usual thermal pressure in interstellar gas. Therefore, assuming that a_0 is constant throughout the disk, we can write

$$\frac{\partial P}{\partial r} \simeq \frac{\partial}{\partial r}\sigma a_0^2 = a_0^2 \frac{\partial \sigma}{\partial r},$$

$$\frac{\partial P}{\partial \theta} \simeq \frac{\partial}{\partial \theta}\sigma a_0^2 = a_0^2 \frac{\partial \sigma}{\partial \theta}.$$

These expressions will be used in (30.21) and (30.22) hereafter.

Problem 30.7. The unperturbed state for a galactic disk has $u = v = 0$. Show that (30.20) and (30.22) are satisfied trivially, and find the unperturbed radial equation. Show that the centripetal acceleration acting on a fluid element is reduced by the turbulent pressure of the gas.

Problem 30.7 shows that turbulence in a gaseous disk supplies additional support against gravitational collapse, when the surface density decreases radially, as is observed. Turbulence can also be shown to counter the tendency of gas in a uniform axisymmetric disk to condense into rings.

The hydrodynamic equations (30.20) to (30.22) relate the fluid variables u, v, and σ to the gravitational potential Φ [the angular velocity $\Omega(r)$ and turbulent velocity of the gas a_0 are assumed known], which is obtained from Poisson's equation. Poisson's equation requires that the perturbed surface density act as the source of the perturbing gravitational field. It is given by (28.1) with ρ replaced by $\sigma_{tot}\delta(z)$, where $\delta(z)$ is the Dirac delta function:

$$\nabla^2 \Phi = 4\pi G \sigma_{tot} \delta(z). \quad (30.23)$$

The gravitational potential of the disk may be written as

$$\Phi(\mathbf{r}, t) = \Phi_0(r, z) + \Phi'(r, \theta, z, t). \quad (30.24)$$

Note that the first term is independent of time. Any axisymmetric disk model may be used for Φ_0.

The nonlinear model (30.20) to (30.23) is difficult to solve; so for the remainder of this and the following three sections, we discuss only the linearized model, which is obtained as follows. Assume that

$$\Phi' \ll \Phi_0, \qquad \sigma' \ll \sigma_0, \quad (30.25)$$

and that all unperturbed quantities (σ_0, Φ_0) satisfy (30.20) to (30.23) identically. Notice that for the unperturbed state, we set $u = v = 0$. To illustrate the procedure, substitute (30.24) and (30.19) into the continuity equation (30.20), and note that the peculiar velocity components u and v are also small quantities (see Section 29.4). The result to first order in the small quantities (u, v, σ', Φ') is

$$\frac{\partial \sigma'}{\partial t} + \frac{1}{r}\frac{\partial}{\partial r}(ru\sigma_0) + \Omega \frac{\partial \sigma'}{\partial \theta} + \frac{\sigma_0}{r}\frac{\partial v}{\partial \theta} = 0. \quad (30.26)$$

In arriving at (30.26), we dropped the two terms

$$\frac{\partial \sigma_0}{\partial t} + \frac{1}{r}\frac{\partial}{\partial \theta}(r\sigma_0 \Omega), \quad (30.27)$$

$$\frac{1}{r}\frac{\partial}{\partial r}(ru\sigma') + \frac{1}{r}\frac{\partial}{\partial \theta}(v\sigma'). \quad (30.28)$$

The first two vanish identically, since σ_0 is independent of time and angle θ, and Ω is dependent only on radius. The last two terms contain a product of two small quantities $u\sigma'$ or $v\sigma'$, and represent higher-order corrections to the linear model. Equation (30.26) is just the continuity equation for first-order perturbations, and expresses the requirement that the time-rate of change of perturbations in the mass of a fluid element ($\sigma' r \Delta r \Delta \theta$) result from perturbations in the mass flux across the fluid element.

Proceeding analogously for the radial component of the equations of motion, we find the first-order result to be

$$\frac{\partial u}{\partial t} + \Omega \frac{\partial u}{\partial \theta} - 2v\Omega = -\frac{a_0^2}{\sigma_0}\frac{\partial \sigma'}{\partial r} - \frac{\partial \Phi'}{\partial r}. \quad (30.29)$$

The radial acceleration in the unperturbed disk depends on the unperturbed pressure gradient and gravitational force. Equation (30.29) shows that the first-order correction to this acceleration arises from small changes in ∇P and $\nabla \Phi$, and in the first-order Coriolis acceleration $2v\Omega$. The θ component of the equations of motion is similarly found, and is

$$\frac{\partial v}{\partial t} + u\frac{\partial r\Omega}{\partial r} + \Omega \frac{\partial v}{\partial \theta} + u\Omega$$
$$= -\frac{1}{r}\left(\frac{a_0^2}{\sigma_0}\frac{\partial \sigma'}{\partial \theta} + \frac{\partial \Phi'}{\partial \theta}\right). \quad (30.30)$$

This may be rewritten by noting that

$$u\frac{\partial}{\partial r}(r\Omega) + u\Omega = 2\Omega u\left(1 + \frac{r}{2\Omega}\frac{d\Omega}{dr}\right) = \frac{u\kappa^2}{2\Omega}, \quad (30.31)$$

where κ is the observed epicyclic frequency defined by (29.74). Using equation (30.31) in place of the second term in (30.30) yields the final form for the θ component of the equation of motion:

$$\frac{\partial v}{\partial t} + \Omega \frac{\partial v}{\partial \theta} + \frac{\kappa^2}{2\Omega} u = -\frac{1}{r}\left(\frac{a_0^2}{\sigma_0}\frac{\partial \sigma'}{\partial \theta} + \frac{\partial \Phi'}{\partial \theta}\right). \quad (30.32)$$

Problem 30.8. Derive the linearized components of the equations of motion (30.29) and (30.32). Interpret physically the terms contributing to the first-order change in the θ component of the acceleration, i.e., the first two terms of (30.32).

Poisson's equation for the perturbed potential in linearized theory is

$$\nabla^2 \Phi' = 4\pi G \sigma' \delta(z). \quad (30.33)$$

To summarize, the linearized hydrodynamic equations for the perturbed density are as follows: (30.26) is the continuity equation; (30.29) is the radial equation of motion; (30.32) is the angular equation of motion. Poisson's equation is given by (30.33). The next section considers physical and observational arguments motivating a set of solutions that have spiral character, and for which the pattern is tightly wound.

Spiral Perturbations

In this section we express the linearized hydrodynamic model in a form suited to tightly wound spiral systems. In fact, the approach to a solution that we will take is to assume that the linearized perturbations σ' and Φ' and the components (u, v) of the peculiar velocity exhibit spiral character, and then show that the assumptions are consistent. First, however, we must express σ', Φ', u, and v in a form that makes their assumed spiral character evident.

A natural approach to the solution of a set of partial differential equations like (30.26) to (30.33) is separation of variables, and the discussion following Problem 30.6 suggests a convenient form to try. Each one of the perturbations may be expressed as the real part of a sum of terms

$$F(r, \theta, t) = \sum_n f_n(r) e^{i(\omega t - n\theta)},$$

where the integer $n \geq 1$. Since we expect $F(r, \theta, t)$ to show spiral character, we expect it to be a function of r and θ such that $n\theta - \Phi(r)$ is constant along spiral contours. Therefore it is convenient to set

$$f_n(r) \equiv \hat{f}_n(r) \exp i\Psi(r)$$

with $\hat{f}_n(r)$ a slowly changing function of r and $\Psi(r)$ defined as in (30.14). The nth term contributing to $F(r, \theta, t)$ represents a spiral structure with n distinct branches. The majority of observed spiral galaxies (including our Galaxy and M31) appear, apart from local irregularities, to be double-armed. Therefore we might expect the $n = 2$ term to dominate in these cases.

The nth component of the perturbations may be written in the form

$$\begin{aligned} \sigma' &= \hat{\sigma} \exp i[\omega t - n\theta + \Psi(r)], \\ u &= \hat{u} \exp i[\omega t - n\theta + \Psi(r)], \\ v &= \hat{v} \exp i[\omega t - n\theta + \Psi(r)], \\ \Phi' &= \hat{\Phi} \exp i[\omega t - n\theta + \Psi(r)]. \end{aligned} \quad (30.34)$$

A subscript n should appear on the amplitudes, but is omitted for convenience.

The perturbing gravitational force $\nabla \Phi'$ contains terms which, according to (30.34), are proportional to

$$-\frac{\partial \Phi'}{\partial r} = -\left(\frac{\partial \hat{\Phi}}{\partial r} + i \frac{d\Psi}{dr} \hat{\Phi}\right) e^{i(\omega t - n\theta + \Psi)}$$

and

$$-\frac{\partial \Phi'}{\partial \theta} = in \hat{\Phi} \, e^{i(\omega t - n\theta + \Psi)}.$$

Define the magnitude of the wave vector k by

$$k = d\Psi/dr. \quad (30.35)$$

Our Galaxy, M31, and others like them are characterized by tightly wound spiral arms with pitch angles typically of order 10° or less. For these systems (30.15) shows that

$$\frac{n}{|k|r} = \tan \alpha \ll 1.$$

For the Galaxy, $n/kr = 2/kr$ and $\alpha \simeq 6.3$ deg, so that $1/kr \approx 0.05$. We see that the condition $n/kr \ll 1$ is equivalent to a tight-winding approximation for the spiral structure, and we expect it to be reasonable for Sa and Sb galaxies. For Sc galaxies, α is in the range 10° to 20° ($\tan \alpha \leq 0.4$). In such cases the tight-winding approximation is not as good, but still yields useful insights into the properties of these systems.

The assumption that the amplitude $\hat{\Phi}$ of the gravitational perturbation is slowly varying is equivalent to $d\hat{\Phi}/dr \ll \hat{\Phi}/r$, or $d \ln \hat{\Phi}/d \ln r \ll 1$. Therefore the tight-winding approximation and the slowly varying character of $\hat{\Phi}$ imply that

$$|k|r \gg |d \ln \hat{\Phi}/d \ln r| \quad \text{if} \quad |k|r \gg n. \quad (30.36)$$

These in turn imply that the θ component of $\nabla \Phi'$ is small relative to the radial one:

$$\frac{1}{r}\left|\frac{\partial \Phi'/\partial \theta}{\partial \Phi'/\partial r}\right| = \frac{n|\hat{\Phi}|}{r[(d\hat{\Phi}/dr)^2 + k^2 \hat{\Phi}^2]^{1/2}} \simeq \frac{n}{|k|r} \ll 1.$$

Furthermore, we may approximate the radial component of $\nabla \Phi'$ by

$$\frac{\partial \Phi'}{\partial r} \simeq -g_r e^{i(\omega t - n\theta + \Psi(r))},$$

where

$$g_r \equiv -ik\hat{\Phi} \quad (30.37)$$

is the amplitude of the radial acceleration.

When equations (30.34) are substituted into the linearized model (30.26), (30.29), (30.32), and (30.33), we obtain three ordinary differential equations that contain the quantities $\hat{\sigma}$, $\hat{u}$, $\hat{v}$, and $\hat{\Phi}$. Equation (30.35) yields a fourth equation, which contains $\hat{\sigma}$ and $\hat{\Phi}$. In addition they contain the quantity k, which, according to (30.35), upon integration gives Ψ. The real part of Ψ determines the spiral geometry through (30.14), and when substituted into (30.34) gives the r, θ dependence of the perturbations. These are indeed spirals, as can be seen by examining any one, say, Φ'. The arguments to follow apply to the other members of (30.34) as well. Recalling that $\hat{\Phi}(r)$ is a slowly varying function of r, we obtain the lines of constant Φ' by holding t constant and requiring that r and θ vary in such a way that the exponential remains constant: this requires that

$$n(\theta - \theta_0) = \Psi(r) - \Psi(r_0). \quad (30.38)$$

The point r_0 is fixed but arbitrary. Since $\Psi(r)$ is a slowly varying, monotonically increasing function of r, (30.38) describes spiral-like behavior. Consequently, the minimum of Φ' will exhibit this same behavior. Specific examples in Problem 30.6 demonstrate these points.

Assuming that the pattern described by (30.38) moves as a rigid body, the pattern speed $r\Omega_p$ may be obtained by holding r fixed and taking the time-derivative of the argument in (30.34). The result

$$d\theta/dt = \omega/n \equiv \Omega_p \quad (30.39)$$

defines the pattern angular velocity. Observational procedures for finding a value for Ω_p are extremely difficult. At present, Ω_p is evaluated from the shape of the spiral arms. In our Galaxy it appears to lie between 11 and 14 km sec^{-1} kpc^{-1}. This means that the spiral features in our Galaxy would complete one revolution around the nucleus in $2\pi/\Omega_p \simeq 5 \times 10^8$ yrs. In other words, the pattern moves about 2.5 times slower than the stars and gas. Theoretical models of M 81 (NGC 3031), which is type Sb and is similar in mass and radius to our Galaxy, indicate that for this system $\alpha \simeq 12°$ and $\Omega_p \simeq 26$ km sec^{-1} kpc^{-1}.

Finally we note that not all modes n entering into the expression of perturbed quantities need have real frequencies ω. In general, $\omega = \omega_1 + i\omega_2$, where ω_1 and ω_2 are real. If $\omega_2 < 0$, then each perturbed quantity will contain a factor $e^{-\omega_2 t}$ that vanishes on a time-scale proportional to ω_2^{-1}. Such modes are unstable, and decay in time. Only those for which $\omega_1 \geq 0$ persist. Those for which $\omega_1 > 0$ build up in time until nonlinear corrections to the model become important. For the following discussion we assume that ω_2 is small or zero, so that the spiral patterns are quasistationary ($\omega_2 \gtrsim 0$) or stationary ($\omega_2 \equiv 0$).

Solutions to the Linear Theory

Solutions to the linear theory are obtained by substituting the expressions for σ', Φ', u, and v given in (30.34) into the equations of motion (30.26), (30.29) to (30.33). This is straightforward, but lengthy. Therefore we outline the procedure. Starting with the continuity equation (30.26), we find the following quantities:

$$\begin{aligned}
\frac{\partial \sigma'}{\partial t} &= i\omega\sigma', \\
\frac{\partial v}{\partial \theta} &= -inv, \\
\frac{\partial \sigma'}{\partial \theta} &= -in\sigma', \\
\frac{\partial}{\partial r} r u \sigma_0 &= \sigma_0 \left(u + iru \frac{d\Psi}{dr} + \frac{u}{\hat{u}} \frac{d\hat{u}}{dr} \right).
\end{aligned} \quad (30.40)$$

In the last expression the term proportional to du/dr may be dropped, since u is a slowly varying function of r, and $u/\hat{u}$ is of order unity. When expressions (30.40) are used in (30.26), one finds, after rewriting,

$$\frac{n(\Omega_p - \Omega)\hat{\sigma}}{\kappa} + \frac{k\sigma_0}{\kappa}\left(1 - \frac{i}{kr}\right)\hat{u} = \frac{\sigma_0 k}{\kappa} \frac{\hat{v}}{kr} n.$$

For tightly wound spirals, $|k|r \gg n$. Therefore the last term on the left-hand side, as well as the term on the right-hand side, can be neglected, and to lowest order in small quantities the continuity equation becomes

$$\frac{n(\Omega_p - \Omega)}{\kappa}\hat{\sigma} + \frac{k\sigma_0}{\kappa} = 0. \quad (30.41)$$

Proceeding in this way, one finds that the lowest-order expressions for the remaining equations, (30.29) and

(30.32), are

$$\frac{ia_0^2 k}{\sigma_0 \kappa} \hat{\sigma} + i\nu - \frac{2\Omega\hat{v}}{\kappa} = g_r/\kappa, \qquad (30.42)$$

$$\frac{\kappa}{2\Omega} \hat{u} + i\hat{v}\nu = 0. \qquad (30.43)$$

The frequency ν appearing in (30.42) and (30.43) is by definition

$$\nu^2 \equiv \frac{n^2(\Omega_p - \Omega)^2}{\kappa^2}. \qquad (30.44)$$

The equations (30.41) to (30.43) and the definition (30.44) represent three equations in the four unknowns $\hat{\sigma}$, $\hat{u}$, $\hat{v}$, and g_r. The latter is given by the solution to Poisson's equation, as we will see shortly. Treating g_r as given, the three remaining unknowns may be obtained as the real parts of

$$\frac{\hat{\sigma}}{\sigma_0} = \frac{-ikg_r}{\kappa^2(1-\nu^2) + k^2 a_0^2}, \qquad (30.45)$$

$$\hat{u} = \frac{i\kappa g_r \nu}{\kappa^2(1-\nu^2) + k^2 a_0^2}, \qquad (30.46)$$

$$\hat{v} = \frac{k^2 g_r/2\Omega}{\kappa^2(1-\nu^2) + k^2 a_0^2}. \qquad (30.47)$$

Notice that the right sides of (30.45) to (30.47) contain known quantities κ, Ω, a_0, the parameters n and Ω_p, and the unknown g_r defined in (30.37). In principle n (number of arms) and Ω_p (pattern frequency) are fixed by observations. Equations (30.45) to (30.47) express the spiral perturbations of the gaseous disk that develop because of the perturbation g_r in the radial force. This, however, must have as its source the density σ', and is determined by Poisson's equation (30.33). As a final step, (30.33) must be solved. This can be done in the tight-winding approximation, though the proof is not simple. Fortunately, the result may be obtained from simple physical arguments. From (30.37) it follows that the amplitude of the potential and the acceleration are related by $\hat{\Phi} \sim ig_r/k$. Since the potential $\hat{\Phi}$, which maintains the density perturbation $\hat{\sigma}$, has this perturbation as its source (self-consistency), we expect $\hat{\Phi}$ to be proportional to $\hat{\sigma}$. Finally, if the density maxima $\hat{\sigma}$ are to be maintained as the interarm spacing $\lambda = 2\pi/|k|$ increases, then $|\hat{\Phi}|$ must increase as λ increases, otherwise there would be an increasing region of low potential between the arms in which the gas could reside, thereby weakening the spiral pattern. These arguments indicate that

$$\hat{\Phi} \simeq -\hat{\sigma} G\lambda = -2\pi G\hat{\sigma}/|k|$$

and, by (30.37) that

$$g_r = 2\pi i G\hat{\sigma} k/|k|, \qquad (30.48)$$

which is identical to the solution in the tight-winding approximation. Using (30.45) and (30.48) to eliminate g_r, we find the spiral density-wave dispersion relation, which relates the wave vector (or wavelength) of the spiral disturbances to frequency,

$$\frac{|k|}{k_0} = (1 - \nu^2) + \frac{k^2 a_0^2}{\kappa^2}, \qquad (30.49)$$

where

$$k_0 = \kappa^2/2\pi G\sigma_0.$$

This may be rewritten in terms of the wavelength by using (30.16).

Equating g_r from (30.48) with g_r in the expression for the density perturbation, we find that (30.45) is a self-consistency condition which guarantees that the spiral perturbations represent solutions to the linear theory.

The dispersion relation (30.49) is a quadratic equation for k, or for the wavelength representing the interarm spacing. Demanding that the wavelength be real imposes the condition

$$\frac{4k_0^2 a_0^2}{\kappa^2}(1 - \nu^2) \leq 1. \qquad (30.50)$$

It follows from (30.44) that if ν is complex, then the angular velocity ω of the spiral perturbation (30.34) will also be complex. For stationary modes ν must be real: $\nu^2 \geq 0$. The solution $\nu = 0$ separates the region of instability from the region of stability. Thus $\nu = 0$ sets a critical value for the local gas turbulent velocity,

$$a_0 < \frac{\kappa}{2k_0} \equiv \frac{\pi G\sigma_0}{\kappa}.$$

In the solar neighborhood, (30.49) gives $a_0 \simeq 60$ km/sec, which is about forty times the thermal velocity at typical interstellar gas temperatures ($T \simeq 100$ K).

The dispersion relation (30.49) may be rewritten as

$$\frac{k_0}{|k|}(1 - \nu^2) = \mathcal{F}_g(\nu), \quad (30.51)$$

where

$$\mathcal{F}_g(\nu) \equiv [2 + y/(1 - \nu^2)]^{-1},$$

and $y \equiv k^2 a_0^2/\kappa^2$. Since $\mathcal{F}_g(\nu)$ is positive, we conclude that $1 - \nu^2 \geq 0$, which, in combination with (30.44), limits Ω_p to the range

$$\Omega - \frac{\kappa}{n} \leq \Omega_p \leq \Omega + \frac{\kappa}{n}. \quad (30.52)$$

Notice that the bounds are determined in linear theory by the structure of the unperturbed disk. In general a spiral pattern contains contributions from all modes ($n > 0$). However, as n increases (contributions from modes with an increasing number of spiral branches), the range of permitted values for Ω_p decreases. Figure 30.5 shows $\Omega(r)$ and $\kappa(r)$ for a model of the Galaxy. Also shown (dotted lines) are the bounds $\Omega \pm \kappa/2$ for the double-armed mode. The spiral pattern presumably begins outside the 3 kpc expanding arm. Therefore, according to Figure 30.5, the $\Omega_p \geq 10$ km sec^{-1} kpc^{-1}. Since the pattern probably extends beyond 18

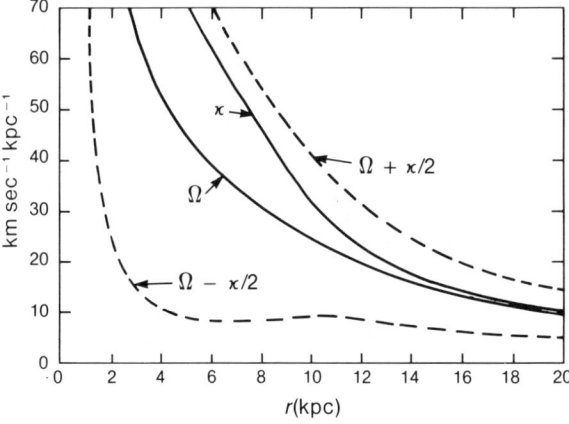

Figure 30.5. Angular velocity and epicyclic frequency for a model of the Galaxy. The pattern angular velocity Ω_p is constrained to lie between the two dashed curves.

kpc, $\Omega_p \leq 14$ km sec^{-1} kpc^{-1}. Thus, a rigidly rotating double-armed spiral wave in our Galaxy should have a pattern angular velocity in the narrow range $10 \leq \Omega_p \leq 14$ km sec^{-1} kpc^{-1}. In the linear model, a three-armed spiral model could contribute only over the range $4 \leq r \leq 13$ kpc. The form of the unperturbed rotation curve $\Omega(r)$ and the epicycle frequency κ are such that two-armed spiral models dominate, and modes with large n are essentially excluded.

Finally we notice that the perturbations $\hat{\sigma}$, $\hat{u}$, and $\hat{v}$, (30.45) to (30.47), are each proportional to $\mathcal{F}_g(\nu)$, and that $\mathcal{F}_g(\nu)$ decreases with increasing turbulent gas velocity a_0. Physically this says that greater turbulence in the disk gas leads to a weaker spiral pattern.

Stellar Component

In the preceding section the possibility of self-consistent spiral density waves propagating with constant angular velocity Ω_p on a differentially rotating gaseous disk was demonstrated. That analysis neglected two features that must be incorporated if the theory is to be applied to typical spiral galaxies. The first is the inclusion of a stellar component, to be discussed in this section. For the moment, we will ignore the presence of a gaseous component, and consider a stellar disk.

The redistribution of stars in the galactic disk because of a small spiral perturbation in the potential is described by the perturbed stellar distribution function, which is obtained as a solution to the collisionless Boltzmann equation. The analysis is significantly more complicated than for the gaseous component. The unperturbed stellar disk, which contains some stars of both populations and of all stellar types, is characterized by the surface stellar mass density σ_{*0}, and by the stellar peculiar velocity components in the plane, c_r and c_θ. These two Cartesian components lie along the radial and angular directions, respectively. The frame of reference in which they are defined moves, at the point r, with velocity $r\Omega$. Therefore, $\Pi = c_r$ and $\Theta = r\Omega(r) + c_\theta$. We are still working within the thin-disk model, so that c_z does not enter explicitly.

A spiral density wave in a stellar disk results in a correction to the surface mass density σ'_* and a shift in the components of the stellar peculiar velocities by v'_r and v'_θ. The three perturbations σ'_*, v'_r, and v'_θ are expressed in forms analogous to (30.34). This yields a linearized model of the stellar disk, which may be solved, along with Poisson's equation relating σ'_* to Φ', for $\hat{\sigma}_*$, $\hat{v}_r$, and $\hat{v}_\theta$. The solutions are found to be strongly decreasing functions of the quantity $k^2 \langle c_r^2 \rangle / \kappa^2$, where

k is the wave vector of the spiral wave propagating in the stellar disk, and $\langle c_r^2 \rangle$ measures the dispersion in stellar velocity along the radial direction in the unperturbed disk. In other words, the stars are effective in maintaining a spiral density wave only if $k^2 \langle c_r^2 \rangle / \kappa^2$ is small. Physically, this means that if a star's peculiar velocity is such that the effective radius of its epicyclic orbit, which is proportional to $\Pi_0/\kappa = c_r/\kappa$, is much larger than the interarm spacing, which is proportional to k^{-1}, then that star would not be expected to respond very much to local spiral perturbations. Instead, this star would contribute primarily to the background axisymmetric field of the disk. But, to order of magnitude,

$$\left(\frac{\text{epicyclic radius}}{\text{interarm spacing}} \right)^2 \simeq \left(\frac{\Pi/\kappa}{k^{-1}} \right) = \left(\frac{kc_r}{\kappa} \right)^2 \simeq \frac{k^2 \langle c_r^2 \rangle}{\kappa^2}.$$

Thus large radial dispersion means that the star has little effect in setting up or responding to spiral structures. Therefore, only a fraction of the disk stars actually are involved in the spiral structure.

Finite Disk Thickness and Complete Theory

For the density-wave theory to be applicable to observed galaxies, the results of the previous two sections must be combined, and we must allow for a finite thickness of the disk. Conceptually the first part of this procedure is straightforward. One starts with the total matter density (30.19), which is now understood to contain stars and gas: the unperturbed parts are

$$\sigma_0 = \sigma_{0g} + \sigma_{0*} \tag{30.53}$$

and the perturbations in density are given by

$$\sigma' = \sigma_*' + \sigma_g'. \tag{30.54}$$

The total mass density now enters as the source for Poisson's equation (30.23), and the response of the gas hydrodynamic and stellar distribution equations is calculated. For a thin disk the solution to Poisson's equation is of the form (30.48) with $\hat{\sigma}_*$ replaced by the total mass density:

$$ik\hat{\Phi} = -2\pi i G (\hat{\sigma}_* + \hat{\sigma}_g) |k|/k. \tag{30.55}$$

The dispersion relation for the gaseous and stellar disk differs from (30.51) for the gaseous disk by less than 10 percent except near $|\nu| = 1$ (these points are discussed in what follows).

Finally, consider the effect of finite disk thickness. The actual thickness of the Galaxy's disk, taken as the height above or below $z = 0$, at which the respective density is one-half its maximum value, is about 0.14 kpc (gas) and 0.31 kpc (stars). This is roughly 1/50 of the diameter. When finite thickness is included, the stars and gas will have velocity components normal to the plane. As a consequence the gravitational field due to the spiral perturbation is weakened, since the average field seen by a star or gas cloud is now less than the $z = 0$ value. When the gas content of a galaxy is small relative to the stellar density, detailed analysis shows that the inclusion of a stellar component and disk thickness produce a model quantitatively similar to the original gas disk model discussed in Section 30.3 if the surface density σ_0 is replaced by a factor of the total density σ_{tot}. For this reason the remainder of the discussion will rely on this relatively simple model as a guide.

Nonlinear Effects

The linear theory of spiral density waves allows both trailing and advancing patterns. The incorporation of nonlinear gas hydrodynamics and stellar dynamics terms leads only to trailing patterns. To appreciate this, consider a typical spiral arm through which a single gas streamline passes. For trailing waves (Figure 30.6), the spiral-wave pattern, which is envisioned as moving into the disk material, encounters regions of decreasing density. The arrow denotes the direction of the wave motion. As the wave moves into a region of lower density, the speed of sound decreases and the amplitude of the disturbance is increased. Eventually a shock front forms, and it will tend to maintain the spiral arm's existence. In Figure 30.6 the galactic center is to the lower left. For advancing arms [Figure 30.6(b)] the wave moves into regions of increasing density. The local speed of sound increases, and the disturbances tend to become smoothed out in time. As a result, any developing pattern is likely to decay. Observational evidence tends to indicate that trailing patterns are, in fact, favored.

The formation of spiral shock waves, existing over a large fraction of the galactic disk, is a significant feature of the nonlinear theory, since, as will be considered in Section 30.4, it offers a mechanism that would aid protostar formation.

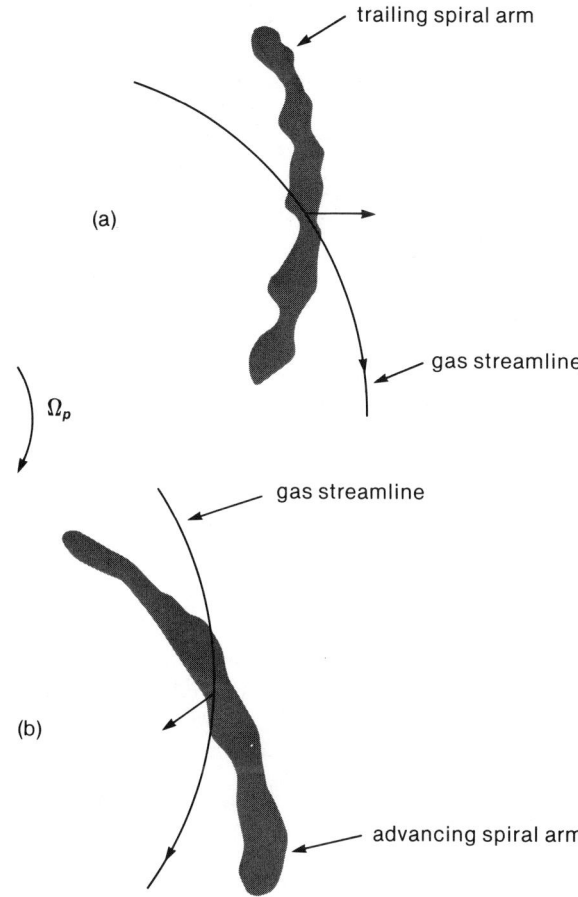

Figure 30.6. Relation of gas streamline (solid) to motion of (a) trailing spiral arm; (b) advancing spiral arm. The pattern rotates clockwise in both.

30.4. OBSERVATIONAL CONSEQUENCES

Equation (30.44) gives the frequency, measured in units of the epicyclic frequency κ, with which a density wave meets matter in the differentially rotating disk, and (30.52) gives the permitted range in the pattern angular velocity Ω_p:

$$\Omega - \frac{\kappa}{n} \leq \Omega_p \leq \Omega + \frac{\kappa}{n}.$$

Given the angular velocity of differential rotation $\Omega(r)$, (30.52) defines the range of values that r may take on and for which spiral density waves propagate as a rigid pattern. Inspection of (30.52) and Figure 30.5 shows that there are generally two values r_0 and r_1 between which (30.52) will be satisfied. The implication is that for $r < r_0$ or $r > r_1$, no spiral pattern is possible, at least in linear theory. These points are the radii of the Lindblad resonances.

Problem 30.9. Consider the angular velocity $\Omega(r)$ shown in Figure 30.5 to be representative of spiral galaxies in general. Using (30.52) show that most galaxies would be expected to have only two spiral arms. How does this compare with observations?

Problem 30.10. Suppose that the angular velocity in the Galaxy is given by

$$\Omega(r) = \frac{\Theta_0}{r} \qquad (30.56)$$

where $\Theta_0 = 250$ km/sec. Taking $n = 2$, find the locations of the Lindblad resonances. How do these values compare with what is known about the Galaxy's spiral structure?

The global spiral pattern is given in linear theory by solving (30.52) for the points of Lindblad resonance, and integrating the wave vector (30.35) between these two limits. Equation (30.38) then relates the angle θ to the radius for points of constant potential Φ', surface density σ, and velocity components (u, v). In other words,

$$\theta - \theta_0 = \int_{r_0}^{r} \frac{k}{n} dr'$$
$$= \int_{r_0}^{r} \frac{\kappa^2 - n^2(\Omega - \Omega_p)^2}{2\pi G \sigma_0(r) n} dr. \qquad (30.57)$$

It is generally convenient to set $\theta_0 = 0$ in (30.57). Notice that $\theta(r)$ in the linear theory does not involve any of the perturbed quantities, and that Ω_p is the only parameter not determined by the unperturbed galactic model. In principle it may be obtained from observational data. For the Galaxy, estimated values of Ω_p are in the range of 10 to 14 km sec^{-1} kpc^{-1}.

Problem 30.11. A simple approximation to $\Omega(r)$ for the Galaxy (1965 Schmidt model) is, with r in kpc,

$$\Omega(r) = \frac{250}{r} \text{ km sec}^{-1} \text{ kpc}^{-1}. \qquad (30.58)$$

Suppose that the projected surface mass density is given by

$$\sigma(r) = \frac{1648}{r} \, M_\odot pc^{-2}. \qquad (30.59)$$

(a) Calculate the total mass of the disk.

(b) Construct and plot the resulting spiral density-wave pattern for an assumed pattern angular velocity $\Omega_p = 20$ km sec^{-1} kpc^{-1}. Assume that only 8 percent of the surface density is effective in responding to the spiral perturbation, and that the galactic radius is 20 kpc when you calculate the mass.

Problem 30.12. Find the interarm spacing in the solar neighborhood and the pitch angle for Problem 30.11.

The density-wave theory explains some of the morphological properties of spiral galaxies in a natural way. Observations indicate that as mass concentrates in galactic nuclei, the gas content of the disk decreases, and the interarm separation decreases. Assume that the total mass M of a spiral galaxy is constant and is given by

$$M = M_N + M_D,$$

where M_D is the total mass of the disk and M_N is the nuclear mass; then Problem 30.13 shows that the density-wave theory accounts for the morphological features distinguishing spiral types Sa, Sb, and Sc.

Problem 30.13. Discuss morphological changes that would result in the density-wave theory as the disk mass is reduced, assuming the density law $\sigma(r) = \sigma_0/r$, and that the total galactic mass M remains constant. Compare the results with properties of Sa, Sb, and Sc galaxies.

Axisymmetric models of the Galaxy, which serve as unperturbed models for the density-wave theory, indicate that the ratio of projected surface density of gas to stars is about

$$\sigma_{0*}/\sigma_{0g} \approx 9 \qquad (30.60)$$

near the Sun. This corresponds to a volume-density ratio $\rho_{0*} \approx 4\rho_{0g}$ and relative gas and stellar-disk thickness given in the discussion of Table 27.2. The ratio (30.60) represents the well-known fact that the bulk of the Galactic mass is in the form of stars. Nevertheless, detailed models show that the induced ratio of perturbed gas to stellar surface densities is of the order

$$\sigma_g' \approx (2/3)\sigma_*' \qquad (30.61)$$

in the solar neighborhood. It is still possible for galaxies that have relatively little gas content to possess spiral structure. Observationally, however, such structure would be exceedingly difficult to detect, since what we normally see are the young hot stars embedded in gas and dust. If these are rare, then the underlying spiral structure of the disk's gravitational field and in the motion of disk-population stars, though present, may not be observed. Finally, stability analysis shows that if stellar dispersion velocities are large, then spiral patterns as described here may not develop. This is consistent with the fact that elliptical galaxies having no observed spiral structure contain stars having large dispersion, and that stellar dispersion in spirals tends to be small.

Problem 30.14. Consider a gaseous disk in which the maximum and minimum surface densities due to a spiral perturbation are

$$\hat{\sigma}_{max} = \hat{\sigma}_0 + \hat{\sigma} \quad \text{and} \quad \hat{\sigma}_{min} = \hat{\sigma}_0 - \hat{\sigma}.$$

(a) Show that the radial gas velocity u, pitch angle i, and pattern speed Ω_p satisfy the relation

$$\hat{u} = \frac{\hat{\sigma}_{max} - \hat{\sigma}_{min}}{\hat{\sigma}_{max} + \hat{\sigma}_{min}} r(\Omega - \Omega_p) \tan i. \qquad (30.62)$$

(b) Suppose that the density contrast $\sigma_{max}/\sigma_{min} = 3$, that $i = 5°$, and that $\hat{u} = 6$ km/sec. What is the pattern speed?

Problem 30.14 suggests at least one way in which a value for the parameter Ω_p could be found observationally. The numerical values used in part (b) are reasonable estimates for the solar neighborhood.

Next, consider the effect that spiral structure should have on the basic rotation curve $\Theta(r)$, which is shown in Figure 30.7 for the unperturbed disk.

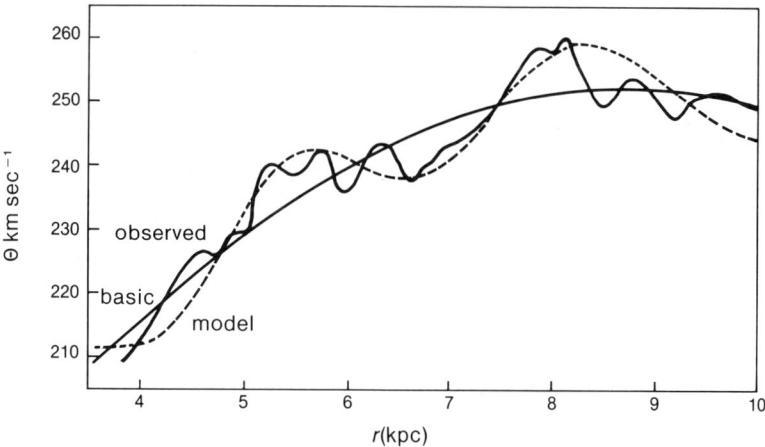

Figure 30.7. Rotation curve in disk with spiral arms. The basic (unperturbed) rotation curve is perturbed to give the dashed curve, as in equation (30.63). The observed rotation curve including spiral and local features is also shown.

According to (30.34) the perturbation in the velocity component v along the θ direction has the general form (in the linear theory)

$$v = \hat{v} e^{i(\omega t - 2\theta + \Phi)}.$$

For simplicity, adopt the equiangular spiral (30.17). Then the perturbed velocity in the θ direction will be

$$\Theta(r) = r\Omega(r) + v$$
$$= r\Omega(r) + v \cos(\Lambda \ln r/r_0). \quad (30.63)$$

The observed velocity curve should consist of the circular velocity $r\Omega(r)$, upon which is superimposed a small periodic variation with slowly increasing period in the r direction. Just such behavior is observed in 21.1-cm absorption studies, as shown in Figure 30.7. The basic curve and the actual curves are shown solid, and a model using density-wave theory analogous to (30.63) yields the dashed curve. The perturbation v represents about a 3 percent effect across the disk, except near the Lindblad radii.

It has already been emphasized that density-wave theory offers a way out of the problems associated with the classical winding dilemma for material arms. There remains the question of whether the model is consistent with the observed existence of stellar associations that have ages of order 10^7 years, and are located near spiral arms, in which they presumably originated. This would present little problem in the material arm model. According to density-wave theory, however, the individual stars move through the spiral pattern. If the theory is to be compatable with observations of associations, the individual stars, despite their motion, must remain near a spiral feature for periods of order 10^7 years. To see that this is indeed the case, consider a young star S formed in the spiral arm segment A, A', as seen by an observer moving with the pattern (angular velocity Ω_p), as shown in Figure 30.8. After time Δt, the star will have moved along its circular orbit (to first approximation) to the new position S'''. In the same time, another star formed at S' when S was formed will, because of differential rotation, have moved a shorter distance, to S'''. Although the stars may have traveled a significant distance in their circular orbits, they are still relatively close to the spiral feature, at least if Δt is not too large. As shown by Problem 30.15, this phenomenon of stellar migration is consistent with what is known about stellar associations.

Problem 30.15. A young star has recently formed out of the relatively dense matter in a spiral arm A, A' (Figure 30.8). Calculate how far it will have moved in time Δt away from its original position S to the new position S' because of galactic rotation. Show that the star is still close enough to be associated with the original spiral feature if $\Delta t = 10^7$ years, $\Omega_p = 11$ km sec^{-1} kpc^{-1}, and the pitch angle $i = 5°$.

30.4. OBSERVATIONAL CONSEQUENCES

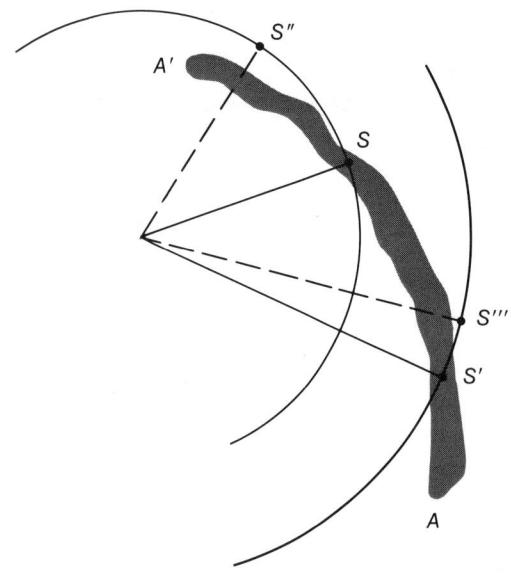

Figure 30.8. Stellar associations and spiral arms. Young stars in stellar associations formed in spiral arms remain near the spiral feature despite having traveled a significant distance in their galactic orbits.

The fact that spiral shock waves develop in association with nonlinear density-wave theory suggests an important triggering mechanism for protostar collapse. For a given density-wave pattern and disk structure, there exists a corotation point, r_c, beyond which the pattern speed exceeds the local velocity of matter given by $r\Omega(r)$. When $r < r_c$, the gas moves into and eventually through the spiral arms. In Section 30.3 it was noted that trailing spiral waves tend to develop density perturbations of steepening amplitude. Numerical calculations indicate that eventually shock waves will develop that have spiral character. Figure 30.9 shows the total gas density (normalized to the unperturbed level), and the perturbing spiral gravitational potential, as functions of angle within the disk. In general, the gas density does not exceed the unperturbed value by much except in a relatively narrow zone immediately behind the shock wave, which appears here as a nearly discontinuous increase in density. The shock front actually develops somewhat ahead of the minimum in the gravitational perturbation. The width of the spiral arm corresponds to the region where $\sigma_g/\sigma_{0g} > 1$, and is seen to extend well beyond the shock wave. The bright HII portion of the arms (shaded portion of the density plot) lies immediately inside the HI region, and is fairly narrow. The picture that emerges is indicated by the schematic view of the disk shown in Figure 30.10. The average density in the spiral arms is only slightly larger than that between arms, and their width is comparable to the interarm spacing. The inner edge of the arm is

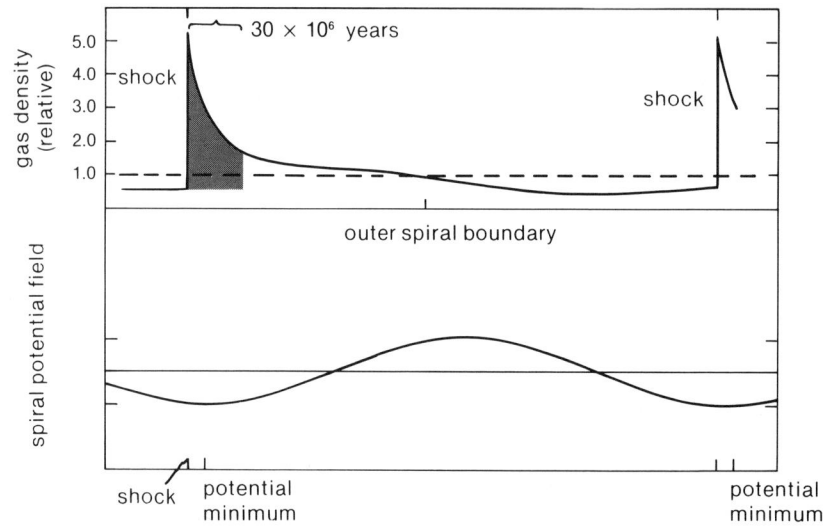

Figure 30.9. Gas density (upper) and spiral gravitational field (lower) in galactic disk versus distance normal to spiral arms. The shock front lies just inside the spiral potential minimum. Gas moves into the shock front (which may trigger star formation). Newly born stars and HII regions lie just behind the shock.

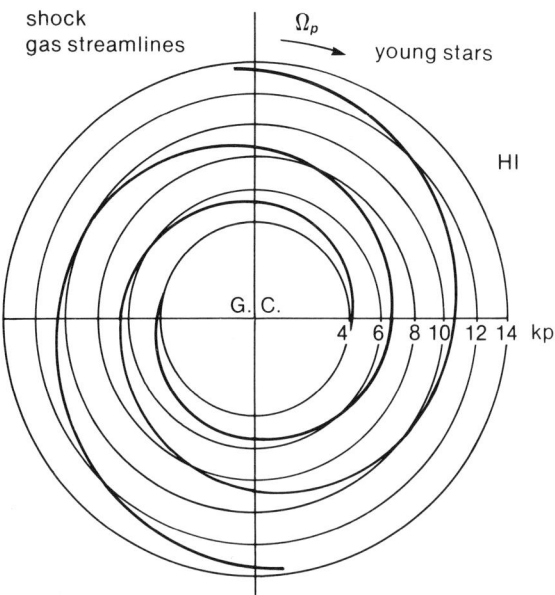

Figure 30.10. Two-armed spiral pattern, showing location of shock front, young stars, and HII regions.

defined by the shock wave, which increases the local gas density by a relatively large factor. Immediately behind the shock front is a region of HII and young stars, extending typically less than half-way into the arm.

A significant prediction of the density-wave model is that bright, young OB stars and associated HII regions, which are the optically prominent delineators of spiral-disk structure, occur along the inner edges of the arms. This is somewhat surprising if one assumes that star formation takes place in the center of spiral arms, where presumably the total mass density (gas plus stars) is a maximum. The explanation, however, is simple. The formation of protostars requires density perturbation in the interstellar gas corresponding to a factor of five or more. This can easily be supplied by the spiral shock wave, which compresses the relatively cool interarm gas to sufficient densities that collapse should be possible. The newly formed protostars then migrate a relatively short distance into the arm before developing into main-sequence stars. For O and B stars, the main-sequence lifetimes are less than 10^8 years. The relatively slow migration rate of these stars and their short lifetime represent the ultimate mechanisms confining them to the inner portions of a spiral arm.

Recently, it has been found that hot, young stars and HII regions in Sc galaxies are preferentially located along the inner edges of the spiral arms. This conclusion is also supported by observations within the Galaxy.

Problem 30.16. How long would it take a newly formed star to migrate across a spiral arm in the Galaxy, assuming that it moves along a circular orbit (the original orbit of the gas cloud from which it condensed). Evaluate for a distance of 10 kpc from the center, a pattern speed is 12.5 km/sec-kpc and a spiral arm width of 2 kpc. Assume a spiral pitch angle $\tan i \approx \sin i \approx 1/7$.

Chapter 31

GALACTIC EVOLUTION

The diversity in type and content of normal galaxies has been noted in Chapter 27. To this diversity must be added the various classes of radio galaxies (Figure 31.1), Seyfert galaxies (Figure 31.2), a more detailed breakdown of irregular galaxies (Figure 31.3), and quasars (Figure 31.4). The continuing accumulation of radio, infrared, and x-ray data about galaxies has revealed even more structure and evidence of more violent activity than had been previously imagined. Some of these phenomena are undoubtedly due to the formation process, but others are probably due to evolutionary processes in the galaxies themselves. The theory of stellar evolution is much more advanced than the theory of galactic evolution. The unresolved issues in stellar evolution are largely concerned with the details of star formation, the outcome of evolution during particular stages (pulsational stars; novae and supernovae outbursts; and the final stages of stellar collapse), or the fate of stellar systems that are intrinsically two-dimensional (rotation or magnetic fields). Galactic evolution, on the other hand, is still attempting to define and identify the fundamental questions to be answered.

Theory and observations have focused attention on several key issues that the theory of galaxy formation must address. For example, did galaxies condense out of the cosmic medium (as was assumed in Chapter 29)? If so, did all galaxies form during the same epoch, or have there been various stages of galactic formation? We would also like to know if galaxy formation is an ongoing process today. Not all types of galaxies need have been formed by a condensation process; instead, they may have been formed by explosive processes in regions of extremely high density in the early universe. If galaxies did condense out of the cosmic background, then we must explain what triggered their formation. The observed diversity in morphological type may not be due entirely (or at all) to evolutionary processes; if it is not, then what are the key initial parameters, such as mass, angular momentum or turbulence, that determine the galactic type? Finally, we might wish to explain why galaxies are generally, though not always, found in pairs or groups.

Once formed, galaxies will evolve, because they are dynamical systems whose members will continue to exchange energy until the system becomes as tightly bound as possible. Unfortunately, we are not certain what this ultimate state is, or whether it is unique. The processes of galactic evolution may differ from disk systems to elliptical systems. Theory must also explain

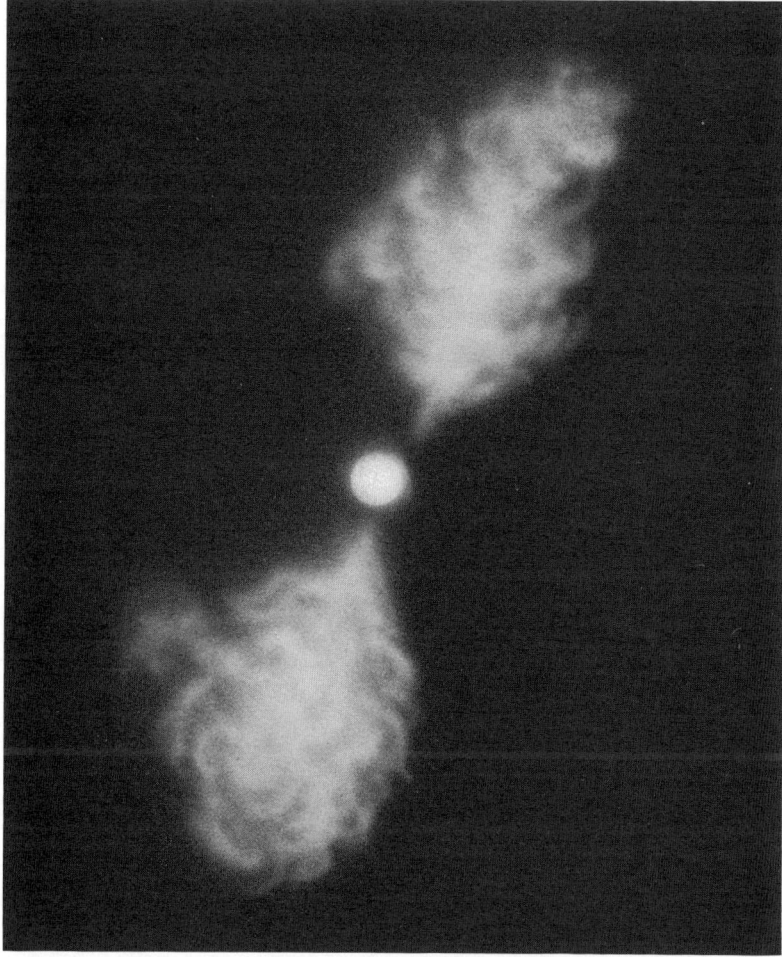

Figure 31.1. Typical radio galaxy, showing central (visible) galaxy and double-lobe structure.
The lobes contain tenuous plasma and magnetic fields that are moving away from the galaxy and that are a primary site of radio emission.

the violent phenomena that are observed in galactic nuclei, their relation (if any) to evolutionary changes in the nonnuclear components of the galaxy, and whether they occur in all galaxies during specific evolutionary stages or only in certain types of galaxies. If they do occur in a specific galaxy, we should find out if they are recurrent.

As improved technology advances the frontiers of the observable universe to greater and greater distances, we expect to see systems that are younger than those nearby. Except perhaps for quasars, the most distant galaxy known to date is 3C 427.1 (Figure 31.5), which has a redshift $z = 1.175$; its light appears very much like that of nearby elliptical galaxies. This galaxy, about 3×10^3 Mpc distant, must have been only about 6×10^9 years old when the light we see from it was emitted. Eventually, we may see galaxies at great enough distances that their appearance corresponds to the epoch of their formation and early evolution. If the redshifts observed for quasars (typically $0.1 \lesssim z \sim 3$) are due to cosmological expansion, then these are the most distant objects known, and may represent the early stages of evolution in galactic nuclei.

Several cosmological epochs can be readily identified as probably being important stages in the formation and evolution of galaxies in Big Bang models of the universe. At $z_{eq} \sim 10^3$ to 10^4, the local energy

Figure 31.2. H_α photograph of the Seyfert galaxy NGC 1275 showing similar wispy structure. Note the similarity with the H_α photograph of the Crab Nebula in Figure 25.6. Although the two objects differ in mass by a factor of 10^{11}, their similar appearance suggests both are explosive. Photograph courtesy of Kitt Peak National Observatory.

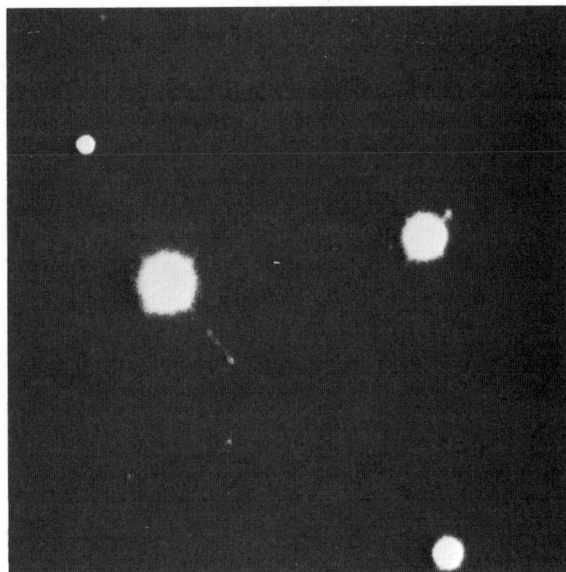

Figure 31.4. This long exposure of the quasar 3C 273 shows a jet of radiating material. The central object is unresolvable on shorter exposures. Photograph courtesy of Kitt Peak National Observatory.

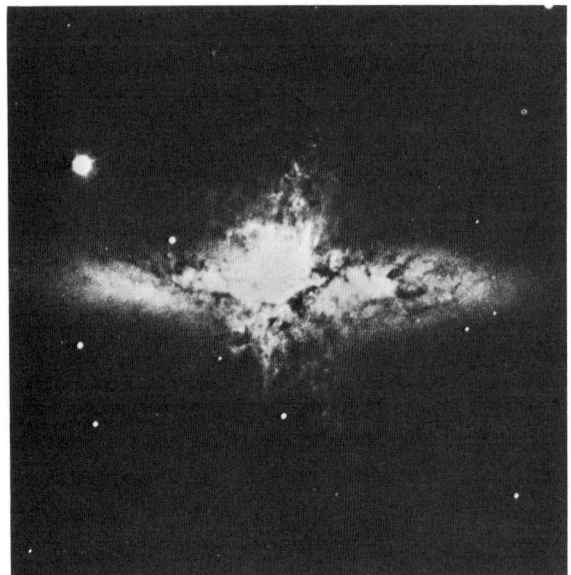

Figure 31.3. H_α photograph of the galaxy M 82. Photograph courtesy of Hale Observatories.

density of matter and radiation are equal. The matter is ionized and is coupled by Thomson scattering to the radiation field. Any protogalactic fluctuation in the matter density will consequently be frozen in magnitude during this epoch. At recombination ($z \sim 10^3$), when matter and radiation decouple, density fluctuations containing more than 10^5 to 10^6 $M_\odot$ will begin to grow in amplitude. Since the universe is expanding, galaxies must presumably have been closer together in the past. Present-day galaxies would overlap at a redshift given by

$$1 + z \approx a_0/a_m = d_g/R_g \approx 10^2,$$

where a typical distance between galaxies, d_g, is one Mpc, and an average galactic radius, R_g, is 10 kpc. Galaxies must have formed *after* this epoch if the standard Big Bang model of the universe is correct. Numerical models of galaxy formation (up to the formation of the first generation of stars) indicate that protogalaxies form within 10^8 to several times 10^9 years after the Big Bang, corresponding to a redshift of about $z_f \sim 3$ to 25. Finally, the observed redshifts of quasars (which may be the nuclei of primeval galaxies) range from epoch $z_Q \sim 0.1$ to about 3.

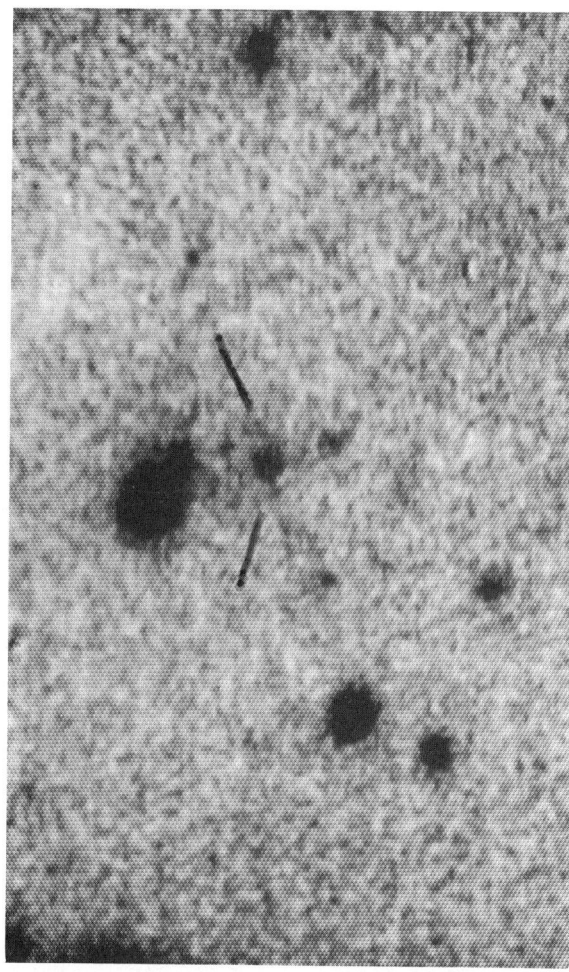

Figure 31.5. Red filter 13-minute exposure of 3C 427.1. The optical galaxy is associated with a radio source. Spectral observations indicate a red shift of 1.175, the highest yet found for a galaxy.

31.1. Formation of Galaxies

Theories of galaxy formation differ in detail, but most assume that galaxies evolved from perturbations present at recombination ($z \sim 10^3$). There is no consensus on the nature of these perturbations, or on how they arose. According to current theories, it is unlikely that the initial perturbations developed out of an otherwise uniform and featureless universe (such as one of the Friedmann models described in Section 26.6). For example, consider a volume of matter V that contains at any instant $\hat{N}$ baryons (instantaneous values of fluctuating variables are denoted by ^). As in any thermodynamic system, there will be continual random fluctuations in the macroscopic variables (such as $\hat{N}$), even though the system as a whole remains uniform and isotropic when averaged over time. In fact, the average fluctuation in density can be shown to be given by

$$\frac{\langle \delta \hat{\rho}^2 \rangle}{\rho^2} = \frac{\langle (\hat{\rho} - \rho)^2 \rangle}{\rho^2} = \frac{kT}{V} \kappa_T, \quad (31.1)$$

where ρ, the average mass density, equals $m_H N/V$, T is the average temperature (the temperature of the matter in the usual sense), and κ_T is the isothermal compressibility. If we assume that the matter can be described by the ideal gas equation of state, (31.1) implies that

$$\frac{\langle \delta \hat{\rho}^2 \rangle^{1/2}}{\langle \hat{\rho}^2 \rangle^{1/2}} = \frac{\langle \delta \hat{\rho}^2 \rangle^{1/2}}{\rho} = \frac{\langle \delta \hat{N}^2 \rangle^{1/2}}{N} = N^{-1/2}. \quad (31.2)$$

A galaxy of mass $M \sim 2 \times 10^{12} M_\odot$ contains about 2×10^{69} baryons; so a density fluctuation large enough to become a galaxy must be of order

$$\frac{\langle \delta \hat{\rho}^2 \rangle^{1/2}}{\rho} \approx 2 \times 10^{-35}. \quad (31.3)$$

The observed density contrast between galaxies and their surroundings is of order $\delta \rho / \rho = 10^6$, indicating that an initial perturbation of order (31.3) must have grown by a factor of about 5×10^{40} in about 10^9 years to become a galaxy (recall that apparently normal galaxies have been observed that are about 10×10^9 light years distant). Growth by such a large factor seems impossible in most models of the universe. A possible exception is the Lemaître universe (which assumes a nonzero cosmological constant). In this model there are two expansion epochs, separated by a period during which little expansion occurs and whose extent can be arbitrarily long. During this intermediate stage, galaxy formation could occur with little difficulty. Other models have also been developed, but most appear to raise more questions than they answer.

Other approaches assume that the perturbations are vestiges of the Big Bang. These theories range from slightly modified Friedmann universes, which may contain primeval turbulence or magnetic fields, to anisotropic cosmologies, or to theories asserting that the very early universe was chaotic, and that all but selected irregularities (those that became galaxies) were damped out prior to recombination. We will next consider a modified version of the Jeans theory, not because it is free from difficulties, but because the others make as yet unverifiable assumptions about the physics during the earliest epochs.

Jeans Mass in an Expanding Universe

The Jeans model of galaxy formation does not explain how protogalactic density perturbations arose. Instead, it attempts to explain when density perturbations become unstable and how rapidly they will grow once they become unstable. As in protostar formation, the Jeans model relies on gravitational instability to initiate collapse; the Jeans mass M_J separates configurations that are stable ($M_J > M$) from those that are unstable, $M > M_J$ (Section 23.2). The situation here is somewhat different, because of possible relativistic effects and because of the expansion of the medium in which the perturbations form. Expansion does not affect the magnitude of the Jeans mass M_J, but relativistic effects can be important. In Newtonian theory, M_J may be written as

$$M_J = (4\pi m_H n/3)(\pi v_s^2/G\rho)^{3/2}, \quad (31.4)$$

where n is the total baryon number density, v_s is the adiabatic sound speed, and ρ is the mass density. When the local energy density exceeds ρc^2, relativistic effects become important; the Jeans mass is then given by (31.4) with ρ replaced by $(\rho + P/c^2)$ (see Section 16.3). Then

$$M_J = \frac{4\pi m_H n}{3}\left(\frac{\pi v_s^2}{G(\rho + P/c^2)}\right)^{3/2}, \quad (31.5)$$

and the sound speed is given by

$$1/v_s^2 = \left(\frac{\partial(\rho + P/c^2)}{\partial P}\right)_s. \quad (31.6)$$

In this situation the square of the sound speed is the change in pressure with respect to total energy density at fixed entropy. In the Newtonian limit, $P/c^2 \ll \rho$, and (31.5) reduces to (31.4). The Jeans mass depends on n and T, which change with expansion. Consider the universe before recombination (10^{10} K $\gtrsim T \gg 4{,}000$ K) during which radiation dominates matter. The principal constituents in this period are photons and nonrelativistic nucleons, and the equation of state of the universe is approximately

$$\begin{aligned} u &= nm_H c^2 + aT^4, \\ P &\simeq P_{\rm rad} = (1/3)aT^4. \end{aligned} \quad (31.7)$$

The energy density includes contributions from matter and radiation, n is the baryon number density, and the pressure of the nucleon gas has been neglected. We also need the entropy in order to evaluate v_s, since the expansion can be considered to be adiabatic as long as there are no energy sources present (quasars, for example, may act as energy sources for epochs around $z = 3$). The entropy per baryon is due predominantly to radiation; from the first law of thermodynamics, it is given by

$$s = \frac{u + P}{nT} \simeq \frac{4aT^3}{3n}. \quad (31.8)$$

During the expansion, s remains constant and T^3 decreases with n, which is consistent with the result of the Friedmann model (Section 26.6). Evaluating v_s^2 from (31.6–8) yields

$$v_s^2 = \frac{1}{3}\left(\frac{Ts}{Ts + m_H c^2}\right). \quad (31.9)$$

Therefore the Jeans mass during the radiation-dominated epoch is given by (31.5), (31.7), and (31.9):

$$\begin{aligned} M_J &= \frac{4\pi}{3m_H^2}\left(\frac{\pi}{3G}\right)^{3/2}\left(\frac{3}{4a}\right)^{1/2}\frac{s^2}{(1 + Ts/m_H c^2)^3} \\ &\simeq 9.0\left(\frac{s}{k}\right)^2\left(1 + \frac{sT}{m_H c^2}\right)^{-3} M_\odot. \end{aligned} \quad (31.10)$$

If we adopt $\sigma = s/k = 10^9$ from (26.98), the Jeans mass becomes

$$\begin{aligned} M_J &= \frac{9 \times 10^{18}\, M_\odot}{(1 + 9.2 \times 10^{-5} T)^3} \\ &\simeq 1.2 \times 10^{13}\, M_\odot/T^3; \\ T &\gg \frac{m_H c^2}{k\sigma} = 10^4\text{ K}. \end{aligned} \quad (31.11)$$

M_J rises rapidly with decreasing temperature [recall $T \sim a(t)^{-1}$ in the radiation-dominated Friedmann models], until the temperature reaches about 10^4 K, at which epoch the effects of baryon pressure become important. Here M_J rolls over, becoming essentially constant, $M_J \simeq 9 \times 10^{18}\, M_\odot$, until recombination. Figure 31.6 shows the qualitative form of M_J as a function of T during the expansion of the universe. Near recombination the radiation pressure becomes less important than the gas pressure, and the equation of state for the universe is approximately that of an ideal nonrelativistic gas:

$$u = nm_H c^2 + (3/2)nkT,$$
$$P = nkT. \qquad (31.12)$$

The speed of sound in this situation is simply

$$v_s^2 = \gamma(P/\rho) = \gamma(kT/m_H)$$
$$= (5/3)(kT/m_H), \qquad (31.13)$$

and the Jeans mass is

$$M_J = \frac{4\pi m_H n}{3}\left(\frac{5\pi kT}{3m_H G}\right)^{3/2}\left(nm_H c^2 + \frac{5}{2}nkT\right)^{-3/2}$$
$$\simeq \frac{4\pi m_H n}{3}\left(\frac{5\pi kT}{3m_H G}\right)^{3/2}\left(\frac{1}{m_H c^2 n}\right)^{3/2}. \qquad (31.14)$$

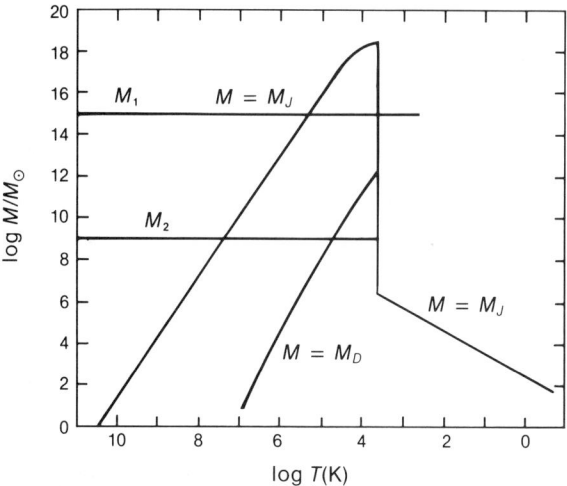

Figure 31.6. Jeans mass as a function of temperature during the expansion of the universe. See text for discussion.

At recombination this can be rewritten as

$$M_J \simeq 10^2\, \sigma^{1/2} M_\odot = 3.2 \times 10^6\, M_\odot. \qquad (31.15)$$

Recombination reduces M_J by about twelve orders of magnitude. After recombination the matter temperature decreases roughly as $a(t)^{-2}$, and

$$M_J \sim (T^3/n)^{1/2} \sim a^{-3/2} \sim T^{3/2} \qquad (31.16)$$

decreases gradually as the universe cools. The behavior of M_J as the universe cooled from 10^{10} K to the present is shown in Figure 31.6. Prior to recombination it rises rapidly with decreasing T, reaches a maximum between $10^{16}\, M_\odot$ and $10^{18}\, M_\odot$ (depending on the assumed value of the present density of the universe, ρ_0) just before recombination. As a result of recombination, M_J plummets to a value between $10^5\, M_\odot$ and $3 \times 10^6\, M_\odot$, and then gradually declines with continuing expansion.

Now consider a disturbance of mass M_1, as shown by the horizontal line in Figure 31.6. Since $M_1 > M_J$ during the epoch $T \gtrsim 10^{5.5}$ K, it is gravitationally unstable, and the amplitude

$$\delta(t) = \delta\rho/\rho \qquad (31.17)$$

will grow with time. During the epoch $10^{5.5}$ K $\gtrsim T \gtrsim$ 4,000 K, $M_1 < M_J$, and further growth will cease. Sufficiently massive disturbances, such as M_1, resemble sound waves during this period, and undergo little change in amplitude until $T \simeq 4,000$ K. The primary stabilizing influence during this period is radiation pressure. At recombination, the opacity of the universe drops as matter and radiation decouple, and the disturbance M_1 begins to grow in amplitude again, possibly becoming a cluster of galaxies, e.g., the Coma cluster. A similar development is expected for disturbances of mass $M \gtrsim 10^{12}\, M_\odot$. The fate of a mass $M_2 < 10^{12}\, M_\odot$ is expected to be different. For it the amplitude grows until $T \simeq 10^7$ K, then becomes a sound wave until $T \simeq 10^{5.5}$ K. At lower temperatures, radiation pressure produces a damping effect on the disturbance, and its amplitude begins to decay. If the mass M_2 is small enough, its amplitude will have decayed enough when recombination occurs that it could not become a stellar system by the present epoch. The boundary separating the regimes of fluctuating and damped disturbances is labeled M_D in Figure 31.6. According to this model, massive perturbations become clusters of galaxies, and the less massive ones become single galaxies. Furthermore, those of sufficiently low mass, perhaps 10^8 $M_\odot$ or less, decay before recombination, thereby setting a lower limit to the galactic mass spectrum.

Problem 31.1. What qualitative effect does the precise value of the present density of the universe ρ_0 have on the form of M_J?

Growth Rate

The masses of gravitationally unstable disturbances predicted by the Jeans model range from values typical of dwarf galaxies (10^9 $M_\odot$) up to clusters of galaxies (10^{16} $M_\odot$). The disturbances associated with protostar formation arise out of the interstellar medium within galaxies, where the effects of expansion are not significant. Their growth rate is exponential, $\delta(t) \simeq \delta(0) e^{\lambda t}$, where λ is positive. Disturbances in an expanding universe need not grow at the same rate. When relativistic effects are small, the growth rate may be found simply from Newtonian dynamics. Relativity, when important, changes the result quantitatively, but not qualitatively. We will therefore consider only the Newtonian case, where the behavior of a fluid element is described by the hydrodynamic equations

$$\frac{\partial \rho}{\partial t} + \nabla \cdot \rho \mathbf{v} = 0, \qquad (31.18)$$

$$\frac{\partial \mathbf{v}}{\partial t} + \mathbf{v} \cdot \nabla \mathbf{v} = -\frac{1}{\rho} \nabla P - \nabla \phi, \qquad (31.19)$$

Poisson's equation, and the Friedmann equations for the scale factor $a(t)$, (26.21), and (26.10). The evolution of an isotropic and homogeneous medium was considered in Chapter 26. Now suppose that a local density fluctuation develops within this medium. The increased density acts as the source in Poisson's equation, generating a gravitational field in its vicinity. As a result, nearby matter elements are accelerated toward the perturbation, and develop a velocity

$$\mathbf{v} = \frac{\dot{a}}{a} \mathbf{r} + \mathbf{u}(\mathbf{r},t). \qquad (31.20)$$

The first term is the characteristic uniform expansion in a Friedmann universe [see (26.7)], and the second term represents the local perturbation. The instantaneous position of a mass element $\mathbf{r}$ is defined as in Section 26.2,

$$\mathbf{r} = a(t)\,\mathbf{r}_0, \qquad (31.21)$$

where $\mathbf{r}_0$ is the Lagrangian coordinate. Because of the perturbation, the density depends on $\mathbf{r}$ and t. We will assume that its amplitude is small, and define the fractional change $\delta(\mathbf{r},t)$ by

$$\rho(\mathbf{r},t) = \rho(t)\,[1 + \delta(\mathbf{r},t)], \qquad (31.22)$$

where $\rho(t)$ is the usual mass density, which would appear in the Friedmann equations. If $\delta \ll 1$, then Poisson's equation becomes

$$\nabla^2 \phi = 4\pi G \rho_0 + \nabla^2 \phi_1 = 4\pi G \rho (1 + \delta)$$

or

$$\nabla^2 \phi_1 = 4\pi G \rho \delta. \qquad (31.23)$$

Next, we assume that the velocity induced by the perturbation is small compared with the expansion rate, or $|\mathbf{u}| \ll |\mathbf{v}|$, and substitute (31.20) and (31.22) into the continuity equation (31.18). Neglecting second- and higher-order terms, we find

$$\left(\dot{\rho} + \rho \nabla \cdot \mathbf{r}\,\frac{\dot{a}}{a}\right) + \left(\dot{\rho} + 3\rho\,\frac{\dot{a}}{a}\right)\delta$$
$$+ \rho\left[\dot{\delta} + \nabla \cdot \mathbf{u} + \frac{\dot{a}}{a}\mathbf{r} \cdot \nabla \delta\right] = 0.$$

The first quantity in parentheses is the continuity equation for the unperturbed medium (which we assume to be satisfied); so it vanishes. The second quantity in parentheses vanishes by (26.10); so we are left with

$$\left(\frac{\partial \delta(\mathbf{r},t)}{\partial t}\right)_\mathbf{r} + \frac{\dot{a}}{a}\mathbf{r} \cdot \nabla \delta(\mathbf{r},t) + \nabla \cdot \mathbf{u} = 0. \quad (31.24)$$

This expression is in terms of Eulerian coordinates $\mathbf{r}$ and t. It will be more convenient to work in terms of the Lagrange coordinate $\mathbf{r}_0$. The Lagrange coordinate is fixed with the mass element; so the time-rate of change in δ in the comoving frame is

$$\left(\frac{d\delta}{dt}\right)_{\mathbf{r}_0} = \left(\frac{\partial\delta}{\partial t}\right)_{\mathbf{r}} + \frac{\dot{a}}{a}\mathbf{r}\cdot\nabla\delta$$

(see the discussion of Lagrange coordinates in Section 21.1). Finally, using (31.21), we find that the divergence

$$\nabla\cdot\mathbf{u} = \frac{\partial u^i}{\partial x^i} = \frac{\partial x_0^k}{\partial x^i}\frac{\partial u^i}{\partial x^k} = a(t)^{-1}\frac{\partial u^i}{\partial x_0^i};$$

so the first-order correction to the continuity equation in comoving coordinates is simply

$$\left(\frac{\partial\delta}{\partial t}\right)_{\mathbf{r}_0} + \frac{1}{a}\frac{\partial u^i}{\partial x_0^i} = 0. \quad (31.25)$$

The same line of attack may be used to obtain the first-order correction to the equation of motion (31.19). In Eulerian coordinates $\mathbf{r}, t$, the result is

$$\left(\frac{\partial\mathbf{u}}{\partial t}\right)_{\mathbf{r}} + \frac{\dot{a}}{a}\mathbf{u} + \frac{\dot{a}}{a}\mathbf{r}\cdot\nabla\mathbf{u} = -\frac{kT}{m_H}\nabla\delta - \nabla\phi_1.$$

Transforming to comoving coordinates as in the preceding, we have, for each component of $\mathbf{u}$,

$$\left(\frac{\partial u^i}{\partial t}\right)_{\mathbf{r}_0} + \frac{\dot{a}}{a}u^i = -\frac{kT}{am_H}\frac{\partial\delta}{\partial x_0^i} - \frac{\partial\phi_1}{\partial x_0^i}. \quad (31.26)$$

Problem 31.2. Derive the equation of motion (31.26) in comoving coordinates. Assume that the fluid is an ideal gas, and that the change in pressure is isothermal.

The two equations (31.25) and (31.26) may be combined to give a single equation for the density fluctuation δ. In the simple case where pressure effects are negligible, we take the divergence of the momentum equation to get

$$\frac{\partial}{\partial t}\frac{\partial u^i}{\partial x_0^i} + \frac{\dot{a}}{a}\frac{\partial u^i}{\partial x_0^i} = -\frac{1}{a}\frac{\partial^2\phi_1}{\partial x_i^2},$$

and use the continuity equation to eliminate $\partial u^i/\partial x_0^i$,

obtaining

$$\left(\frac{\partial^2\delta}{\partial t^2}\right)_{\mathbf{r}_0} + 2\frac{\dot{a}}{a}\left(\frac{\partial\delta}{\partial t}\right)_{\mathbf{r}_0} - 4\pi G\rho(t)\delta = 0. \quad (31.27)$$

This may be solved once the form of the unperturbed scale factor $a(t)$ is known. If there were no expansion ($\dot{a} = 0$), growth would be exponential, with time constant: $\tau_0 \approx (\rho G)^{-1/2}$. An initial fluctuation $\delta \simeq 2 \times 10^{-35}$ would grow rapidly enough to appear today as a typical galactic density ($\delta \approx 10^6$). In general, the expansion rate is characterized by $(a/\dot{a}) = H(t)^{-1} \simeq (8\pi G\rho/3)^{-1/2}$, which is comparable to the growth rate τ_0, and the term proportional to $\dot{a}/a$ must be retained. In an Einstein–de Sitter universe, for example, $\dot{a}/a = (2/3t)$, and (31.27) reduces to

$$\ddot{\delta} + \frac{4\dot{\delta}}{3t} - \frac{2\delta}{3t^2} = 0;$$

the solution is easily seen to be

$$\delta(t) = \alpha t^{2/3} + \beta/t, \quad (31.28)$$

where α and β are constants. The solution to (31.27) may also be obtained analytically for the other ($k \neq 0$) Friedmann models, in which the pressure term is negligible. Analysis of these models shows that for $0.014 \lesssim q_0 \lesssim 2$, the ratio of the density constant now to its value at recombination satisfies

$$10^2 \lesssim \delta(t_0)/\delta(t_R) \lesssim 3\times 10^3. \quad (31.29)$$

Solutions can also be found that include the pressure term during early times [$a(t) \ll a(t_0)$], or that include the effects of general relativity. In these cases the density contrast varies essentially as a power of the time, $\delta(t) \approx t^\alpha$, with α of order unity.

Problem 31.3. Obtain the following equation for the density fluctuation when the pressure term can not be ignored:

$$\delta(\mathbf{r}, t) = \sum_{\mathbf{k}} \delta_{\mathbf{k}}(t) e^{i\mathbf{k}\cdot\mathbf{r}/a(t)},$$

$$\ddot{\delta}_{\mathbf{k}} + 2\frac{\dot{a}}{a}\dot{\delta}_{\mathbf{k}} = \left(4\pi G\rho(t) - \frac{4\pi^2 kT}{m_H\lambda^2}\right)\delta_{\mathbf{k}}, \quad (31.30)$$

where $\lambda = 2\pi k/a$ is the wavelength associated with the wave vector **k**. Note that the Jeans length is unaffected by the expansion.

When the pressure term is included, the differential equation for the density contrast is given by (31.30), where δ_k is the amplitude of the mode whose wave vector is **k** and whose wavelength is λ. As shown in Problem 31.3, λ is a function of the expansion parameter $a(t)$. A stability analysis of (31.30) shows that modes whose wavelengths exceed

$$\lambda_J = (\pi kT/\rho Gm_H)^{1/2}$$

are unstable toward continual growth, but those with $\lambda < \lambda_J$ oscillate. In linear theory, the mass associated with the unstable modes is of order

$$M_J = \frac{4\pi}{3} nm_H \lambda_J^3 = \frac{4\pi}{3} nm_H \left(\frac{\pi kT}{\rho Gm_H}\right)^{3/2},$$

which is just (31.4).

The analysis above is strictly applicable only in the linear regime ($\delta \ll 1$). When $\delta(t)$ is of order unity, nonlinear effects become important. It may be supposed that a galaxy forms not long after the beginning of the nonlinear phase, and that the subsequent growth is rapid. Supposing that the epoch $z \simeq 2$, when quasars appear, signals the onset of galaxy formation, an initial density increase should have grown at least to order unity by then. Since $\delta(t_0)$ as predicted by linear theory is comparable to δ at $z = 2$, we may use (31.29) with $\delta(t_0) \approx 1$ to estimate the density contrast at recombination; the result is

$$3 \times 10^{-4} \lesssim \delta(t_R) \lesssim 10^{-2}. \quad (31.31)$$

Whether or not fluctuations of this magnitude exist at recombination is an open question, but their existence might be observable. Suppose that the fluctuation is adiabatic; then the change in density is accompanied by a change in radiation temperature given by

$$\frac{\delta T_R}{T_R} \simeq \frac{1}{3}\frac{\delta n}{n} = \frac{1}{3}\frac{\delta \rho}{\rho} = \frac{1}{3}\delta(t). \quad (31.32)$$

Evaluating this at $T = T_R$, and noting the bounds in (31.31), we should find the fluctuation in radiation temperature to be in the range

$$3 \times 10^{-3} \lesssim (\delta T_R/T_R) \lesssim 10^{-4}$$

and to appear as local irregularities in the microwave background. As noted in Section 26.6, none have appeared down to several times 10^{-3} K, but satellite observations in the near future may yield sufficient resolution to test this model.

Problem 31.4. The angular size θ of a protogalactic fluctuation can be obtained from the results of Section 26.5. Show that

$$\theta \approx \frac{H_0(1 + z_R)}{2c}\lambda_J \quad (31.33)$$

for $q_0 = \frac{1}{2}$, where the angular diameter of the fluctuation is λ_J.

In fact, if $q_0 = 1$, and if the mass density ρ_0 is assumed to be 10^{-29} g/cm^3, then a fluctuation at recombination should have an angular size of order 30".

31.2. STELLAR POPULATIONS

After a volume of increased density separates dynamically from the expanding background, it will collapse essentially on a free-fall time-scale (about 10^8 year). What sort of galaxy results may depend on the initial density, the protogalaxy's rate of rotation, and its interaction with neighboring protogalaxies. The dissipationless collapse models of elliptical galaxies (Chapter 29) assume that star formation exhausted most of the initial gas content of the protogalaxy at the onset of collapse. In protogalaxies destined to become spirals, a significant amount of gas must have remained to form a disk after the collapse was complete. This suggests that star formation occurs throughout the dynamic collapse stage; each epoch of star formation is followed by mass loss from the most massive stars, with accompanying enrichment of the gaseous medium. In this way successive stellar populations form, each of increasing heavy-metal abundance, and each increasing the heavy-metal abundance of the interstellar medium. Observations of the ultraviolet excess, peculiar velocity, and angular momentum of subdwarfs in the solar neighborhood indicate that our Galaxy probably evolved in this way.

An important assumption in analyzing the characteristics of stars in the solar neighborhood involves the relation between ultraviolet excess and stellar age. Ultraviolet excess measures the underabundance of heavy metals relative to a normal star (one having the same $B - V$ color as the sample star). Heavy metals produce absorption lines, mostly at short wavelengths. These lines remove energy primarily from the ultraviolet portion of the stellar spectrum; so the star appears to emit less ultraviolet radiation than would a black body at the same effective temperature (that is, of the same spectral type). The young stars in the Hyades cluster (recently formed from heavy-element-enriched interstellar gas) are usually taken as standards. A star having the same $B - V$ color as a member of the Hyades, but lower Z, will appear to emit more energy in the ultraviolet. Since stars (such as subdwarfs) that formed earlier in the evolution of the galaxy are deficient in heavy metals, they should exhibit more ultraviolet excess than subdwarfs formed during the later stages in the galaxy's evolution. It is generally assumed, therefore, that the older a subdwarf is, the larger its ultraviolet excess should be.

Recall that stellar relaxation times for galaxies greatly exceed the age of the universe; thus stellar orbits determined by violent relaxation during the initial collapse of the protogalaxy should retain characteristics of their formation to the present day. In general, the orbits of newly formed stars will be governed by the galactic (or protogalactic) potential field and will not be closed; many will occupy noncircular orbits that precess around the galactic center. The definition of eccentricity for such orbits is arbitrary, but one useful choice is

$$e = \frac{R_A - R_P}{R_A + R_P}, \quad (31.34)$$

where R_A is the furthest distance traveled by the star from the galactic center (apogalacticum), and R_P is the radius of closest approach to the center (perigalacticum). The result is correct for an orbit that is a focal ellipse. If the galactic formation process is dynamic, then the orbits will tend to be highly noncircular. For example, consider a circular orbit at the onset of dynamic collapse of the protogalactic gas cloud. As mass falls through the star's orbit, the gravitational force acting on the star will increase, pulling it out of circular orbit into an orbit of greater effective eccentricity. If the collapse time is comparable to the orbital period, then the star will not be able to adjust its motion to establish a new circular orbit of smaller radius; instead it will continue to increase its effective eccentricity until the collapse is complete. These arguments suggest two possible correlations between effective orbital eccentricity and stellar age, depending on the relation between the collapse time t_c and the orbital period P. If galaxy formation was gradual ($t_c \gg P$), then there would be large numbers of old stars moving in essentially circular orbits about the galactic center. If formation was dynamic ($t_c \lesssim P$), then the oldest stars would lie in orbits of greatest effective eccentricity e.

Finally, if the protogalaxy was axisymmetric, and if angular momentum exchange did not occur between concentric shells of matter during the formation process, then the angular momentum distribution with mass should have been preserved. In fact, the angular momentum distribution of the inner part of M 31 is similar to that of a uniform mass distribution in rigid rotation. These observations suggest that the angular momentum h of a star about the galactic center reflects the angular momentum per unit mass of the gas from which it formed, and is unchanged by the formation process.

The preceding arguments may be applied to stars in the solar neighborhood to deduce characteristics of our Galaxy's early evolution. Dwarf stars are chosen for the analysis, since their lifetimes are comparable to the age of the Galaxy (several times 10^{10} years). The observed velocity components of dwarfs near the Sun can be used, in conjunction with a model Galactic potential, to construct orbits. Figure 31.7 shows segments of the orbits of three typical dwarf stars constructed in this way (the solar neighborhood is denoted by the circle through which all orbits pass, and the Galactic center by the cross). An eccentricity can be assigned to each orbit by (31.34), of roughly $e \sim 0.2$, 0.6, and 0.9 for the stars shown. The ultraviolet excess $\delta(U - B)$ observed for each star is 0.05, 0.17, and 0.26, respectively. Figure 31.8 extends the analysis of e versus $\delta(U - B)$ to 221 dwarfs in the solar neighborhood, and shows that the stars moving in the most eccentric orbits exhibit the greatest ultraviolet excess. Since heavy-metal abundance is believed to be lowest in the oldest stars, Figure 31.8 is evidence for a correlation between stellar age and orbital eccentricity, at least in dwarf stars. The absence of stars with low e and large $\delta(U - B)$ is believed to be real,

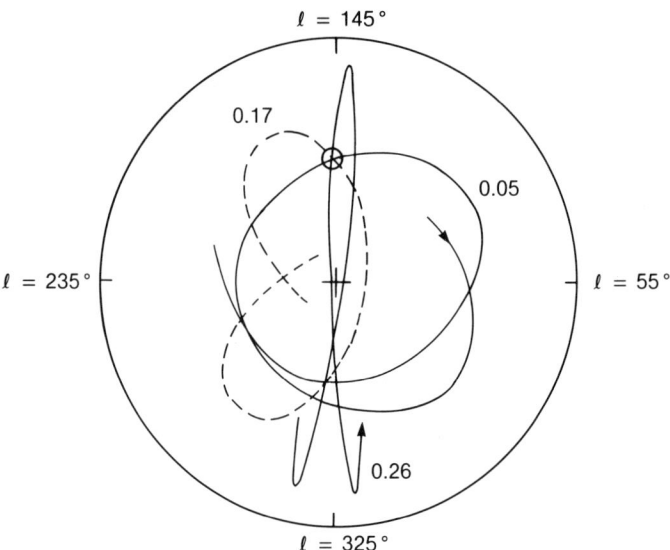

Figure 31.7. Segments of orbits of dwarfs near the Sun. See text for discussion.

and has been taken as evidence that the Galaxy's collapse occurred on a time-scale comparable to the orbital period at the Sun's Galactic radius, that is, $t_c \sim 10^8$ years. The lack of stars in the lower right-hand corner of Figure 31.8 (young stars with eccentric orbits) is consistent with the hypothesis that star formation occurs in disk gas that is essentially in circular motion about the galactic nucleus.

Figure 31.9 shows the angular momentum h of the stars versus ultraviolet excess. If we assume that stellar angular momentum was conserved during the Galaxy's formation, this figure is simply a different representation of the correlation shown in Figure 31.8. In the units at left, $h_\odot = 25 \times 10^2$ kpc km/sec, assuming $r_\odot = 10$ kpc. The angular momentum of young stars is large, as would be expected if they formed out of gas near $r_\odot$ that was moving in an essentially circular orbit about the Galactic center. The absence of young stars with low h is also to be expected. Consider the oldest stars, for which $h \simeq 12 \times 10^2$ kpc km/sec. If the gas from which these were formed were in circular orbit when the stars were born, they would have had circular orbits of radius less than about 5 kpc, and could not be in the solar neighborhood. Therefore they must have formed from gas that was not yet in equilibrium in the Galactic potential. This offers further support for the hypothesis that the protogalaxy's collapse was dynamic, producing highly eccentric orbits for the oldest stars. Furthermore, we see that between the formation of the oldest dwarfs and the youngest (which are at least 10^9 years old), the collapse reached its present equilibrium state. This amounts to about several times 10^8 years, which is comparable to the collapse time t_c. A collapse time of about 10^8 years is

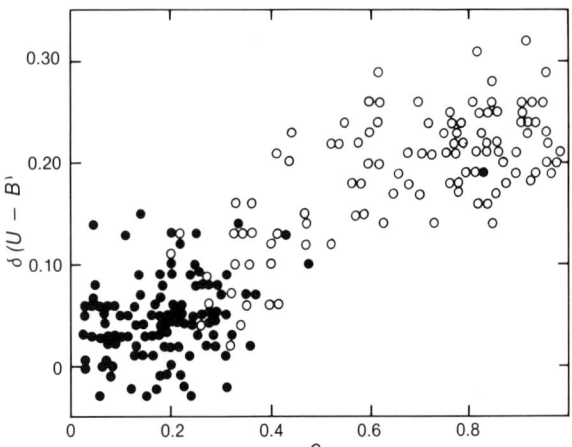

Figure 31.8. The correlation between the ultraviolet excess, $\delta(U-B)$, and the orbital eccentricity, e, for a sample of 221 stars. The filled and open circles represent stars from a first and a second catalogue, respectively. From O. J. Eggen, D. Lynden-Bell, and A. R. Sandage, *Astrophysical Journal*, **136** (1962), 748.

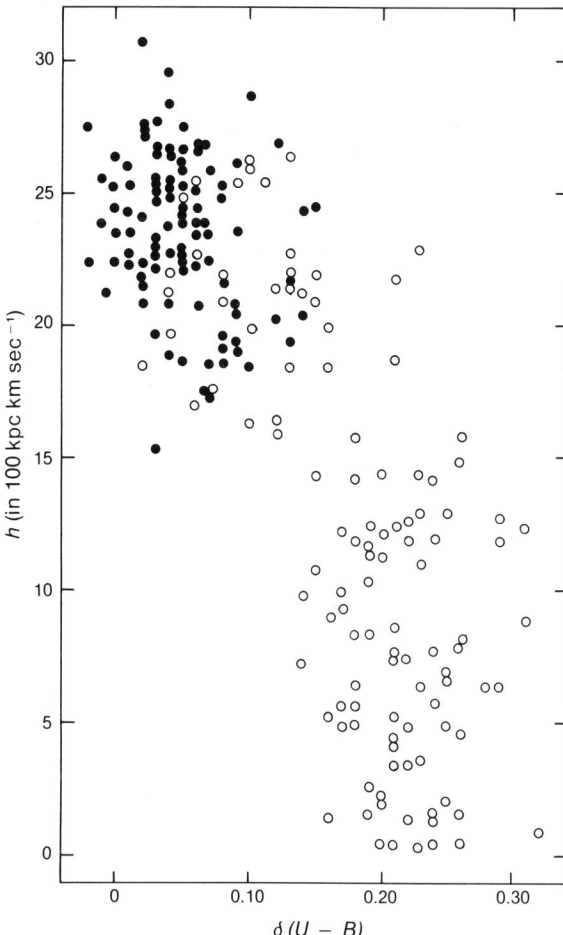

Figure 31.9. The correlation between the angular momentum, h, in units of 10^2 kpc km/sec and the ultraviolet excess of the stars shown in Figs. 31.8 and 31.10. From Eggen et al., op. cit.

consistent with the narrow range in composition variation observed in halo globular clusters, which is taken as evidence that they are of essentially the same age.

Problem 31.5. Show that a star in circular orbit that has the same angular momentum as the oldest dwarfs in the Galaxy must lie within 5 kpc of the Galactic center. Assume a Galactic mass distribution $m(r) \sim r$.

The distribution of ultraviolet excess with stellar velocity normal to the Galactic plane $|w|$ for the same set of dwarf stars is shown in Figure 31.10. On the right axis is given the maximum distance above the galactic plane, Z_{max}, that a star with velocity w can reach. We see that the youngest dwarfs [low $\delta(U-B)$] are confined to a region extending no more than 0.4 kpc above or below the Galactic plane. These stars evidently formed out of the disk material after the Galaxy attained its present state. The oldest dwarfs [large $\delta(U-B)$], on the other hand, occupy a region extending to about 10 kpc above and below the disk. According to the assumptions we have stated, these stars formed early in the collapse; so their maximum distance from the Galactic center sets a lower limit to the protogalaxy's extent when star formation began. The envelope bounding the unoccupied portion of the $\delta(U-B), |w|$-plane gives the height above the disk of the region below which star formation was possible at epochs corresponding to specific values of $\delta(U-B)$. The ratio of the greatest Z_{max} for young dwarfs, 0.4 kpc/10 kpc = 0.04, is a measure of the extent by which the Galaxy contracted normal to the plane during the epoch of star formation.

A lower limit may be placed on the degree of contraction in the plane of the Galaxy. Among dwarfs of the highest eccentricity, there are a few for which apogalacton exceeds 50 kpc. These dwarfs, being among the oldest, formed when the gas in the protogalaxy was in dynamic collapse and was streaming almost radially inward. After a first-generation star formed, the remaining gas from which it formed continued its infall, exchanging energy with neighboring gas streams, and finally cooling to a state of centrifugal equilibrium. We can assume that the angular momentum per unit mass h of the gas was conserved during this process; so it and the first-generation dwarf formed during the initial collapse would have the same h. Furthermore, a dwarf recently formed from the gas in centrifugal equilibrium would also have the same h. It follows from Figure 31.9 and reasonable models of the Galaxy's angular momentum distribution that such a newly formed star's circular orbit would lie within 5 kpc or so of the Galactic center. This implies that the Galaxy has contracted by at least a factor of ten since star formation began.

The evolutionary picture revealed by the analysis of the kinematic parameters and chemical composition of dwarf stars near the Sun may be summarized as follows. The first generation of stars condensed out of the protogalactic gas during the onset of dynamic collapse a few times 10^{10} years ago. Within about 10^8 years, collapse was halted, and the gas that had not

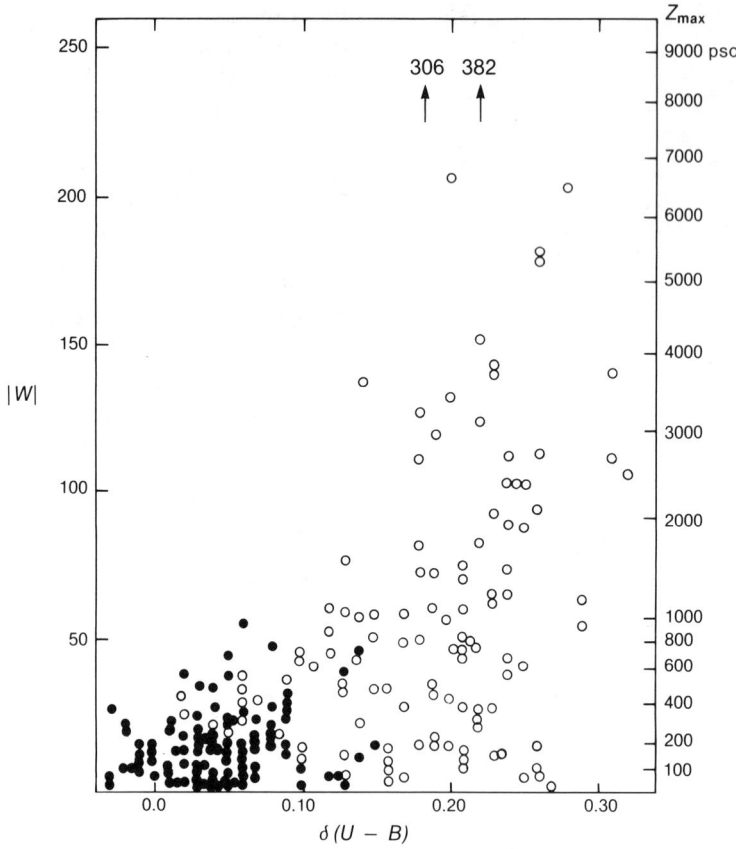

Figure 31.10. The correlation between the W-velocity, perpendicular to the galactic plane, and the ultraviolet excess for the 221 stars in sample. Filled and open circles as in 31.8. From Eggen et al., op. cit.

been consumed by star formation settled into an equilibrium determined by the balance between rotation and gravitation. In attaining this state, the protogalaxy shrank by a factor of about 25 normal to the disk, and by at least a factor of 10 in the radial direction. The subsequent evolution has involved star formation from disk gas.

Throughout earlier discussions we have classified stars as either Population I or Population II. The smooth distribution of e, h, and $|w|$ with $\delta(U - B)$, and thus with stellar age, during collapse shows that this is an oversimplification. The classification can be extended to include, for example, extreme Population II, intermediate Population II, and so on.

Appendix 1

CONSTANTS AND UNITS

Mathematical Constants

$\pi = 3.1416$
$e = 2.7183$
1 radian = 57.296 degrees
1 arc-sec = $1'' = 4.848 \times 10^{-6}$ radians
$\log e = 0.4343 = (\ln 10)^{-1}$

Physical Constants

Speed of light	$c = 2.9979 \times 10^{10}$ cm/sec
Gravitational constant	$G = 6.670 \times 10^{-8}$ dynes cm^2 g^{-2}
Planck's constant	$h = 2\pi\hbar = 6.626 \times 10^{-27}$ erg sec
	$\hbar = 1.055 \times 10^{-27}$ erg sec
Electric charge	$e = 4.803 \times 10^{-10}$ esu
Mass of the electron	$m_e = 9.110 \times 10^{-28}$ g
Mass of the proton	$m_p = 1.673 \times 10^{-24}$ g
Mass of the pion	$m_\pi = 2.49 \times 10^{-25}$ g
Boltzmann's constant	$k = 1.381 \times 10^{-16}$ erg deg^{-1}
	$= 8.617 \times 10^{-5}$ eV deg^{-1}
Avogadro's number	$N_0 = 6.022 \times 10^{23}$ mole^{-1}
Bohr radius	$a_0 = h^2/m_e e^2$
	$= 5.292 \times 10^{-9}$ cm
Bohr magneton	$\mu_B = he/4\pi m_e c = 9.274 \times 10^{-21}$ erg gauss^{-1}
Electron magnetic moment	$\mu_e = 1.001\, \mu_B$
Proton magnetic moment	$\mu_p = 1.521\, \mu_B$
Radiation constant	$a = 8\pi^5 k^4/15 c^3 h^3$
	$= 7.565 \times 10^{-15}$ erg cm^{-3} deg^{-4}
Stefan-Boltzmann constant	$\sigma = ac/4$
	$= 5.670 \times 10^{-5}$ erg cm^{-2} sec^{-1} deg^{-4}
Fine-structure constant	$\alpha = e^2/\hbar c = 1/137.04$
Classical electron radius	$e^2/m_e c^2 = 2.818 \times 10^{-13}$ cm
Compton wavelength of electron	$\lambda_e = \hbar/m_e c$
	$= 3.861 \times 10^{-11}$ cm
Thomson cross section	$\sigma_T = (8\pi/3)/(e^2/m_e c^2)^2$
	$= 6.652 \times 10^{-24}$ cm^2
1 eV	$= 1.16 \times 10^4$ K
	$= 1.602 \times 10^{-12}$ ergs
1 Rydberg	$m_e e^4/2\hbar^2 = 13.606$ eV
	$m_p/m_e = 1836$

Astronomical Constants

Astronomical unit	A.U. $= 1.496 \times 10^{13}$ cm
Parsec	pc $= 3.086 \times 10^{18}$ cm
	$= 3.261$ light years
Solar mass	$M_\odot = 1.989 \times 10^{33}$ g
Solar radius	$R_\odot = 6.960 \times 10^{10}$ cm
Solar luminosity	$L_\odot = 3.862 \times 10^{33}$ erg/sec
Solar absolute magnitude	$M_{b,\odot} = +4.77$
Solar effective temperature	$T_\odot = 5{,}800$ K
Sidereal year	$= 3.156 \times 10^7$ sec

Appendix 2

ATOMIC MASS EXCESSES

Z	Element	A	M − A, Mev
0	n	1	8.07144
1	H	1	7.28899
	D	2	13.13591
	T	3	14.94995
	H	4	28.22000
		5	31.09000
2	He	3	14.93134
		4	2.42475
		5	11.45400
		6	17.59820
		7	26.03000
		8	32.00000
3	Li	5	11.67900
		6	14.08840
		7	14.90730
		8	20.94620
		9	24.96500
4	Be	6	18.37560
		7	15.76890
		8	4.94420
		9	11.35050
		10	12.60700
		11	20.18100
5	B	7	27.99000
		8	22.92310
		9	12.41860
		10	12.05220
		11	8.66768
		12	13.37020
		13	16.56160
6	C	9	28.99000
		10	15.65800
		11	10.64840
		12	0
		13	3.12460
		14	3.01982
		15	9.87320
7	N	12	17.36400
		13	5.34520
		14	2.86373
		15	0.10040
		16	5.68510
		17	7.87100
8	O	14	8.00800
		15	2.85990
		16	−4.73655
		17	−0.80770
		18	−0.78243
		19	3.33270
		20	3.79900
9	F	16	10.90400
		17	1.95190
		18	0.87240
		19	−1.48600
		20	−0.01190
		21	−0.04600
10	Ne	18	5.31930
		19	1.75200
		20	−7.04150
		21	−5.72990
		22	−8.02490
		23	−5.14830
		24	−5.94900

Z	Element	A	M − A, Mev	Z	Element	A	M − A, Mev
11	Na	20	8.28000	19	K	36	−16.7300
		21	−2.18500			37	−24.8100
		22	−5.18220			38	−28.7860
		23	−9.52830			39	−33.8033
		24	−8.41840			40	−33.5333
		25	−9.35600			41	−35.5524
		26	−7.69000			42	−35.0180
12	Mg	22	−0.14000			43	−36.5790
		23	−5.47240			44	−35.3600
		24	−13.93330			45	−36.6300
		25	−13.19070			46	−35.3400
		26	−16.21420			47	−36.2500
		27	−14.58260	20	Ca	38	−21.6900
		28	−15.02000			39	−27.3000
13	Al	24	0.1000			40	−34.8476
		25	−8.9310			41	−35.1400
		26	−12.2108			42	−38.5397
		27	−17.1961			43	−38.3959
		28	−16.8554			44	−41.4596
		29	−18.2180			45	−40.8085
		30	−17.1500			46	−43.1380
14	Si	26	−7.1320			47	−42.3470
		27	−12.3860			48	−44.2160
		28	−21.4899			49	−41.2880
		29	−21.8936	21	Sc	40	−20.9000
		30	−24.4394			41	−28.6450
		31	−22.9620			42	−32.1410
		32	−24.2000			43	−36.1740
15	P	28	−7.6600			44	−37.8130
		29	−16.9450			45	−41.0606
		30	−20.1970			46	−41.7557
		31	−24.4376			47	−44.3263
		32	−24.3027			48	−44.5050
		33	−26.3346			49	−46.5490
		34	−24.8300			50	−44.9600
16	S	30	−14.0900	22	Ti	42	−25.1230
		31	−18.9920			43	−29.3400
		32	−26.0127			44	−37.6580
		33	−26.5826			45	−39.0020
		34	−29.9335			46	−44.1226
		35	−28.8471			47	−44.9266
		36	−30.6550			48	−48.4831
		37	−27.0000			49	−48.5577
		38	−26.8000			50	−51.4307
17	Cl	32	−12.8100			51	−49.7380
		33	−21.0140			52	−49.5400
		34	−24.4510	23	V	46	−37.0600
		35	−29.0145			47	−42.0100
		36	−29.5196			48	−44.4700
		37	−31.7648			49	−47.9502
		38	−29.8030			50	−49.2158
		39	−29.8000			51	−52.1989
		40	−27.5000			52	−51.4360
18	Ar	34	−18.3940			53	−52.1800
		35	−23.0510			54	−49.6300
		36	−30.2316	24	Cr	48	−42.8130
		37	−30.9509			49	−45.3900
		38	−34.7182			50	−50.2490
		39	−33.2380			51	−51.4472
		40	−35.0383			52	−55.4107
		41	−33.0674			53	−55.2807
		42	−34.4200			54	−56.9305

Z	Element	A	M − A, Mev	Z	Element	A	M − A, Mev
		55	−55.1130			62	−66.7480
		56	−55.2900			63	−65.5160
25	Mn	50	−42.6480			64	−67.1060
		51	−48.2600			65	−65.1370
		52	−50.7020			66	−66.0550
		53	−54.6820	29	Cu	58	−51.6590
		54	−55.5520			59	−56.3590
		55	−57.7048			60	−58.3460
		56	−56.9038			61	−61.9840
		57	−57.4800			62	−62.8130
		58	−55.6500			63	−65.5831
26	Fe	52	−48.3280			64	−65.4276
		53	−50.6930			65	−67.2660
		54	−56.2455			66	−66.2550
		55	−57.4735			67	−67.2910
		56	−60.6054			68	−65.4100
		57	−60.1755	30	Zn	60	−54.1860
		58	−62.1465			61	−56.5800
		59	−60.6599			62	−61.1230
		60	−61.5110			63	−62.2170
		61	−59.1300			64	−66.0003
27	Co	54	−47.9940			65	−65.9170
		55	−54.0140			66	−68.8810
		56	−56.0310			67	−67.8630
		57	−59.3389			68	−69.9940
		58	−59.8380			69	−68.4250
		59	−62.2327			70	−69.5500
		60	−61.6513			71	−67.5200
		61	−62.9300			72	−68.1440
		62	−61.5280	31	Ga	63	−56.7200
		63	−61.9200			64	−58.9280
28	Ni	56	−53.8990			65	−62.6580
		57	−56.1040			66	−63.7060
		58	−60.2280			67	−66.8650
		59	−61.1587			68	−67.0740
		60	−64.4707			69	−69.3262
		61	−64.2200			70	−68.8970

SOURCE: D. Clayton, *Principles of Stellar Evolution and Nucleosynthesis* (New York: McGraw-Hill, 1968)

Bibliography

General

G. O. Abell, *Exploration of the Universe,* 2nd ed. (New York: Holt, Rinehart & Winston, 1975).

C. W. Allen, *Astrophysical Quantities* (London: Athlone Press, 1973).

F. H. Shu, *The Physical Universe* (Mill Valley, Calif.: University Science Books, 1982).

M. Harwit, *Astrophysical Concepts* (New York: Wiley, 1973).

F. Hoyle and J. Narlikar. *The Physics-Astronomy Frontier* (San Francisco: W. H. Freeman, 1980).

W. K. Rose, *Astrophysics* (New York: Holt, Rinehart & Winston, 1973).

A. Unsold, *The New Cosmos* (New York: Springer-Verlag, 1977).

VOLUME I

PART 1: INTRODUCTION

Chapter 1: An Overview of Stellar Structure and Evolution

I. S. Shklovskii, *Stars: Their Birth, Life and Death* (San Francisco: W. H. Freeman, 1978).

Chapter 2: Properties of Matter

H. B. Callen, *Thermodynamics* (New York: Wiley, 1961).

F. Reif, *Fundamentals of Statistical and Thermal Physics* (New York: McGraw-Hill, 1965).

G. H. Wannier, *Statistical Physics* (New York: Wiley, 1966).

Ya. B. Zeldovich and I. D. Novikov. *Relativistic Astrophysics* (Chicago: University of Chicago Press, 1971).

Chapter 3: Aspects of Observational Astronomy

G. O. Abell, *Exploration of the Universe,* 2nd ed. (New York: Holt, Rinehart & Winston, 1975).

E. E. Salpeter, "The Luminosity Function and Stellar Evolution." *Ap. J.,* **121** (1955): 161.

T. L. Swihart, *Astrophysics and Stellar Astronomy* (New York: Wiley, 1968).

R. Trumpler and W. Weaver, *Statistical Astronomy* (Berkeley, Calif.: University of California Press, 1953).

PART 2: STELLAR STRUCTURE

General

J. P. Cox and R. T. Guili, *Stellar Structure,* Vol. I (New York: Gordon and Breach, 1968).

H. Y. Chiu, *Stellar Physics* (Waltham, Mass.: Blaisdell, 1968).

L. Motz, *Astrophysics and Stellar Structure* (Waltham, Mass.: Ginn, 1970).

E. Novotny, *Introduction to Stellar Atmospheres and Interiors* (Oxford, England: Oxford University Press, 1973).

Chapter 4: Static Stellar Structure

S. Chandrasekhar, *An Introduction to the Study of Stellar Structure* (Chicago: University of Chicago Press, 1939).

M. Schwarzschild, *The Structure and Evolution of the Stars* (Princeton, N.J.: Princeton University Press, 1958).

H. Y. Chiu, *Stellar Physics* (Waltham, Mass.: Blaisdell, 1968).

Chapter 5: Radiation and Energy Transport

S. Chandrasekhar, *Radiative Transfer* (New York: Dover, 1960).

T. G. Cowling, "Magnetic Stars," in *Stellar Structure*, edited by L. H. Aller and D. B. McLaughlin (Chicago: University of Chicago Press, 1965).

L. Mestel, "Meridian Circulation in Stars," in *Stellar Structure*, edited by L. H. Aller and D. B. McLaughlin (Chicago: University of Chicago Press, 1965).

D. Mihalas, *Stellar Atmospheres*, 2nd ed. (San Francisco: W. H. Freeman, 1978).

E. A. Spiegel, "Convection in Stars. I: Basic Boussinesq Convection," *Ann. Rev. Ast. Ap.*, **9** (1971): 323.

———, "Convection in Stars. II: Special Effects," *Ann. Rev. Ast. Ap.*, **10** (1972): 197.

A. Unsold, *Physik der Sternatmosphären* (Berlin: Springer-Verlag, 1955).

Chapter 6: Atomic Properties of Matter

A. N. Cox, "Stellar Absorption Coefficients and Opacities," in *Stellar Structure*, edited by L. H. Aller and D. B. McLaughlin (Chicago: University of Chicago Press, 1965).

T. R. Carson, "Stellar Opacity," *Ann. Rev. Ast. Ap.*, **14** (1976): 95.

D. G. Hummer and G. Rybicki, "The Formation of Spectral Lines," *Ann. Rev. Ast. Ap.*, **9** (1971): 237.

J. T. Jeffries, *Spectral Line Formation* (Waltham, Mass.: Blaisdell, 1968).

PART 3: STELLAR EVOLUTION

General

D. D. Clayton, *Principles of Stellar Evolution and Nucleosynthesis* (New York: McGraw-Hill, 1968).

J. P. Cox and R. T. Guili, *Stellar Structure,* Vol. II (New York: Gordon and Breach, 1968).

H. Gursky, "Neutron Stars, Black Holes and Supernovae," in *Frontiers of Astrophysics,* edited by E. H. Avrett (Cambridge, Mass.: Harvard University Press, 1976).

I. Iben, Jr., "Normal Stellar Evolution," in *Stellar Evolution,* edited by H. Y. Chiu and A. Murriel (Cambridge, Mass.: MIT Press, 1972).

S. L. Shapiro and S. A. Teukolsky, *Black Holes, White Dwarfs and Neutron Stars* (New York: Wiley, 1983).

R. F. Stein, "Stellar Evolution: A Survey with Analytical Models," in *Stellar Evolution,* edited by R. F. Stein and A. G. W. Cameron (New York: Plenum Press, 1966).

Ya. B. Zeldovich and I. D. Novikov, *Relativistic Astrophysics* (Chicago: University of Chicago Press, 1971).

Chapter 7: Nuclear Energy Sources

D. D. Clayton, *Principles of Stellar Evolution and Nucleosynthesis* (New York: McGraw-Hill, 1968).

Chapter 8: Introduction to Stellar Evolution

R. J. Taylor, *The Stars: Their Structure and Evolution* (New York: Springer-Verlag, 1972).

Chapter 9: The Main Sequence

L. Motz, *Astrophysics and Stellar Structure* (Waltham, Mass.: Ginn, 1970).

Chapter 10: Evolution Away from the Main Sequence

I. Iben, Jr., "Post Main Sequence Evolution of Single Stars," *Ann. Rev. Ast. Ap.*, **12** (1974): 215.

L. Motz, *Astrophysics and Stellar Structure* (Waltham, Mass.: Ginn, 1970).

Chapter 11: Deviations from Quasistatic Evolution

N. Baker, "Stellar Stability and Stellar Pulsations," in *Stellar Evolution,* edited by H. Y. Chiu and A. Murriel (Cambridge, Mass.: MIT Press, 1972).

J. P. Cox, "Nonradial Oscillations of Stars: Theory and Observations," *Ann. Rev. Ast. Ap.*, **14** (1976): 247.

R. F. Christy, "Variable Stars—Realistic Models," in *Stellar Evolution,* edited by H. Y. Chiu and A. Murriel (Cambridge, Mass.: MIT Press, 1972).

K. J. Fricke and R. Kippenhahn, "Evolution of Rotating Stars," *Ann. Rev. Ast. Ap.*, **10** (1972): 45.

P. Ledoux, "Stellar Stability," in *Stellar Structure,* edited by L. H. Aller and D. B. McLaughlin (Chicago: University of Chicago Press, 1965).

Chapter 12: Final Stages of Stellar Evolution

W. D. Arnett, G. J. Hansen, J. W. Truran, and A. G. W. Cameron, eds., *Nucleosynthesis* (New York: Gordon and Breach, 1968).

D. D. Clayton, *Principles of Stellar Evolution and Nucleosynthesis* (New York: McGraw-Hill, 1968).

E. E. Salpeter, "Stellar Evolution Leading up to White Dwarfs and Neutron Stars," in *Relativity Theory and Stellar Structure, Lectures in Applied Mathematics,* Vol. 10, American Mathematical Society, 1967.

Chapter 13: Weak Interactions in Stellar Evolution
J. N. Bahcall and R. L. Sears, "Solar Neutrinos," *Ann. Rev. Ast. Ap.,* **10** (1972): 25.
Z. Barkat, "Neutrino Processes in Stellar Interiors," *Ann. Rev. Ast. Ap.,* **13** (1975): 45.
H. Y. Chiu, *Stellar Physics* (Waltham, Mass.: Blaisdell, 1968).

Chapter 14: Degenerate Stars
J. R. P. Angel, "Magnetic White Dwarfs," *Ann. Rev. Ast. Ap.,* **16** (1978): 487.
L. Mestel, "The Theory of White Dwarfs," in *Stellar Structure,* edited by L. H. Aller and D. B. McLaughlin (Chicago: University of Chicago Press, 1965).
J. P. Ostriker, "White Dwarfs," in *Stellar Evolution,* edited by H. Y. Chiu and A. Murriel (Cambridge, Mass.: MIT Press, 1972).
R. Sexl and H. Sexl, *White Dwarfs—Black Holes* (New York: Academic Press, 1979).

Chapter 15: Supernovae
G. E. Brown, H. A. Bethe, and G. Baym, "Supernovae," *Nucl. Phys.,* **A375** (1982): 481.
D. N. Schramm, *Supernovae* (Dordrecht, Holland: Reidel, 1977).
F. Zwicky, "Supernovae," in *Stellar Structure,* edited by L. H. Aller and D. B. McLaughlin (Chicago: University of Chicago Press, 1965).

Chapter 16: Compact Stellar and Relativistic Objects
G. Baym and C. Pethick, "Physics of Neutron Stars," *Ann. Rev. Ast. Ap.,* **17** (1979): 415.
R. Giaconi and R. Ruffini, *Physics and Astrophysics of Neutron Stars and Black Holes* (Amsterdam: North Holland, 1978).
H. Gursky and R. Ruffini, *Neutron Stars, Black Holes and Binary X-ray Sources* (Dordrecht, Holland: Reidel, 1975).
W. H. Press and K. S. Thorne, "Gravitational-Wave Astronomy," *Ann. Rev. Ast. Ap.,* **10** (1972): 335.
C. Misner, K. S. Thorne, and J. A. Wheeler, *Gravitation* (San Francisco: W. H. Freeman, 1973).
R. N. Manchester and J. H. Taylor, *Pulsars* (San Francisco: W. H. Freeman, 1977).
J. V. Narliker, *Introduction to Cosmology* (Boston: Jones and Bartlett, 1983).
M. Ruderman, "Pulsars: Structure and Dynamics," *Ann. Rev. Ast. Ap.,* **10** (1972): 427.
R. Sexl and H. Sexl, *White Dwarfs—Black Holes* (New York: Academic Press, 1979).
S. Weinberg, *Gravitation and Cosmology,* (New York: Wiley, 1972).
Ya. B. Zeldovich and I. D. Novikov, *Relativistic Astrophysics* (Chicago: University of Chicago Press, 1971).

Chapter 17: Close Binary Systems
B. Paczynski, "Close Binaries," in *Stellar Evolution,* edited by H. Y. Chiu and A. Murriel (Cambridge, Mass.: MIT Press, 1972).
J. S. Gallagher and S. Starrfield, "Theory and Observation of Classical Novae," *Ann. Rev. Ast. Ap.,* **16** (1978): 171.
W. K. Rose, "Novae," in *Stellar Evolution,* edited by H. Y. Chiu and A. Murriel (Cambridge, Mass.: MIT Press, 1972).

Bibliography

General

B. Bok and P. F. Bok. *The Milky Way* (Cambridge, Mass.: Harvard University Press, 1981).

L. Spitzer, *Physical Processes in the Interstellar Medium* (New York: Wiley, 1978).

VOLUME II

PART 4: THE INTERSTELLAR MEDIUM

Chapter 18: Interstellar Matter

H. Alfven and C. Falthamnan, *Cosmic Electrodynamics,* 2nd ed. (London: Oxford University Press, 1963).

H. Goldstein, *Classical Mechanics* (Reading, Mass.: Addison-Wesley, 1981).

S. A. Kaplan, *Interstellar Gas Dynamics,* (London: Pergamon Press, 1966).

L. Spitzer, *The Physics of Fully Ionized Gases* (New York: Interscience, 1962).

Chapter 19: Interstellar Dust Grains

S. A. Kaplan and S. B. Pikelner, *Interstellar Medium* (Cambridge, Mass.: Harvard University Press, 1970).

E. E. Salpeter, "Formation and Destruction of Dust Grains," *Ann. Rev. Ast. Ap.,* **15** (1977): 267.

Chapter 20: Gaseous Nebulae

R. L. Brown, "Galactic Nonthermal Continuum Emission," in *Galactic and Extragalactic Radio Astronomy,* edited by G. L. Verschuur and K. I. Kellermann (New York: Springer-Verlag, 1974).

E. J. Chaisson, "Gaseous Nebulae and Their Interstellar Environment," in *Frontiers of Astrophysics,* edited by E. H. Avrett (Cambridge, Mass.: Harvard University Press, 1976).

A. Dalgarno and R. A. McCray, "Heating and Ionization of H II Regions," *Ann. Rev. Ast. Ap.,* **10** (1972): 375.

G. G. Fazio, "Infrared Astronomy," in *Frontiers of Astrophysics,* edited by E. H. Avrett (Cambridge, Mass.: Harvard University Press, 1976).

T. K. Merron, "Lectures on Radio Frequency Absorption Line Studies of Galactic Structure," in *Galactic Astronomy,* Vol II, edited by H. Y. Chiu and A. Murriel (New York: Gordon and Breach, 1970).

J. M. Moran, "Radio Observations of Galactic Masers," in *Frontiers of Astrophysics,* edited by E. H. Avrett (Cambridge, Mass.: Harvard University Press, 1976).

D. E. Osterbrock, *Astrophysics of Gaseous Nebulae* (San Francisco: W. H. Freeman, 1974).

P. A. G. Scheuer, "Radio Galaxies and Quasi-Stellar Sources," in *Plasma Astrophysics,* edited by P. A. Sturrock (New York: Academic Press, 1967).

Chapter 21: Hydrodynamics

F. F. Chen, *Introduction to Plasma Physics* (New York: Plenum Press, 1974).

L. D. Landau and E. M. Lifshitz, *Fluid Mechanics* (Reading, Mass.: Addison-Wesley, 1959).

Chapter 22: The Virial Theorem
J. P. Cox and R. T. Guili, *Stellar Structure,* Vol. I (New York: Gordon and Breach, 1968).

Chapter 23: Star Formation
L. Mestel, "Stellar Magnetism and Rotation," in *Stellar Evolution,* edited by H. Y. Chiu and A. Murriel (Cambridge, Mass.: MIT Press, 1972).
J. L. Tassoul, *Theory of Rotating Stars* (Princeton, N.J.: Princeton University Press, 1978).
P. R. Woodward, "Theoretical Models of Star Formation," *Ann. Rev. Ast. Ap.,* **16** (1978): 555.

Chapter 24: Supersonic Flow and Shock Waves
J. C. Brandt, *Introduction to the Solar Wind* (San Francisco: W. H. Freeman, 1970).
L. D. Landau and E. M. Lifshitz, *Fluid Mechanics* (Reading, Mass.: Addison-Wesley, 1959).
Ya. B. Zeldovich and Yu. P. Raizer, *Physics of Shock Waves and High Temperature Phenomena,* Vols. I and II, edited by W. D. Hayes and R. F. Probstein (New York: Academic Press, 1967).

Chapter 25: Diffuse Supernova Remnants
L. Woltjer, "Supernova Remnants," *Ann. Rev. Ast. Ap.,* **10** (1972): 129.

PART 5: GALAXIES AND THE UNIVERSE

Chapter 26: The Expanding Universe
M. Davis, "Galaxies and Cosmology" in *Frontiers of Astrophysics,* edited by E. H. Avrett (Cambridge, Mass.: Harvard University Press, 1976).
J. V. Narlikar, *Introduction to Cosmology* (Boston: Jones and Bartlett, 1983).
P. J. E. Peebles, *Physical Cosmology* (Princeton, N.J.: Princeton University Press, 1971).
——— *The Large Scale Structure of the Universe* (Princeton, N.J.: Princeton University Press, 1980).
D. W. Sciama, *Modern Cosmology* (Cambridge, England: Cambridge University Press, 1973).

Chapter 27: Galaxies
S. M. Faber and J. S. Gallagher, "Masses and Mass to Light Ratios of Galaxies," *Ann. Rev. Ast. Ap.,* **17** (1979): 135.
A. Sandage, *The Hubble Atlas of Galaxies* (Washington, D.C.: Carnegie Institute of Washington, 1961).

Chapter 28: Dynamics of Stellar Systems
S. Chandrasekhar, *Principles of Stellar Dynamics* (Chicago: University of Chicago Press, 1943).

G. B. Field, "The Mass of the Universe: Intergalactic Matter," in *Frontiers of Astrophysics,* edited by E. H. Avrett (Cambridge, Mass.: Harvard University Press, 1976).
——— "Intergalactic Matter," *Ann. Rev. Ast. Ap.,* **10** (1972): 277.
I. King, "The Structure of Round Stellar Systems: Observations and Theory," in *Dynamics of Stellar Systems,* edited by A. Hayli (Dordrecht, Holland: Reidel, 1975).

Chapter 29: Axially Symmetric Galaxies
D. Mihalas, *Galactic Astronomy* (San Francisco: W. H. Freeman, 1968).
M. Schmidt, "Rotation Parameters and Distribution of Mass in the Galaxy," in *Galactic and Extragalactic Radio Astronomy,* edited by G. L. Verschuur and K. I. Kellermann (New York: Springer-Verlag, 1974).

Chapter 30: Spiral Structure
W. B. Burton, "The Large-Scale Distribution of Neutral Hydrogen in the Galaxy," in *Galactic and Extragalactic Radio Astronomy,* edited by G. L. Verschuur and K. I. Kellermann (New York: Springer-Verlag, 1974).
C. C. Lin, "Theory of Spiral Structure," in *Galactic Astronomy,* Vol. II, edited by H. Y. Chiu and A. Murriel (New York: Gordon and Breach, 1970).
W. W. Roberts, "Shock Formation and Star Formation in Galactic Spirals," in *Galactic Astronomy,* Vol. II, edited by H. Y. Chiu and A. Murriel (New York: Gordon and Breach, 1970).
C. Yuan, "On the Comparison Between the Density-Wave Theory and Observations," in *Galactic Astronomy,* Vol. II, edited by H. Y. Chiu and A. Murriel (New York: Gordon and Breach, 1970).

Chapter 31: Galactic Evolution
J. Andouze and B. M. Tinsley, "Chemical Evolution of Galaxies," *Ann. Rev. Ast. Ap.,* **14** (1976):43.
K. Brecher, "Active Galaxies," in *Frontiers of Astrophysics,* edited by E. H. Avrett (Cambridge, Mass.: Harvard University Press, 1976).
G. R. Burbidge and E. M. Burbidge, *Quasi Stellar Objects* (San Francisco: W. H. Freeman, 1967).
D. J. Eggen, D. Lynden-Bell, and A. R. Sandage, *Ap. J.,* **136** (1974):748.
C. Hazard and S. Mitton, eds. *Active Galactic Nuclei* (London: Cambridge University Press, 1979).
J. H. Oort, "The Galactic Center," *Ann. Rev. Ast. Ap.,* **15** (1977):295.
D. W. Weedman, "Seyfert Galaxies," *Ann. Rev. Ast. Ap.,* **15** (1977):69.

Index

Volume I ends on page 344. Volume II begins on page 345.

absolute luminosity. *See* Luminosity
absorption, 69, 99
   line, 130–131, 349
   neutrino, 240
   radio, 363–369
   starlight, 349
   true, 71, 79–82
absorption coefficient, 113–114, 132
   bound-free, 99–100
   dust grains, 371
   free-free, 119
   gaseous nebulae, 389–390
   lines, 115
   radio, 398
   weak line, 136
absorption edge, 122
absorption lines, 39, 349
   chemical abundances, 39
   cosmic background radiation, 349
   excitation and ionization, 39
   hydrogen, 98–99
   mass transfer indicator, 337
   molecular, 349
   opacity, effect on, 85
   profile, 86
   spectral classification, 106
absorptivity by dust grain, 352
abundances
   deuterium, 484–485
   gaseous nebulae, 393, 396–398
   HI regions, 351
   heavy metals, 203, 371, 377, 378, 584
   helium, primeval, 483–484, 485
   light elements, primeval, 483–484
   line absorption, 39, 85
   mass fractions, 55
   metallic, 104
   primordial, 15
   relative, 20, 138, 393, 396–398
   solar, 106, 231
   solar atmosphere, 104
   stellar, 130
   supernovae, 272, 274
accretion, 11, 321–322, 327, 340–343, 447–450
   adiabatic, 448
   black holes, 447
   compact objects, 11, 327, 447–450
   gamma rays, 11, 447
   luminosity from, 449–450
   neutron stars, 447
   novae, 340–343
   rate, 449
   spherical, 447–448
   stellar winds, 448
   white dwarfs, 447
   x rays, 11, 331, 334, 447
accretion disk, 321–322, 327, 334, 337, 340
   eclipse of x-ray source, 334
   Her X-1, 334–335
   novae, 340–343
action integral, 98
adiabat, 435
   see Shock adiabat
adiabatic indices, 18, 21, 411
   average, 411
   general relativistic modifications, 298
   mixture of matter and radiation, 21–22
   stability and, 214, 259
adiabatic processes, 18
   and convection, 89
   and instabilities, 18
advanced potentials, 425
Alfvén surface, 337
Alfvén waves, 424–425
   angular momentum loss, 424–426
   stellar evolution, 207
Algol-type stars, 324
Andromeda galaxy, 492
   red shift, 468
   spiral structure, 566
   x-ray luminosity, 333
angular diameter, 478
angular momentum
   binary system, 317
   dwarf stars and galactic evolution, 584–585
   galaxy formation, 511
   loss from rotating magnetic clouds, 423–425
   and mass loss, 321
   planetary systems, 48
   quantum, 99–100
   stellar angular momentum reduction, 207
   and stellar rotation, 207, 284, 423
   Sun, 47
   and ultraviolet excess, 587
antiparticle, 24
Ap stars, 103, 430
   magnetic fields in, 207
astronomical statistics, 48–52
astronomical unit, 4
atmosphere
   gravitational acceleration in, 57
   grey atmosphere, 78–79, 83, 85
   hydrostatic equilibrium in, 57, 78, 87

atmosphere *(continued)*
  line formation in, 85
  main-sequence type B0.5, model, 89
  main-sequence type G0, model, 88
  plane–parallel, 57
  scale height, 58, 307
  solar, opacity, 101
  solar, temperature structure, 83
  stellar models, 41
  surface gravity, 41, 87
  white dwarfs, 262
atom, three-level, 396
atomic structure
  magnetic field effects on, 290–291
  see Hydrogen atom

Balmer continuum, 100
Balmer discontinuity, 34
Balmer jump. *See* Balmer discontinuity
Balmer series, 98–99
barred galaxies, 490–493, 497
  spectrum of bar, 505
  spiral density waves, 561
baryon number, 150
Bernoulli's law, 404, 435, 449
beta decay, 240
  neutron, 240
  neutron-rich nuclei, 288
  nuclear, 234
Big Bang, 17, 475, 481
  galaxy formation, 578
  relics, 486–487
binary galaxies, 502
binary pulsar, 338–340
  general relativity test, 338
  orbital parameters, 338
  radio emission, 338
  transverse Doppler effect, 338
binary systems, 42–45
  absolute orbit, 43
  amplitude of radial velocity, 318
  angle of inclination, 44
  angle of nodes, 317
  angular momentum of, 317
  apparent orbit, 44
  black holes, 42, 326–327
  disruption of, 327–329
  Doppler effects, 41, 42
  double-lined spectroscopic, 41, 43
  eccentricity, 317, 327
  eclipse, primary and secondary, 42
  eclipsing, 42, 45, 317
  Euler angles of orbit, 317

evolving, 322–330
general relativity, 42
globular clusters, 519
gravitational radiation, 42
heating effects, 318, 336
light curve, 43
longitude of periastron, 317
mass function, 44, 317
mechanics of binary systems, 316–318
neutron stars, 42
orbital elements, 44, 316–318
orbital inclination, 317
orbital motion, 42
orbital period, 44, 317, 329
pulsar mechanism, 305
radial velocity, 44
relative velocity of two stars, 317
rotation, 283–284
rotational effects on stars, 58–59, 95
semimajor axis, 44, 317
single-lined spectroscopic, 41, 43, 317
spectroscopic, 42, 44
spectrum binary, 41, 42–43
stellar evolution effects, 42, 327–329
supernovae in, 327–329
tidal effects, 58–59, 318, 327, 337, 338
true orbit, 43
velocity curve, 317–318
visual, 41, 42
x-ray sources, 42, 58
see Close binary systems; X-ray binary systems
birthrate function. *See* Stellar birthrate function
black body, 4
black-body radiation, 20, 65, 74
  energy density, 74
  flux, 74
  galactic nuclei, 509
  mean intensity, 74
  Planck function, 74
  thermodynamic equilibrium, 74
black hole, 10, 11, 15, 267, 291–301, 337–338, 451, 530
  active galactic nuclei, 301
  binary systems, 42, 326–327, 337–338
  collision with galaxy, 530–531
  Cyg X-1, 301, 337–338
  effects on nearby stars, 301, 337–338
  event horizon, 295, 297
  formation, 11, 237, 267, 285, 297–298

galactic mass, 11
galactic nuclei, 11
giant, 530
Kruskal coordinates, 296–297
magnetic fields, 285
mini black holes, 301
quasars, 11
search for, 301
spacetime near a black hole, 296–297
spherically symmetric, 295
tidal effect on nearby galaxy, 530
x-ray sources, 331–332, 337–338
Bohr radius, 98
  magnetic field effect on, 290–291
Bok globules, 415
bolometric correction, 30
  table, 31
bolometric luminosity. *See* Luminosity
Boltzmann distribution function, 514
  averages over, 544
Boltzmann equation, 514, 544–548
  collision-free, 514, 544, 546–547
  collision term, 514
  cylindrical coordinates, 546
  moments of, 544–545
Boltzmann's law, 102
Bremsstrahlung. See Free-free transition
brightness
  measurements, 28
  surface, 64
buoyancy forces, 89

carbon flash, 169
Cassiopeia A, 459, 460
Cen X-3, 333
centrifugal force, 321, 419
Cepheid variable, 36, 45, 46, 47
  classical, 218–223
  magnitude versus time, 220
  period-luminosity relation, 218–219
  position in HR diagram, 214
  radius versus time, 220
  temperature versus time, 220
Chandrasekhar mass limit, 13, 167, 255, 257
  hot white dwarf, 282
chemical composition
  effect on stellar evolution, 203–207
  evolutionary changes, 176
  inhomogeneities, 185–186, 206, 207
  star clusters, 34

stellar atmospheres, 79
stellar evolution, 54
chemical elements
abundances. *See* Abundances
primordial, 15
chemical equilibrium, 24
chemical inhomogeneities, 3
rotational mixing, 96, 206
chemical mixing, 3
chemical potential, 18, 22
antiparticles, 24–25
chemical equilibrium, 109
electron, 163
external potential, 163
fermion, 23–24
ideal gas, 109
matter near absolute zero, 24
photons, 24–25
relativistic, 25
circulation, 95–96
currents set up by, 96
meridional, 96
velocity, 96
circumstellar gas rings, 319, 321
close binary systems, 36, 42, 45, 316, 318–322
accretion disk formation, 321–322
angular momentum loss, 321
contact, 319
detached, 319
emission lines from, 319
equipotential surfaces, 319–320
evolving, 322–330
Hertzsprung–Russell diagram for member stars, 320
inner Lagrange point, 319
Lagrange points, 319–320, 321
mass exchange, 319, 320–322, 323–325, 337, 340–343
mass-luminosity relation for member stars, 321
mass transfer from convective envelopes, 325
mass transfer from radiative envelopes, 325
morphology of, 319
novae, 340–343
observed properties of stars in, 320–321
outer Lagrange point, 321
radius of Roche lobe, 319, 323–324
relative orbit during mass exchange, 322
Roche lobes, 319–320, 340
semidetached, 319
stellar evolution in, 322–330
stellar winds in, 327, 332–333

supernova explosions in, 327
tidal distortion, 337
time-scale for mass transfer, 325
x-ray sources, 327
clusters of galaxies, 478, 488, 503, 524–531
cluster mass, 503–504
Coma cluster, 525
evolution, 528
intergalactic gas in, 527, 528
irregular cluster, 525
isothermal sphere models, 64
mass-to-luminosity ratio, 525
projected x-ray emission, 529
regular clusters, 524
Virgo cluster, 525
virial theorem, 503
x-ray luminosity, 528
x-ray surface brightness, 528
CNO cycle, 13, 155–156
upper main sequence, 175, 176, 181
collision strength, 387
collisional cooling rate, 388
collisional coupling of dust grains and gas, 378–379
collisional damping constant, 134–135
collisional de-excitation, 101
cross section, 133, 387–388
line damping, 133
rate, 388
collisional drag force, 379
collisional excitation, 79–80, 101, 347, 357–359, 386–389
cross section, 387
rate, 388
trace elements in nebulae, 382, 386
collisional relaxation, 353
deflection time, 356
energy exchange time, 356
time-scale, 353–357
color-color diagram, 33–34
reddening line, 35
color indices, 30, 31
black body, 31
elliptical galaxies, 499
globular clusters, 499
interstellar absorption, 34–35
surface temperature, 34
table, 31
UBV system, 30, 38
color-magnitude diagram, 33
galactic cluster, 37, 38
M 5, 37
for Population I and Population II, 35
and stellar mass, 34

Coma cluster, 525
intergalactic gas content, 528
luminosity, 525
mass, 525–527
compressibility, isothermal, 579
Compton scattering, 71, 116, 447
cosmology, 482
energy exchange between radiation and matter, 482
Compton wavelength, 24, 116, 252
condensation
central mass condensation, 185, 186
nuclei and interstellar grain formation, 377
conduction, 95
heat flow by, 95
radiative, 95
thermal, 95
conductivity
electrical, 406
electrical, in neutron star matter, 308
electron, 123
radiative, 95, 209
thermal, 95, 124, 209, 253, 256, 437
continuum intensity, 40
convection, 3, 88–95
energy flux by, 93, 128
energy transport by, 76, 92, 199
luminosity, 76, 228
opacity, effect on, 89
stellar atmospheres, 92
stellar surface, 94
temperature gradients, 89
time-scale for, 89, 93
turbulent, 95
convective core, 126
convective envelopes, 264
convective mixing, 169
convective stars, 127–130
convective zone, 92
cooling rates, 24–25
interstellar gas in protogalaxies, 538
nebulae, 351–352, 357, 373–374, 383, 386, 388–391
neutron stars, 24, 289
white dwarfs, 24
Cooper pairs, 288–289
core collapse. *See* Stellar core collapse
core convergence, 276
Coriolis force, 321
coronal gas, 350
cosmic background radiation, 349
electromagnetic background radiation, 485–487

cosmic background radiation *(continued)*
  line absorption by interstellar gas, 349–350
  isotropy, 487
  neutrino background radiation, 483, 487
  spectrum, 486
cosmic rays, 312, 447
  energy spectrum, 455–456
cosmological constant, 472, 473
cosmological models, 473–476
  age of the universe, 475, 476, 478
  closed universe, 474–475
  collapse time, 475
  critical mass density of the universe, 474, 475, 476
  current density of the universe, 474, 487–488
  early universe, 481–482
  Einstein–de Sitter model, 474, 475, 477, 478
  Friedmann models, 472–473, 474, 476, 480–481, 533–537, 579
  galaxy formation, 533–537
  general properties, 473–476
  light travel, 476–478
  matter-dominated universe, 473, 474, 485
  maximum radius of the universe, 475
  open universe, 475
  radiation-dominated universe, 473, 477, 580
  redshift due to expansion, 477
  spatial curvature, 473
  static universe, 472
  see Big Bang
cosmological nucleosynthesis, 483–485
cosmological principle, 470
cosmological redshift. *See* Redshift
cosmology, 17, 314, 468–469
  closed universe, 475
  comoving coordinates, 471
  cosmological principle, 470
  deceleration parameter, 473
  distance scales, 478–481
  early universe, 481–482
  general relativity, 470
  homogeneity of space, 470
  isotropy of space, 470
  missing mass, 474, 488, 504
  world map, 470
  world picture, 470
  see Newtonian cosmology
Coulomb barrier, 144–145, 151, 155
Coulomb effects, 26

Coulomb energy, 25
  screened, 162–163
Coulomb interactions, 257–258, 353–357
  pressure due to, 258
Coulomb logarithm, 355
Coulomb scattering, 354–357
  angle, 354
  impact parameter, 354
  logarithm, 355
Crab nebula, 15, 451, 463–465
  blue light, 461
  electromagnetic radiation from, 464
  energetics and pulsars, 312
  filamentary structure, 463
  gaseous structure, 463
  lifetime of synchrotron electrons, 464
  line spectra, 464
  mass, 465
  optical polarization of light, 462
  polarization of light, 464
  pulsar as power source, 464
  radio polarization of light, 462
  red light, 461
  spectrum, 455–456, 459
  structure, 461, 463–464
  synchrotron energy-loss rate, 464
  x-ray component, 464
Crab pulsar, 286, 301
  characteristics, 304
  Crab nebula's power, 312
  energy spectrum, 312–313
  pulse structure, 302, 312–313
  slowdown rate, 307
critical density of the universe. *See* Cosmological models
cross section, 70
  absorption, 114, 141
  atomic electron capture, 120–121
  collisional, 133, 347, 354
  Coulomb, 124, 354, 356
  de-excitation, 387
  dust grain, 371
  excitation, 387
  ionic capture on dust grain, 375
  neutrino, 240, 244, 249, 279
  nuclear reactions, 150–154
  photoionization, 121, 383
  radiative recombination, 384
  Rayleigh scattering, 117–118
  screened, 162–163
  Thomson scattering, 117, 440
crystal lattice, 27, 258, 288, 290–291
  melting, 258
curve of growth, 137–138

cyclotron frequency, 361
Cygnus A, 365
Cygnus X-1, 15, 42, 301, 319, 327, 337–338
cylindrical coordinates, 407

damping broadening of lines, 137
damping constant, 112
damping profile. *See* Lorentzian profile
dark clouds. *See* Interstellar clouds
Debye length, 162–163, 354
Debye screening, 162
  electron, 353
deceleration parameter, 473, 475–476, 481, 485
  galactic evolution, 481
deflection time scale, 356, 517
degeneracy, 12, 26
  degree of, 253
  effect on neutrino pair production, 245
  effect on nuclear burning, 145–146, 164
  electron, 12, 257
  electron screening, 164
  neutrino, 22, 23
  neutron, 236
  relativistic electrons, 26
degenerate matter. *See* Equations of state
  evolved stars, 252
  properties, 252–254
  in stars, 252
  temperature in degenerate matter, 285
density discontinuities, 20
density fluctuations, 579, 583
density wave theory. *See* Spiral density waves
detailed balancing, 109–115, 365
deuterium production
  cosmological, 484
  nuclear fusion in stars, 147–148
  proton-proton chain, 155
dielectric constant, 245, 247, 360, 422
differential rotation, 414
  effects on star formation, 419–420
diffuse background radiation
  stellar, 371
  see Cosmic background radiation
diffuse clouds. *See* Interstellar clouds
diffusion, 65, 86, 115
  approximation, 3, 74, 86
  coefficient, 74

diffusion (cont.)
  magnetic diffusion in plasma, 422
  thermal energy transport, 422
diffusion equation, 86
dispersion for normal distribution, 500
dispersion measure, 304, 360–361
dispersion relation, 416
  density waves, 568
dissipative processes and stellar pulsations, 215
distance scales
  angular-diameter, 478–480
  cosmological, 478–481
  extragalactic, 36, 42, 46, 269
  galactic, 34, 304, 362
  globular clusters, 518
  light travel time, 468–469
  luminosity, 478–479
  metric, 478–479
  stellar, 28–29
Doppler broadening, 133, 368
  equivalent width of line, 136
Doppler shift, 41, 132
  apparent wavelength, 41
  binary systems, 41, 338
  radial velocity, 41
  stellar radial velocity, 41
  transverse Doppler effect, 338
Doppler velocity, thermal, 132
Doppler width, 132
dynamic collapse, 11
  electron capture induced, 259
dynamical friction, 517

early universe
  entropy in, 580
  equation of state for matter in, 580–581
  expansion, 481–482
  expansion time-scale, 482
  galaxy formation, 485
  matter in, 482
  neutrino decoupling, 483
  neutrino-matter coupling, 482–483
  radiation decoupling, 485–486
  temperature, 483
  temperature fluctuations, 584
  see Big Bang, cosmology
Earth, 7
Eddington limit, 12–13, 75, 226, 342, 447
  for neutrinos, 280
Eddington standard model, 62–63
effective temperature, 4, 76, 77–78
  relation to atmospheric temperature, 77–78
  table, 31

Einstein–de Sitter universe, 474, 475, 477, 478–480, 537
  protogalactic perturbations in, 583
Einstein transition probabilities. *See* Transition probabilities
elastic collisions, 353
electric current density, 406, 410
electric dipole
  absorption by, 112
  field, 309–310
  radiating, 112
electric field, energy density in, 117
electrodynamics of rotating magnetized conductor, 308–312
electromagnetic radiation, 65, 298
  circular polarization, 361
  and pair production, 24
  see Synchrotron radiation
electromagnetic spectrum, 21, 66
  early universe, 485–487
  expanding universe, 485
  Planckian, 21, 74
  radio, 399
electromagnetic wave, 360
  attenuation, 360
  speed of propagation in plasma, 360
electron capture, atomic
  cross section, 120–121
electron capture, nuclear
  by heavy nuclei, 236, 249, 259, 276, 278
  stellar core collapse, 214–215, 276
electron conduction, 123–124
  conductivity, 123
  degeneracy, 124
  energy transfer, 123, 124
  opacity, 123
electron screening
  degeneracy effects, 164
  nuclear reactions, 162–164
  potential energy, 162–163
  reaction cross section, 162
elliptical galaxies, 489, 532–538, 584
  classification, 489–490
  composition gradients in, 533
  evolution, 533–538
  formation, 532
  gas dynamic effect, 538
  luminosity distribution, 495
  mass distribution, 538
  masses, 500–501
  rotation in, 500, 537
  spiral structure, 572
  surface brightness, 532–533, 538

  time-scales, 518
  velocity dispersions, 500
  virial theorem, 500
emission, 69
  bands from novae, 340
  free-bound, 99–100
  gaseous nebulae, 396, 398
  induced. *See* Stimulated
  interstellar hydrogen, 363–364
  line, 130, 319, 396–398
  neutrino, 240
  radio, 363–369, 398–399
  rate, 386
  scattering emission, 80
  spontaneous, 71, 386, 388
  stimulated, 71, 110, 114, 115
  true, 71, 80
  two-photon, 101
emission coefficient, 80, 363
emission measure, 398
emissivity, 69–71
  dust grain, 352
  free-free, 527
  line, 369
  volume emissivity, 71
energy conservation across shock fronts, 435
energy conservation for fluids, 404
energy exchange time scale, 356
  radiation-matter coupling, 482
energy generation, 3
  rates, 147–154
  stellar interiors, 76
energy levels
  continuum depression, 103
  coolant atoms in nebulae, 387
  degenerate, 102
  half-life, 109
  hydrogenic. *See* Hydrogen atom
  mean life, 109
  nitrogen atom, 387
  oscillator, 348
  oxygen atom, 387
  relative populations, 103
  statistical weight, 102, 103
  zero-point, 164
energy-momentum density tensor, 291
energy transfer, 3
energy transport, 3
  convection, 88–95, 94, 199
  efficiency of convective, 94
  lines, 130
  thermal, 223
enthalpy, 434
entropy, 20, 434
  conservation equation, 452
epicyclic frequency, 551, 553–554, 560

Volume II begins on page 345

epicyclic motion, 551, 553–554
   retrograde, 554
equations of state, 8, 16–27
   chemical potential, 18, 22, 23–24, 25
   Coulomb interactions, 258
   degenerate electrons, 256–258
   degenerate fermions, 23–24, 254–255
   degenerate matter, 8, 13, 20, 22–24, 285
   degenerate neutrinos, 24
   degenerate neutrons, 285
   ideal gas, 8, 17
   low-temperature electrons, 261
   matter in the universe, 473
   mixture of ideal gases, 19–20
   neutrinos, 24, 250
   neutron stars, 8, 285, 289
   photon gas, 20
   polytropic, 18, 59, 257
   radiation pressure, 8
   real fluids, 25–27
   white dwarfs, 8
equilibrium process (e-process), 234, 236
equivalence of mass and energy in general relativity, 297
equivalence principle, 299
escape velocity, 516
Euler's equations, 293
event horizon, 11, 297
   physics near the event horizon, 296–297
evolution. *See* Stellar evolution; Galactic evolution
excitation temperature, 138
expansion of the universe, 468–469, 473–476
exponential integral, 85
extinction
   coefficient, 70, 371
   efficiency, 371
   of starlight, 370–371

Faraday rotation, 361, 379
Fermi–Dirac distribution, 23, 24
   neutrinos, 249–250
Fermi energy, 22, 23, 145
   relativistic electron, 245
Fermi momentum, 22, 23, 145
Fermi sea, 24
fermion, 22
filamentary structure. *See* Supernova remnants
fine structure, 99–100
   spin-orbit interaction, 100

flare stars, 340
forbidden transitions, 100–101, 102, 141, 272
   galactic disk matter, 502
   gaseous nebulae, 382–383, 386–387
   interstellar matter, 347
Fraunhofer lines, 39
free-free transition, 100, 438
   energy loss rate, 389
   hydrogen, 358
   thermal spectrum, 389
Friedmann equations, 472–473, 474
   closure of the universe, 475
   development angle, 475
   parametric form, 475
fusion. *See* Thermonuclear fusion

Galactic coordinates, 539
galactic clusters, 35
   age, 38
   color-magnitude diagram, 37
   time-scales, 518
galactic disk
   galactic clusters, 35
   neutral hydrogen emission, 101, 365, 501
   and Population I stars, 35
galactic evolution, 481, 576–579
   effect on deceleration parameter, 481
   effect on luminosity, 481
   recombination, 577
galactic nuclei, 17, 518
   black holes, 11
galactic gravitational potential, 547–548
galaxy, 11, 15
   active, 42, 423
   Andromeda, 333–334
   binary encounters between galaxies, 526
   characteristics, 508–511
   color-aperture relations, 498
   color distributions, 497–499
   disk component, 496, 498
   double. *See* Binary galaxies
   ellipsoidal, 17
   forbidden line emission from, 502
   gas content, 510
   Hubble type, 489, 492–493
   hydrogen distribution in, 365
   isothermal sphere models, 64
   luminosity, 15, 508, 510
   luminosity class, 42
   luminosity distribution, 495, 497
   Magellanic irregulars, 492

   magnitude, absolute, 495
   magnitude, photographic, versus radius, 496–497
   mass, 499–504, 508, 510
   mass-to-light ratio, 507–508
   morphology, 489–494
   nucleus, 17, 423, 496, 508
   obscuring matter in, 499
   quasar, 42
   recessional motion, 468–471, 487
   rotational motion in, 500
   spheroidal component, 498
   spiral component, 496
   star formation, 415
   stellar content, 42, 504–508
   stellar population synthesis, 504, 508
   surface brightness, 494–496
   synthetic spectrum, 42
   tidal radius, 530
   x-ray luminosity, 333
   see Barred galaxies, radio; Elliptical galaxies; Galactic evolution; Galaxy formation; Irregular galaxies; Lenticular galaxies; Spiral galaxies
Galaxy, the, 346, 539
   center, 365
   characteristics, 509
   circular velocity, 539
   coordinates, 539–540
   density contrast of spiral pattern, 561
   differential rotation of disk, 414, 543
   disk composition, 346
   disk gas temperature, 561
   disk structure, 349
   epicyclic motion, 551, 553–554
   expanding rings, 542–543
   force laws, 547–548
   gravitational potential, 548
   halo stars, 541
   kinematics, 539–541
   magnetic field, 361–362, 379
   mass density, disk, 548
   mass density, solar neighborhood, 548
   mass distribution, 549–552
   models, 549–551
   noncircular orbits, 551–554
   nuclear mass loss, 559
   Oort's constants, 543
   Poisson's equation, 548, 549
   projected surface density of gas and stars, 572
   rotation curve, 542–543
   rotation velocity in disk, 414
   Schmidt model, 552

spiral pattern angular velocity, 567
spiral structure, 365, 566
Sun's location, 542
turbulent velocities in disk, 561
x-ray background, 350, 352
galaxy formation, 478, 485, 533, 579–584
- density fluctuations, 579
- density of surrounding matter, 579
- elliptical, 532
- interstellar gas effects, 538
- Jeans model, 580–584
- protogalactic perturbations, 533, 582
- random fluctuations and, 579
- relativistic effects, 580
- rotation, 537
- spiral galaxy, 510, 538–539
- tidal interactions, 537

Gamow peak, 152–153, 232–233
gaseous disks, 563–569
- hydrodynamic equations for, 564, 565
- linearized model, 565
- Poisson's equation for, 564
- spiral perturbations, 566–567
- surface density, 563
- thin disk models, 563
- turbulent pressure, 564

gaseous nebulae, 382–383
- chemical abundances, 383, 398
- cooling processes, 383
- cooling rates, 386, 388–391
- electron number density, 395–397
- emission line relative strength, 393, 396–398
- forbidden processes, 382–383
- heating process, 386
- heating rates 385, 391
- ionization state, 390, 392–393
- ionization structure, 392–394
- mass, 399
- models, 391–398
- optically thick models, 392
- optically thin models, 392
- statistical equilibrium in, 393
- steady state structure, 389
- structure equations for, 389–391
- temperature, 389, 390–391, 393–395, 396–397
- temperature measurements, 397
- thermal equilibrium, 390, 393–396
- thermal stability, 391

Gaunt factor
- bound-bound transitions, 113
- bound-free, 121
- free-free, 119

Gaussian distribution. *See* Normal distribution

general relativity, 10–11, 291–294, 338–340
- binary pulsar, 338–340
- curvature of spacetime, 292–294, 472
- distance scales, 478–481
- energy density, 297
- event horizon, 295, 297
- galactic nuclei, 509
- geodesics, 292, 293–294
- geometric interpretation, 292–294
- gravitational mass, 297
- gravitational redshift, 11, 296, 338
- hydrostatic equilibrium, 297
- light cone, 292–294, 296–297
- light propagation in curved spacetime, 476–478
- mass-energy equivalence, 297
- metric tensor, 292
- neutron star structure, 285–286
- proper distance, 294
- proper time, 294
- spacetime, 292
- spacetime diagram, 296–297
- spatial curvature, 473
- time dilation, 295
- weak field approximations, 293
- world lines, 292
- *see* Black hole; Cosmology; Gravitational radiation

globular clusters, 17, 35, 518–523
- absolute magnitude, 518
- age estimates, 38, 206, 518
- black hole, 523
- cluster radius, 520
- core contraction, 521, 523
- energy distribution, 519
- evolution, 518–519, 520, 521, 523
- halo, 521
- horizontal branch, 47
- isothermal sphere models, 64
- M 5 color-magnitude diagram, 37
- mass distribution, 523–524
- mass function, 48
- relativistic effects, 519
- steady state, 519
- stellar content, 518
- stellar loss from, 521–522
- tidal force, 520
- tidal radius, 520
- surface brightness distribution, 520, 521
- time-scales, 518
- x-ray emission, 523

gravitation
- Einstein's theory, 292
- force in general relativity, 292
- general relativistic model, 291–297
- geometric interpretation, 292–294
- Newtonian, 291, 298, 410
- universal nature, 471

gravitational charge, 298
gravitational collapse, 2, 291–301, 291
- asymmetric, 300
- interstellar clouds, 420–422
- quantum gravity, 295
- rotational effects, 419–420
- uniform matter distribution, 297–298

gravitational constant, time-variation of, 161
gravitational contraction, 12, 13, 57
- red giant stage, 196–197, 201
- stellar core, 184, 186, 194

gravitational field, 58
gravitational mass, 297
gravitational potential, 58, 291
- equipotential surfaces, 58, 319

gravitational potential energy, 410
- black hole, 10
- disk, 410
- neutron star, 14
- spherical mass distribution, 410
- Sun, 10

gravitational radiation, 11, 291, 298–300, 338–340
- angular distribution, 299–300
- binary pulsar, 338–340
- binary systems, 42, 299–300, 301, 338–340
- effect on binary orbital period, 339
- energy loss due to, 299
- galactic nuclei, 300
- multiple star systems, 300
- pulsar slowdown, 306
- stellar collisions and coalescence, 300–301

gravitational radius, 10–11, 295
- *see* Event horizon
gravitational redshift. *See* Redshift
gravitational wave. *See* Gravitational radiation
guillotine factor, 123

HI regions, 350, 365
- composition, 350

HI regions *(continued)*
  cooling mechanism, 353, 357–359
  forbidden processes, 353
  free-electron number density, 352
  heating, 357–359
  mass, 350
  neutral H emission from, 501
  relative abundances, 350
  rotation curves for spiral galaxies, 501
  spiral arms, 574
  temperature, 353
HII masers, 366
HII regions, 350, 382, 438–447
  brightness temperature, 366
  density, 442
  dust grains, 350, 374
  emission lines from, 397
  free-electron number density, 350, 399
  infrared emission, 377
  masers, 366
  mass, 399
  models, 389
  spiral arms, 574
  structure equations for, 389–391
  temperature, 350, 399, 442
Hayashi track, 171–173
heat capacity, 17
  low-temperature fermions, 24
heat conduction, 75
heat conductivity, 75, 209
heat engine, 215
heat-loss rate, 18
heat transfer
  conduction, 75
  perturbation of heat transfer equation, 209
  rate, 405
  stellar structure equations, 76
heating rates
  interstellar gas, 538
  nebulae, 352, 357, 373–374, 385–386, 389–390
helium abundance, primeval, 483–484
helium flash, 169, 196–197
helium-rich stars, 203
Her X-1, 15, 319, 334–335, 336
  accretion disk, 334
Hertzsprung gap, 198
Hertzsprung–Russell diagram, 33, 88
  Cepheid strip, 219
  close binary members, 320
  cooling white dwarfs, 266
  evolutionary path of 5 $M_\odot$ star, 168

evolutionary path, post-main-sequence, 178
evolutionary path, pre-main-sequence, 172
instability band, 223
M 31, 507–508
main-sequence position in, 161
red giant region, 184, 193
stellar evolution, 34
subgiant region, 320
variable stars, 47, 214
Holtsmark distribution, 139
homology transformation, 158–159
horizontal branch, 47
HR diagram. *See* Hertzsprung–Russell diagram
Hubble classification, 489–493, 494
Hubble constant, 468, 475, 478, 481
  galactic evolution, 481
Hubble diagram, 469
Hubble parameter, 470
  time rate of change, 473
Hubble relation, 468–469
Hugoniot. *See* Shock adiabat
Hyades cluster, 229
hydrodynamics, 54
  advection, 401
  body force, 402
  comoving coordinates, 400
  continuity equation, 403, 416, 545, 557, 564
  cylindrical coordinates, 407, 546–547
  energy conservation, 404
  energy equation, 404–405
  equations in Lagrange form, 405
  equations of motion, 402–406, 544–545
  Eulerian coordinates, 401, 403, 582
  Eulerian time derivative, 401
  generalized Gauss's theorem, 402
  incompressible fluid, 402, 403
  irrotational fluid, 402
  Lagrangian coordinates, 400–401, 403, 470, 582
  linearized equations, 415–416
  magnetic effects. *See* Magnetohydrodynamics
  momentum, 403
  momentum equation, 402–406, 544–545, 564
  nonviscous fluid, 402
  perturbations of hydrodynamic equations, 209, 415–416
  rarefaction, 432–433, 444
  reference frames, 400–401
  rotating coordinates, 407
  shearing stress, 402

shock front. *See* Shock waves
shock waves. *See* Shock waves
stellar systems, 544–546
supersonic flow, 432, 434
work done on fluid element, 404
hydrogen atom, 97–102
  Bohr magneton, 101
  Bohr radius, 98
  continuum, 98
  electronic transitions, 98–99
  energy levels, 97–102
  ground state, 98
  Landé g factor, 101–102
  perturbed energy levels, 101
  photoionization cross section, 383
  spectrum, 98–99
hydrogen burning, 12
  CNO cycle, 155–156
  main sequence, 154–156
  proton-proton chain, 155
  in stellar cores, 12, 36
hydrostatic equilibrium, 6, 7, 11, 54–59
  boundary conditions, 56, 127
  degenerate matter, 254–256
  density-temperature relation, 26
  equations, 56, 78
  impossible inside event horizon, 11
  mass limits, 13
  perturbations away from, 208–210
  pressure, central, 56
  rotating stars, 58, 95
  stars in binary systems, 58
  stellar atmosphere, 78, 87
  vector form, 58
  virial theorem for, 411
hydroxide ion, 349
hyperfine structure, 99–100
  galactic disk structure, 101, 365
  spin-spin interaction, 100
  transitions, 347
hyperon, 288

impact approximation, 133, 138
inelastic collisions and heating of dust grains, 373
infrared excess, 343
  interstellar dust grains, 376–377
infrared radiation, 347, 366, 389
  dust grains, 350, 370, 376–377, 382
  free-free transitions, 389
  novae, 343
  protostellar, 173
  pulsars, 312
initial luminosity function, 52

initial mass function, 48–49, 52
instabilities, 18, 115
    convective, 18, 90–92
    dynamic, 18, 298
    pulsational, 18, 115
    thermal, 115
intensity
    black-body, 30
    radio, 363
    saturation, 367
intergalactic matter, 365, 487–488
    clusters of galaxies, 527, 528
    evolution of spiral galaxies, 539
    x-ray emission from, 527
internal energy, 9, 17, 411
    matter near absolute zero, 24
interstellar absorption, 31–32, 347
    absorption of radiation, 31
    clouds, 347
    coefficient, 32
interstellar clouds, 347, 357–360
    absorption by, 347
    collapse in galactic magnetic field, 422
    collapse of, 420–421
    collisions, 432
    dark, 350, 423
    dark globules, 371
    diffuse, 350
    evolution, 421
    fragmentation during collapse, 425–429
    magnetic fields in, 413
    properties, 350
    star formation, 413, 414–415, 419–421
    velocities, 347
interstellar dust grains, 31, 32, 350, 359, 370–373, 382
    absorptivity, 352
    accretion of interstellar atoms, 378
    alignment in galactic magnetic field, 372, 379–381
    collisional coupling to gas, 378–379
    collisional drag force, 379
    composition, 350, 371, 377, 378
    condensation nuclei, 377
    condensation temperature, 378
    cooling rate, 373–374
    cross section for ion capture, 375
    cross section, geometric, 371
    dirty ices, 372
    dynamics, 378
    electric charge, 375
    emissivity, 352, 371
    evaporation, 374
    evolution, 377–378
    extinction of starlight, 370–371
    formation, 376–377
    grain-grain collisions, 378
    grain size, 370, 378
    growth, 378
    heating rate, 373–374
    infrared emission from, 350
    infrared excess, 376–377
    ionic capture, 375–376
    iron spheres, 372
    magnetic moment, 381
    molecule formation on, 370, 373–374, 379
    novae, 343
    paramagnetic, 381
    paramagnetic relaxation time, 381
    photoelectric emission, 359, 375
    polarization of starlight, 350, 371–372, 379
    reddening of starlight, 371–372
    rotation in magnetic fields, 379–381
    scattering of light, 370
    structure, 378
    temperature, 352, 373–374
    vaporization, 378
    work function, 375
interstellar electrons, 360–363
    number density, 360–363
interstellar gas, 12, 31, 346
    atomic processes, 347
    collisional excitation, 347
    composition, 347
    cooling rates, 357–360
    elliptical galaxy formation, 538
    emission coefficient, 363
    forbidden atomic transitions, 347
    heating rates, 357–360
    ionization state, 349
    molecular processes, 347
    radio emission, 363
    spiral galaxies, 556
    stationary absorption features, 347
    temperature, 351–356
    thermal equilibrium, 357
    thermal stability, 357
    vibrational atomic states, 347
interstellar matter, 11, 15, 17, 31, 130, 346, 419, 584
    composition, 346
    density, 346
    enrichment by heavy elements, 237
    evolution, 346
    forbidden transitions, 100–101, 141, 347, 382–383
    in galaxies, 42
    magnetic field, 346
    see Interstellar dust; Interstellar gas
interstellar reddening, 31, 32, 34, 370
    dust grains, 371–372
    effect on color indices, 35
    measurement, 38
ionization
    degree of, 107–108
    photoionization, 383–386
ionization equilibrium, effect of metallic abundances on, 104
ionization fronts, 440–447
    D condition, 446
    energy conservation, 444
    formation, 446
    Jouquet point condition, 444
    junction conditions across, 443–444
    motion of, 445–447
    R condition, 446
    shock waves and, 443
ionization potential, 98
    chemical elements, 104–105
    hydrogen-like atoms, 98
ionization rate, 384
ionization state, 102
    gaseous nebulae, 392–393
    interstellar gas, 349
    notation, 102
ionization temperature, 18
ionization zone, 18, 107–108, 129–130
    convective instabilities, 92
    convective stars, 127, 129–130
    gaseous nebulae, 390
    helium, 107, 217, 219
    hydrogen, 107, 217, 219
    pulsational instabilities, 214
    stability, 412–413
IR maser, 366
iron core, 276
iron-peak elements, 14, 233
    formation, 233
irregular galaxies
    Magellanic irregulars, 492
    star formation, 415
irreversible processes, 18
isothermal changes, 18
isothermal core, 169, 181, 182–184
    temperature of, 183
isothermal gas sphere, 63

Jeans length, 417–419, 420, 426, 584
Jeans mass, 419, 580
    matter-dominated epoch, 581

Jeans mass *(continued)*
  radiation-dominated epoch, 580–581
  recombination, 581
  relativistic effects on, 580
  versus temperature, 581
Jeans model, 416–417
  density fluctuations, 417
  expanding medium, 580
  galaxy formation, 580
  marginal stability, 417
  perturbation growth rate, 582
  stability, 416–417
  star formation, 416–419
Jeans radius, 12
Jouquet point condition, 444
junction conditions, 435
Jupiter, 12

Kepler's second law, 318
Kepler's third law, 5, 42, 44, 316, 542
Kirchhoff's law, 81, 363, 371
Kramers' opacity, 120, 126
Kruskal coordinates, 296–297

Ladenburg's relation, 114
Landé *g* factor, 101–102
Lane-Emden equation, 60–63
  analytic solutions, 61
  boundary conditions, 61
  functions, 62–63
Laplace transform, 83
lenticular galaxies, 490–491
  barred, 490
  structure, 495
light curves
  novae, 340–343
  pulsational, 220
  stellar rotation, 59, 95
  stellar tidal distortion, 59
  supernova, 268–269, 272
light cylinder, 307, 310
limb darkening, 82–83
  coefficient of, 83, 134
  eclipsing binary stars, 83
  function, 83
Lindblad resonance, 571
line
  absorption, 130–132
  absorption coefficient, 115
  blanketing, metallic, 105
  broadening, 101, 112, 114, 130–135, 138–141
  collisional broadening, 133
  core, 40, 85
  damping half-width, 133
  damping profile, 133–135
  damping wings, 134, 137
  depth, 40
  Doppler, 132–135, 136–137
  emission, 130, 319
  emission from galaxies, 502
  equivalent width, 40, 136–137
  formation, 130–131, 274
  intensity, 40, 135–138
  metallic, 104–106
  natural broadening, 131–132
  opacity, 40, 109–115, 130–135
  profile, 40, 136
  radio (21.1 cm), 365
  rectified profile, 135–136
  relative line strength, 39, 103
  residual intensity. *See* Rectified profile
  saturation, 136
  strength, 107–108, 135
  width, 112
  wings, 40, 85, 134
line-broadening function, 114
Local Group of Galaxies, 468
  relative motion, 487
local standard of rest (LSR), 539
  dynamical, 539
  kinematic, 539
local thermodynamic equilibrium (LTE), 72, 75, 352
  moderate LTE, 81
  stellar atmospheres, 77
  strong LTE, 77, 79
  weak LTE, 81
Lorentz force, 361, 410
Lorentzian profile, 112–114
LTE. *See* Local thermodynamic equilibrium
luminosity, 2, 76
  absolute, 4, 28
  accretion, 449–450
  apparent, 28
  bolometric, 4
  convective, 76, 228
  core, 194, 198
  energy generation, relation to, 76
  galaxy, 15, 481
  neutrino, 15, 22, 280, 281
  novae, 342
  radiative, maximum stable, 92
  shell, 190, 198
  supernova, 15, 268
  white dwarfs, 263, 265
  x-ray, 11, 331, 333
luminosity class, 33
  galaxy, 42
  stellar, 39, 107
luminosity distance. *See* Distance scales
luminous shock waves, 437–440
  density and temperature behind, 439–440
  energy balance in, 438, 439
  HII regions, 438
  radiant flux from, 440
  width, 440
Lyman alpha radiation, 374
Lyman continuum, 100
Lyman series, 98

M 13, 521
M 31. *See* Andromeda galaxy
Magellanic clouds, 36, 492
  large, 492
  small, 492
  spectrum of bar in, 505
magnetic dipole
  energy loss from rotating, 305–306
  field, 309–310
  radiation from, 306
  rotating model for pulsar, 305–306
magnetic fields, 282, 421–426, 430
  effect on atomic energy levels, 101, 290–291
  effect on star formation, 421–425, 426, 430, 431
  energy density of, 411
  Galactic, 379
  interstellar medium, 346, 424
  pulsars, 305–312
  relaxation time in plasma, 422
  spiral galaxies, 561
  stellar, 102, 207, 284
  stellar core collapse, 284–285
  stellar evolution, 284–285
  supernova remnants. *See* Supernova remnants
  x-ray pulsars, 337
magnetic moment, 380
  dust grain, 381
  electronic, 100–101
  orbital electronic, 100
magnetic susceptibility, 381
magnetohydrodynamics, 400, 406–407
  equations of motion, 406, 415, 544–545
  linearized equations, 415–416
magnetosphere, 307–312
  charged particle flow, 311
  critical field line, 311
  electric and magnetic fields, 309
  energy radiated by, 311–312
  light cylinder, 307, 310–311
  momentum equation, 561

polar cap, 311
wind zone, 311
magnitude system, 29
  absolute, 29
  absolute visual (table), 31
  apparent, 29
  bolometric, 30
  bolometric table, 31
  measurements, 29
  sensitivity function, 30
  uncertainties, 32
main sequence, hydrogen burning, 12, 33, 36, 95, 161–162
  contraction onto, 95
  evolution, 176–184
  lifetime, 13, 51, 52, 176, 414
  lower, 155–156, 177–181
  luminosity–temperature relations, 160–161
  mass-luminosity relation, 51–52, 160–161, 218
  mass-radius relation, 51–52, 160–161, 260
  mass-visual magnitude relation, 51–52
  nuclear burning, 154–156
  in Population I and II systems, 35
  termination of evolution, 167, 184
  turn-off point, 38, 229, 518
  upper, 155–156, 181–182
  zero-age, 12, 167, 175–176
main sequence, helium burning, 200
maser, 363, 366–369
  brightness temperature, 368
  gain, 368
  HII, 366
  infrared sources, 366
  IR, 366
  line shape, 368
  molecular, 366
  population inversion, 366–367
  pump process, 366–367
  saturation, 368
  saturation intensity, 367
  transfer equation, 367
mass excess in nuclear reactions, 149–150
mass exchange
  absorption lines and, 337
  compact x-ray sources, 322, 337
  novae, 322
  see Close binary systems
mass fraction, 55
mass function. *See* Binary systems
mass limits
  Chandrasekhar, 13, 167
  degenerate stars, 255, 260
  magnetic interstellar cloud, 413, 423
  minimum for nuclear burn, 145–147
  neutron stars. *See* Neutron stars
  rotating stars, 283–284
  Schönberg–Chandrasekhar, 183
  stars, 12–13
  Tolman–Oppenheimer–Volkoff, 255, 286
  white dwarfs. *See* White dwarfs
mass loss
  radiation pressure, 226
  red giants, 225
  solar, 205
  from stars in close binary systems, 319, 321–322, 323–325, 332–334
  stellar, 15, 54, 205
  rate, 325, 332–334
mass-luminosity relation, 13, 48, 414
  homologous stars, 160–161
  main-sequence relation, 51–52, 218
  stars in close binary systems, 321
mass-to-light ratio
  clusters of galaxies, 525
  galaxies, 507–508
  stars, 507
mass transfer. *See* Mass exchange
Maxwell's equations, 406
  linearized, 415–416
  plasma, 422
Maxwellian distribution, 19, 353
  collisional equilibrium state, 353, 384
  elliptical galaxies, 537
  energy distribution, 144
  truncated, 519
  velocity distribution, 19, 48, 64, 355, 515
mean free path, 32, 65
  absorption mean free path, 70
  collisional, 347
  Planckian, 86
  scattering mean free path, 70
mean molecular weight, 8, 19–20, 108, 186
  per electron, 20, 257
metastable levels, 101, 347, 366
microwaves, 363
Milne–Schwarzschild equation, 79
missing mass, 474, 488, 504, 527
  clusters of galaxies, 530–531
  neutrinos, 487
mixing length, 92, 94
mixing-length theory, 92–95
molecular dissociation, 170, 348
  energy of, 170
  protostar formation, 170
  pulsational instabilities, 214
molecular masers, 366
molecular transitions, 115, 363
molecules, 347
  complex, 374
  formation on dust grains, 370, 373–374, 379
  ground-state energy, 348
  hyperfine transitions, 348
  photodissociation, 348
  rotational states, 348
  vibrational dissociation, 348
  vibrational states, 347–348
moment of inertia, generalized, 167, 409, 412
momentum conservation across shock front, 435
momentum equation. *See* Hydrodynamics
multiplicity of states, 19

NGC 3031 (M 81), spiral structure, 567
nebulae, 17, 346
  composition, 371
  energy loss by lines, 130
  homogeneous model, 86
  reflection, 370
  Rosetta, 379
  see Gaseous nebulae; Supernova remnants
negative hydrogen ion, 101, 105
  opacity due to, 127
  red giants, 199
neutrino, 66, 68, 276, 482–483
  absorption, 240
  antineutrinos, 240
  Bremsstrahlung, 158
  coherent scattering off heavy nuclei, 249–251
  cooling of white dwarfs, 261–262
  cosmology, 482–483, 488
  cross section, 240, 244
  decay, 243
  degeneracy, 24
  diffusion, 280
  electron neutrino, 240
  emission, 240, 280
  energy deposition, 279–281
  energy loss from nuclear fusion, 157–158
  energy loss rates, 243–249
  interactions, 240–241
  luminosity, 280, 281
  mean free path, 240, 241, 249

neutrino *(continued)*
  momentum deposition, 279–281
  muon neutrino, 240
  from nuclear fusion, 147–148
  opacity, 277, 280
  pair production, 244–245, 280, 482
  photoneutrinos, 158, 244, 245–246
  plasma neutrinos, 158
  plasma process, 244, 246–247
  radiation force, 280–281
  red giant evolution, 197, 201
  shock damping by, 280
  stellar core collapse, 214–215, 239
  stellar evolution, 239
  supernovae, 276
  tau neutrino, 240
  thermalization, in stellar cores, 240, 249
  transport, 276, 280
neutrinosphere, 277, 280
neutron
  capture, 237
  decay, 240, 288
  degeneracy, 236
neutron star, 2, 10, 267, 285–291, 451
  atmosphere scale height, 307
  binding energy, 278–279
  composition, 287–289
  crystalline crust, 312
  degeneracy in, 253
  electrical conductivity, 308, 422
  formation, 14–15, 237, 267, 277
  gravitational potential energy, 14
  gravitational radius, 10
  hot neutron star, lifetime, 250–251
  magnetic properties, 289, 307–312, 430
  mass, 286, 289–290, 296, 330, 338–340
  mass limits, 15, 286
  moment of inertia, 286, 312–314
  parameters, 287
  polar cap, 311
  rotating magnetic. *See* Radio pulsar
  seismological effects, 312–314
  solid, 287–288, 312
  star quakes, 312, 314
  superconductivity in, 287–289
  superfluidity in, 287–289, 314
  tidal distortion of neutron star in binary system, 338
  x-ray sources, 331–332
neutron-to-proton ratio, 483

neutronization, 288
Newtonian cosmology, 470–473
Newtonian gravitation, 291, 513
  force, 294, 316
normal distribution, 500, 515
novae, 223, 322, 340–343, 346
  accretion, 340
  binary nature, 340
  dust grains, 343
  ejected shell, 454
  emission bands from, 340
  energy released by eruption, 343
  expansion velocity, 342
  infrared emission, 343
  infrared excess, 343
  light curves, 340–343
  luminosity, 342–343
  mass exchange in close binary systems, 322
  mass loss by, 399
  prenova star, 340
  recurrence, 223
  rise time, 343
  spectrum, 342
  ultraviolet dwarfs, 340, 382
  unstable nuclear burning, 340–342
  white dwarfs, 340
nuclear burning. *See* Thermonuclear burning
nuclear dissociation, 280
nuclear matter, 285–286, 287–289
  phase transition in, 288
nuclear potential energy, 144–146
nuclear statistical equilibrium. *See* Stellar nucleosynthesis
nuclei
  bound states in nuclear matter, 288–289
  clusters in neutron stars, 288
  neutron-rich, 288
nucleosynthesis, 15
  rotational mixing, 96
  stellar. *See* Stellar nucleosynthesis; Cosmological nucleosynthesis

OB runaway stars, 329–330
observational selection effects, 49
Ohm's law, 308
on-the-spot approximation, 392
Oort's constants, 543
opacity, 3, 13, 69–71, 105
  absorption opacity, 70
  bound-free, 120–123
  conductive, 95
  continuous, 99, 101, 105–106, 115–125
  electron conduction, 123–124

  electron scattering, 13, 75, 117
  free-free, 100, 119, 398
  frequency dependence, bound-free, 123
  Kramers', 120
  line, 130–131
  line center, 114
  mean opacity, 73
  negative hydrogen ion, 101, 105, 127
  neutrino, 155, 158, 277, 280
  optically thin matter, 86
  per atom, 70
  per particle, 70
  photospheric, 127
  radiative, 95
  Rosseland mean, 74, 75, 116
  scattering, 80
  solar atmosphere, 101
  specific opacity, 70, 115–116
  stellar atmospheres, 77
  total, 116, 122, 124–125
  true, 80
  upper-main-sequence, 181
  volume opacity, 70
optical depth. *See* Optical thickness
optical thickness, 32, 70, 72
Orion association, 414
Orion nebula, 11
oscillator strength, 113, 137
  Balmer series, 113
  Kramers' formula, 113

pair production, 24–25
  cosmological, 482
  neutrinos. *See* Neutrinos
  photons, 482
  stellar stability, 413
parallax
  dynamical, 44
  trigonometric, 28–29
paramagnetic relaxation, 380–381
parsec, 28
partial pressure, 19
particle emission rate from nuclei, 233
partition function, 19, 102–103
Paschen continuum, 100
Paschen series, 98
Pauli exclusion principle, 12
peculiar velocity of stars, 539
  galactic evolution, 584
  Sun, 540
period-luminosity relation, 46, 211
  Cepheids, 46, 211
  RR Lyrae, 47
perturbation, 208, 209–210
  adiabatic, 405, 416

perturbation theory, 415
  applied to hydrodynamic equations, 415–416
phase space, 23, 103, 513–514
  statistical weight, 103
photoejection of electrons from dust grains, 375
photodisintegration, 14, 232–233, 276, 278
  helium nuclei, 14, 236
  iron nuclei, 14, 109, 236
  nuclear, 280
photoelectron, 352
photoionization, 99, 120, 352–353, 382
  cross section, 121, 383
photon, 21
  absorption and emission, 99
photon gas, 20
photosphere, 127
  opacity in, 127
  supernova, 272–274
pion decay, 240
Planck function, 74
Planck relation, 98
Planckian opacity, 86
Planckian spectrum, 21
planetary nebulae, 36, 382
  central star in, 382
  dust grains in, 370, 374
  emission lines from, 397
  free electron number density, 399
  globular cluster, 518
  infrared emission, 376–377
  mass, 399
  models, 389, 391–398
  and stellar pulsations, 214, 225
  structure equations for, 389–391
  temperature, 399
planetary systems, 48
planets, 7, 22, 48
plasma, 354, 422
  dielectric constant, 360, 422
  magnetic field, 422
  radio emission, 398
plasma frequency, 247, 360
plasmon, 246–247
plasmon decay, 247
Pleiades cluster, 35, 229
Poisson adiabat. *See* Adiabat
Poisson's equation
  electron screening, 163–164
  galactic disk, 548, 549
  gravitation, 406, 512–513
  linearized, 415–416
  thin gaseous disk, 564–565
Poisson's law, 95
polarization field, 360

polarization of starlight. *See* Interstellar dust grains
  by galactic disk matter, 556
polytrope, 60, 257
polytropic processes, 17
population inversion, 366
post-main-sequence evolution, 185, 194–195
potential energy, gravitational, 6
Poynting vector, 311
pp chain. *See* Thermonuclear burning
precession
  dust grains in magnetic field, 380
pressure, 23
  broadening of lines, 101
  Coulomb, in white dwarfs, 258
  degeneracy, 253, 257
  ionization, 256
  magnetic, 411, 421–422
  mixture of ideal gases, 19–20
  tensor, 545
primeval fireball, 481–487
protogalactic cloud, 511
  collapse and fragmentation, 511
  evolution in expanding universe, 533–537
  final collapse, 537
  specific angular momentum, 538
protogalactic perturbations, 533, 535–536
  at recombination, 578
protogalaxy, 485, 526, 579–584
  evolution, 584
  formation, 578, 579–584
protostar, 12, 25
  central temperature, 170
  evolution, 170–174
  evolutionary track in HR diagram, 171
  formation from collapsing clouds, 428
  luminosity, 171
  maximum stable radius, 170
  opacity in, 171
  spiral galaxies, 574–575
protostellar cloud, 170
pulsar, 2, 272, 301–315
  age of radio pulsars, 306
  binary pulsar, 301
  classes, 314
  gamma-ray emission, 312
  infrared emission, 312
  integral profile, 301
  milliseconds pulsars, 315
  optical emission, 312
  period, 301, 304
  period distribution, 301, 303, 304
  pulse intensity, 301

pulse structure, 302, 312–313
radio pulsars, 301, 312
rate of increase of period, 301, 303
rotating neutron stars, 47, 290, 301, 305–307
spinup, 314
subpulse structure, 301–302
x-ray pulsars, 301, 312
  see Radio pulsar; X-ray pulsar
pulsating stars, 45, 210–213
  Cepheids, 45, 208, 210, 219
  Cepheid, type II, 46
  period-luminosity relation, 46
  pulsation period, 45
  RR Lyrae stars, 46, 208, 210, 219
  W Viriginis stars, 46
pycnonuclear reactions, 164–165

quadrupole moment of mass distribution, 229
quantization, 97–98
  action integral, 98
  angular momentum, 348
  bound states, 98
  energy levels, 97–98, 99
  ground state, 98
  hydrogen, 97–102
  oscillator, 348
  principal quantum number, 98
quantum gravity, 295
quantum mechanics, 19
  zero-point energy, 163, 253, 348
quantum number
  angular momentum, 100
  magnetic, 101
  orbital, 100
  oscillator, 348
  principal, 98
quasar, 469
  black holes, 11
  and galactic nuclei, 42
  redshift, 578
quasistellar object. *See* Quasar

RR Lyrae variables, 35, 46–47, 218–223
  distance to globular clusters, 518
  ionization zones, 219
  model, 219–222
  period-luminosity relation, 218
  position in HR diagram, 214
radiation
  acceleration of matter, 75, 378
  anisotropic field, 67
  coupling to matter, 73, 75, 81
  depth of formation, 85

radiation *(continued)*
  energy density, 66, 67–68
  field description, 66, 67
  field in stellar atmospheres, 69
  flow in rotating stars, 58
  flux, 68, 73
  flux at stellar surface, 84
  force, 12
  intensity, 66
  isotropic field, 67–68
  kinetic-theory description, 69
  mass limits, effect on, 75
  mean free path, 32
  momentum, 67
  momentum flux, 67
  optical thickness, 32, 70
  polarization, 69
  pressure, 8, 13, 20, 67–68, 73
  solar atmosphere, 82–86
  synchrotron. *See* Synchrotron radiation
  work on matter, 73
radiation field, 298
  electromagnetic, 298
  electromagnetic dipole, 298
  gravitational, 299
radiative damping force, 112
radiative equilibrium, 75–76
  grey atmosphere, 79, 83
  nonspherical star, 96
radiative recombination, 382, 383–386
  coefficient, 384–386
  cooling rate, 386
  cross section, 384
  heavy elements, 386
radiative recombination rate, 384, 438, 442
radiative transfer equation, 71–73
  gaseous nebulae, 389
  integral form, 72
  interstellar hydrogen, 363–364
  masers, 367
  solar disk, 82–86
  stellar interiors, 73, 74
radiative transition, 79–80, 109
  bound-bound, 115
  bound-free, 115, 120–123
  forbidden, 100–102
  free-free, 115, 118–120
radiative transport, 65
  convective instability, 92
  moments, 68
radio emission, 347, 389, 398–399
  brightness, 455
  contours from galaxies, 528
  cooling of interstellar gas, 347
  extragalactic, 459
  free-free transitions, 389
  galactic disk structure, 101
  galaxies, 527
  intensity, 398
  neutral hydrogen, 101, 364, 501, 555
  nonthermal, 312, 455–462
  pulsars. *See* Radio pulsars
  spectra, 454, 455–456
  spectral index, 454, 456, 458
  supernova remnants, 454
  thermal, 398–399
radio galaxies, 527
  radio emission contours, 528
  radio lobes, 527
radio pulsar, 301–312, 370
  age, 306
  braking index, 306–307
  brightness temperature, 304
  characteristics, 305
  coherent emission, 304, 307
  cosmic-ray source, 312
  Crab nebula, powered by pulsar, 312
  dispersion measure, 304, 360–361
  distances, 304, 362
  electric field at surface, 309
  energy radiated by, 311–312
  energy spectrum, 312–313
  light cylinder, 307, 310
  location in the Galaxy, 304
  magnetic field strength, 307
  magnetosphere, 307–312
  neutron star model, 305–314
  nomenclature, 301
  nonaligned rotating magnetic dipole, 305–306
  nonthermal radio emission, 312
  period, 301, 304
  polarization, 303, 361
  power-law spectrum, 303
  pulse emission mechanism, 305–312
  pulse intensity, 301
  pulse structure, 302, 312–313
  radio emission, 307–312, 360
  relation to supernovae, 307
  rotation measure, 361
  slowdown, 306
  space velocities, 304
  spectral index, 303
  spectrum, 303–304, 306, 312–313
radio sources, 398
radioactive decay, 109–110
  barrier penetration, 145
random walk, 65
Rayleigh line, 436–437
Rayleigh scattering, 117–118, 370
Rayleigh–Jeans limit, 363
recombination, 485–487
  epoch, 485–487, 533, 581
  see Radiative recombination
red giant, 33, 195–203
  core luminosity, 198
  core structure, 228
  mass loss from, 225–228
  model of a 5 $M_\odot$ star, 199–201
  shell luminosity, 198
red giant branch, 190, 191, 193, 195–203
  evolution up, 190–203
  tip of, 199
redshift
  Andromeda, 468
  cosmological, 41, 476–478
  Doppler, 41
  Einstein–de Sitter universe, 477
  expansion of the universe, 468–470
  Friedmann model, 474
  galactic evolution, 577–578
  galaxies, 468
  gravitational, 11, 296, 338
  gravitational redshift of spectral lines, 296
  quasars, 41
  relativistic, 41, 338
  versus isophotal angular diameter of galaxies, 480
  versus luminosity, 481
  in white dwarfs, 41
red supergiant
  core structure, 274–276
  supernova progenitor, 274–275
reddening line, 34
relative line strengths. *See* Lines
relativistic effects, 10
  in stars, 10, 57
relativity. *See* General relativity; Special relativity
retarded potentials, 425
reversing layer, 135
Rosetta nebula, 379
Rosseland mean opacity, 74, 75, 77, 86, 116
  bound-free opacity, 122
  free-free opacity, 119
rotating stars, 47–48, 283–284
  angular momentum, 47–48
  Doppler broadening, 47
  giants, 283–284
  hydrostatic equilibrium in, 58
  main-sequence, 283–284, 429
  pulsars, 47, 301, 305–307
  rotational velocity, 47, 48, 283
rotation
  effect on star formation in galactic disks, 419–420, 429
  in elliptical galaxies, 500–501

Volume I ends on page 344

gravitational collapse, 419
synchronous in binary, 58
rotation curves, 501–502, 542–544
    observational determination, 542
    spiral perturbations, 572
rotation measure, 304, 361
rotational break-up, 48
rotational mixing, 96
    in the Sun, 96

Saha equation, 104, 107–109, 121, 484
scale height, 58, 262, 307
scattering, 69, 79–82, 353–357
    coherent scattering of neutrinos off heavy nuclei, 249–251, 277, 280, 281
    coherent scattering of photons off electrons, 71
    Compton scattering, 71
    Coulomb, 124, 354–357
    electron-electron, 353, 356
    electron-photon, 12–13, 71, 115, 116–118
    incoherent, 71
    inelastic, 71
    ion-ion, 356
    neutrino-electron, 240, 277, 280
    neutrino-nucleon, 240
    pure scattering, 71, 81
Schmidt model of the Galaxy, 552
Schönberg–Chandrasekhar mass limit, 183–184
Schwarzschild radius. *See* Gravitational radius
shell sources, 187–189, 276
    active nuclear, 185, 189
    envelope expansion, 186, 188–189
    evolution of, 189–195
    flash, 169, 196–197, 202, 226
    inactive nuclear, 185, 188–189
    luminosity, 190
    thermostatic nature of, 186, 187–189
    unstable, 223–225, 226–228, 340–343
shock adiabat, 435–436
shock waves, 406, 432–437, 451
    density increase across, 436
    expanding shock front, 452
    formation, 433
    generated by convection, 95
    heating by, 434, 436–437
    heating of interstellar gas, 350
    interstellar matter, 437
    junction conditions, 435
    mass loss, 226, 228
    shell flash, 226, 228
    shock front, 433–434
    solar corona, 95
    speed, 436
    spiral density waves, 433, 570, 574–575
    star formation, 420
    supernovae, 277–279, 280
    thermal conductivity, 437
    viscosity, 437
    width, 437
    see Ionization fronts; Luminous shock waves
Sirius, 2
solar constant, 4
solar disk, 82
solar neutrinos, 241–243
    detection, 241
    flux rate at Earth, 243
    as probes of solar core, 242–243
    production, 241–242
    solar neutrino unit, 241
solidification, 258
    neutron star matter, 287
sound speed, 54, 405, 432
    adiabatic, 432, 434
    relativistic matter, 580
sound waves, 405–406
source function, 72
    linear model, 83
    pure absorption, 81
    pure scattering, 81
spacetime. *See* General relativity; Special relativity
special relativity, 292
    geodesics, 292
    Minkowski metric, 292
    light cone, 292
    sound speed, adiabatic, 580
specific energy, 17
specific heat, 17
    convective instabilities, 91
    ideal gas, 17, 265
spectra, 2, 84
    general properties of stellar spectra, 85
    hydrogen, 98
    stellar, 103
    Type I supernova, 269–272
    Type II supernova, 272–274
spectral classification, 106–107
    metallic lines, 106
    stellar, 39
spectral index, 454, 456, 458
spectral type, 31, 40
    effective temperatures, 40
    metallic lines, 106
spectrophotometry, 30
    color indices, 30, 31
    sensitivity function, 30
    UBV system, 30
spectroscopic notation, 99–100
spectroscopy, 38
spectrum
    continuous, 130
    galactic, 505–506
    rectified, 40
    stellar, 39, 84
    synthetic, 42
spectrum binary. *See* Binary systems
spherical galaxies. *See* Elliptical galaxies
spin
    angular momentum, 100
    electronic, 100
    intrinsic, 100
    quantum mechanical, 19
spin-orbit interaction, 100
spin-spin interaction, 100
spiral curves, 562–563
    equiangular (logarithmic) spiral, 562
    interarm spacing, 562
    pitch angle of spiral, 562
spiral density waves, 408, 560–571
    angular velocity of spiral pattern, 558, 560, 567, 569
    arms, 563
    disk thickness, 570
    dispersion relation, 568
    formulation of theory, 561–567
    gaseous component, 563
    gravitational potential, 541
    initial perturbation, 561
    linear theory, 564–570
    nonlinear effects, 570
    magnetic fields, 561
    observational consequences, 571–575
    parameters for spiral systems, 558
    physical basis for, 560
    spiral density wave, 560
    spiral gravitational potential perturbation, 560
    spiral pattern, 560
    spiral shock waves, 570
    and star formation in galactic disks, 420, 433
    stellar component, 563, 569–570
    tight winding approximation, 566
    see Gaseous disks
spiral galaxies, 11, 490–492, 532, 538–542
    angular momentum, 538, 539
    angular velocity of spiral pattern, 544

spiral galaxies *(continued)*
  barred spirals, 561
  differential rotation of disk, 543, 556
  disk, 538, 555
  disk temperature, 561
  equations of motion for disks, 407–408
  evolution, 539
  force laws in disk, 544–549
  formation, 539
  gravitational field of disk, 560
  HI regions, 574
  HII regions, 574, 575
  halo, 538, 555
  interarm spacing, 562
  interstellar gas in, 556
  Lindblad resonances, 571
  luminosity distribution, 495
  magnetic fields, 556, 561
  mass distribution, 506
  metallicity gradients, 538
  morphology, 572
  nuclear bulge, 538
  nuclei, 506, 508, 538
  protostar formation, 575
  rotation curves, 501–502, 542–544, 572–573
  spectrum of nuclei, 506
  spheroidal component, 532
  spiral arms, 490, 555, 566
  spiral pattern, 556
  spiral shock waves, 570
  star formation, 415, 561
  stellar association, 573
  stellar component, 555
  stellar migration, 573
  streaming motion, 547, 556–560, 559
  turbulence effects on spiral pattern, 569
  winding dilemma, 556–557, 561
  see Gaseous disks; Galaxy, the
sputtering, 378
stability, 18, 88, 342
  degenerate stars, 260
  infinite disk, 417–419
  ionization zones, 412
  Jeans model, 416–417
  macroscopic, 412
  marginal, 417, 418
  molecular dissociation, 412
  oscillations, 211
  pair formation, 413
  and perturbations, 208
  pulsational, 208, 215–217
  radial perturbations, 210
  radiation pressure, 88
  stellar, 213–215, 260

thermal. *See* Thermal instability
star clusters, 34
  ages, 38, 206
  chemical composition, 34
  color-color diagram, 34
star formation, 11–12, 15, 346, 414–415
  angular momentum loss during, 423–425, 429
  Bok globules, 415
  differential rotation of galactic disk, 419–420
  fragmentation of collapsing interstellar clouds, 425–426
  from interstellar matter, 413, 419–420, 428
  rotation, 419–420
  rotational effects on, 419–420, 423–425, 429–430
  shock wave induced, 420, 433
  sites, 414–415
  spiral density waves, 420, 561, 574
  spiral galaxies, 11, 419
  stellar associations, 419
  stellar rotation, 429
  see Stellar birthrate function
Stark effect, 135, 138–141
  linear, 139
  quadratic, 141
stars, 2
  binding energy, 61, 278–279
  blue supergiants, 225, 326–327
  bolometric corrections, 31
  Cepheids, 36, 45
  chemical composition, 34, 79
  color indices, 31
  convective, 95, 127–130, 171
  degenerate, 254, 260
  dwarf Cepheids, 47
  dwarfs, 39
  effective temperatures, 31
  formation, 11
  galactic clusters, 50
  gas pressure, 9
  giants, 39, 376
  homogeneous, 177
  long-period variable, 36, 47
  luminosities, 31, 171
  magnetic, 102
  magnitudes, 31
  mass limits, 9, 12, 13
  masses, 31, 61
  mean density, 31, 61
  Mira variables, 47, 376
  neutron, 9, 10, 14, 22
  novae, 340–343, 376
  pressure, central, 56, 61
  protostars, 170–174

pulsating, 45
radiation pressure, 9
radii, 31, 61
red giants, 33, 195–203
red supergiants, 47
relativistic, 10, 57
rotating stars, 47
RR Lyrae variables, 35, 46–47
semiregular variable, 47
spectral type, 31
subdwarfs, 106–107, 585
subgiants, 320
supergiant, 202–203, 376
supernovae, 267–282
ultraviolet, 226
ultraviolet excess, 585
variable, 6, 36
white dwarfs, 9, 10, 14, 22, 202, 256–266
static universe, 472
statistical fluctuations and dust grain formation, 377
statistical line broadening, 135
statistical Stark effect. *See* Stark effect
stellar associations, 414, 416, 426
  spiral density waves and, 573
stellar atmospheres, 2, 76–79
  absorption lines, 39, 85
  boundary conditions for model, 77, 86–88
  convective zones in, 77
  gravitational acceleration in, 57
  hydrostatic equilibrium in, 57
  model atmospheres, 55, 77
  plane-parallel, 57
  radiative equilibrium, 77
  scale height, 58, 77, 307
stellar birthrate function, 49, 415
stellar core collapse, 14, 23, 186, 274–282
  energetics, 278
  energy released by, 278–279
  halted by neutron degeneracy, 236
  magnetic fields in, 284–285
  neutrino luminosity from, 250–251
  photodissociation induced, 14, 214–215, 276
  supernovae, 14, 274–282
  time-scale for, 250–251
stellar distribution functions, 48–52
  stars in solar neighborhood, 50
  statistical distribution function, 513–515
stellar dynamics, 512–515
  axially symmetric, 546–549
  collision-free hypothesis, 514

deflection time, 517
distribution function, 513–515
dynamical evolution time-scale, 518
dynamic friction, 517
dynamic relaxation time, 517
equilibrium under gravitational scattering, 351
hydrodynamic treatment, 544
relaxation time, 516–518
sling-shot effect, 516
stellar encounters, 515, 517
stellar mixing time, 517–518
three-body effects, 516
velocity ellipsoid, 540
vertex deviation, 541–542
stellar envelope ejection, 226–227
novae, 342–343
stellar evolution, 3, 13–15, 36, 54–55, 403
binary systems, 42
dynamical stages, 403
effect of composition, 203–207
effect on close binary systems, 322–330
final stages, 205, 229–230
main sequence. *See* Main sequence, evolution
massive stars, 14
neutrino emission and, 239
pair processes, 25
post-main-sequence. *See* Post-main-sequence evolution
quasistatic, 55, 57, 76
radius versus time of 5 $M_\odot$ star, 322–323
red giant branch. *See* Red giant branch
rotational mixing, 96, 206
stages of, 166–169
and stellar statistics, 48, 49, 50–51
supernovae and neutron stars, 14, 15, 267
time-scale for, 55
stellar interiors, 3
opacity, 74
radiative transfer in, 74
stellar lifetime, 13
main sequence, 52, 414
stellar luminosity function, 48, 50
stellar mass function, 48, 50
stellar models, 54–55, 59–64, 125–130
evolutionary sequences, 169
homologous, 158–161
initial for main sequence, 177
linear, 59–60, 177
neutron stars, 254, 289–290

polytropic, 59–60, 257
radius versus time for 5 $M_\odot$ star, 322–323
white dwarfs, 254
stellar nucleosynthesis, 230–238
e-process, 234, 236
explosive, 275
heavy elements, 237
intermediate mass elements, 232, 276
iron-peak elements, 233
light elements, 230–232
neutron enrichment, 234
nuclear statistical equilibrium, 234
r-process, 237
s-process, 237
supernovae, 272, 276
stellar populations, 35–36, 548, 584–588
abundances, 55, 104
disk population, 548
formation, 584–585
galaxies, 504
globular clusters, 523
halo population, 548
metal abundance, 584
Population I, 35
Population II, 35, 47, 523
properties, 36
synthesis of, 504, 508
stellar pulsation, 210–213
adiabatic, 210–213
frequency, 214
fundamental mode, 213–214
homologous, 218
nonradial, 218
novae, 342
nuclear energy generation and, 216–217
opacity and, 216–217
overtones, 214
period, 211
period-density law, 211
period-luminosity law, 211
pulsar mechanism, 305
radial, 210
valve mechanism, 217
stellar structure, 54–55
boundary conditions, 55, 127
closed, 474
core in highly evolved stars, 230–231, 274–276
degenerate stars, 254–256
envelope, post-main-sequence, 192
equations for, 55, 76, 87, 160, 169
homologous, 159–160
initial conditions, 55

linear model 59–60, 177, 180, 181–182, 183
main-sequence model, 90, 179–180
model stellar atmospheres, relation to, 55, 87
polytropic models, 60, 257
quasistatic, 76
radiative, summary, 86–88
radius versus time for 5 $M_\odot$ star, 322–323
white dwarfs, 256–266
stellar systems
axially symmetric, 546–549
characteristics, 513
eccentricity, 585–586
orbits, 551–554, 586
statistical approach, 512
stellar time-scales, 4, 193
stellar wind, 205, 327, 332–333, 432
in close binary systems, 327
Strömgren sphere, 382, 440–447
evolution of, 441–443
HII region, 441
strong interactions, 7, 25, 285, 288–289
nucleon-nucleon bound states, 288
sun, 3, 4, 9, 16
flares, 3
internal temperature and pressure, 9
luminosity, 4
solar constant, 4
solar granulation, 3
sunspots, 3, 102
superadiabatic temperature gradient, 92, 94
superconductivity, 287–289
superfluidity, 287–289, 314
supermassive stars, stability and general relativity, 298
supernovae, 2, 14, 158, 229, 283, 412, 451
absolute visual magnitude, 268
angular size of, 272
in binary systems, 327–329
comparison to novae, 272
core bounce, 277–280
distance to, 272–273
ejecta, 15
ejecta velocity, 451
envelope ejection, 270, 272–274, 279–280
formation rate, Type I, 282
historical, 267–268
kinetic energy of ejected envelope, 281
light curve, 268–269, 272

supernovae *(continued)*
  line emission and absorption, 274
  luminosities, 15, 268
  mass ejection, 279–282
  neutrino damping, 280
  neutrino luminosity from, 280–281
  neutrinos, 239, 276, 279–280
  neutron stars, 307
  observed properties, 267–274
  photosphere, 272–274
  progenitor, 229–230, 268, 274–276
  rate in the Galaxy, 268–269
  remnant mass, 451
  spectra, 269–274
  stellar core collapse, 23, 214–215
  thermonuclear burning, 279, 281
  Type I, 268–270, 281–282, 451
  Type II, 237, 268–273, 451
supernova remnants, 283–285, 346, 451
  age, 453
  blast-wave model, 451
  brightness temperature, 455
  charged particles in, 459–460
  compact. *See* Black hole; Neutron star
  cosmic rays, 455–456
  diffuse, 451
  electron temperature, 455
  evolution, 451–454, 460
  expansion rate, 453
  filamentary structure, 451, 454–455
  high-energy electrons, 459
  initial energy, 453
  lifetime of charged particle populations, 459–460
  magnetic fields in, 423, 455, 459
  nonthermal radio component, 451, 455–463
  radio intensity from, 458
  radio sources, 455
  radius, 453
  relativistic electrons in, 459
  spectra, 454, 455–456, 458, 459
  temperature, 455
supersonic flow, 432, 434
surface brightness, 83
  stellar intensity, 84
surface gravity, 57, 78
  effective, in atmospheres, 87
surface temperature, 4, 78
  stellar atmospheres, 77
synchrotron radiation, 455–459
  charged particle populations, 459–460
  energy loss by relativistic electrons, 455, 464
  fundamental frequency, 455
  high-frequency cutoff, 459
  lifetime of high-energy electron population, 459, 464
  polarization, 455, 457
  power law, 455
  radio galaxies, 527
  radio intensity, 458
  radio source, 455
  self-absorption, 458, 459
  spectrum, 455, 458
  see Cassiopeia A; Crab nebula; Synchrotron radiation; Veil nebula
synchrotron self-absorption. *See* Synchrotron radiation

T Tauri variables, 174
temperature, 351
  brightness, 363, 364, 366, 368, 455
  color, 351
  condensation, 378
  electron, 440
  excitation, 352
  ion, 440
  ionization, 351
  kinetic, 351, 365, 384, 393–394
  radiation, 351, 365
  saturation, 377
  spectral. *See* Color
  spin, 363, 365
thermal equilibrium, 16, 25, 351–352
  interstellar gas, 357
  relaxation rate, 357
thermal excitation, 102
thermal instability, 223–225
  novae, 342
thermal ionization, 102
thermal runaway, 196, 225
  and Type II supernovae, 281
thermalization, 352–353
thermodynamics, 17
  equilibrium states, 351
  first law, 17, 215, 404
  phase transitions, 360
thermonuclear burning, 12
  carbon burning, 157, 201, 248
  carbon flash, 169, 202
  energy generation rate. *See* Thermonuclear reactions
  helium burning, 13, 14, 156–157, 196, 199–201
  helium flash, 169
  hydrogen burning, 12, 13, 14, 147, 149, 193–194
  network, 157, 232
  oxygen burning, 157
  pre-main-sequence, 156
  shell source, 168, 181, 182, 184, 185, 187–195, 223–225
  silicon burning, 232–236, 247, 276
  supernovae, 279
  unstable, 223–225, 340–343
  unstable helium shell, 226–228
thermonuclear energy, 3, 7, 144–149
thermonuclear fusion, 144
  ignition temperature, 145–147
thermonuclear reactions, 13
  energy generation rate, 147, 149–154, 156, 157, 234, 249
  energy release, 147–149, 150
  half-life, 148
  high temperatures, 233
Thomson scattering, 440
  cross section, 117
tidal acceleration, 299, 327, 338
  effect on globular clusters, 520
  elliptical galaxy formation, 537
time dilation, gravitational, 295
time-scales, 4, 16
  deflection time, 517
  dynamic relaxation time, 517
  Einstein, 7
  evolutionary, 6, 7, 55, 150
  free-fall, 5
  Kelvin–Helmholtz, 6, 225, 323, 325
  mass limit, 9
  mass transfer in close binary systems, 325
  nuclear, 150, 176
  relaxation time, 516–518, 526
  stellar, 4, 193
  thermal. *See* Time-scales, Kelvin–Helmholtz
  violent relaxation, 526–527
Tolman–Oppenheimer–Volkoff mass limit, 255, 286
transition probability, 109–115
  classical, 112
  Einstein, 109–111
  optical, 112
  radio, 112
transition rate, 111
  bound-bound transitions, 113–114
turbulent motion
  galaxy formation, 511
  macroturbulence, 133
  microturbulence, 133

microturbulent velocity, 133
spiral structure, 569
in stellar atmospheres, 133
turbulent pressure, 564
two-color diagrams. *See* Color-color diagram

U Geminorum, 340
ultraviolet dwarf, 340, 382
ultraviolet excess, 584–585
  angular momentum of stars in galaxies, 587
  eccentricity of stellar orbits, 586
  stellar age, 585
  stellar velocity, 587
ultraviolet stars, 226
uncertainty principle, 112–113, 164
  and degenerate matter, 252–253, 254
universe
  age, 475, 476
  current mass density, 474, 487, 488
  early stages in Big Bang, 481–482
  Euclidean, 474
  expanding. *See* Cosmology
  mass density in galaxies, 485, 487
  mass density of, 474, 487, 488
  maximum expansion, 474
URCA process, 281

van der Waals field, 101
variable stars, 5
  T Tauri, 174
Veil nebula, 15, 437, 451, 453
  age, 453
Vela pulsar, 301
  energy spectrum, 313
  spinup, 314
velocity dispersions, 500, 541
velocity ellipsoid, 540
vertex deviation, 542
violent relaxation, 523, 526, 537
Virgo cluster, 525

virial, 410
  magnetic, 410
virial theorem, 9, 56–57, 183, 409–410
  clusters of galaxies, 503, 527
  dynamic, 166–167
  elliptical galaxies, 500
  embedded gas, 420
  in hydrostatic equilibrium, 56–57, 411
  missing mass in the universe, 488, 504
  stellar systems, 515–516
viscosity, 437
Voigt function, 134
Von Zeipel's theorem, 58
vortex lines in superfluids, 314

W Ursae Majoris stars, 319
W Viriginis stars, 46
WW Aurigae, 43
wave equation, 405, 416
  advanced and retarded potentials, 425
  Alfvén, 424–425
  dispersion relation, 416
  mass term, 416
weak interactions, 7, 14, 25, 240–241
  coupling constant, 244
  deuterium production in stars, 155
white dwarfs, 8, 9–10, 14, 229, 256–266
  convective envelopes, 264
  cooling of, 261–262, 266
  core structure, 256–261
  envelope structure, 261–265
  evolving, 265–266
  formation rate, 269, 281
  luminosity, 253, 263, 265
  magnetic fields in, 284, 430
  mass limit, 9, 282
  masses, 260, 296
  mass-radius relation, 260
  novae, 340–343

  observed parameters, 260
  pulsars, 305
  radii, 260
  spectra, 264
  supernovae, Type I, 281–282
  surface temperatures, 14
  thermal energy content, 253
work function, 375

x-ray binary systems, 330
  eclipsing, 331
  mass accretion rate, 331
  neutron star masses, 332, 336–337
  properties, 330
x-ray blanketing, 332, 334
x-ray pulsars, 301, 335–338
  accretion disk, 337
  binary character, 336–337
  Doppler-shifted pulse period, 336–337
  emission mechanism, 337
  heating of companion, 336
  integrated pulse profile, 335
  magnetic fields, 337
  subpulses, 335–336
x-ray sources, 15, 327, 330–338, 488
  accretion, 447
  accretion disk, 337
  accretion rate, 334
  clusters of galaxies, 527
  compact, 319
  eclipse of, 334
  galactic nuclei, 509
  luminosities, 11, 331
  magnetic fields, 337
  mass exchange in close binary systems, 322
  soft, 454
  variability, 331
  see X-ray binary systems

Zeeman effect, 102, 284
zero-age main sequence. *See* Main sequence